Combined Proceedings International Plant Propagators' Society 2024

IPPS Volume 74

Held 2024

Editors:

Robert L. Geneve
Ranjith Pathirana
William Barnes Jr.
Masanori Tomita
Jill Reader
Elsa du Toit
Fred T. Davies Jr.
Ryan Stewart

ISBN: 979-8-3313-2700-2

Printed from e-media with permission by:

Curran Associates, Inc.
57 Morehouse Lane
Red Hook, NY 12571

Some format issues inherent in the e-media version may also appear in this print version.

Printed with permission by Curran Associates, Inc. (2025)

For permission requests, please contact International Plant Propagators' Society (IPPS) at the address below.

International Plant Propagators' Society (IPPS)
174 Crestview Drive
Bellefonte, PA 16823-8516
USA

Phone: +01 814-355-2467

secretary@ipps.org

Additional copies of this publication are available from:

Curran Associates, Inc.
57 Morehouse Lane
Red Hook, NY 12571 USA
Phone: 845-758-0400
Fax: 845-758-2633
Email: curran@proceedings.com
Web: www.proceedings.com

THE INTERNATIONAL PLANT PROPAGATORS' SOCIETY

Combined Proceedings

Volume 74

2024

IPPS

Sharing Plant Production Knowledge Globally

Published 2024

Combined Proceedings International Plant Propagators' Society (ISSN 0538-9143) is a yearly publication of the International Plant Propagators' Society (IPPS) headquartered in Pennsylvania, U.S.A.
Email: Secretary@ipps.org
Website: www.ipps.org

Business Office. Memberships are processed at the regional level, with dues being allocated to both international and regional expenses. Contact the regional secretary / treasurer in your area of the world for membership information and applications.

Table of Contents

Australian Region

New Zealand Region

Southern Region – USA

Eastern Region - USA

Combined Proceedings IPPS 74:1-9. 2024.

Our Breeding of Endemic Southern African Plants

Andy de Wet

De Wet Plant Breeders, Johannesburg, South Africa

hello@cndnursery.co.za

Keywords: *Aloe*, *Agapanthus*, plant improvement, new cultivars

Summary

The development of a successful plant breeding program targeting South African native plants is described. *Aloe* was the major plant improved for ornamental use but additional genera including *Agapanthus* have also been highly successful.

INTRODUCTION

My journey started in 1973. As a teenager I found a passion for aloes, and whilst studying Botany at university I realised that the different species sometimes naturally hybridised in the wild. I thought that if I did the selection of the parents myself, I could do a better job.

Our rich diversity of amazing Southern African plants has always been improved in the Northern Hemisphere, generally rendering them less suitable for our warm sunny climate. I decided to change that, as it's OUR proud heritage and we can do better because we know and have access to the broader gene pool.

IPPS Vol. 74 – 2024

Starting with nothing, it was a long, tedious road, but today I have a breeding company called De Wet Plant Breeders which is based at The ALOE FARM and CND Nursery in South Africa (**Fig. 1**). De Wet Plant Breeders aims to improve mainly Southern African ornamental plants that are heat and drought resistant. I also have an extremely capable junior partner, Quinton Bean, and our cultivars have received numerous awards worldwide - undoubtedly the best known is the prestigious 'Plant of the year 2023' at the Chelsea International Flower Show in the UK.

Figure 1. Andy at The ALOE FARM, in one of the display gardens filled with De Wet cultivars.

Ornamental plant breeding is a slow, expensive process which starts with studying a potential genus and then growing and collecting suitable specimens to work with. After a few years of tedious hand pollination, you may achieve some success (and lots of failure!), but you could end up with a gene pool that can serve as basis to build your breeding upon. It is important to keep a record of your hybrids so that you know the hidden characteristics of each seedling. Even if you select the best parents, you never know exactly what you're going to get. All the progenies differ slightly from each other, and the recessive and dominant characteristics are at first generally unknown to you. Successful plant breeding is a scientific numbers game (**Fig. 2**).

We currently work on approximately 22 different selected genera. With only *Agapanthus*, we do an average of 12,000 seedlings annually from selected, hand pollinated parents (**Fig. 3**). They are then grown in the best possible conditions in order to bring them to flower as soon as possible, so that we can evaluate and select the best and then discard the majority to create space for the next batch. We destroy the rejects because we have to protect our genetics, selling them would give the opposition shortcuts into our expensive work.

Figure 2. Colourful drought and heat-resistant gardens filled with De Wet aloe cultivars.

Figure 3. New hybrid *Agapanthus* seedlings, grown from selected hand pollinated parents, each pot containing a different combination. This is typical of one season's work and the parentage of each combination is kept on tags that will always stay with the plants.

Plant breeding is fascinating work, because we continuously raise seedlings and each one harbours promise of great expectations.

The real fun begins when they start flowering! This is when we get lots of surprises, and then there are often individual plants which exceed expectations or open up new avenues to follow. Quinton and I do all the selections ourselves. On first flowers we normally pick a soft selection, which is the "best" 3-4%. Mistakes are unavoidable, but this is where our experience and instinct matter. These selected few then get tested further, usually potted up and left to flower again when they are more mature plants. The selection is very strict and the final product must fit several criteria.

There is no such thing as a perfect plant and a good plant breeder is never satisfied - that is what keeps us going! The search for the proverbial "holy grail" is never ending. One can only release a plant to the horticultural community if it is distinctly better and different to anything that came before it. On the level at which we are breeding today, it very often happens that a plant can have only one really outstanding characteristic, but it's not good enough if it isn't balanced out with overall excellence. Our massive compost heaps are littered with plants that have some amazingly unique characteristics.

A new cultivar must satisfy the consumer, people are always wanting something new and better, that is the nature of mankind....and womankind! But most importantly, we breed for the growers, like the distinguished members of IPPS. We have only succeeded when we supply the market with a new release that has outstanding new features, is easy to grow, flowers in record time and is disease and heat resistant. Most importantly, it must appeal to the retailers and consumers. A perfect example of this is *Agapanthus* 'Blackjack' - apart from fitting the above criteria, its uniqueness has created a buzz worldwide (**Figs. 4 to 7**).

Figure 4. *Agapanthus* 'Blackjack' winner of the prestigious "Plant of the Year 2023" at the Chelsea International Flower Show in the UK.

Figure 5. De Wet Plant Breeders, Quinton Bean and Andy de Wet, celebrating the Chelsea win.

Figure 6. Some mature *Agapanthus* from our diverse trailing section. It is often very difficult to select which one to release.

Figure 7. *Agapanthus* 'Fireworks', won a third at The Chelsea International Flower Show, popular for its distinctive drooping bicolor flowers and being extremely floriferous.

De Wet Plant Breeders try to release as few cultivars as possible, in order to create memorable plants and not to confuse the industry and public with meaningless multitudes. There is no room for sentiment in plant breeding, but sometimes during the selection process, we stumble on plants that may not fit the commercial requirements but are just too good to destroy. This happens especially with aloes. Fortunately, we have The ALOE FARM with ever-expanding gardens that are continuously fed with the best new and unique aloe plants in the world! Visitors from many countries and all over South Africa visit us annually during our winter Aloe Festival season.

Plant breeding works very much like the music and film industry, we create a unique product that becomes patented as our intellectual property. We only start getting a return for our efforts when our products are sold in the stores and royalties are calculated and shared with our agent. That is why the protection of our intellectual property is paramount to the success of our endeavours.

We generally do not sell directly to retailers, but regularly send clean samples for testing and reproduction purposes to trusted associates. We are represented by Plantipp, who are excellent agents worldwide and communicate, distribute, and monitor the wholesalers who grow our plants under license. It sounds extremely lucrative, but once we've identified a new release it takes 7-8 years before the supply chain is satisfied and saturated to the point where the plant starts selling in the stores. Plant breeders must have deep pockets, or as in our case, stay poor and patient for a long time. I have been told this is not a good business plan …. but it's a lot of fun and we're establishing a lasting legacy.

The goals in our breeding are varied and not always profit orientated. One important goal is conservation - by making hybrids of popular but endangered species like *Aloe peglerae* (and several others), we have created similar looking hybrids that grow and flower better. The aim here is to alleviate the illegal collection pressure on the natural plant population.

Figure 8. A new cultivar should surpass all previous similar plants in its overall appearance. To be a good cultivar, many characteristics like vigour, diseases and heat resistance, are also essential.

Often there are unusual, attractive characteristics in an obscure species that need to be enhanced, e.g. from *Aloe perrierii* we created *Aloe* 'Marilyn'. Alternatively, there is the easy growing and architectural *Aloe* 'Samson' - it's a practical lookalike of the extremely difficult and very slow growing but popular Aloe dichotoma.

Another target is the landscape industry. Plants like *Aloe* 'Hedgehog' and *Aloe* 'Peri Peri' (and several others) have proved to fulfil that need. It is however difficult to get landscape architects and garden designers to change their habits.

Mostly, however, we breed to satisfy the nursery industry with ornamental plants that will become available in local retail stores.

The practicality of use is also very important in a cultivar. The size and neatness must make handling, packing and transportation as easy and cost effective as possible, without any damage to the plants.

An important aspect which we always keep in mind during the breeding and selection process, is the environmental impact of our creations. A successful new plant must be tough in many situations - easy to grow and resistant to most plant diseases, but also add to the surrounding environment by attracting and feeding beneficial insects and birds. Fortunately, it's sometimes an automatic thing - when a cultivar flowers more prolifically and for longer periods, it will also generate more nectar and pollen for our fellow creatures. The ALOE FARM is very popular with nature lovers, and we also have a very active Facebook group, BIRDS OF THE ALOE FARM.

Our Southern African flora has the potential to take an even bigger place in world gardens, but it has to be improved by enhancing the colours, number of flowers produced and lengthening the flowering season. Good examples of this are *Agapanthus* 'Buccaneer', *Aloe* 'Rocket', and *Aloe* 'Starstruck', and a plant like the smaller *Aloe* 'Goldfish' will flower randomly throughout the year.

Finding the correct name for new cultivars is very important as a bad name can sink a good plant. A name like Blackjack is perfect as it's short, strong, descriptive and easy to remember. Two of my favourite aloes are named Charles and Sannie after my parents, and as a tribute to our family's love for them, there is one of each planted on their grave....and they flower beautifully every winter.

Our international *Agapanthus* cultivar distribution was kickstarted by *Agapanthus* 'Twister' which was widely accepted worldwide. We have set the new benchmark for the genus *Agapanthus* because our cultivars have much more attractive flowers and longer flowering seasons than traditional agapanthus. Good examples of this are *Agapanthus* 'Poppin' Purple', 'Bingo White' (Syn. 'Ever White'), 'Buccaneer' (Syn. 'Poppin' Star'), and many other award-winning varieties. We also have a breeding programme for cut flower *Agapanthus*.

Some breeding programmes have come to a dead end, like *Albuca* that seemed to have such promise at first. The negative characteristics just wouldn't go away, so we dropped our *Albuca* breeding programme after ten years of dedicated work. A fantastic, improved *Plumbago* which we created is just too slow growing to ever put out to the market.

However, we have many beautiful *Gazania*, *Osteospermum*, *Dietes*, *Plectranthus*, *Salvia*, day lilies, *Bulbine*, *Kalanchoe*, *Arctotis*, *Tecomaria*, *Cotyledon*, *Coleus* and many other exciting hybrids that we have either released or have them slowly moving through our extensive secret pipeline (**Fig. 9**).

Figure 9. *Gazania* 'Suncarpet', a new De Wet *Gazania* cultivar, unique large flowers on short, neat stems. It's a improvement as it's heat and drought tolerant, sterile. Being self-cleaning, keeps the plants looking fresh, which means it reduces the amount of labour involved by growers and retailers.

Combined Proceedings IPPS 74:10-15. 2024.

Auxin Use in Propagation – Then and Now

Robert L. Geneve

Department of Horticulture University of Kentucky, Lexington, Kentucky 40515, USA

Rgeneve@uky.edu

Keywords: IBA, IAA, plant hormones, application methods, foliar auxin, total immersion, Bonzi

Summary

Until the 1930s and 1940s, cutting propagation was limited to those species that were easy to root. With the discovery that auxin significantly enhanced rooting in cuttings, the number of vegetatively propagated species available to greenhouse and nursery producers significantly increased. It is interesting to see the early adoption of "hormones" in commercial practice and how the delivery methods for treating cuttings with auxin evolved. With an increased emphasis on propagation efficiency for modern greenhouse and nursery production, alternative methods are resurfacing as potential ways to deliver auxin to cuttings.

INTRODUCTION

The greenhouse and nursery industry has seen significant changes in the past twenty years. Annually, there is an increase in the diversity of plants offered to an international market. This may be most dramatic in the annual bedding plants where there has been an increase in cutting propagation in an industry that was exclusively seed propagated a few decades ago. Breeding efforts have extended the number of interspecific and intergeneric hybrids that necessitate vegetative propagation. Stock plant management has moved from being predominantly local to being consolidated to larger specialty growers. Local greenhouse and nurseries are increasingly receiving unrooted cuttings to produce plugs and liners. With increased emphasis on efficiencies in cutting propagation, there is renewed interest in auxin and its various application methods. This paper traces the early adoption of auxin for cutting propagation and its current application practices.

Historical background

Fritz Went in 1928 building on the initial research of Charles Darwin (Darwin, and Darwin, 1880) and Boysen Jensen in 1911 developed the first bioassay for detecting hormones in plants based on the bending of grass and oat seedlings to light. Went placed agar blocks containing suspected hormones asymmetrically on decapitated oat seedlings and measured the bending of the coleoptile. Kögl and Haagen-Smit, between 1933 and 1935, found that substances in human urine and various plant extracts were active in Went's coleoptile bioassay. This led to the chemical isolation of "heteroauxin" [indole-3-acetic acid (IAA)] identified as the first plant hormone. Soon after this, Went and Thimann in 1934 developed another bioassay based on the discovery that auxin-induced adventitious rooting in etiolated pea cuttings.

The first specific report of IAA being used to stimulate rooting in cuttings was by William Cooper at the Boyce Thompson Institute in 1935. He applied IAA in lanolin paste to stimulate rooting in lemon (*Citrus*), lantana (*Lantana*), and chenille plant (*Acalypha*) stem cuttings. By 1935, synthetic auxins were developed that were shown to promote rooting in cuttings (Thimann, 1935; Zimmerman and Wilcoxon, 1935). These included the familiar α-naphthaleneacetic acid (NAA) and indolebutyric acid (IBA) compounds still used by modern propagators.

The potential commercial importance of auxin to cutting propagation became evident as researchers at the Boyce Thompson Institute showed the efficacy of auxin in stimulating rooting in cuttings of over 85 genera of plants, including woody plants that had proven difficult to propagate in the past (Zimmerman, 1935).

In 1936, the soak method replaced lanolin paste for auxin delivery. Auxin was made soluble in alcohol and diluted in water. Typical soak durations were 10 to 24 hours (Zimmerman and Hitchcock, 1935). Grace in 1937, developed the method of incorporating auxin in talc to deliver auxin to cuttings that would eventually become the standard commercially. In 1938, Chadwick and Kiplinger (1938) treated over 100 types of woody cuttings with IBA in a soak solution or as commercially available dusts with positive results. The quick-dip method for treating auxin was developed by Hitchcock and Zimmerman in 1939 and later refined by Cooper in 1944.

The Boyce Thompson Institute was granted a patent for use of auxins in rooting and subsequently licensed Merck to distribute Hormodin A for commercial application. By 1947, four commercial companies were offering synthetic auxin formulations in talc for application to cuttings. These included Hormodin (Merck), Rootone (American Chemical Paint Co.), StimRoot (Plant Products Co.) and Quick-Root (Dow Chemical).

In an excellent early review of research in cutting propagation by Avery et al. (1947), they provide a table with references on experimental rooting for over 600 different kinds of woody plants. This provides a great overview of the impact of auxin on rooting cuttings in the era prior to the use of mist by the greenhouse and nursery industry.

Early adoption of auxin in propagation

Although the research on auxin use to improve cutting propagation was very positive, the industry could be reluctant to accept this new technology. In the preface for Wilfrid Sheat's 1948 book on propagation of woody plants, he is skeptical about the use of auxin in cuttings. "I think it is true to say that to date no real commercial advantage has yet been gained by the use of the substance (auxin) for the production of plants by cuttings. As a practical propagator, I would add a word of warning. The use of chemical root-producing materials is no substitute for the exercise of intelligent practice in the art of propagation." (Sheat, 1948).

The early Proceedings of the Plant Propagator's Society provide insight into early adoption of auxin and cuttings by commercial nurseries. In the first issue of the Proceedings in 1951, James Wells discusses auxin as "HORMONE TREATMENTS – There are growers who say that there are no results obtained by the use of hormones which the skilled propagator cannot develop without them. This is an argument to which we do not subscribe. We believe that used intelligently the plant hormones have a most definite place in modern plant propagation and we use them extensively. For our easily rooted varieties we use a powder containing 6 mg/g of indole butyric acid. This is the strongest commercially available powder in this country."

In the same Proceedings issue, Richard Fillmore commented on auxin use in his review of woody plant propagation. "The use of synthetic hormones is a well established and often beneficial practice in rooting cuttings. Assuming that one is thoroughly familiar with the most suitable hormone and the optimum concentration for the species under consideration, hormone treatments will unquestionably promote improved results with a wide variety of plants. When the requirements of this assumption cannot be met, the indiscriminate use of hormones may do more to inhibit than to promote rooting. I do not wish to be misunderstood. I am a pro-hormone man and I have successfully used hormones on dozens if not hundreds of species."

In a wonderfully frank discussion of auxin use by an established nursery propagator, Peter Vermeulen defers to the next generation. "The use of hormones! You know, I never went to high school. I never went to college, and I am awfully dumb in chemicals. I am quite thick-headed as a Dutchman and I didn't want to try a lot of things. I probably should have, but we finally got to the point we are trying chemicals. This year for the first time, we tried some indolebutyric acid, one and two percent. I don't know whether it will work out. I am not alone down there anymore, so they tell me I shouldn't be so old-fashioned, I should try some new things. We have to give in once in a while to the younger generation. So we have!"

Transition from dilute soak to talc and quick dip methods

As auxin treatment of cuttings was becoming accepted, the delivery method was evolving from the original dilute soak method to the talc (dust) and quick dip methods. Again, the Proceedings of the Plant Propagator's Society provides insights into this transition. In 1956, Henry Kirkpatrick provides a very good summary of this transition with his work in lilac (Kirkpatrick, 1956). Indolebutyric acid (IBA) and naphthaleneacetic acid (NAA), alone and in 50-50 mixtures, were used in a range of concentrations. The chemicals were applied to the cuttings by the 24-hour solution soaking method, by the talc powder method, and by the concentrated dip method. The solution soaking method required a 24-hour treatment in solutions containing from 40 to 80 mg active chemical in one liter of water. The solution soaking method has been largely replaced by the talk powder method because of the work and time involved in preparing solutions. The concentrated dip method required concentration from 10 to 20 mg of active chemical per ml of solution (20 to 30 times stronger than used for solution soaking). In preparing concentrations of IBA or NAA in this range, 95 per cent ethyl alcohol must be used to dissolve the chemicals since they are not water soluble."

It is apparent by 1959, that the quick dip method had become a preferred method for treating woody plant cuttings. In the 1959 Volume 9 of the Proceedings there were nine papers discussing the use of the quick dip method including a paper by Charles Hess comparing the quick dip with the powder method (Hess, 1959).

In 1959, the first edition of the reference textbook "Plant Propagation: Principles and Practices" by Hudson Hartmann and Dale Kester was published. It had detailed information on auxin use for cutting propagation. It is interesting to note the understanding of plant hormones at that time. They state that "There are several groups of substances considered plant hormones. These are (a) auxin, (b) traumatic acid, (c) caulines, and (e) vitamins. Auxin appears to act as a sort of "master hormone."

Auxin application methods

Selected auxin delivery systems are outlined in Table 1. The two methods that are most commonly used in greenhouse and nursery propagation are the talc and quick dip methods (Davies et al., 2018). For the talc application, the basal end of the cutting(s) are placed into the talc for a brief period until the powder adheres to the cutting. The available talc preparations come in a range of concentrations that are predetermined by the manufacturer. It is advisable to remove a small portion from the original container for daily use rather than dipping directly into the larger container. A dibble hole in the substrate may prevent talc loss when inserting the cuttings.

The quick dip method is often preferred by larger operations because it tends to give more uniform results as well as the flexibility to control the auxin concentration. Auxin solutions used for quick dips are available as concentrated stock solutions. The concentrated stocks are diluted to achieve the appropriate concentration to treat cuttings. Concentrated stock solutions use a solvent to keep the auxin in solution. Stock solutions for K-IBA are prepared in water. The quick dip application involves dipping cutting bundles into the solution for three to five seconds. The solution is rapidly absorbed into most cuttings prior to and after sticking. An alternative is to include a gel (carboxymethyl cellulose) with the auxin solution to increase the time auxin is in contact with the cutting base (Dip'N Grow, 2024). There are also premade preparations that include this gel and IBA at 3,000 ppm.

With an increasing emphasis on efficiency in cutting propagation, there has been renewed interest in alternative methods for delivering auxin. These include foliar auxin sprays and total cutting immersion. Foliar K-IBA sprays have been shown to be an efficient, cost-effective, labor-saving delivery method for treating cuttings (Dranm, 2007; Martindell, 2019). K-IBA in an aqueous solution is sprayed on cuttings following sticking. It is also a useful way to treat cuttings after being stuck by a robot. An entire day's sticking can be treated with auxin in a backpack sprayer.

Total immersion involves submersing a batch of cuttings for a short period of time. Total immersion can be used to rehydrate cuttings with or without a wetting agent, treat with a biopesticide (like BotaniGard or SuffOil-X), or as an auxin delivery system. It is also possible to combine these treatments before sticking cuttings. For example, cuttings can be first submerged in a wetting agent (like Uptake) for rehydration plus SuffOil-X for insect control followed by immersion for one to three minutes in K-IBA.

There is also interest in combining either wetting agents or growth regulating substances along with K-IBA during spray applications to cuttings. To date with a limited species list, there is little evidence that including a wetting agent to a K-IBA spray improves rooting (Bowden et al., 2022; Geneve, 2023). There are also limited studies investigating growth regulator combinations and root formation in cuttings. In a study using angelonia cuttings, Bonzi was combined with K-IBA to investigate whether this combination would have an impact on rooting and a carry-over effect on plant height or branching post-rooting (Baloh et al., 2022). The combination showed a slight improvement in rooting but there was no impact of plant growth post-rooting. Future studies will determine if batch spray applications combining growth regulating compounds have a positive impact on cutting production.

LITERATURE CITED

Avery, G. S., Jr. and Johnson, E.B. (1947). Hormones and Horticulture. McGraw-Hill Book Co. N.Y.

Baloh, A.G., Owen, W.G, and R.L. Geneve. (2022). Impact of foliar applied paclobutrazol in combination with auxin on rooting and subsequent shoot growth in *Angelonia* cuttings. Comb. Proc. Intl. Plant Prop. Soc. *72*:340–344.

Bowden, A.T.,. Knight, P.R.. Ryals, J.B.. Coker. C.E.H., Langlois. S.A., Broderick. S. R., Blythe. E.K., Sakhanokho. H.F. and Babiker. E.M. (2022). Evaluation of one-time applications of foliar applied auxin co-applied with surfactant for use in commercial cutting propagation. Agron. *12*:2243. /doi.org/10.3390/agronomy12102243

Boysen Jensen, P. (1911) La transmission de l'irritation phototropique dans l' *Avena*. K. Danske Videnskab. Selskab., Forh. *3*:1-24.

Chadwick, L.C. and Kiplinger, D.C. 1938. The effect of synthetic growth substances on the rooting and subsequent growth of ornamental plants. Proc. Amer. Soc. Hort. Sci. 36:809-816.

Cooper, W.C. (1935). Hormones in relation to root formation on stem cuttings. Plant Physiol. *10*:789-794.

Cooper, W.C. (1944). The concentrated-solution-dip method of treating cuttings with growth substances. Proc. Amer. Soc. Hort. Sci. *44*:533-541.

Darwin, C. and Darwin, F. (1880). The Power of Movement in Plants. London.

Davies, F.T. Jr, Geneve, R.L. and Wilson, S.B. (2018). Hartmann and Kester's Plant Propagation: Principles and Practices (9th ed). Prentice-Hall, Boston, MA, USA.

Dip'N Grow Bulletin, (2024). https://www.dipngrow.com/wp-content/uploads/2010/08/Dip-Gel-Bulletin.pdf Accessed March 2024.

Drahn, S. R. (2007). Auxin application via foliar sprays. Comb. Proc. Intl. Plant Prop. Soc. *57*:274–77.

Fillmore, R. (1951). General review of woody plant propagation. Comb. Proc. Intl. Plant Prop. Soc. 1:40-50.

Geneve, R.L. (2023). Foliar application of K-IBA to the abaxial or adaxial leaf surface with or without a surfactant does not impact root formation in buttonbush and burning bush euonymus cuttings. Journal of Environmental Horticulture *41*:97-100.

Grace, H.H. (1937). Physiologic curve of response to phytohormones by seeds, growing plants, cuttings, and lower plant forms. Can. J. Res. *15*:538-546.

Hess, C. (1959). Comparison between the quick dip and powder methods for growth substance application to cuttings. Comb. Proc. Intl. Plant Prop. Soc. 9:45-46.

Hitchcock, A.E. and Zimmerman, P.W. (1939). Comparative activity of root-inducing substances and methods for treating cuttings, Contrib. Boyce Thompson Inst. *8*:63-79.

Kirkpatrick, H. (1956). propagation of hybrid lilacs from cuttings. Comb. Proc. Intl. Plant Prop. Soc. 6:81-83.

Kögle, F., Haagen-Smit, A.J., and Erxleben, H. 1933. Über ein phyohormon der zellstreckung. Reindarstellung des auxins aus menschlichem harn. Hoppe-Seyler's Z. Physiol. Chemie. *214*:241-261.

Martindell, J.C. (2019). Foliar infole-3-butyric acid rooting hormone application and cost analysis. Comb. Proc. Intl. Plant Prop. Soc. *69*:165–168.

Thimann, K.V. (1935). On an analysis of activity of two growth-promoting substances on plant tissues. Proc. Kon. Akad. Wetensch. *38*:896-912.

Vermeulen, J. (1954). The propagation of *Taxus* by cuttings. Comb. Proc. Intl. Plant Prop. Soc. 4:76-81.

Wells, J. (1951). Propagating Rhododendrons from stem cuttings. Comb. Proc. Intl. Plant Prop. Soc. 1:12-14.

Went, F.W. (1938(. Specific factors other than auxin affecting growth and root formation. Plant Physiol. *13*:55-80.

Wilfrid G. Sheat. 1948. Propagation of Trees, Shrubs and Conifers. London.

Zimmerman, P.W. and Hitchcock, A.E. 1935. Several chemical growth substances which cause initiation of roots and other responses in plants. Contr. Boyce Thompson Institute 7:209-229.

Zimmerman, P.W. and Wilcoxon, F. (1935). Several chemical growth substances which cause initiation of roots and other responses in plants. Contrib. Boyce Thompson Inst. 7:209-229.

Combined Proceedings IPPS 74:16-19. 2024.

Thinking Outside the Box(wood)

Tom Saunders

Saunders Nursery, Piney River, Virginia 22964 USA

tom@saundersbrothers.com

Keywords: business, nursery, farm market, boxwood, Saunders Brothers

Summary

Saunders Brothers is a multi-disciple business including a wholesale nursery, fruit orchard, farm market and garden center. After a brief history of Saunders Brothers, key aspects of business practices were described.

INTRODUCTION

Saunders Brothers was started in 1915 by five brothers who believed that a handshake and a verbal contract were binding. Working together, they raised cattle, tobacco, and apples. During the Great Depression they caught and sold rabbits to keep their business afloat.

The business was continued by 2nd generation Paul Saunders who had a civil engineering degree and a surveying business. His 4-H project in 1947 was rooting boxwood. With the help of his science teacher, he turned growing plants into a successful business. He grew boxwood mainly in the field and, in a limited fashion, some in containers. The container nursery expanded tremendously after Hurricane Camille in 1969 as over 30" of rainfall fell over an 8-hour period. Rich bottomland of field-grown plants were washed away, and his business approach changed. Paul and Tatum had 7 boys to meet the labor needs on his thriving farm.

Starting in 1980, his sons started returning from college. Six of the 7 biological sons returned to work at one point or another in the business. Also, Paul and Tatum took in a "son" from Canada and one from Russia. Four of the boys stayed on to work in the business along with their Canadian brother. The '80's were good for business as Ronald Reagan was President and interest rates were down.

Today, Saunders Brothers produces over 1100 plant varieties on their 100+ acre container nursery. An adjacent 500-acre farm was purchased in 2021. The first company auto vent greenhouses were built in 2023. Sixteen 30' x 200' houses were built with the layout for 10 more. These greenhouses will be covered 24/7 for *Buxus* production. The goal is to keep dew and rainfall off the crops thus reducing the possibility of those plants developing Boxwood Blight.

Production at Saunders Brothers includes an extensive line of woody plants, annuals, perennials, and some shade and ornamental trees. The goal is to be a one-stop shop for independent garden centers and landscapers. The company is part of a five nursery LLC called Synrgy; all five share common business philosophies. The LLC shares business ideas and, through breeders, their own plant genetics. The businesses, on a rotating basis, open their operations to the other partners and their employees for a few days of "seeking and sharing".

Boxwood are the number one item produced at Saunders Brothers and constitute nearly 30% of the corporate sales. Field plants are grown on raised beds to improve drainage. The company owns patents on several cultivars and has its own breeding program and an off-site Boxwood Blight testing site. The NewGen™ line of Buxus is the property of Saunders Brothers and currently has licensed 32 growers around the world.

Product is shipped into 13 states which are mainly in the northeast Atlantic area. The product is mainly shipped on rolling carts to improve plant presentation and reduce employee touches. It also makes it easier for independent delivery drivers to deliver the company product. In 2023, the company signed a contract with a Lean Flow company to improve the efficiency of its shipping system.

In addition to the container operation, there are 159 acres of field production. The rolling topography has been a challenge and requires heavier horsepower tractors to accomplish the work. Fruit production includes apples, peaches, Asian pears, and nectarines. Also, the company produces over 50,000 one-year bench-grafted apple trees.

Retail speaking, the company has a Farm Market that is open for 9 months annually and sells their fresh fruit, and plants. Jams, jellies, salsa, beef, pork, and ice cream are also sold through the Farm Market. In the winter months, the company sells Christmas trees and makes high-end wreaths for their customers.

What has it taken for the company to succeed? Certainly, *Buxus* is the crop on which they hang their hat, notably providing plants to the White House Rose Garden on two occasions.

To keep the business going for future generations, the company hired independent succession planning contractors who have helped steer and structure the growing company. At present, five of the fourth-generation offspring have returned with roles in the company. Saunders Brothers has 66 full-time employees, 13 part-time, and 108 H2A workers.

Job descriptions are the key for an employee to understand his role. The description must be detailed enough for the employee but broad enough so that an employee can assist in other capacities. This is especially true with a company that has seasonal workloads. Training an employee requires a manager to spend the essential amount of time for a worker to understand the expectations for them.

Also, employees must be provided with the right tools for success whether it is a computer, vehicle, or anything else that makes them more efficient. One method of improving efficiency that the farm still utilizes is incentive pay. This type of piece rate work can reduce a job's per unit cost but requires management to set quality standards. Holding onto good workers is done by providing the right company culture, paying them comparable wages, offering good benefits, and treating workers with respect. Challenging workers is also important to make workers reach new levels.

Communication is always important in any relationship and Saunders Brothers believes in the sandwich approach when addressing an issue. Approach the topic with something positive, present the issue at hand, and wrap up by touching on something positive.

As mentioned before, take time to train a worker. When you feel they are ready to move ahead, give them a longer leash. Not training a worker about basic tasks results in a worker not meeting the owner's satisfaction. Be patient during the training process and understand that employees will make some mistakes and not do some things the way you expected. These small mistakes can turn into teaching opportunities.

Training workers includes letting them go to industry events. Making them accountable includes them writing up a summary of what they learned and saw while away. Sometimes other staff members can benefit from their notes.

At Saunders Brothers, we believe in detailed recordkeeping. Workers punch in on time clocks associated with profit centers. These detailed time records help us determine which products are making money and which are not.

Also, SOP's (standard operating procedures) are written by seasoned staff members to avoid going through a learning curve twice. SOP's can be written for even the simplest tasks. We also develop "plant recipes" that include when to plant a crop, what the soil type should be, when to prune it, and possible pests. If crops require extra attention, we try to determine how to tweak the recipe to eliminate them.

As with anything, good numbers are critical. We developed a spreadsheet years ago that includes all the production expenses associated with a crop. This spreadsheet helps us determine the profitability per plant and the return per square foot.

Growers are taught that spacing can be reduced, inconsistent liners can be thrown away, and percentage to market can be improved to result in a better margin.

In recent years, we have invested in some technology that makes us better stewards and better producers. Examples include web relays in the place of thermostats, a Vapor Pressure Deficit (VPD) misting system in propagation and an Evapotranspiration (ET) based irrigation system. The ET based irrigation system alone saved us millions of gallons of water and reduced our fertilizer usage.

We lock in our gas prices up to 3 years down the road with a local company and have saved tremendously.

What does the future have in store for us? We certainly can't predict it but we want to be more sustainable. We hope to move our Farm Market to an area with higher traffic patterns. We will introduce new genetics and hope to build a gutter-connect propagation greenhouse. With all this, we continue to think outside the box.

Most importantly, at SBI we remember that we didn't achieve what we've done alone. It's been a great journey, and we believe the best is still ahead.

Combined Proceedings IPPS 74:20-21. 2024.

Lessons Learned: My 43 Year Journey with Commercial Micropropagation

Gayle R.L. Suttle

CEO-Retired, Microplant Nurseries, Inc., Gervais, Oregon 97026 USA

gayle@microplantnurseries.com

Keywords: tissue culture, business, philosophy

Summary

Gayle Suttle the long-time CEO of Microplant Nurseries, Inc, provides insights into running a successful micropropagation business learned over four decades.

INTRODUCTION

Microplant Nurseries, Inc., is one of the largest and longest running commercial plant micropropagation facilities in the world. Gayle provided a history of Microplant and also told some stories of what it was like walking into the project 43 years ago. She then shared some of the most important lessons she has learned along the way. Here is a summary:

1) You learn most from your mistakes. So ... give yourself a break! Don't waste time beating yourself or others up about it. Figure out what went wrong, fix it the best you can, set safeguards in place so you don't repeat the mistake, and then move on.

2) Simplify the process. Make it easier to do things right and harder to do things wrong.

3) It is easy to build a lab, but very hard to make one work.

4) Know where you are going ... and ... believe you will get there. This means you should know what the goal is. Declare where you are going! Know what you want! Keep your eyes on the prize.

5) It's not a success unless it is repeatable. Document the process so you know what you did.

6) Calculators are fabulous toys. The author uses a horseshoeing story to illustrate the power of exponential growth, calculating the value of each of 32 nails doubled every time. The first nail is worth one penny. The last one is worth over $21million. It looks so easy, but EVERYTHING must be doubled, and EVERYTHING must go as planned.

7) Under-promise and over-deliver. Customers rely on their orders to arrive on time and to be filled in full. When planning, fudge factors are our friends. Work backwards on the calendar, set benchmarks to keep on track and have the grace to renegotiate the delivery with your customers if things aren't on track. Let them decide if they want to take the crop late or take less. But really, plan to use so many fudges that you hit the goal right on time, maybe even early! The most important thing of all in developing a business is to recognize that whatever you do, whatever you produce, whatever you are selling MUST be what the customer wants!

8) Be relentless in the pursuit of excellence. Robert Townsend, in his great book 'Up the Organization' said "If you don't do it excellently, don't do it at all. Because if it's not excellent, it won't be profitable or fun, and if you are not in business for fun or profit, what the hell are you doing there?

9) Clean plants are easier to manage. Plants are not sterile. Build defenses against bad guys, make it inhospitable for them. Also, build early detection systems. Be relentless about it.

10) Plan on growing. This means – plants, space and people.

11) Document the process. Everything. Hard copy and digitally. This will help you do things right and avoid repeating past mistakes. Use technology, build books and use them to teach and remember. Update regularly. It is exhausting but be relentless about it. Computers are terrific.

12) Share the wealth. Knowledge, ideas, triumphs, challenges, fun, dollars and promote from within.

13) Look around. Get out there! Be curious! Make friends ... you need them.

14) Enjoy the journey. Have fun with – fellow employees, customers, compatriots and competitors. All are friends and life is short.

15) Books on my reading list include:

The Goal by Eliyahu M. Goldratt

Seven Habits of Highly Effective People by Steven R. Covey

Up the Organization: How to Stop the Corporation from Stifling People and Strangling Profits by Robert Townsend

Who Moved My Cheese by Spencer Johnson, M.D.

Combined Proceedings IPPS 74:22-27. 2024.

Investigation of the Propagation of Pink-flowered *Haemanthus* by Inflorescence Culture

Ayaka Ohuchi and Hiroaki Ohashi

Graduate School of Agriculture, Ehime University, 3-5-7, Tarumi, Matsuyama, Ehime, 790-8566, Japan

ohashi@agr.ehime-u.ac.jp

Keywords: tissue culture, organogenesis, geophyte, bulb

Summary

Adventitious bud formation in H*aemanthus* tissue culture was successful using floret peduncles as the original explant. It was also possible to get organogenesis from callus cultures originally taken from ovary tissue after exposing cultures to low temperatures.

INTRODUCTION

Haemanthus species is bulbous plant of the Amaryllidaceae family and about 60 species have been identified, mainly in South Africa. In Japan, *H. albiflos* Jacq. is the most popular species of the genus, which is widely cultured and distributed. On the other hand, red to pink-flowered *Haemanthus* are expensive and distribution amount is few, because it takes a long time to propagate through division. The more efficient propagation method for this species is needed to meet demand. *Haemanthus* is bulbous plant, so it has dwarf stem inside of the bulb. The inflorescence of *Haemanthus* plants is enclosed in bract leaves. These are specialized leaves with axillary buds in the leaf axils that become flowers. For example, *Haemanthus* has about sixty florets present inside the bract leaves, blooms with the bract leaves slightly open. *Haemanthus albiflos* propagates by division and seeds.

Flower stalk and inflorescence culture are a beneficial method, in the plants that have dwarf stem which is underground or at the surface of the earth, and where the flower stalk is the only elongating stems. The reasons are as follows: low rate of contamination, high rate of organogenesis and callus formation, cause little damage to mother plant (Ohashi *et. al.*, 2009, Matsumoto and Ohashi, 2015).

In this study, we aim to create the propagation method of pink-flowered *Haemanthus*. We used the inflorescence because could have the availability of adventitious bud differentiation.

Allium fistulosum L. var. *viviparum* Makino and *A. macrostemon* Bunge, which is Amaryllidaceae family same with *Haemanthus*, set bulbils at inflorescence. Especially, *A. macrostemon* sets florets and bulbils at inflorescence in the same timing. Based on *Allium* plants, inflorescence of *Haemanthus* could make bulbils by inflorescence culture.

MATERIAL AND METHODS

Pink-flowered *Haemanthus*, presumed to be derived from interspecific hybrids, were used in the experiment. Inflorescences were collected in late September 2023, when the inflorescences elongated about 4 cm from the scale (**Fig. 1**).

Figure 1. Inflorescence of pink flowered *Haemanthus* before cutting.

The inflorescences were collected with the flower stalks attached about 2 cm from the bract leaf base (BLB) by using box-cutter, and the cut ends were immediately sealed with melted wax to prevent the entry of bacteria. The bract leaves were cut at a position that would not damage the BLBs, and the florets were cut off perianths etc., leaving the ovaries (**Fig. 2**). Then, the BLBs were sterilized with sodium hypochlorite solution supplemented with agrochemical spreader for 8 minutes for surface sterilization. After that, BLBs are rinsed in sterile distilled water. After that, flower stalks and florets detached from the BLBs, leaving about 2 mm above and below the BLBs.

BLBs which were divided into three pieces, and ovaries of florets were placed on the culture medium, which were MS medium (Murashige and Skoog, 1962) supplemented with 30g $\cdot$ L^{-1} sucrose, 2.5 g $\cdot$ L^{-1} Gellan Gum. Plant growth regulators (PGRs) combination was shown Table 1, which is consisted with 0, 1, 2 mg$\cdot L^{-1}$ 6-Benzylaminopurine (BA), 0, 1, 2 mg$\cdot L^{-1}$ 1-

Naphthaleneacetic acid (NAA). The medium pH adjusted to 5.8 prior to autoclaving at 121°C for 15 minutes. The culture condition kept under a 16/8h light/dark regimen at 21±2°C.

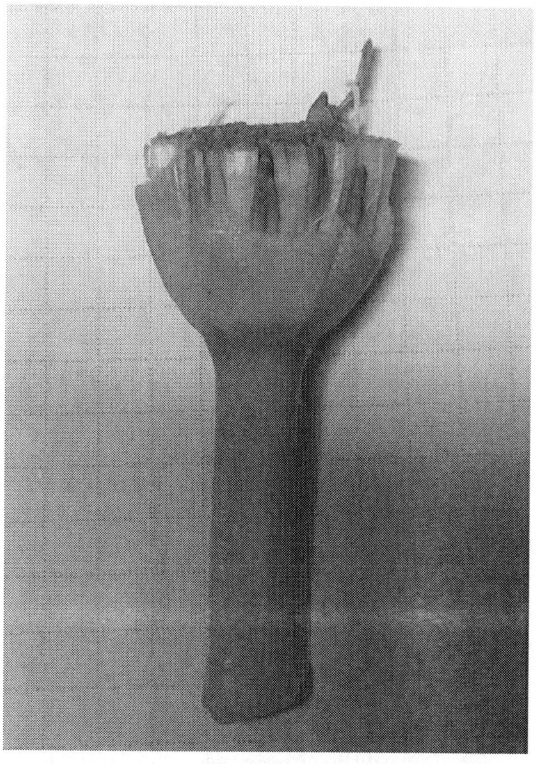

Figure 2. The inflorescence before surface sterilization. It was collected with the flower stalks attached about 2 cm from the bract leaf base, and the cut ends were immediately sealed with melted wax. The bract leaves were cut at a position that would not damage the BLB, and the florets were cut off perianths etc.

BLBs culture was conducted with 6 repetitions in 5 treatments; 3 repetitions in control (**Table 1**), and ovaries culture conducted with 9 repetitions in 5 treatments (**Table 2**).

RESULTS AND DISCUSSION

After placement, survival, contamination, callus formation, and organogenesis recorded periodically. After 56 days, organogenesis observed in all PGRs combination BLBs culture, including the control section (**Table 1**). Organogenesis observed from the base of the cut end of floret peduncles (**Fig. 3**), and although number of organogenesis tended to be higher in the PGRs combination, there was also a higher tendency for the organogenesis to be malformation.

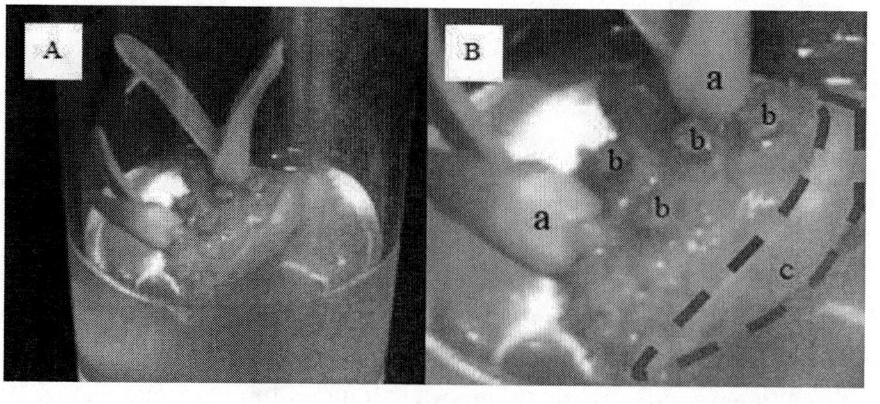

Figure 3. Organogenesis on control on bract leaf base culture. A) Whole explant with adventitious bud formation. B) Enlargement of A. (a) Adventitious buds. (b) The base of the floret peduncles. (c) Cutting surface of bract leaf.

Table 1. Effects of plant growth regulators for organogenesis on bract leaf base explants of pink flowered *Haemanthus* (after 56 days incubation at 21±2°C).

PGRs ($mg \cdot L^{-1}$)		Explants for testing (No.)	Survived explants (No.)	Explants with observed organogenesis (No.)	Rate of organogenesis (%)
NAA	BA				
0	0	3	2	2	100
1	1	6	4	3	75.0
1	2	6	5	3	60.0
2	1	6	4	3	75.0
2	2	6	4	4	100

PGRs: Plant growth regulators (PGRs). NAA: 1-Naphthaleneacetic acid, BA: 6-benzylamino-purine.

In addition, the contamination rate was about 30% in total. It might be because the bract leaves were slightly open when the inflorescences were collected, which increased the contamination rate of the inflorescence interior by some insects and fungi, as well as the fact that the surface fungicide did not sufficiently penetrate to the culture area (BLB) due to the morphological characteristics of the inflorescences.

After 56 days, no organogenesis observed in all PGRs combination ovaries culture, but callus formation observed around cut end of peduncle side (**Table 2**).

Table 2. Effects of plant growth regulators for organogenesis and callus formation on ovary explants of pink flowered *Haemanthus* (after 56 days incubation at 21±2°C).

PGRs ($mg \cdot L^{-1}$)		Explants for testing (No.)	Survived explants (No.)	Explants with observed organogenesis (No.)	Explants with observed callus (No.)	Callus formation (%)
NAA	BA					
0	0	9	5	0	1	20.0
1	1	9	9	0	5	55.6
1	2	9	9	0	4	44.4
2	1	9	9	0	5	55.6
2	2	9	8	0	2	25.0

PGRs: Plant growth regulators (PGRs). NAA: 1-Naphthaleneacetic acid, BA: 6-Benzylamino-purine. Rate of callus formation = (No. of explants observed callus / No. of survived explants) x 100.

Flowering timing of *Haemanthus* sp. is from late September to early October, after that, vegetative growth stage has come, incubation at low temperature would occur organogenesis. After 5 months incubation at 21±2°C, we moved ovaries to growth chamber, which was set 15°C, 20°C and 25°C. After 230 days incubation, we got organogenesis from ovaries via callus at PGRs combination of 15°C and 20°C; no organogenesis was observed at 25°C (**Table 3**).

Table 3. Effects of plant growth regulators and temperature for organogenesis on ovary explants of pink flowered *Haemanthus* (after 230 days incubation).

Incubation temperature	PGRs (mg·L^{-1})		Explants for testing (No.)	Explants observed organogenesis (No.)	Rate of organogenesis (%)
	NAA	BA			
	0	0	1	0	0
	1	1	2	0	0
15	1	2	4	2	50.0
	2	1	3	0	0
	2	2	2	0	0
	0	0	-	-	-
	1	1	3	2	66.7
20	1	2	3	1	33.3
	2	1	3	1	33.3
	2	2	3	0	0
	0	0	1	0	0
	1	1	3	0	0
25	1	2	2	0	0
	2	1	3	0	0
	2	2	2	0	0

PGRs: Plant growth regulators (PGRs). NAA: 1-Naphthaleneacetic acid, BA: 6-Benzylaminopurine. After 5 months incubation at 21±2°C, we moved ovaries to 15°C, 20°C and 25°C.

In conclusion, we got organogenesis which include adventitious buds from inflorescences of *Haemanthus* sp. BLBs culture at 21±2°C, we got adventitious buds directly from the base of the floret peduncles. The number of organogenesis was higher at PGRs combination, there was also a higher tendency for the organogenesis to be malformation.

In ovaries culture, it was difficult to get organogenesis at 21±2°C, but by changing to low-temperature culture at 15°C and 20°C, we got organogenesis via callus from ovaries.

We will study the effect of low temperature on BLBs culture to establish more efficient propagation method of *Haemanthus*. In addition, we will conduct multiplication by using adventitious buds which were obtained from this study.

LITERATURE CITED

Ohashi, H., K. Akaboshi, and Y. Ochi. (2009). Obtaining plantlets by flower bud culture and scape culture in *Primula kisoana*. Comb. Proc. Intern. Plant Prop. Soc. *59*:133-136.

Matsumoto, Y., and H. Ohashi. (2015). *In vitro* shoots formation by inflorescence apex culture of *Primula ×polyantha*. Comb. Proc. Intern. Plant Prop. Soc. *65*.97-100.

Murashige, T., and F. Skoog. (1962). A revised medium for rapid growth and bioassays with tobacco tissue culture. Physiol. Plant *15*:473–497.

Combined Proceedings IPPS 74:28-33. 2024.

Polyploidy Induction by Colchicine Treatment in Kenaf (*Hibiscus cannabinus* L.)

Masaki Ochiai, Kazumasa Okuda, Yuta Kawahara, Yuki Matsumoto and Hirokazu Fukui

Faculty of Applied Biological Science, Gifu University, 1-1, Yanagido, Gifu, Gifu, 501-1193, Japan

ochiai.masaki.h3@f.gifu-u.ac.jp

Keywords: breeding, tetraploid, phenotype, flow cytometry

Summary

Polyploid breeding improves the ornamental value and efficiency of interspecific crosses. Kenaf (*Hibiscus cannabinus* L.) is expected to be a breeding resource for blue flowers among ornamental plants of the genus *Hibiscus*. We investigated the optimal conditions for polyploidy induction and the morphological changes induced by polyploidization in kenaf. The most efficient conditions for tetraploidy induction in blue-flower-type kenaf were soaking of the seedlings in 3.0×10^{-3} M colchicine solution for 24 h. Colchicine soaking at 1.0×10^{-3} M for 12 to 24 h was suitable for white-flower-type kenaf. In either type of kenaf, leaflet length to width ratio, guard cell length, petal length to width ratio, petal thickness, pollen diameter, and seed fresh weight were greater in tetraploids than in diploids.

INTRODUCTION

Some tropical and temperate species of the genus *Hibiscus* (Malvaceae family) are popular as ornamental plants because of their beautiful flower color and shape and their ease of cultivation. There is a wide range of flower colors, including white, yellow, peach, orange, and red, but few blue flower cultivars exist. Kenaf (*Hibiscus cannabinus* L.) is used mainly as a raw material for fiber, but it has not only white and yellow flowers but also blue flowers, which are rare among both wild species and horticultural cultivars of the genus *Hibiscus*. Therefore, kenaf with blue flowers is expected to be useful as a breeding resource to establish blue horticultural cultivars of the genus *Hibiscus*.

In interspecific crosses, sterility of the hybrid progeny is an obstacle to further breeding. Use of polyploids—mainly tetraploids—as hybrid parents can improve the success rate of interspecific hybridization and the fertility of the hybrid progeny. In this study, we tried to induce polyploids in kenaf as a preliminary step to interbreeding between kenaf and other *Hibiscus* species that are used as ornamental plants.

MATERIALS AND METHODS

Two strains of *H. cannabinus* L. were used (blue flower type and white flower type). Seeds were collected from plants grown in a greenhouse at the Gifu Field Science Center, Gifu University.

Seeds were dipped in sulfuric acid for 60 min for blue flower type and 30 min for white flower type to scarify the impermeable seed coat and then washed in running water for 1 h. They were then placed on moist filter paper in sealed plastic cases at 25 °C until germination. Germinated seeds with about 2 mm of root were subjected to polyploidy induction treatment with colchicine.

The seedlings were soaked for 12 to 48 h in 3.0×10^{-3} to 1.0×10^{-2} M colchicine solution containing 10% dimethyl sulfoxide. After the colchicine treatment, the seedlings were rinsed in running water for 1 h and then planted in plastic pots. They were grown under natural daylight in a glasshouse that was heated only in the winter.

Ploidy analysis was conducted by flow cytometry (Partec, Ploidy Analyser PAII, Germany). Fully expanded leaves were prepared for flow cytometry by using the chopping method (Galbraith et al., 1983).

Leaflet length to width ratio, guard cell length, petal length to width ratio, petal thickness, pollen diameter, and seed fresh weight were measured in diploids and tetraploids. Leaflet length to width ratio was measured in terminal leaflets. Petal thickness was evaluated from the ratio of petal fresh weight to petal area. Seeds were acquired through artificial pollination between plants of the same ploidy.

RESULTS AND DISCUSSION

Seedlings of blue-flower-type kenaf were treated for 24 or 48 h with 1.0×10^{-3}, 3.0×10^{-3}, or 1.0×10^{-2} M colchicine. Diploids, tetraploids, and octaploids, as well as chimeras in various combinations of 2x, 4x, and 8x, were observed after the colchicine treatment (**Fig. 1**).

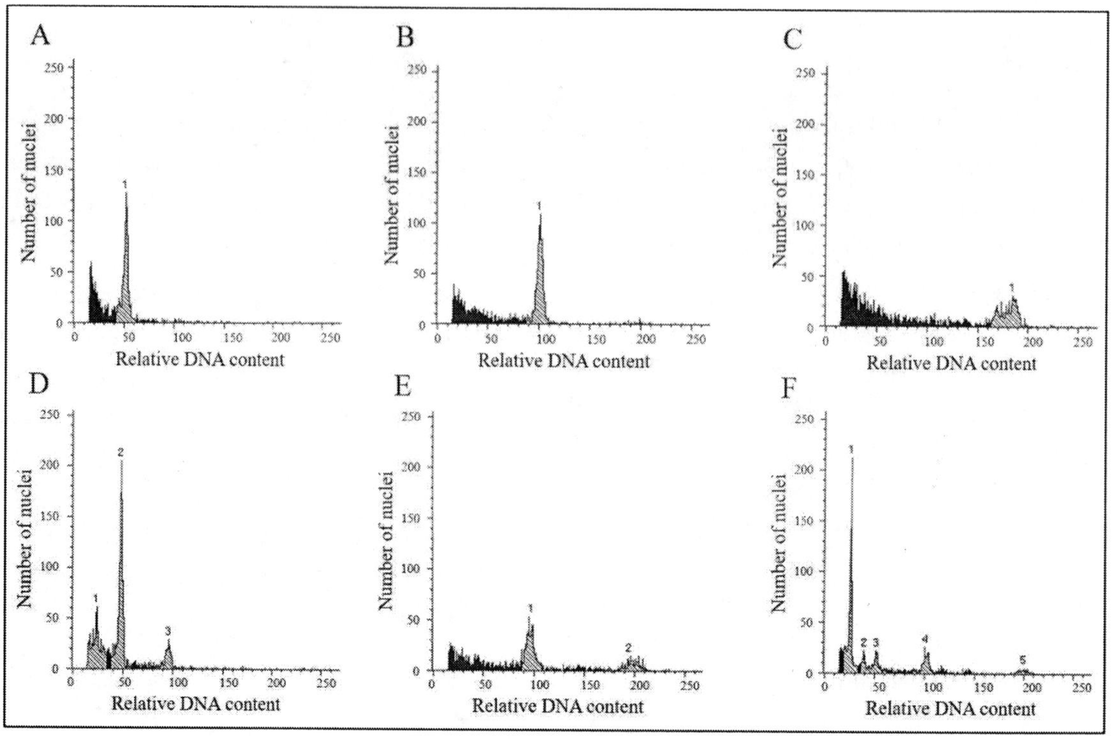

Figure 1. Histogram of flow cytometry after colchicine treatment in blue-flower-type kenaf. A) diploid; B) tetraploid; C) octaploid; D) chimera of 2x and 4x; E) chimera of 4x and 8x; F) chimera of 2x, 4x, and 8x. Numbered and shaded areas indicate estimated peaks.

RESULTS AND DISCUSSION

Tetraploids were obtained under all conditions (**Table 1**). About 300 seedlings were treated under each set of conditions. The greatest number of tetraploids was obtained with 1.0×10^{-3} M colchicine for 48 h, followed by 1.0×10^{-3} M colchicine for 24 h. The number of tetraploids decreased with increasing colchicine concentration. The numbers of other ploidy states were greater with 1.0×10^{-3} M colchicine than with 3.0×10^{-3} or 1.0×10^{-2} M.

The tetraploidy rate in surviving plants was highest with 3.0×10^{-3} M colchicine for 24 h, followed by 3.0×10^{-3} M colchicine for 48 h. The most efficient conditions for polyploidy induction in blue-flower-type kenaf were therefore 3.0×10^{-3} M colchicine for 24 h.

Based on the results in blue-flower-type kenaf, we tested a smaller range of optimal conditions for polyploidy induction in white-flower-type kenaf (1.0×10^{-3} or 3.0×10^{-3} M colchicine for 12 or 24 h). The survival rate was 48.4% or 34.4% with 1.0×10^{-3} M colchicine, and it was almost 0% with 3.0×10^{-3} M colchicine (**Table 2**). In the case of blue-flower-type kenaf, the survival rates were about 30% and 15%, respectively. High concentrations of colchicine inhibit root and shoot elongation and organ differentiation by inhibiting cell division. As we had observed no difference in germination rates between the two types of kenaf without colchicine in prior tests, we concluded that the white flower type was more susceptible than the blue flower type to the negative effects of colchicine.

The optimal conditions for polyploidy induction therefore differed within the same species. Comparison with other *Hibiscus* species revealed that the optimal colchicine concentration for polyploidy induction in the two types of kenaf was higher than that for *H. mutabilis* L. (Ogasawara et al., 2010).

Table 1. Polyploid induction by colchicine in blue-flower-type kenaf.

Concentration of colchicine (M)	Duration of treatment (h)	Number of treatment	Number of survival	Number of 2x	Number of 4x	Number of other ploidy state	Survival rate (%)	Rate of 2x (%)^Z	Rate of 4x (%)^Z
1.0×10^{-3}	24	298	84	1	65	18	28.2	1.2	77.4
1.0×10^{-3}	48	300	95	9	80	6	31.7	9.5	84.2
3.0×10^{-3}	24	304	47	0	46	1	15.5	0	97.9
3.0×10^{-3}	48	283	43	0	41	2	15.2	0	95.3
1.0×10^{-2}	24	290	15	3	9	3	5.2	20.0	60.0
1.0×10^{-2}	48	294	44	2	34	8	15.0	4.5	77.3

^{Z}Ratio of total number of plants to number surviving.

Table 2. Polyploid induction by colchicine in white-flower-type kenaf.

Concentration of colchicine (M)	Duration of treatment (h)	Number of treatment	Number of survival	Number of 2x	Number of 4x	Number of other ploidy state	Survival rate (%)	Rate of 2x (%)^Z	Rate of 4x (%)^Z
1.0×10^{-3}	12	64	31	0	27	4	48.4	0	87.1
1.0×10^{-3}	24	64	22	0	22	0	34.4	0	100.0
3.0×10^{-3}	12	64	1	0	0	1	1.6	0	0
3.0×10^{-3}	24	64	0	0	0	0	0.0	0	0

^{Z}Ratio of total number of plants to number surviving.

We then compared the organ shape between diploids and tetraploids. Both the mature leaves and the petals of the tetraploids seemed wider than those of the diploids in the blue-flower-type kenaf (**Fig. 2**). In fact, the leaflet width to length ratio and the petal width to length ratio were significantly greater in the tetraploids than in the diploids (**Fig. 3**). Petal thickness, pollen diameter, guard cell length, and seed weight were also significantly greater in tetraploids than in diploids.

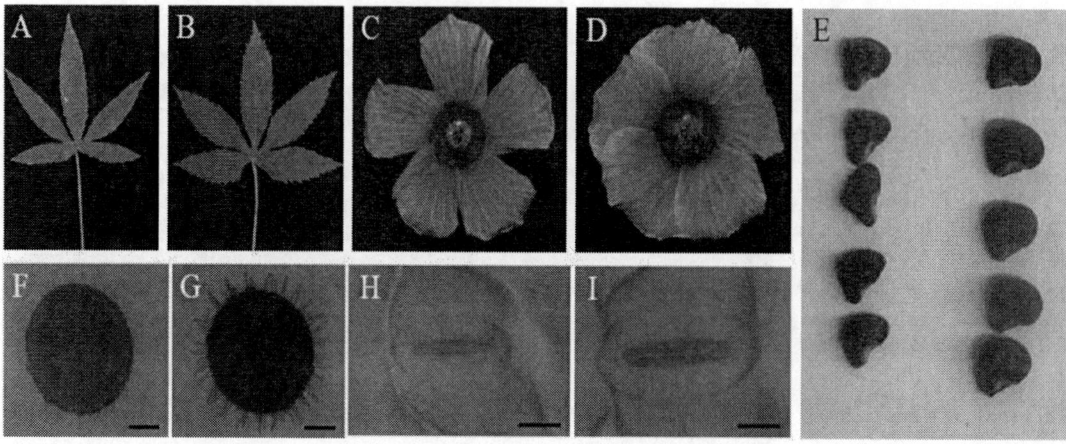

Figure 2. Characteristics of diploids and tetraploids of blue-flower-type kenaf. Mature leaves - A) diploid; B) tetraploid). Flowers - C) diploid; D) tetraploid. Seeds - E) left: diploid; right: tetraploid. Pollen - F) diploid; G): tetraploid, bars = 40 µm. Guard cells -H) diploid; I) tetraploid, bars = 10 µm.

Similar morphological changes were also observed in the white flower type (**Figs. 3** and **4**), and the trends and degrees of morphological change were roughly the same between the blue flower type and the white flower type. Increased organ width and larger size are common morphological changes caused by polyploidization in many plant species (Adachi et al., 2016; Niazian and Nalousi, 2020; Sugimoto et al., 2010), and these morphological changes were conserved in kenaf. In the genus *Hibiscus*, similar morphological changes with polyploidization have been observed in *H. moscheutos* L. and *H. mutabilis* L. (Li and Ruter, 2017; Ogasawara et al., 2010).

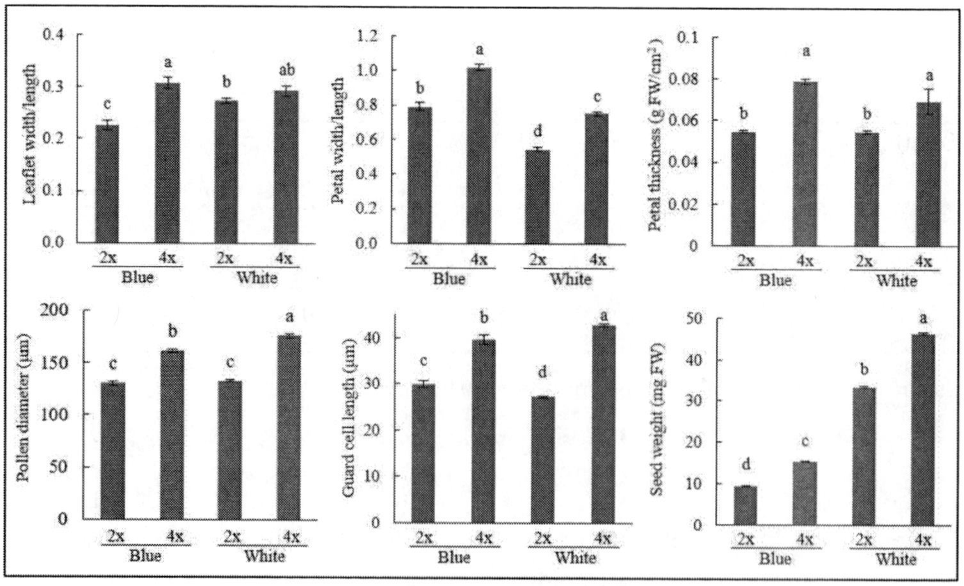

Figure 3. Characteristics of diploids and tetraploids. 2x: diploid; 4x: tetraploid; Blue: blue-flower-type kenaf; White: white-flower-type kenaf. Different lower-case letters in the same graph indicate a significant difference at $P < 0.05$ according to the Tukey–Kramer test.

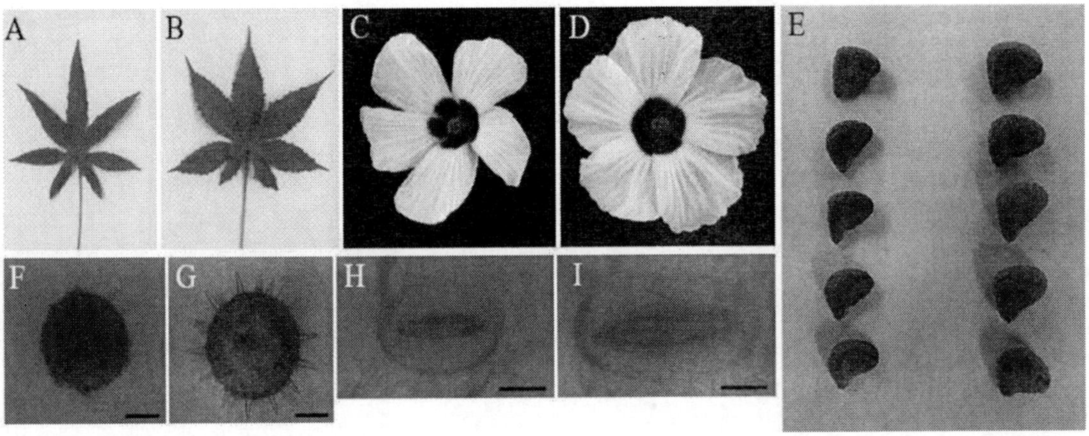

Figure 4. Characteristics of diploids and tetraploids of white-flower-type kenaf. Mature leaves - A) diploid; B) tetraploid. Flowers - C) diploid; D) tetraploid. Seeds - E) left: diploid; right: tetraploid. Pollen - F) diploid; G) tetraploid, bars = 40 μm). Guard cells - H) diploid; I) tetraploid, bars = 10 μm.

LITERATURE CITED

Adachi, Y., M. Ochiai, and H. Fukui. (2016). Induction of polyploidy plants through colchicine treatments in balloon vine (*Cardiospermum halicacabum*). Proc. Int. Plant Prop. Soc. *66*:101–104.

Fukui, H., and T. Yokota. (2007). Tetraploid induction by colchicine and oryzalin in *Rosa multiflora*. Acta Hortic. 751:313–322.

Galbraith, D. W., K. R. Harkins, J. M. Maddox, N. M. Ayres, D. P. Sharma, and E. Firoozabady. (1983). Rapid flow cytometric analysis of the cell cycle in intact plant tissues. Science, *220*:1049–1051.

Li, Z., and M. Ruter.(2017). Development and evaluation of diploid and polyploid *Hibiscus moscheutos*. HortSci. 52: 676–681.

Niazian, M., and A. M. Nalousi. (2020). Artificial polyploidy induction for improvement of ornamental and medicinal plants. Plant Cell Tiss. Org. Cult. *142*:447–469.

Ogasawara, R., Y. Kawahara, and H. Fukui. 2010). Tetraploids induction by colchicine treatment in *Hibiscus mutabilis*. Proc. Int. Plant Prop. Soc. *60*:636–641.

Sugimoto, H., H. Fukui, Y. Aoki, T. Tatematsu, and M. Hayashi. (2010). Tetraploid induction by colchicines in *Rosa banksiae*. Acta Hortic. *870*:147–152.

Combined Proceedings IPPS 74:34-38. 2024.

Induction of Adventitious Buds from Unopened Flower Buds and Expanded Petals of African Violet

Chinatsu Kaneko and Wakanori Amaki

Department of Agriculture, Tokyo University of Agriculture, 1737 Funako, Atsugi, Kanagawa 243-0034, Japan

amaki@nodai.ac.jp

Keywords: *Streptocarpus*, *SaintPaulia*, tissue culture, organogenesis

Summary

Streptocarpus ionanthus (H.Wendl.) Christenh. (common name: African violet, saintpaulia), a member of the Gesneriaceae family, is native to the mountainous regions of tropical Africa and is an ornamental plant that is popular as an interior plant because it grows well even in low light interior conditions. Plant regeneration has been reported from petal culture of carnations, and chrysanthemums, but there have been no reports of petal culture of African violet, so this was attempted in this study. Two flowering stages, fully expanded petals and unopened flower buds were set up, and four plots were set up with a combination of α-naphthaleneacetic acid (NAA) at 1.0 or 2.0 mg/L and thidiazuron (TDZ) at 0.25 or 0.5 mg/L as PGRs. It was possible to induce adventitious buds from both of unopened flower bud and fully expanded petal explants. When adventitious buds that had grown to a length of about 5 mm were removed from the test tube and subcultured on a PGR-free medium, rooting was confirmed after 3 weeks and after 8 weeks they had become plantlets that could be acclimatized.

INTRODUCTION

Streptocarpus ionanthus (H.Wendl.) Christenh. (common name: African violet, saintpaulia), a member of the Gesneriaceae family, is native to the mountainous regions of tropical Africa and is an ornamental plant that is popular as an interior plant because it grows well even in low light interior conditions (Tamai, 1989). It can be propagated by leaf cuttings, and various forms of propagation by tissue culture have been attempted, with regeneration of plantlets reported from leaf fragments (Cooke, 1977; Mii and Ohashi, 1988), leaf petioles (Bilkey and Cocking, 1981), and protoplasts derived from leaf blades (Hoshino et al., 1995; Winkelmann and Grunewaldt, 1995).

There are many varieties of African violet, but the bicolored varieties are chimeras, so leaf cutting propagation results in single-colored flowers. In order to propagate the plant while maintaining the chimeric structure, division using apical or axillary buds is performed, but the propagation efficiency is not high. A method of inducing shoots from the axillary buds of the involucre just below the inflorescence has been attempted to maintain the chimeric structure, but there are few research examples. On the other hand, there have been attempts to fix the flower color by tissue culture of the chimeric multicolored parts by color. Plant regeneration has been reported from petal culture of carnations (Nugent et al., 1991), *Hypericum perforatum* (Don Palmer and Kellar, 2011) and chrysanthemums (Takahashi et al., 2024), but there have been no reports of petal culture of African violet, so this was attempted in this study.

MATERIALS AND METHODS

A commercially available potted African violet 'Wood Trail' (blue mono-color flowers) was purchased and used for the experiment. The culture vessel used was a φ20×120 mm glass test tube, into which 10 mL of medium was dispensed and closed with double layer aluminum foil, and then sterilized in an autoclave at 121°C for 15 minutes. The basal medium was MS (Murashige and Skoog, 1962) composition added with 30 g/L sucrose and 8 g/L agar, and the pH was adjusted to 5.8 before autoclaving.

Confirmation of sterilization method

Fully expanded petals (corolla) were collected with the calyx attached and sterilized in 70 % ethanol for 30 seconds and in 1 % aqueous solution of sodium hypochlorite containing one drop of Tween 20 for 5 minutes, then washed three times with sterilized pure water. The petals were cut into 1 cm square pieces and placed on the basal medium. As a result, there was almost no contamination, so subsequent unopened flower buds and petals were sterilized in the same procedure.

Effects of flowering stage and plant growth regulators (PGRs) on callus and adventitious bud formation

With reference to previous studies (Nugent et al., 1991; Nakano et al., 1994; Don Palmer and Keller, 2011), two flowering stages [fully expanded petals; 1 cm square segments placed adaxial side up and unopened flower buds (bud length approx. 4 mm; cut in half longitudinally and placed cut surface down)] were set up, and four plots were set up with a combination of α-naphthaleneacetic acid (NAA) at 1.0 or 2.0 mg/L and thidiazuron (TDZ) at 0.25 or 0.5 mg/L as PGRs.

Rooting of adventitious buds and subsequent plantlet cultivation

Adventitious buds formed on the explants that reached about 5 mm in size were subcultured on PGR-free medium. After that, the two adventitious buds that were confirmed to have sufficient roots were removed from the test tubes, washed carefully with tap water so as not to damage the roots, and transplanted into 3 cm pots filled with vermiculite (fine grains). After acclimatization under maintaining high humidity for 2 weeks, they were continued to be cultivated and potted up into 6 cm and then 9 cm pots according to their growth.

The test tubes after the explants were placed on the medium, the acclimation period for the young plantlets removed from the test tubes, and the subsequent pot cultivation were all carried out under the same conditions: 23 ± 1 °C, 16 hours of light (20 $\mu mol \cdot m^{-2} \cdot s^{-1}$ PPFD) under white fluorescent lamps (FLR40S·EX-N/M-H, Toshiba Lighting and Technology Co., Ltd.) and 8 hours of darkness.

RESULTS AND DISCUSSION

Although omitted in this report, in a similar experiment using the PGR-free medium, the petals of unopened flower buds were expanded but after that no other changes were observed. The expanded petal explants browned and died within 1-2 weeks without any morphological changes. The addition of PGRs was essential for inducing morphological formation in petal culture of African violet.

Although the fully expanded petal explants formed callus, it took longer time for the callus to form than the unopened flower bud explants, and callus formation began 5 to 7 weeks after placement on the medium. When the culture was continued thereafter, adventitious buds formed on the surface of the callus, although not in large numbers. On the other hand, using the unopened flower bud explants, callus formation was confirmed after 3 weeks, and adventitious bud formation was confirmed after 8 weeks from the start of culture (**Fig. 1**). Although there was no significant difference in callus formation between the different types and concentrations of PGRs used in this study, adventitious bud formation was the fastest in the 1.0 mg/L NAA + 0.25 mg/L TDZ.

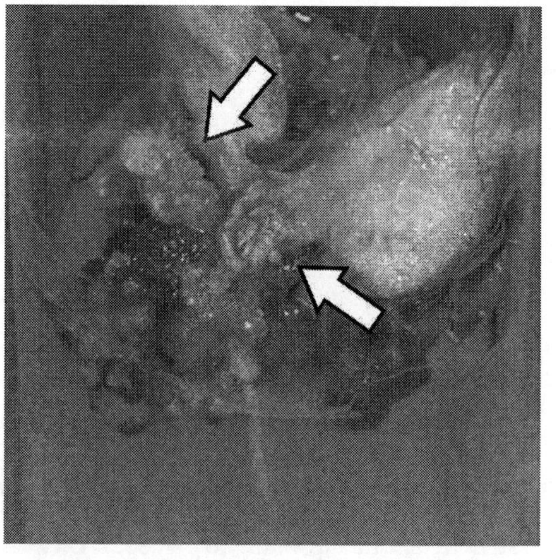

Figure 1. Adventitious bud formation via callus in unflowered flower bud culture (Positions are indicated by two arrows). Eight weeks after culturing in medium containing 1.0 mg/L NAA + 0.25 mg/L TDZ.

When adventitious buds that had grown to a length of about 5 mm were removed from the test tube and subcultured on a PGR-free medium, rooting was confirmed after 3 weeks and after 8 weeks they had become plantlets that could be acclimatized (**Fig. 2**).

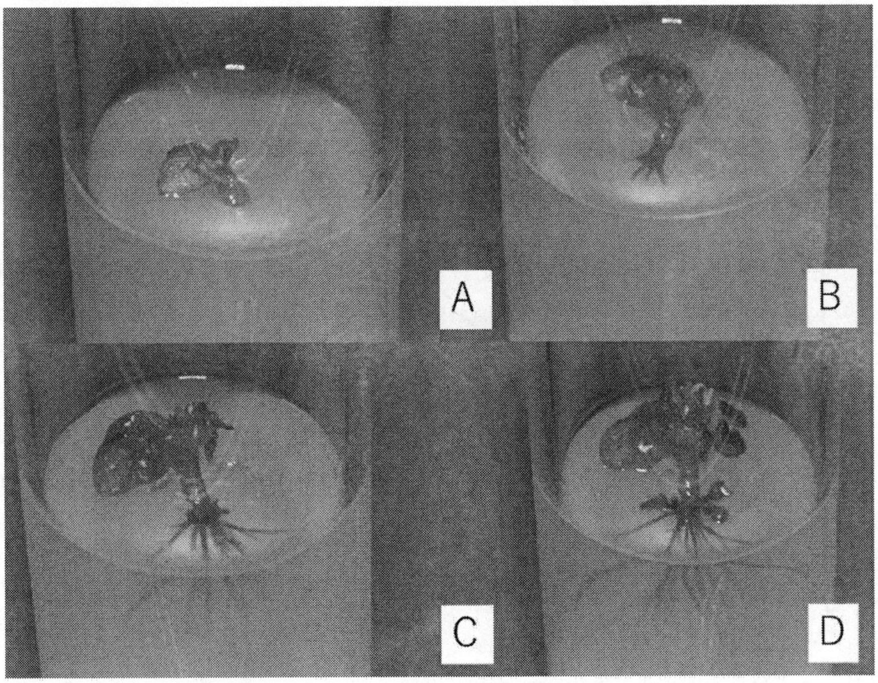

Figure 2. The formed adventitious bud was subcultured to PGR-free medium to induce rooting. A) Adventitious bud when inoculated on the medium, B: 3 weeks after, C) 6 weeks after, D) 9 weeks after subculture.

The plantlets were removed from the test tube, acclimatized and potted to 3 cm pots in the usual way, and grew well. They were re-potted into 6 cm pots 2 months after the start of acclimatization, and 9 cm pots 3.5 months later, and continued to grow vigorously (**Fig. 3**).

Figure 3. Growth status of the plants 5 months after the start of acclimation.

Unfortunately, they had not yet bloomed at this point, so we were unable to check for changes in flower color, etc. In this experiment, unopened flower buds were used that had been cut longitudinally, so the explants contained multiple organs (e.g., petals, sepals, and ovary), and callus was induced from all of them. Although the organ from which the plantlets derived from adventitious buds shown in **Figure 2** originated can be determined only by visual confirmation of the position of the callus at the time of development, it is determined that they are most likely derived from petals. In the future, when using unopened flower buds, it will be necessary to separate them by organ and use them as test materials.

As described above, it was possible to induce adventitious buds from both of unopened flower buds and fully expanded petals in African violets. It was shown that by confirming the flower color from the petals of a chimeric bicolor flower and then taking explants, it may be possible to obtain a strain exhibiting new traits.

LITERATURE CITED

Bilkey, P. C. and Cocking, E.C. (1981). Increased plant vigor by *in vitro* propagation of *Saintpaulia ionantha* Wendl. from subepidermal tissuse. HortScience *16*: 643-644.

Cooke, R. C. (1977). Tissue culture propagation of African violets. HortScience 12: 549.

Don Palmer, C. and Keller, W.A. (2011). Plant regeneration from petal explants of *Hypericum perforatum* L. Plant Cell Tiss. Organ Cult. *105*:129-134.

Hoshino, Y., Nakano, M. and Mii, M. (1995). Plant regeneration from cell suspension-derived protoplasts of *Saintpaulia ionantha* Wendl. Plant Cell Reports 14: 341-344.

Mii, M. and Ohashi, H. (1988). Saintpaulia. pp. 192-193. In: H. Higuchi (ed.) In-vitro culture of horticultural plants. Shibata Hario Glass Co., Ltd., Tokyo, Japan (In Japanese)

Murashige, T. and Skoog, F. (1962). A revised medium for rapid growth and bioassays with tobacco tissue cultures. Physiol. Plant. 15:473-497.

Nugent, G., Wardley-Richardson, T. and Lu, C.-Y. (1991). Plant regeneration from stem and petal of carnation (*Dianthus caryophyllus* L.). Plant Cell Reports *10*:477-480.

Takahashi, K., Ouchi, A., Ogawa, R., Nagasawa, R., Kishi, K., Koseki, Y., Muto, R., Amemiya, N., Moriyama, H., Takahashi, K. Sato, K., Miura, M. and Suzuki, S. (2024). Influence of blooming stage, different parts within the petal and dark treatment on adventitious bud formation in chrysanthemum petal tissue culture. Hort. Res. (Japan) *23*:99-108. (In Japanese with English abstract)

Tamai, Y. (1989). *Saintpaulia*. pp. 88-95. In: T. Aiga (ed.) The grand dictionary of horticulture. Vol. 3. Shogakukan, Tokyo, Japan (In Japanese)

Winkelmann, T. and Grunewaldt. J. (1995). Genotypic variability for protoplast regeneration in *Saintpaulia ionantha* (H. Wendl.). Plant Cell Reports *14*:704 -707.

Combined Proceedings IPPS 74:39-43. 2024.

Effect of Cytokinin in Tissue Culture in the Ornamental Aquatic Plant – Pearl Grass

Mitsuteru Goto and Wakanori Amaki

Department of Agriculture, Tokyo University of Agriculture, 1737 Funako, Atsugi, Kanagawa 243-0034, Japan

amaki@nodai.ac.jp

Keywords: thidiazuron, organogenesis, *Micranthemum*

Summary

In recent years, interest in aquariums has increased in Japan, and the demand for aquatic plants used to decorate aquariums is on the rise. Many of them are grown by tissue culture and sold commercially as sterilized, pesticide-free products. The production of these tissue culture transplants is exclusively carried out by venture companies. As the results, there is little research information published about the growth of aquatic plants by tissue culture. We believe that there is still a lot of room for optimization, including the composition of the culture medium, and have presented the results of tissue culture experiments of three aquatic plants. In previous experiments plant growth regulators were not added to the medium due to the risk of tissue culture mutations, but this time, cytokinin was added to promote growth and improve propagation efficiency. Without the addition of cytokinin, shoots developed and elongated from the axillary buds of both species of pearl grass and rooting occurred, but the addition of cytokinin suppressed leaf formation and rooting, especially in thidiazuron. In the cytokinin-added medium where leaf formation was inhibited, green globular masses that are thought to be derived from shoot axillary buds were formed.

IPPS Vol. 74 – 2024

INTRODUCTION

In recent years, interest in aquariums has increased in Japan, and the demand for aquatic plants used to decorate aquariums is on the rise. The aquatic plants used are native to Southeast Asia, Central and South America, Australia, etc., and many of them are grown by tissue culture and sold commercially as sterilized, pesticide-free products. The production of these tissue culture transplants is exclusively carried out by venture companies. As the results, there is little research information published about the growth of aquatic plants by tissue culture.

We believe that there is still a lot of room for optimization, including the composition of the culture medium, and have presented the results of tissue culture experiments of three aquatic plants of the Scrophulariaceae family at the 20th IPPS-J Gifu annual meeting (Niki and Amaki, 2014) and the 24th IPPS-J Okinawa annual meeting (Minamiyama et al., 2017). In previous experiments plant growth regulators (PGRs) were not added to the medium due to the risk of tissue culture mutations, but this time, cytokinin was added to promote growth and improve propagation efficiency.

Here, we report the formation of green globular masses that can be used as intermediate propagules in the process of tissue culture of pearl grass, which is a member of the Linderniaceae family, in a medium supplemented with cytokinin.

MATERIALS AND METHODS

New large pearl grass (*Micranthemum umbrosum* (J.F. Gmel.) S.F. Blake) and Cuban pearl grass (*Micranthemum callitricoides* (Griseb.) C. Wright) of the Linderniaceae family, which are commercially available and produced by Aqua Design Amano Co., Ltd. (Niigata, Japan), were purchased and used in this experiment. The culture vessels were flat-bottomed glass test tubes measuring φ40 × 130 mm, and 30 mL of medium was dispensed into each tube. The medium was a 1/2 concentration MS (Murashige and Skoog, 1962) composition with 20 g/L sucrose and 2 g/L gellan gum (Wako Pure Chemical Industries Ltd., Oosaka, Japan) added, the pH was adjusted to 5.8, and the medium was sterilized in an autoclave at 121°C for 15 minutes. The apical part of each shoot (cutting) was prepared to about 1 cm long, and 5 cuttings were cut (inserted) into the one test tube respectively. Those test tubes were plugged with double layer aluminum foil for multiplication of shoots, and the in vitro shoots after subcultures for multiplication were used in subsequent experiment.

The cytokinin used were benzyladenine (BA; Wako Pure Chemical Industries Ltd., Oosaka, Japan) and thidiazuron (TDZ; Wako Pure Chemical Industries Ltd., Oosaka, Japan), and two levels of addition concentration were 0.5 and 1.0 mg/L. All culture conditions, including multiplication by subculture, were 23 ± 1°C, 16 hours of illumination (40 $\mu mol \cdot m^{-2} \cdot s^{-1}$ PPFD) under cool white-fluorescent lamps (FLR40S ·EX-N/M-H, Toshiba Lighting and Technology Co., Ltd., Yokosuka, Japan) / 8 hours of darkness. Two months after the start of culture, the fresh weight of 5 explants per test tube was measured and morphological observation was carried out.

RESULTS AND DISCUSSION

Without the addition of cytokinin, shoots developed and elongated from the axillary buds of both species of pearl grass and rooting occurred, but the addition of cytokinin suppressed leaf formation and rooting, especially in TDZ (**Table 1, Figs. 1 and 2**).

Figure 1. Effect of cytokinin on the growth of new large pearl grass. From left: no addition (0 mg/L), 0.5 mg/L BA, 1.0 mg/L BA, 0.5 mg/L TDZ, 1.0 mg/L TDZ.

Figure 2. Effect of cytokinin on the growth of Cuban pearl grass. From left: no addition (0 mg/L), 0.5 mg/L BA, 1.0 mg/L BA, 0.5 mg/L TDZ, 1.0 mg/L TDZ.

Table 1. Effect of cytokinin on the growth of pearl grass (n = 5).

Plant name	Cytokinin name	Concentration (mg/L)	Total fresh weight (mg/explant)
New large pearl grass	No addition	0	708 ± 72
	Benzyladenine	0.5	432 ± 56
		1.0	694 ± 46
	Thidiazuron	0.5	384 ± 42
		1.0	382 ± 58
Cuban pearl grass	No addition	0	964 ± 0.0
	Benzyladenine	0.5	424 ± 36
		1.0	288 ± 32
	Thidiazuron	0.5	248 ± 26
		1.0	190 ± 16

In the cytokinin-added medium where leaf formation was inhibited, green globular masses that are thought to be derived from shoot axillary buds were formed (**Fig. 3**). In many general plants, cytokinin promotes shoot development from axillary buds, but it was inhibited in pearl grass. A similar effect was confirmed in ferns (Higuchi et al. 1987; Amaki, 1997), and it has been shown that the tissues formed can be used as intermediate propagules for tissue culture propagation as GGBs (green globular bodies) (Suneetha and Hegde, 2022). In tissue culture transplants of aquatic plants, endophytic contaminating microorganisms that are not detected during the micropropagation process often become apparent after sale. It may be possible to use green globular masses induced by cytokinin as a new micropropagation method, because it is easy to observe the presence of endophytic contaminating microorganisms.

Figure 3. Green globular masses formed from the axillary buds of explants. Cuban pearl grass, medium with 1.0 mg/L TDZ added.

LITERATURE CITED

Amaki, W. (1997). Ferns: Production of tissue culture transplants of ornamental plants. Nogyo-Gijutu-Taikei Ornamental Flowers Vol. 5, pp.525-529. Rural Culture Association, Tokyo, Japan (In Japanese)

Higuchi, H., Amaki, W. and Suzuki, S. (1987). *In vitro* propagation of *Nephrolepis cordifolia* Prsel. Sci. Hortic. *32*: 105-113.

Minamiyama, M., Noguchi, A. and Amaki, W. (2017). Micropropagation of ornamental aquatic plants, *Glossostigma*, *Microarpaea* and *Limnophila*. 2. Effects of CaCl $\cdot 2H_2O$, KH_2PO_4, Fe-EDTA concentrations on the growth of explants. Combined. Proc. Intern. Plant Prop. Soc. *67*:387-390.

Murashige, T. and Skoog, F. (1962). A revised medium for rapid growth and bioassays with tobacco tissue cultures. Physiol. Plant. *15*:473-497.

Niki, T. and Amaki, W. (2014). Micropropagation of ornamental aquatic plants, *Glossostigma*; *Microarpaea* and *Limnophila*. Combined. Proc. Intern. Plant Prop. Soc. *63*:369-373.

Combined Proceedings IPPS 74:44-48. 2024.

Effect of Light Quality on the Tuber Sprouting of *Pinellia ternata* (Thunb.) Makino

Masamichi Torii and Wakanori Amaki

Department of Agriculture, Tokyo University of Agriculture, 1737 Funako, Atsugi, Kanagawa 243-0034, Japan

amaki@nodai.ac.jp

Keywords: aroid, LED, irradiance

Summary

Pinellia ternata (Thunb.) Makino, a member of the Araceae family, is a medicinal plant, and the tubers are peeled and dried to be used as herbal medicine called "Han-ge". *Pinellia ternata* is used in many traditional Chinese medicine prescriptions and is in high demand as herbal medicine ingredient. *Pinellia ternata* basically forms one leaf from the tuber, and the photosynthetic products form the next generation of tubers underground. In this experiment, we investigated the influence of the light quality of the cultivation light on the sprouting of the tubers, and the extent to which the tubers in the soil recognize and respond to the amount of light. The quality of the irradiated light clearly affected the sprouting of *Pinellia ternata* tubers. In the red LED light irradiation area, the tubers sprouted earlier than other light conditions and the cumulative number of sprouts was the highest. On the other hand, the other treatment areas sprouted later than the dark area, and the cumulative number of sprouts was the lowest in the blue LED irradiation area. The measurement results showed that approximately 0.5% of the light reaching the soil surface penetrates to a depth of about 5 mm from the soil surface.

IPPS Vol. 74 – 2024

INTRODUCTION

Pinellia ternata (Thunb.) Makino, a member of the Araceae family, is a medicinal plant, and the tubers are peeled and dried to be used as herbal medicine called "Han-ge". *Pinellia ternata* is used in many traditional Chinese medicine prescriptions, and is in high demand as herbal medicine ingredient, but currently it is almost entirely imported from China (Harashima, 2012). *Pinellia ternata* grows wild throughout Japan and is considered a weed in the protected cultivation fields, that is rarely actively cultivated.

In order to explore the possibility of domestic self-sufficiency, the authors borrowed facilities from Tamagawa University and attempted to produce it in a plant factory (Amaki et al., 2015). It was revealed that the life cycle of *Pinellia ternata* was repeated about once a year in the open field, but about four times a year under white-fluorescent lights in an environmentally controlled facility, but the total yield during four- time cycles was not profitable due to the production costs. In a simultaneous experiment using monochromatic light irradiation with light emitting diodes (LEDs), the sprouting of tubers was suppressed, especially under blue light irradiation. *Pinellia ternata* basically forms one leaf from the tuber, and the photosynthetic products form the next generation of tubers underground.

At the 27th IPPS-J Gifu annual meeting, we reported that six months of low temperatures (4°C) are necessary for *Pinellia ternatea* to flower, and that when grown under white-fluorescent lamps, sprouting of tubers was significantly delayed under short-day conditions compared to long-day conditions (Torii et al., 2022). In this report, we investigated the influence of the light quality of the cultivation light on the sprouting of the tubers again, and the extent to which the tubers in the soil recognize and respond to the amount of light.

MATERIALS AND METHODS

Pinellia ternata growing wild in the open field was dug up, and the tubers were planted in a 9 cm plastic pot filled with the medium of Akadama soil (small grain): Metro-mix 360 (Hyponex Japan Ltd., Oosaka, Japan) = 1:1, and cultivated and propagated in a glass greenhouse with a minimum night temperature of 16 °C until they were used for experiments. Some of the dug-up tubers were placed in a plant box (Magenta Box G7, Magenta, Chicago, USA) and stored in a refrigerator at 4 °C.

Sprouting of tubers under different light quality environments

Tubers cultivated in a greenhouse and tubers stored in a refrigerator were planted in groups of five in 9 cm plastic pots at about 1 cm depth (distance between the surface of the medium and the top of the tuber bud) filled with the medium of Akadama soil (small grain): Metro-mix 360 = 1:1 and cultivated in a fully enclosed artificial light irradiation room. The cultivation lights were white fluorescent lamps (FLR40S・EX-N/M-H, Toshiba Lighting and Technology Co., Ltd., Yokosuka, Japan), white LED (peak wavelength: 450nm + 550nm), blue LED (470nm), green LED (525nm), and red LED (660nm), and each of light was irradiated at 16 hours (8:00 – 24:00) and 8 hours of darkness (24:00-8:00). Additionally, a completely dark treatment area was also set up. The cultivation temperature was 23°C. The PPFD of the irradiation light was 80 $\mu mol \cdot m^{-2} \cdot s^{-1}$ on the soil surface. Ten pots (total of 50 tubers) were used for each experimental treatment.

Incidence of light into the medium used

The medium used in this experiment was filled to a specified thickness in a container lined with Saran Wrap (polyvinylidene chloride film; Asahi Kasei Home Products Corporation, Tokyo, Japan), and an illuminance sensor was attached to a hole in the bottom of the container. In a dark environment, the amount of light passing through the medium was measured. An illuminance meter (T-1M, KONIKA MINOLTA JAPAN Inc., Tokyo, Japan) was used.

RESULTS AND DISCUSSION

Sprouting of tubers under different light quality environments

The quality of the irradiated light clearly affected the sprouting of *Pinellia ternatea* tubers (**Fig. 1**). In the red LED light irradiation area, the tubers sprouted earlier than other light condition area and the cumulative number of sprouts was the highest. On the other hand, the other treatment areas sprouted later than the dark area, and the cumulative number of sprouts was the lowest in the blue LED irradiation area. This delay in tuber sprouting due to monochromatic light irradiation was consistent with a previous report (Amaki et al., 2015).

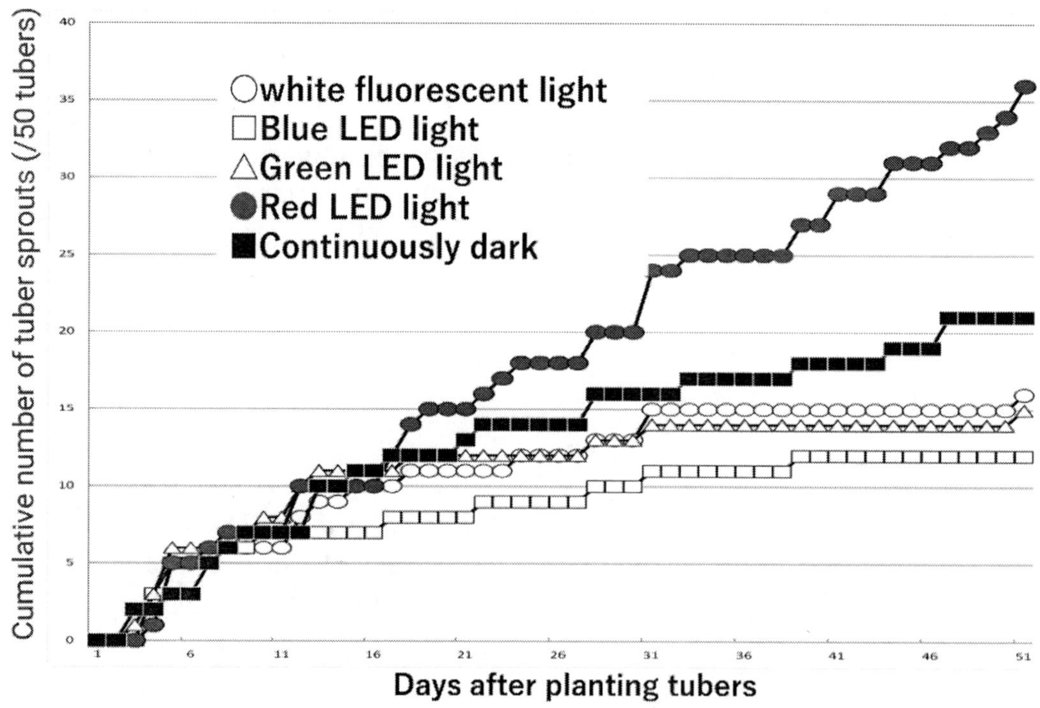

Figure 1. Effect of light quality conditions during cultivation on the sprouting of *Pinellia ternata* tubers.

Incidence of light into the medium used

However, there have been no previous studies on whether the tubers in the soil recognize the irradiated light or the limiting illuminance. The measurement results showed that approximately 0.5% of the light reaching the soil surface penetrates to a depth of about 5 mm from the soil surface (**Table 1**), and it is believed that the quality of this light is involved in the photomorphogenesis of tubers. Light other than blue light is incident up to a depth of 20 mm from the surface of the medium (**Table 1**), and it is believed that this light is recognized by the bud of *Pinellia ternata* tubers. In the cultivation experiment, the PPFD of 80 μmol・m^{-2}・s^{-1} at the soil surface is equivalent to 6,000 lx in white fluorescent light, and even if the incidence rate is 0.1%, 6 lx of light reaches a depth of 20 mm. It is known that the release of the seedling hook is a physiological response of plants to the low amount of light, and the minimum light required for this is said to be 40 μlx (Bickford and Dunn, 1972), and the light irradiation conditions used in this experiment are believed to be sufficient to have a physiological effect on *Pinellia ternata*.

Table 1 Percentage of incident light intensity by depth in the medium for cultivating *Pinellia ternata*.

Moisture condition of medium	Measurement depth (mm)	Relative light intensity (%: soil surface is 100)			
		Light quality of irradiated light			
		Mixed White LED	Blue LED	Green LED	Red LED
Dry	1	7	0.6	5.5	5.7
	5	0.6	0.1	0.4	0.5
	10	0.3	0.1	0.2	0.3
	15	0.2	0.1	0.1	0.2
	20	0.1	0	0.1	0.1
Wet	1	－*	－	－	－
	5	1.9	0.2	4.3	2.8
	10	0.3	0	0.2	0.3
	15	0.3	0	0.1	0.2
	20	0.1	0	0.1	0.1

* Did not measure at a depth of 1 mm in wet condition.

Similar experiments have been conducted for several years in our laboratory, but the numerical results of the light response seem to vary depending on the time since digging, the size of the tuber, and whether it was exposed to low temperatures. In the future, it will be necessary to examine these factors and the response to light. However, from the our results so far, it is clear that low temperature treatment causes the tuber to flower, which is a negative factor for tuber production, and that cultivation under white light in a constant environment such as a plant factory causes the tuber's secondary buds to form secondary tubers on top of the parent tuber, resulting in a shape that is unsuitable for use as a herbal medicine (Amaki et al., 2015; Torii et al., 2022). Since it is also clear from multiple experiments that blue light suppresses tuber sprouting, it may be possible to suppress secondary spherical sprouts and cultivate tubers with a shape suitable for herbal medicines by irradiating blue light when the leaves wither.

LITERATURE CITED

Amaki, W., Yamanishi, Y. and Watanabe, H. (2015). Effect of light quality on the growth of the medicinal plant, *Pinellia ternata*. Hort. Res. (Japan) *14* (Supple.1): 387.

Bickford, E. D. and Dunn, S. (1972). Lighting for plant growth. pp.48. Kent State University Press, USA.

Harashima, H. (2012). Syoyakutan -Pharmacognocy- (revised ed.). pp.212-213. NTS-BOOK, Tokyo.

Torii, M., Noguchi, A., Nakatsuka, H. and Amaki, W. (2022). Effect of day length and tuber storage temperature and duration on sprouting, enlargement and flowering of tubers in *Pinellia ternata* (Tunb.) Makino. Combined Proc. Intern. Plant Prop. Soc. *72*:13-18.

Combined Proceedings IPPS 74:49-51. 2024.

Consideration of Planting Materials for *Phalaenopsis* Seedlings Taken Out of the Bottle.

Ayano Shiraki and Masahiro Hayami

Shizuoka Prefectural Agricultural and Forestry College, 678-1Tomigaoka Iwata city Shizuoka Pref. 436-8577 Japan

hana1214hayami@outlook.jp

Keywords: orchid, transplanting, tissue culture, acclimatization

Summary

Survival in Phalaenopsis planting material was successful using polystyrene foam. It significantly reduced time, cost and improved uniformity.

INTRODUCTION

When planting *Phalaenopsis* seedlings which were obtained by sterile sowing or culturing, it is common to wrap them tightly in sphagnum moss, as this allows for good rooting and growth. However, in addition to the rising cost of sphagnum moss, wrapping it tightly is a lot of work, and the survival rate varies greatly depending on the skill of everyone.

Therefore, we conducted a survey of the survival rate, growth, and planting time, using foam polystyrene as an alternative material to sphagnum moss, for the following reasons, (1) it is inexpensive and easy to obtain; (2) it is highly durable; (3) it suits the characteristics of Phalaenopsis, which is to spread its roots on tree branches and trunks in well-ventilated places; (4) by clamping it firmly, it has the same effect of fixing the roots as tightly wrapping sphagnum moss; (5) it does not require much work; and (6) there is little difference in skill between people.

MATERIAL AND METHODS

Phalaenopsis (cultivar unknown) were grown in the greenhouse at Shizuoka Prefectural Agricultural and Forestry College. There were four treatments with 40 plants in 2 replicates per plot. Treatments include sphagnum moss or polystyrene foam substrates with or without mist.

Phalaenopsis seedlings were taken from the bottles on July 2, 2023, and planted in sphagnum moss or foam polystyrene in hard black polyethylene pots. On November 13, 2013, after checking for survival, the areas without mist were moved to the greenhouse. At that time, Hyponex (6-10-5) fertilizer at 1,000th dilution was sprayed on all surviving plants. Final evaluation was on February 7, 2014.

RESULTS AND DISCUSSION

The growth of the *Phalaenopsis* was visually observed. We surveyed four students majoring in potted plants who are familiar with planting from bottles, compared with the time when they were planted in sphagnum moss and polystyrene foam.

The survival rate was 90% for plants grown in sphagnum moss without mist, while all other treatments showed 100% survival.

The time it took for students to transplant 25 *Phalaenopsis* seedlings using polystyrene foam took 1/4 the time it took to wrap them tightly in sphagnum moss (**Table 1**).

Table 1. Comparison of times to transplant 25 *Phalaenopsis* seedlings in sphagnum moss vs. polystyrene foam substrates.

Planting materials	Student	Time
Sphagnum moss	A	39min. 14 sec.
	B	42min. 28 sec.
	C	44min. 45 sec.
	D	45min. 15 sec.
Foam polystyrene	E	11min. 34 sec.

Although the growth of plants in polystyrene foam were generally inferior to that in sphagnum moss, the difference in growth were small and the growth was uniform (**Fig. 1**). Growth in plants in polystyrene foam that were misted was better than those that were unmisted (**Fig. 1 B and D**).

Conclusion

The reason why growth in the polystyrene foam area was poorer than that of the sphagnum moss area is thought to be because polystyrene foam does not have water retention, and therefore there was a lack of nutrients from the weekly fertilization. Therefore, it is thought that growth in plants with mist, which was consistently moiste, was better than those that were not misted.

In conclusion, it seems that polystyrene foam can be used as a planting material for *Phalaenopsis*, as it significantly reduces time, cuts costs, and eliminates differences between individuals due to the reduced effort required, and has a good survival rate.

In the future, we would like to try methods such as sandwiching sphagnum moss between sheets of polystyrene foam to make up for nutrient deficiencies.

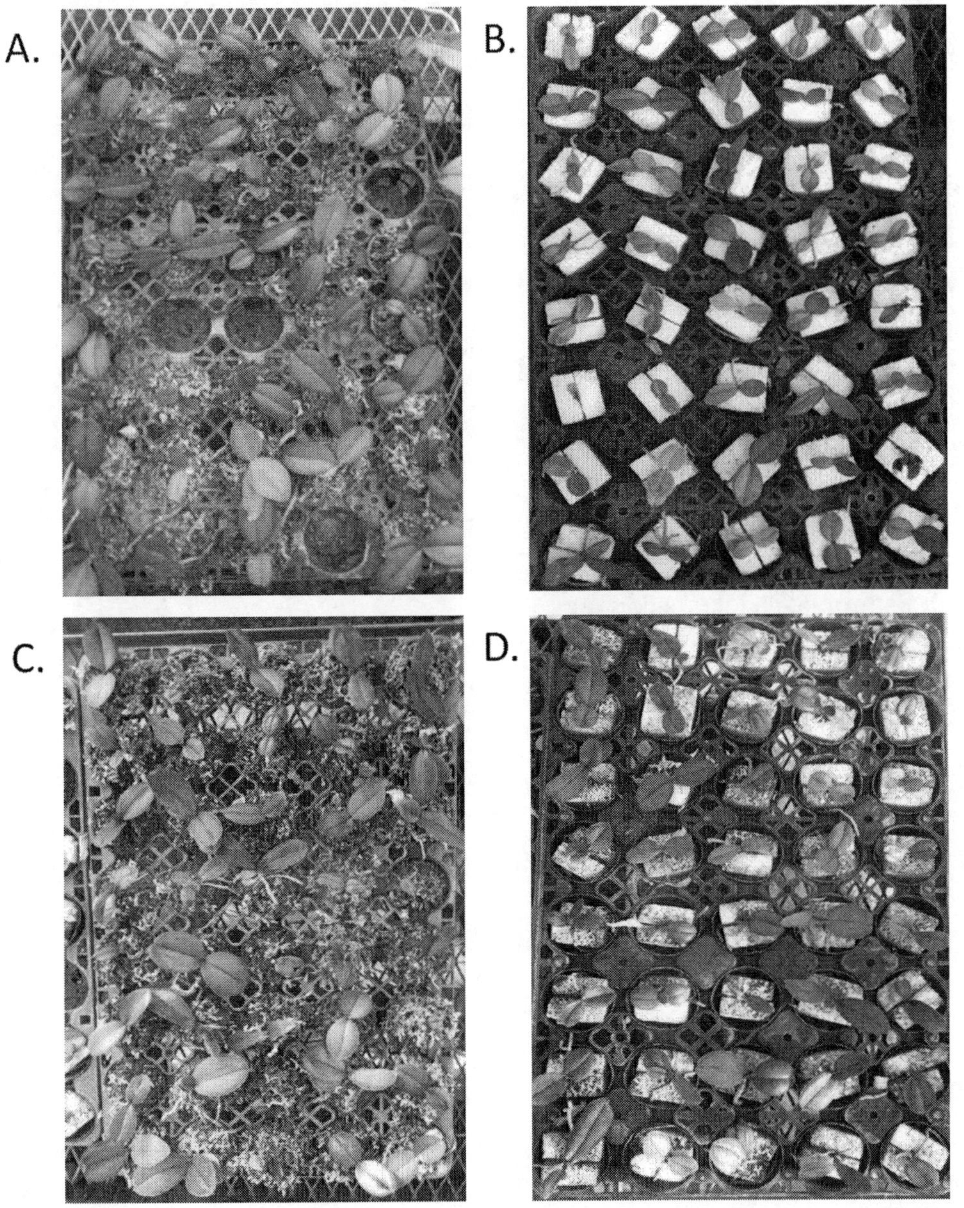

Figure 1. *Phalaenopsis* transplant growth in A) sphagnum moss without mist, B) polystyrene foam without mist, C) sphagnum moss with mist and D) polystyrene foam with mist.

Combined Proceedings IPPS 74:52-75. 2024.

A Breath of Fresh Air: 52nd Conference of the International Plant Propagators' Society – Australia Region

Ranjith Pathirana[1], Bruce Higgs[2] and Tim Hills[3]

[1]School of Agriculture, Food & Wine, University of Adelaide, Urrbrae, SA 5064, Australia
[2]Darvill Nursery, 54 Darvill Road, Orchard Hills NSW 2748, Australia;
[3]82, Fraser Road, Dunoon, NSW 2480, Australia

Ranjith.Pathirana@adelaide.edu.au

Keywords: awards, propagation, grafting, nursery tours, New South Wales

Summary

"A Breath of Fresh Air' was the theme for the 52nd Annual Conference of the International Plant Propagators' Society (IPPS) Australia Region held during 22 – 25 May 2024 at RSL conference Centre in Ballina, New South Wales. The conference was packed with different activities, including a pre-conference tour, traditional golf competition, trade displays, award session during the gala dinner, nursery tours, grafting demonstrations and of course many interesting presentations. The traditional six pack of young energetic people selected by the Executive Committee helped in running the conference. Some highlights from these activities are presented in this Editorial to the Proceedings of the IPPS Australia Region.

IPPS Vol. 74 – 2024

INTRODUCTION

The theme of the 52nd Annual Conference of the Australian Region of IPPS "A breath of Fresh Air" was evident from the very beginning of the conference, during the pre-conference tour, and throughout the conference held at the spacious RSL conference facility at the edge of the scenic Richmond River. The Ballina Coast and Hinterland are the traditional Nyangbul Country of the Bundjalung nation.

The conference was organised by Ray Parker, supported by President Bruce Higgs, Clive Larkman and the dedicated Secretary Pam Berryman and was attended by 135 delegates. It was packed not only with interesting presentations from a variety of scientists, academics, hands-on nursery managers, practitioners, company representatives and students, but also with on-stage interviews, nursery tours, the traditional golf competition and a variety of trade displays. This editorial is meant to cover the activities associated with the conference as well as giving a brief run-down of the conference proceedings and to introduce IPPS Australia award winners in 2024.

PRE-CONFERENCE TOUR

For the preconference tour, the delegates were picked up from the Gold Coast Airport and then travelled to Mt. Tamborine for the evening. There was a total of 25 delegates on the tour.

Pre-conference Tour – Day 1

The tour began at Tamborine Mountain Wholesale Nursery. Despite having been affected by a tornado around Christmas of 2023, the retail and wholesale areas were both in full swing. Tour attendees were shown around by Alicia who took over the nursery within the last couple of years. Ebb and flow irrigation systems were used across the entire nursery.

Second stop was Mt Nathan Propagation. They had endured 3 major floods in recent years. The nursery is located on two cites close to the scenic Coomera River and easily accessible from Gold Coast. A wide range of exotic and native trees and shrubs are grown across the two sites (**Fig. 1**), with a focus on wholesale production. Operating for over 25 years, Mt. Nathan Nursery produces an extensive selection of both native and exotic plant stock, currently over 200 species (Mtnathan, 2025).

Next stop was Crystal Waters Nursery in Gold Coast. It stocks excellent quality indoor and outdoor landscaping ornamental stock ranging from 100 mm pots to 100 L bags (**Fig. 2**).

The final stop was Boyd's Bay Nursery who are well established and grow for the landscaping trade and their own gardening business projects (**Fig. 3**). They had begun treating some potting mix batches with bifenthrin as part of their fire ant requirements. This can add up to $20 per cubic metre and was a topic of discussion throughout the tour (as well as the topic of a presentation) as some nurseries visited were within the fire ant zone. The Boyd's Group's rehabilitation, restoration and ecological services are impressive as they have a range of restoration services.

The delegates spent the evening at Mantra on Salt Beach, Kingscliff.

Figure 1. Despite three major floods in recent years, Mt Nathan Propagation has recovered to continue the propagation of native and exotic species. Photo credit – Pam Berryman.

Figure 2. Crystal Waters Nursery in Gold Coast stocks excellent quality indoor and outdoor landscaping ornamental stock.
Photo credit – Pam Berryman.

Figure 3. Boyd's Bay Nursery was the last visited on day 1 of the pre-conference tour. Located in the Gold Coast, the nursery stocks landscaping plants. Photo credit – Pam Berryman.

Pre-conference Tour – Day 2

The inspiring Limpinwood Botanic Gardens and nursery was the first stop on day 2. Russell Costin had recently passed away, but his wife Sharon and Mary guided the delegates through extensive gardens with many interesting native plants such as the smooth Davidson's Plum, *Davidsonia johnsonii*. Established in 1977, the nursery stocks Australian bushfood, camellias and vireyas (tropical rhododendrons) **(Fig. 4)**. They also specialise in *Grevillea* spp.

The final visit was to the Gondwana Nursery, where Joy and Gahan took us on a native plant frenzy. Gondwana leads the way with a well thought out nursery design and impressive range of plants including grafted natives and new selections **(Fig. 5)**.

Figure 4. Limpinwood Botanic Gardens and Nursery stocks a variety of Australian natives.

Figure 5. The final visit in the pre-conference tour was to the Gondwana Nursery stocking an array of native plants for gardens and revegetation projects. Photo credit Pam Berryman.

AWARDS

IPPS Australia has a suite of awards for members and even non-members who excel in propagation-related activities or have contributed to the Society. They are selected by committees appointed by the Board and are honoured at the annual conference.

IPPS Conference Youth Initiative (Six Pack) and South African Exchange

Year 2024 was the 22nd year since the Six Pack award commenced in 2003 at Coffs harbour. There have been 20 conferences with 120 awardees. It is inspiring what this has meant for many of these young horticulturists in their careers. This year's Six Pack (**Fig. 6**) was sponsored by Garden City Plastics. The Six Pack has an honourable task of assisting the Organizing Committee in running the conference smoothly and they learn from the best while on the 'job'. In addition to the Six Pack Support Team, South African Exchange is another program that is designed to offer two exciting opportunities to new propagators. Joshua Taylor (Australia Awardee) and Nosipho Ndlovu (South Africa Awardee) shared their experiences visiting the two countries at the Conference and are published in these Proceedings.

Figure 6. The 'Six Pack' of young horticulturists selected for each IPPS Australia Region conference has been a tradition for the last 20 years. 2024 Ballina Conference was no exception. From left to right: Nicoletta Centofanti, Royal Botanic Gardens, VIC (RBGV); Emily Smith, RBGV Cranbourne Gardens, VIC; Willow Sawyer, Glasshouse TubeStock, QLD; Nosipho Ndlovu – South Africa Exchange, Gabriella Lee, Westlands Nurseries, TAS and Indie Keenan, Ellenby Tree Farm, WA.

Rod Tallis Memorial Youth Award and Anita Boucher Award for the Best Presentation

IPPS recognizes outstanding achievements by the younger members of the industry. One way the Australian Region of the Society achieves this is through the Rod Tallis Memorial Youth Award which is presented annually to the most commendable achievement by a person under 30 years who is working within the horticulture industry or studying horticulture in Australia. The award is named after one of the Society's most respected members, Rod Tallis. Rod was a committed nurseryman with a passion for plant propagation and had a keen interest in the youth of the industry.

The recipient of 2024 Rod Tallis Memorial Youth Award was Lisa Wightwick from Peninsula Growers, Victoria (**Fig. 7**) for her achievements in developing protocols for micropropagation of *Grevillea* spp. In her presentation, she demonstrated how a growth retardant paclobutrazol can help in attaining better acclimation and survival of Grevillea 'Bonnie Prince Charlie' microshoots in the rooting and acclimation stages of micropropagation. For her outstanding presentation, Lisa was also awarded the Anita Boucher Award for the Best Presentation at the 2024 Conference as selected by a panel of judges.

Figure 7. Lisa Wightwick from Peninsula Growers, Victoria receiving Rod Tallis Memorial Youth Award for her research in micropropagation of *Grevillea* spp.

Ed and Mary Bunker Award

The focus of the Edward and Mary Bunker Award is to recognize an outstanding contribution from someone who has demonstrated the IPPS motto 'To Seek and To Share' for the betterment of the industry at large. The Edward and Mary Bunker award is a relatively new award with the person not necessarily being a member of IPPS. The initial awardee was John MacDonald in 2019 at Twin Waters, QLD followed by Gabe Mostafa in 2022 at Leura, NSW and Jane Edmondson at Geelong, VIC in 2023.

The winner of the 2024 award Karen Smith (**Fig. 8**) is the epitome of the IPPS motto "To Seek and Share". Based in Sydney Karen has worked in the horticultural industry for over 30 years as an employee and business owner and is the editor of the national horticultural industry magazine Hort Journal. She is a recognised Horticultural Trainer, Presenter, Master of Ceremonies, Writer and Podcaster.

Figure 8. Karen Smith (centre) receiving Ed and Mary Bunker award, 2024.

Karen is active in the industry and has served on many committees including the Hort Media Association and the Interior Plantscape Association and is also a member of various horticultural organisations including IPPS. She currently is an Executive Board Member of the Interior Plantscape Association. In November 2017 she was awarded the Allan Seale Award by Nursery Garden Industry NSW & ACT. In 2020 Karen was awarded the Australian Institute of Horticulture's Golden Wattle Award for her work in the media, and for her raising the awareness of the importance of horticulture in today's world.

Steve Vallance - Pewter Tankard Award

In 1979 the Great Britain & Ireland Region of IPPS, presented this pewter tankard to our region, to be used as an annual award to recognise the contributions of one of our members to the society. It was regularly awarded until 1991. After that time, it was paused until 2010. In that year, at the Freemantle conference, it was awarded to Steve Vallance. Steve really embodied the ideals of the award, contributing without fanfare, but with commitment. In honour of Steve, and the way he went about his 'seeking & sharing', in 2017 the award was renamed the *Steve Vallance Tankard.*

The 2024 awardee Dermot Molloy **(Fig. 9)** has been involved in horticulture his whole working life. Prior to employment with the Royal Botanic Gardens Victoria (RBGV) he worked in a nursery, owned a garden design business and was head gardener at the historic Invergowrie property in Hawthorn. His membership of IPPS started in 2005 and he has been a board member since 2018. He has travelled widely promoting the values of IPPS and spoken at many IPPS conferences. With a Diploma of Horticulture obtained in 2006 from Oakleigh College of Horticulture in Victoria, his propagation and growing experience now especially includes Cycads, *Clivia, Agathis* and *Araucaria* as well as native and exotic species from around the world

Figure 9. Dermot Molloy receiving Steve Vallance - Pewter Tankard Award from Natalie Vallance (right) and speaking to the audience after receiving the award (left).

He enjoys mixing horticulture and travel to discover the world's plants and people. He is the Senior Curator of Horticulture at the RBGV, where he has worked for over 22 years.

The RBGV have awarded him the Guilfoyle Award for excellence in team achievement in July 2010 and for excellence of Individual achievement in August 2011.

Award of Honour

The Australian Region of the IPPS has been awarding individuals who have made a significant contribution to the society for many years. The trophy is designed using an individually selected free-form piece of rare Australian native timber such as red cedar, teak, blackwood etc. onto which the cast of the IPPS logo is fixed. The recipient's name is cast or engraved onto the award.

David Hancock from Natural Area Holdings, awardee for 2024 (**Fig. 10**) has made an exceptional contribution to the Society and horticulture in general. Additionally, David has accounting and economics qualifications and spent 22 years of his working life in finance and banking at general management level. His love for native plants and restoration of mine-sites and other disturbed areas fostered a change in career to horticulture, founding a business in 2001. He is currently a major shareholder and contributing consultant to the business with a nursery, contracting and environmental consulting divisions. His workforce varies from 115 to 150 based upon seasonal demand. His business services Government at all levels, property developers, mining companies and private landowners for their ecological requirements.

Figure 10. David Hancock from Natural Area Holdings (right) receiving the Award of Honour for 2024 from the President of IPPS Australia Region Bruce Higgs.

In his presentation at this conference (available in these Proceedings), David described how important provenance of seeds he collects for his nursery. He has been involved in the training and support for many new nursery industry people in both personal and business skills while encouraging a willingness to research and experiment in plant development. His restoration focus has been on recalcitrant species propagation and developing methods to overcome seed dormancy. According to David, "Working with researchers and practitioners in hot spot locations has been an abiding interest".

He has been a long serving member of the IPPS Australian Region board, having convened a conference in Perth in 2017, as well as regional meetings in Western Australia and more recently at Boomeroo nursery in Queensland. He also serves on the management committees for the Revegetation Industry Association of WA (RIAWA), The Society for Ecological Restoration Australasia (SERA) and Australian Institute for Horticulture. As part of the additional community support provided by the business, he is active in the operations of the Dieback Working Group (DWG), the Association of Mining & Exploration Companies (AMEC) and the Australian Network for Plant Conservation (ANPC).

He is recognised for the Plant Breeders Rights filing in 2014 for *Hibbertia spicata* Ocean Reef. In 2002 the City of Joondalup recognised him as Citizen of the Year, and he received the Premiers Award for the state of Western Australia and Volunteer of the Year by the Department of Environment & Conservation in the same year.

Like IPPS motto of "seek and share" is the common saying of "you get out of life what you put into it". David Hancock, the recipient of the most prestigious award of IPPS Australia Region truly lives up to both with his energetic approach to life.

Honorary Life Membership

IPPS Australia Region awards honorary memberships to its members who have immensely contributed to the horticulture industry in any field, including but not limited to, education, research, promotion, industry etc. Additionally, to earn the honorary membership one must have a demonstrated history of sharing their knowledge and belief in the principles of the Society and had been a member for a minimum period of 10 years.

Michael Gleeson was recognised as a person who has more than lived up to these and for many was the face of IPPS. In his absence, the President will be presenting the certificate to Michael at a separate meeting.

Peter Smith Perpetual Golf Trophy

It has been a tradition of the IPPS Australia to have a golf tournament among willing participants before the conference. The 2024 Trophy was won by James Gardner from Organic Crop Protectants.

THE SPONSORS AND TRADE DISPLAYS

The conference would not have been affordable to many without the support of the sponsors and the trade displays. The conference had many sponsors, namely Garden City Plastics, Growth Technology, Hort Journal, Norwood, KW Automation, Proptec Horticultural, Syngenta, Greenlife Gro, Croft Structures, Enviro Tec, Australian Plant Production and Transplant Systems.

In 2023, a new award for the best trade display was introduced and Garden City Plastics were voted by the participants as the best trade display in 2024.

PRESENTATIONS

There were seven sessions spread over two days with 23 presentations covering a wide range of areas related to the mandate of the Society – from personal experiences to nursery automation, plant propagation and healthy environments for both plants and humans.

Propagation and Breeding

In Vitro Technologies in the Nursery Industry

There were four presentations in the applications of tissue culture technologies for the nursery industry including the Rod Tallis Memorial Youth Award and Anita Boucher Award for the Best Presentation by Lisa Wightwick from Peninsula Growers in Victoria. Lisa lives in Melbourne and grew up on the Mornington Peninsula. She has worked in horticulture for 10 years, specialising in plant tissue culture and micropropagation. She feels lucky to have been selected for the Six-Pack initiative in 2017 and attended her first IPPS conference in Perth. Her award-winning presentation was on the improvement to the rooting and acclimation stages of popular Grevillea cultivar 'Bonnie Prince Charlie'. As the plants were growing spindly and wilting during acclimation, she decided to test paclobutrazol, a gibberellic acid inhibitor that helps to shorten internodes and make plants stouter. Testing different concentrations, she optimised the concentration to 2 mg/L for this variety and she is now testing other varieties and species with similar problems to ascertain if there are similar improvements to be achieved.

The second presentation on micropropagation was by Dr. Puthiyaparambil Josekutty, (Jose), an agricultural biotechnologist with 37 years of experience in research and research management. In Australia, Jose managed commercial plant tissue culture laboratories at Yuruga Nursery and Fleming's Nursery before joining Skybury farms as Research Manager. He has micropropagated over 100 plant species and many crop varieties from Australia, USA, New Zealand, Micronesia, South Africa and India. Dr Josekutty's presentation was on the development of a micropropagation protocol for three valuable table grape cultivars for rapid and reliable cloning of virus-free stock material for orchard establishment. The work was undertaken at the laboratories of Skybury Farm where he is the Research Manager. His presentation also touched upon various in vitro methods applied in grapevine improvement, virus eradication and conservation.

Another presentation on micropropagation protocol development was on a new fruit crop for Australia - Chinese bayberry or Yangmei (*Myrica rubra* – Myricaceae family). It was presented by one of our new IPPS members Dr. Jayeni Hiti-Bandaralagé, Co-founder and Director of J&S Plant Biotech and STC Link Pvt. Ltd, an expert in plant biotech tools for horticulture. With expertise in tissue culture, genetic improvement, and cryopreservation spanning academia and industry, Jayeni has lately focused on industry outcomes benefiting environment and community. Thus, this project is partly funded by Agrifutures Australia through an Emerging Industries Business Grant 2023 and was conducted in collaboration with several industry partners; Microplants Pty Ltd, Australian Nurserymen's Fruit Improvement Company (ANFIC), Wide Bay Seedlings, and Australian Horticultural Traders ensuring that the outcomes benefit the horticultural industry in the spirit of IPPS. In her presentation, Jayeni explained how she optimized plant material selection, seasonal timing, and media composition for successful red bayberry micropropagation. The developed protocol will facilitate improved growth and mass propagation of red bayberry for commercial applications within and beyond Australia. It is expected that in the future this novel fruit with many nutritional and medicinal benefits will be available in the shelves of Australian fruit traders thanks to the untiring work of Dr Hiti-Bandaralagė.

The fourth presentation on in vitro applications was by Dr Ranjith Pathirana, Editor, IPPS Australia Region. His presentation covered multitude of applications of in vitro technologies, mostly with examples from his own research. In addition to micropropagation, the presentation covered an array of in vitro applications in crop improvement and conservation. These included eradicating viral and bacterial diseases infecting clonal plant material to produce high-health planting material by combining different methods such as meristem culture, thermotherapy, chemotherapy, electrotherapy and cryotherapy, development and deployment of new cultivars to the industry much faster and efficiently than traditional field-based plant breeding methods using in vitro mutagenesis, genetic transformation, gene editing, distant hybridization, polyploid and haploid induction, and increasing the proportion of hybrid seeds in apomictic species. The application of in vitro technologies in ex situ conservation through in vitro and cryo repositories was explained using his personal experience in setting up the cryo-genebank of horticultural species in New Zealand including the rescue of thousands of kiwifruit accessions and breeding lines through in vitro gene banking after the incursion of bacterial canker of kiwifruit in New Zealand.

A Breath of Fresh Air, Healthy Soils and Plants and Clean Irrigation Systems

Inspired by the theme of the conference, there were several presentations on keeping our nurseries, plants, irrigation systems and soils in good health in addition to ascertaining effect of plants on the air we breathe and on our own well-being.

Plants in the Classroom Improve Student Performance

Many studies in the past have shown that plants and growing media (as a biofilters) maintained indoors improve air quality, ambiance and mood of workers resulting in improved staff productivity, performance, job satisfaction and reduced sick leave absence, stress, depression and negative mood states (Sadick & Kamardeen, 2020; Hähn et al., 2021). However, only few studies on the effect of plants on classroom performance of school children have been conducted so far. To fill this gap, in collaboration with the Plants and Indoor Environmental Quality Group of the Centre for Environmental Sustainability, Faculty of Science at the University of Technology, Sydney, John Daly of Eco-Environment in Queensland conducted a study to understand the performance of students in classrooms with and without potted plants. It involved 360 students in grades six and seven in 16 classes in three schools in Queensland, Australia. Student performance was tested across three curriculum course streams: Numeracy, Literacy and Science. The results indicated that the presence of plants and long-term specialist growing media in the classroom consistently led to improved performance in spelling, mathematics and science – i.e., across the curriculum. The results were statistically significant with 10 to 14% improvement in all but one of the five sets of scores in two schools. In the third school where results were not significant between groups with and without plants' presence, the students were already involved in an active gardening program, involving both ornamental and vegetable species.

John Daly who initiated these trials has more than 48 years of experience as a specialist horticulturist, specialising in landscape design and developing soils and potting media for landscape, nursery broadacre and indoor, Plantscape and green building environments across Australia, Singapore, Qatar and Saudi Arabia. His research and treatment of problematic soils such as acid sulphate and calcic soils has led to manufacturing soil amelioration solutions, converting desert sands and marine mud into healthy fertile soils and potting mixes that act as biofilters of air and water to grow biophilic landscapes.

Significance of Vegetation Solutions for Sustainable and Productive Workspaces and Built Environment

Mark Thompson, a registered Queensland Architect who founded Eco Effective Solutions - a consulting organisation working in the design, education and research sectors was the next speaker on the theme of plants in built environment. He co-authored the book "The Environmental Brief: Pathways to Green Design" and co-developed the Queensland Government's Ecological Fitout Guidelines (QldGov, 2023). For 10 years Mark was a member of Greenstar Faculty, the Expert Teaching Panel of the Green Building Council of Australia and as an Adjunct Professor at QUT, assisted in the establishment of the Centre for Subtropical Design. He is an Honorary Member of NIPA (National Indoor Plant Association) and is a passionate advocate of vegetation integrated in the built environment.

Mark reviewed the research findings in multiple research projects he was involved in, demonstrating the significance of integrating vegetation within the indoor environment and the resulting improvement of air quality, reduction of pollutants, and enhancement of occupant well-being. The Revitaliser he designed, and other biophilic design strategies show quantifiable advantages from lowering volatile organic compound levels and stabilising CO2 which improves cognitive function and reduces stress in occupants. Using data from biochemical measurements he further demonstrated the significance of strategic implementation of biofiltration and sustainable construction practices and methods to creating healthy workplaces and urban areas. While indoor environments are vital to occupants, the results of research support incorporating green spaces into proposed urban development approval requirements of cities around the world. Fostering collaboration between architects, scientists, urban planners and legislators will be crucial in developing built environments into healthy and self-sustaining ecosystems. This collaboration will reinforce human and environmental health while paving the way for a sustainable urban future with proven good indoor air quality and increased indoor wellbeing for building occupants.

Plant Quality Control in the Nursery

Natural Area Nursery in Western Australia run by the recipient of the IPPS Australia Award of Honor, David Hancock, has stringent control on quality of their stock. A qualified veterinarian Dr Sabine Suess changed her profession to horticulture and plant nursery management due to her keen interest in plants. Thanks to her eye for detail, she has been selected for the role of Plant Yield Coordinator of Natural Area Consulting and Management in 2020. In her presentation, Dr Suess shared her experience and best practices she put in place for improving plant quality in the nursery. The key areas described ranged from monitoring and identifying areas of concern, water management, determining and implementing actions, teamwork, documentation, to research and development. The focus of the presentation was on the role's integration with the nursery team, and practical ways adopted in the overall management of key functions and areas. After describing the work in improving quality of plants in detail, Dr Suess reflected over the past operations and growth, emphasizing multiple benefits of targeted nursery quality control. The benefits are both of short and long-term nature and include:

- Reduced plant losses, improved stock quality, increased profitability, happier clients.
- Early recognition of problem areas for the production and sales perspective.
- Increased efficiency in dispatch grading, weeding and all maintenance tasks
- Optimised water efficiency and quality
- Improved record keeping and documentation.
- Motivation for procedure improvement and innovation.
- Pride in all staff for the combined efforts as a self-driven team to efficiently grow a quality product.

Green Technology for Sustainable Management of Irrigation Systems

Irrigation lines, drippers and sprinklers will operate inefficiently when fouling occurs due to scale, sediment and/or biofilms. Inefficient irrigation leads to lower production, higher operating costs and lower water use efficiency. Sediment removal is a function of flocculation/filtration while scale is caused when metallic ions are oxidised to inorganic salts and accumulate in-situ. Iron and calcium salts are the most prevalent in irrigation systems. Biofilms on the other hand are complex, can be formed by bacteria, algae, fungi, protozoans or combinations thereof. Common phyla in irrigation systems include *Proteobacteria, Actinobacteria, Chloroflexi, Mycobacterium* etc.

Gary Murdoch-Brown, a senior executive with 25 years of regenerative agriculture experience has provided a platform of knowledge and practical solutions to assist growers in achieving sustainable production systems. As part of this endeavour, Gary described the sustainable solutions developed to manage biofilms. Biofilms are problematic in all production regions. Biofilm fouling is traditionally carried out using oxidising agents, chlorine dioxide, hypochlorite, hydrogen peroxide etc. Most of these products have very high environmental impacts as well as toxicity to humans, water systems and soil and are inefficient in that the biofilms will recur. The modern environmentally friendly 'green' alternatives use either physical forces such as hydrodynamics (flushing through high water flows) or the use of substances that are capable of interfering with the matrix structure of biofilms. In the driest continent Australia, first option is not sustainable and therefore the latter group is being promoted. The latter group includes biocatalysts (enzymes, phages) and organocatalysts (organic non-enzymes). AquaMate® is a patented organocatalyst developed by Advanced Nutrients (Advanced Nutrients, 2025) that causes destruction of the biofilm matrix by breaking it into simple sugars and flushing out from the system whose repeated use prevents reoccurring of biofilm and is low-cost, non-toxic, and non-hazardous.

Managing Red Imported Fire Ant for Soil Health

The production nurseries in Southeast Queensland (SEQ) have been battling Red Imported Fire Ant (RIFA) since 2001 when the first incursion was detected in Richlands suburb of Brisbane. The costs associated with the pest are high due to market access, treatment, site management, lost markets, etc., costing the sector in SEQ more than $20 million annually. RIFA is a pest of the environment, more so than horticulture, where if established it dominates the invertebrate world, out competing or killing native insects and small animals. Greenlife Industry Australia National Biosecurity Manager/Director Research Development & Extension (RDE) John McDonald with 35 years' experience in production horticulture has been involved at state and national levels addressing industry development, plant protection, biosecurity, policy and RDE. In his presentation, he pointed out shortfalls in funding the RIFA control (and eradication) program and emphasized that it is critical that Australia eradicates RIFA to ensure our way of life for future generations is unaltered and that our environment is protected. RIFA is one of the most invasive species known and it has been known to 'farm' plant pests such as aphids within cropping systems which add to growers' pest pressures and further enhances worker exposure and subsequent health impacts from stings. It is estimated that 43,000 - 174,000 people in Australia get medical assistance due to allergic reactions to RIFA stings annually.

After describing the history of RIFA eradication/control programs and the current situation John concluded that RIFA is continuing to spread, and it is expected to move into other Australian jurisdictions in future. For production nurseries, his message is to keep RIFA off the site and trade with businesses that are RIFA aware and have their own risk mitigation programs in place. Incorporating bifenthrin granular into the growing media is one measure. Inspecting consignments, property surveillance, crop monitoring and dispatch inspections have been more effective than simple chemical treatment. He suggested using the pest management resources available to industry including RIFA specific technical information plus a plant protection program (BioSecure HACCP) that provide all the guidance on surveillance, inspections and crop monitoring.

Propagation, Breeding and Nursery Infrastructure

Apart from the four presentations on tissue culture-based propagation, *in situ* plant propagation in nurseries, plant development, adaptation, breeding and developing machinery and infrastructure were also addressed in several presentations.

As only 1% of the Subtropical rainforests (STRF) are left in eastern Australia after their decimation in the 1800s and early parts of 1900, planting is the only solution for restoration. Momentum that gathered in the 1970s saw annual plantings increasing from mere hundreds in the 1980's to over half a million by 2020's. Over the years it was realised that not only the quantity of plants but also its diversity is important. Hence seedlings have become the planting material for rainforest restoration. As a result, seed collection, storage, propagation and growing techniques of seeds of over 450 rainforest species became the cornerstone of research and a multimillion-dollar industry in recent years. Species composition and functional trait representation in these forests is of utmost importance. It is no exaggeration to state that Mark Dunphy is the most qualified person to discuss Australian rainforest seed propagation as he has worked on over 100 restoration projects as project manager or as a consultant in Australia and in the Pacific. His Firewheel Nursery has been operating since 1988 and produces over 250,000 trees annually from over 300 species. Mark is the lead author of the CSIRO published "Australian Rainforest Seeds. A Guide to Collecting, Processing and Propagation" (Dunphy et al. 2020), considered the handbook for any nursery dealing with rainforest seeds and has sold thousands of copies. Mark passionately described the reasons for increase in demand for seedlings, importance of diversity of species in restoration projects, skills, knowledge and experience needed for seed collection (species identification, masting issues, viability, handling etc.), reasons and techniques for processing STRF fruit and seeds, recalcitrance and viability issues, sowing and germinating rainforest seeds and techniques for storing seedlings (as seeds cannot be stored).

It is heartening to note that, thanks to the efforts of dedicated seed collectors and nurseries such as Mark's Firewheel Rainforest Nursery (FRN, 2025), the trend of reducing land and species diversity under rainforests found in many countries has reversed in the northern New South Wales. The species diversity, number of trees planted, and area of rainforest established is increasing annually in eastern Australia. Collecting rainforest seed, propagating it, growing it on, then planting it and maintaining the trees to form a young rainforest is a relatively new science. Since 1980's this science has taken momentum, and we have been moving ahead in leaps and bounds since then. However, there is much more to learn and discover ways to improve and speed germination, growing on and establishing rainforest. Funding this research is therefore critical for restoring our degraded rainforests.

Des Boorman has had a life-long interest in plants, especially flowering natives. After completing a degree in Production Horticulture in 1991, Des has worked in the horticultural industry in various roles. Plant breeding has always interested Des, and he gave an update on the *Brachychiton* hybridisation and selection program he has been undertaking since 1998. He described the methods of pollination and grouping of hybrid plants and the outcomes from different hybrid combinations. With time, the initial three groups had to be crossed with other genotypes and species producing interspecific and poly specific (four species in one of the groups) hybrids. The multiple crosses have yielded unusual genotypes of breeding interest and some already promising selections are being bulked up for commercialisation. Despite the time it takes to breed ornamental tree species, Des encouraged more people to take up breeding as the results are very satisfying.

Propagation and breeding of another two native Australian species, *Anigozanthos* and *Macropidia* popularly known as kangaroo paw (some species and hybrids also called cat's paw) were the theme of the presentation by Angus Stewart (GWA, 2025), who co-operates with a network of horticultural businesses and individuals to create a unique blend of information that includes tried and trusted topics as well as the new and experimental in Australian horticulture. Angus works with a range of specialist Australian native plant nurseries. He coauthored the book "Grow Your Own– How to be an Urban Farmer" with Simon Leake of Sydney Environmental and Soils Laboratory (Stewart and Leake, 2020). The book features excellent new growing systems for urban farmers and it won a Laurel for best gardening book at the Horticultural Media Association of Australia in 2020. Angus is organising the 2026 IPPS Australia Region conference in Hobart, Tasmania.

Angus' presentation on kangaroo paws covered the brief history of domestication of the two species as ornamental crops, methods of propagation including tissue culture, using seed and using parts of divided rhizomes. He emphasised the importance of breakthrough research by Kingsley Dixon and co-workers of Kings Park and Botanic Gardens in Perth, WA where they showed the role of smoke water in breaking seed dormancy (Dixon et al, 1995). In his presentation, he vividly illustrated the possibilities of developing cultivars by intra and interspecific hybridisation of the most adapted *Anigozanthos flavidus* with short stature and red and green coloured species for urban areas and as potted plants with spectacular colours. Tissue culture techniques such as embryo culture have played a distinct role in developing some spectacular interspecific hybrids.

Natural Area Nursery in Western Australia started from humble beginnings in 2005 and grew into an 800,000 annual turnover of tubestock from 80,000. Over the 15 years, the Government leased land of the nursery quadrupled in area as well. At the beginning of Covid pandemic, in January 2020, the Government of Western Australia asked for the return of the land for a new train station complex. The founder of the nursery and the recipient of Award of Honor David Hancock described the identification of land, logistics, designing, building and relocation of the massive operation within the two-year timeframe allocated by the Government. In addition to all other factors, the dedication and hard work of the staff of the Natural Area Nursery made this relocation possible.

Automation is key to successful plant nursery business in developed countries, as manual labour is costly. Moreover, automation helps achieve uniformity of the product and better efficiency. Equipment and machinery production for nurseries requires both engineering skills and knowledge of plant biology and physiology. KW automation is a leading provider of nursery automation solutions founded in 1979 by Kurt Weisenberger. KW Needle Seeder set a new standard for efficiency and productivity. Today, as a third-generation family-owned business, KW Automation continues Kurt's legacy of quality and innovation. Their range of machinery includes soil mixers, hoppers, conveyors, & elevators, pot & bag fillers, needle seeding equipment, tray fillers, potting machines, tray & pot washers, watering tunnels and customizable equipment etc. Thus, their expertise includes a broad range of automation technologies from seeding lines, soil mixing lines, pot and tray fillings solutions plus customised solutions.

Luke Weisenberger is a third-generation member of KW Automation, a dynamic sales technician emerging at KW Automation. He Joined the family business in 2016 and transitioned from Purchasing to Sales in 2024 and presented the profile of the company in his talk.

Plant Biology, Physiology and Evolutionary Biology

Biology of plants from cell to species levels were discussed at the conference.

In his presentation, Carl Barry, co-founder of Growth Technology Pty Ltd discussed the different types of meristematic cells in trees and how they contribute to the growth. He described the differences in apical meristems in shoot and root and how these in turn differ from lateral meristems that contribute to girth of massive trees. Then he went on to discuss the differences between the two lateral meristems – cork cambium and vascular cambium. He suggested that a thirty-meter-high tree trunk with a diameter of one meter would have a lateral meristem with an area of over 90 square meters and a thickness of tissue paper. Then he moved onto describe how the meristematic cells act, differences in apical, axillary and root meristem division. A mention was also made of the dedifferentiation of cells and redifferentiation, not only in tissue culture, but also when we use external auxins to induce roots in cuttings in nursery industry. Barry's knowledge in the fundamentals of plant physiology and development made Growth Technology a successful business with novel products for the hydroponic community and the development of plant hormone products for the plant propagation industry. Under his management Growth Technology went from a two-man operation in South Fremantle WA to a business with factories in Australia and the UK. Exports now account for nearly half of their production.

Clive Larkman elected the President of IPPS Australia at the Ballina Conference in 2024 is a qualified botanist and nursery person with over thirty years managing a major propagation nursery. He developed an interest in the lavender industry and is now recognised as a world expert in the growing and breeding of commercial lavender varieties. He is a passionate collector of new, different, old and rare plants from all parts of the globe and his presentation was about the importance of understanding the evolutionary biology of a species for a successful nursery industry using lavender as an example. He first described the mediterranean climate in southern France where *Lavandula* evolved and how his own experience and field trials in Australia helped grow and breed lavender successfully and transform it into a business with a large market share.

Personal Experiences

IPPS is all about personal experiences, research and sharing the knowledge so acquired. Hence, every IPPS Australia Conference has a renowned speaker or two talking about their experiences in the gardening/nursery industry. These speeches motivate young people attending the conference. In fact, the first two inaugural presentations of the Conference were about the journey of two contrasting but equally successful people in horticulture/nursery industry.

The inaugural speech was by Samantha Birkwood from Bamboo World – a 15-acre nursery specialising in clumping bamboo in the picturesque Northern Rivers region of New South Wales where the participants had the chance of visiting during the Conference. Samantha described how her character was moulded by the Army in her early days as an Officer in the Australian armed forces, then transitioning into a travel writer, while working in project management roles in leading international brands. She then went onto describe how she transformed into managing the Bamboo World, after acquiring it in 2019 combining her project management skills with her husband Matt's extensive landscaping experience in Macau, including the world's largest interior vertical garden. Their key strategies include innovation, sustainability, community engagement, and adaptability. They use e-commerce platforms, automated processes, and online ordering while prioritising sustainability by minimising their environmental footprint and promoting conservation efforts. They also stay prepared for unexpected challenges, having navigated droughts, fires, a global pandemic, floods, storms, and more in just 5 years!

Samantha's presentation was followed by Des Boorman's story, about his journey in horticulture. He had a life-long interest in plants, especially flowering natives that made him complete a degree in Production Horticulture in 1991 and has worked in the horticultural industry his whole career in various roles. Des has many skills including native plant pollination and breeding, grafting and nursery management. He says, "being able to weld, build and fix things may not seem like a horticultural skill but it certainly makes you think about a lot of different industries as you apply their tools of the trade". Des described his school and university days when his enthusiasm in exotic plants made him establish the first tea plantation in New South Wales and selecting 'Mineola Tangelo' by tasting and sugar analysis of a range of selections. He went on to describe the different native species he worked on and how he learned their biology while working with them to unlock their potential as ornamentals. He did not forget to mention people and teachers who inspired him throughout his horticultural journey.

The conference also had an interesting on-stage interview of a young and emerging horticulturist Zoe Williams, also IPPS Australia Social Media Editor, interviewed by a renowned radio/media personality and 2023 IPPS Australia Edward and Mary Bunker Award winner Jane Edmanson (Pathirana and Williams, 2023) (**Fig. 11**). The detailed papers on all the abovementioned presentations are published in these Proceedings including the experiences of the two South Africa Exchangees Nosipho Phiwokuhle Ndlovu (from South Africa to Australia) and Joshua Taylor, Australia Region awardee to visit South Africa.

Figure 11. Zoe Williams, a successful young horticulturist was interviewed by award winning media personality Jane Edmanson on stage during the Conference.

Nursery Tours During the Conference

On 24 May, the second day of the conference, all the participants had the opportunity to visit four nurseries in the Northern Rivers region of New South Wales where conference venue was also located. Bamboo World (Bamboo World, 2025) was the first nursery to visit, and the participants already had an idea about its operation as Samantha Birkwood from the nursery had already given the inaugural presentation about her involvement. At the nursery the participants were taken through the plantation and the nursery (**Fig. 12**), and Matt described how to distinguish runner bamboos from clumping bamboos as it is important to select clumping bamboos for home gardens. In addition to being the largest bamboo collection supplying bamboos Australia-wide (except Western Australia and Tasmania), Bamboo World provides garden design and consultation, landscaping and maintenance, holds workshops and garden tours as well as hires plants.

Next, Ray Parker, Convenor and Organiser of the Conference and Diane hosted the participants at their Parkers Place Nursery (Parkers Place, 2025), operating as a wholesaler, for a stopover and lunch. During the lunch, Des Boorman who has two manuscripts in these Proceedings gave a demonstration of grafting techniques (**Fig. 13**).

Figure 12. Participants visited the Bamboo World, the leading bamboo nursery in New South Wales during the Conference.

Figure 13. Des Boorman demonstrating grafting techniques during the lunch at Parkers Place.

After lunch at Parkers Place, the participants visited Alstonville Plants, another wholesale nursery. Alstonville specialises in indoor plants, tropical foliage and landscaping plants including Aspidistras, Cordylines, Waterhousia, Heliconia, Hoyas, Magnolias, Dracaenas, Strelitzia, Sansevierias and Palms (Alstonville, 2025). Production Manager Josh Duncan and

Chief Plant Officer Lynne Sutherland showed the stunning collections of landscaping and indoor plants they produce **(Fig. 14).**

The last nursery to visit was Pearce's Nurseries, a wholesale nursery supplying potted plants to garden centres, florists, landscapers, city councils etc. (Pearce's, 2025). Their collection of succulents was diverse and stunning **(Fig. 15).**

Figure 14. Alstonville Plants specialising in indoor and landscaping plants was the third nursery visited by the Conference participants.

Figure 15. Pearce's Nursery visited by the Conference participants had a stunning collection of succulents.

Acknowledgement

The authors wish to thank Pam Berryman, Executive Officer of the IPPS Australia Region for reading the manuscript and suggesting improvements and for supplying some images used in the figures.

LITERATURE CITED

Advanced Nutrients. (2025). AquaMate®. Available at: https://advancednutrients.com.au/aquamate%C2%AE-1 Downloaded on 23 January 2025.

Alstonville. (2025). Alstonville Plants. Available at: https://www.alstonville-plants.com.au/services-5 Downloaded on 15 January 2025.

Bamboo World. (2025). Bamboo World, Available at: https://bambooworld.com.au/ downloaded on 24 January 2025.

Dunphy, M., McAlpin, S., Nelson, P., Chapman, M. and Nicholson, H. (2020). Australian Rainforest Seeds. A Guide to Collecting, Processing and Propagation, *CSIRO Publishing*, Clayton South, Victoria, Australia.

Dixon, K.W., Roche, S. and Pate, J.S. (1995). The promotive effect of smoke derived from burnt native vegetation on seed germination of Western Australian plants. Oecologia *101*:185–192.

GWA. (2025). Gardening with Angus, Available at: https://gardeningwithangus.com.au/about-angus-2/ Downloaded 22 January 2025.

Hähn, N., Essah, E., and Blanusa, T. (2021). Biophilic design and office planting: a case study of effects on perceived health, well-being and performance metrics in the workplace. Intelligent Buildings International *13*: 241-260.

Mtnathan. (2025). About Mt Nathan Nursery. Available at: https://www.mtnathannursery.com.au/about/ Downloaded 20 March 2025.

Natural Area Nursery. (2024) .Natural Area Perth Native Nursery. Available at: https://naturalarea.com.au/perth-natandive-nursery/ Accessed 23 October 2024.

Parkers Place. (2025). Parkers Place Wholesale Nursery. Available at: https://www.parkersplace.com.au/about-parkers-place-wholesale-plant-nursery/ Downloaded 23 January 2025.

Pathirana, R. and Williams, Z. (2023). Propagation – Essential to Life on Earth: 51st Conference of the International Plant Propagators' Society – Australia Region. Comb. Proc. Intern. Plant Prop. Soc. *73*:46-66.

Pearce's. (2025). Pearce's Nurseries. Available at: https://pearcesnurseries.com.au/ Downloaded 23 January 2025.

QldGov. (2023). Office Accommodation Management Framework Guideline 3: Fitout. Available at: https://www.forgov.qld.gov.au/property-land-and-infrastructure/manage-government-buildings-and-assets/office-accommodation-management-framework/guideline-3-fitout Downloaded 22 January 2025.

Stewart, A. and Leake, S. (2020). Grow Your Own – How to be an Urban Farmer, available at: https://gardeningwithangus.com.au/product/grow-your-own/ Downloaded 23 January 2025.

Sadick, A. M. and Kamardeen, I. (2020). Enhancing employees' performance and well-being with nature exposure embedded office workplace design. Journal of Building Engineering *32*:101789.

Combined Proceedings IPPS 74:76-81. 2024.

Kangaroo Paws – From Wildflowers to World Market

Stewart Angus

Premaydena, Tasmania, Australia

auangus@gardeningwithangus.com.au; stewart.angus00@gmail.com

Keywords: *Anigozanthos,* Australian natives, pot plant, breeding, embryo rescue, tissue culture, propagation, ornamental plants

Summary

Kangaroo paws (*Anigozanthos* and *Macropidia* species) have been successfully domesticated over the last few decades through various genetic improvement programs and the development of propagation methods suited to mass production. This paper summarises the various strategies and techniques employed by the author and other researchers to achieve the goal of domesticating this crop for landscape and garden use as well as for cut flower production.

IPPS Vol. 74 – 2024

INTRODUCTION

Kangaroo paws (*Anigozanthos* and *Macropidia* species) are one of Australia's most distinctive native plants. Growing wild only in the southwest corner of Western Australia (**Fig. 1**), the furry flowers bear a resemblance to the paw of the iconic kangaroo, hence their common name.

The bright colours of the flowers added to the distinctive and unique shapes created instant horticultural interest when early European botanists and plant collectors sent propagation material back to Europe where it was primarily grown from seed.

Figure 1. Kangaroo paws growing wild in western Australia

It was not until the 1960's that the first report of genetic improvement of kangaroo paws emerged from Bob Dixon, a horticulturist who was growing them at Perth Zoo. This appears to be the start of the journey of kangaroo paws from a difficult to propagate 'wildflower' to a domesticated plant that has taken its place as a significant albeit minor floricultural crop in the world market (**Fig. 2**). This paper examines some of the breeding and propagation advances that have supported this journey.

SEED GERMINATION

A breakthrough by researchers at Kings Park and Botanic Gardens in the 1990's, led by Kingsley Dixon, discovered that a chemical from smoke could enhance germination in a number of *Anigozanthos* species where germination had previously been erratic and unreliable. Smoke was found to be a trigger for breaking seed dormancy for a range of plant species from natural habitats that are prone to bushfires. It has long been observed that regeneration of kangaroo paws in these landscapes was dependent on fire.

Figure 2. Domesticated kangaroo paw found its niche market – growing in the garden adjacent to the famous Sydney Opera House and the Sydney Harbour Bridge.

The action of smoke in breaking seed dormancy has led to the development of various treatments to apply the active ingredient in smoke to seed propagation in kangaroo paws. Smoke impregnated water or vermiculite is the most common and practical way of treating seed on a commercial scale **(Fig. 3)**. Other methods involve using smoke in specially created chambers to expose seed planted in containers to prolonged exposure to smoke. A tool used by beekeepers (called a 'bee smoker') to create smoke to clear bees from their hives has proven useful in generating smoke to treat seeds sown *in* situ. The materials burnt to create the smoke can come from a wide array of substances. There are also commercially available products available such as smoke impregnated water or vermiculite that can be found through internet searches.

Figure 3. Smoke impregnated water or vermiculite is the most common and practical way of treating kangaroo paw seed on a commercial scale.

Micropropagation of *Anigozanthos* and *Macropidia*

The first published report on micropropagation of *Anigozanthos* was by Roger Ellyard in 1978 at the Australian National Botanic Gardens in Canberra (Oliver, 1991). His work outlined a protocol whereby very small parts of the apical meristem were excised from vegetative shoots and grown on a sterile nutrient medium. This protocol enabled cost effective mass propagation of *Anigozanthos* cultivars for the first time and it is now a routine practice in commercial horticulture around the world **(Fig. 4)**. Much of the micropropagation production of *Anigozanthos* has shifted from Australia to offshore laboratories in countries such as Sri Lanka and Indonesia. There is considerable variability in the multiplication rates of various species and cultivars within *Anigozanthos* and *Macropidia* genera and cultivars leading to some clones being not commercially viable by this propagation method.

Figure 4. Cost effective mass propagation of *Anigozanthos* cultivars through tissue culture is now a routine practice in commercial horticulture around the world, particularly in offshore laboratories in countries such as Sri Lanka and Indonesia.

Division of *Anigozanthos* and *Macropidia*

Division of kangaroo paw rhizomes is an alternative method of vegetative propagation that can be used where micropropagation has proven to be not commercially viable. In the experience of the author the best time to divide is in the autumn after the plants have finished flowering. Good results have been obtained by cutting the rhizome into sections that have several healthy green shoots as part of the rhizome section. Reducing the length of foliage by about 50% by cutting it back reduces water stress on the newly divided section of rhizome, thereby reducing the stress on it while new roots are forming. Planting divided rhizome sections into pasteurised general purpose potting mix provides a suitable growing medium with optimal drainage conditions for establishment **(Fig. 5)**.

Figure 5. Division of kangaroo paw rhizomes is an alternative method of vegetative propagation. Planting divided rhizome sections into pasteurised general purpose potting mix provides a suitable growing medium with optimal drainage conditions for establishment.

BREEDING ADVANCES IN KANGAROO PAWS

The species *Anigozanthos flavidus* is unlike the other 11 species of kangaroo paws in that it has proven to be a long-lived perennial plant in cultivation that has proven to be very adaptable to a wide range of soil and environmental conditions. It has been a key species in breeding programs in creating improved growing performance in intra- and interspecific hybrids. Other species of kangaroo paws such as *A. humilis* (cats paws – **Fig. 6**) and *A. bicolor* (dwarf red and green kangaroo paw) have been used to bring spectacular colours and dwarf growth habits to hybrids with *A.* flavidus. Breeding along these lines has been particularly effective in creating compact cultivars for pot plant production that also often flower continuously in frost free conditions. Whilst *A. flavidus* is useful in breeding to create adaptable relatively long-lived cultivars, its tall height (flower stems between 1.5 to 3 metres tall) is not conducive to the logistics of pot plant production and transport.

Figure 6. Cat's paws (*Anigozanthos humilis*) (above) and *A.bicolor* have been used in hybridization to bring stunning colors and dwarfism to new hybrids.

The challenge in breeding for the future is to create cultivars in a wide range of colours with more compact height that are adaptable and long flowering in cultivation for both pot and cut flower production (**Fig. 7**).

Figure 7. Hybrids of *Anigozanthos flavidus* (tall but adaptable and long-lived) with *A. humilis* can create cultivars suitable as potted plants with stunning colours. Shown above is such a cultivar 'Bush Volcano'.

Embryo culture has been a useful *in vitro* method to create new hybrid seedlings that have not been viable by conventional seed germination. The microscopic hybrid embryos are extracted under a dissecting microscope and placed onto a standard multiplication medium. As well as rescuing embryos that would not normally germinate through standard seed propagation, the hybrid embryos rapidly multiply and enable the production of clonal plants for evaluation, thus shortening the breeding cycle (Worrall, 1995).

LITERATURE CITED

Oliver, K.R. (1991). Hybrid kangaroo paws. Australian Plants *16*: 53-61. https://anpsa.org.au/wp-content/uploads/Australian-Plants-Vol16-126.pdf Downloaded on: 27 July 2024.

Worrall, R. (1995). Breeding of dwarf kangaroo paw (*Anigozanthos*) for use as flowering pot plants. Acta Hortic. *397*:189-196. DOI: 10.17660/ActaHortic.2009.812.51

Combined Proceedings IPPS 74:82-86. 2024.

From a Miniscule Ball of Cells to a Giant: How Meristems are Involved in Producing Giant Trees

Carl Barry

Growth Technology Pty Ltd, 1/45 Stockdale Road, O'Connor, Western Australia 6163

carl@growthtechnology.com.au

Keywords: apical meristem, cell differentiation, cell division, cell growth, cork cambium, lateral meristems, totipotency, vascular cambium

Summary

This article is about how some trees grow to be so large. It talks about the various types of meristematic tissue, their similarities and differences in form and function and how meristems can be of relevance to plant propagators.

INTRODUCTION

How do some things get so big? An African Bush Elephant is not an animal to cross. They are huge. The Blue Whale is thought to be the largest animal to have ever lived. Larger even than the dinosaurs. There are squids with eyes the size of basketballs. These, however, pale into insignificance compared with the majesty of tall trees.

IPPS Vol. 74 – 2024

In Tasmania, there is an *Eucaluptus regnans* that is just a touch over 100 meters (330 feet) tall. It even has a name: Centurion. It is amongst the tallest of all the flowering plants. Centurion must be one of the largest organisms to have ever lived in the history of our planet. Another *E. regnans* with the not so majestic name of Still Sorrow has an above ground volume of 400 cubic meters. It is one of the most massive living hardwood trees in existence.

How do some trees get so large? What is the mechanism that causes this growth?

MERISTEMS

Meristems aren't the most talked about parts of plants. Indeed, as a conversation starter they pale into insignificance compared to flowers and foliage and the like, but they have an interest of their own and are worth more consideration than they usually get around the dinner table.

Considering how small most meristems are, it is quite easy to spot where one could be found. Meristems are found wherever there is new growth on a plant. So, we are talking about for example growing tips of stems and branches. To be more accurate, at the very tip of the growing tips. Meristems are the ultimate source of all new growth in plants. This is their function. This ensures continual growth throughout the plants' life by providing a continual supply of new plant cells.

So, what is a meristem? Meristems are made up of very specialized cells that are different from anything in the rest of the plant. They have a superpower called totipotency. This power allows meristem cells to change just about everything about themselves like their size, shape and function. They are at the start of a biological assembly line that forms the leaves and flowers. Meristems are where new growth comes from. The cells that make up a meristem are very small with their interiors consisting mainly of the cell's nucleus which contains the genetic information coded into genes which control the structure and operation of a plant's growth and development. Meristematic cells divide rapidly. That is their job. When a meristematic mother cell divides into two daughter cells something happens that sets off new plant growth. One of the daughters stays within the meristem so maintaining the number of meristematic cells inside the meristem. The other daughter cell forms on the outside of the meristem and it is these cells and their daughters and their daughters that differentiate, divide and grow that ultimately produce new growth that is evident at growing tips on the shoots and roots.

Where would we find a meristem. They are small, microscopically small. The shoot apical meristem or SAM as it is known by generally ranges in size between 50 and 200 micrometers in diameter and the root apical meristems are smaller again at between a few tens to around 100 micrometers. For reference, human hair is around 100 micrometers in diameter. So, because of their size, finding a meristem can be daunting to say the least. They are also not all that obvious a structure since they are under a surface layer of cells which makes spotting them difficult. However, there is a hint given in the name apical meristem. The term "apical" refers to the apex of a branch or stem. This is the absolute tip of the growing tip. All other parts of the plant are behind this point. It represents the newest of the new growth found at this place. As the meristem produces new cells behind it, the meristem moves upward and the plant extends upward in the case of the SAM or downward in the case of root apical meristems, which are, not surprisingly, known as the RAMs.

DIVIDING, GROWING AND DIFFERENTIATING

Shoot and root growth is more than just the apical meristems producing new cells. These new cells have three jobs to do.

Firstly, they themselves need to divide and divide repeatedly to get the cell number up to what is required to produce mature leaves and branches. The meristem itself produces new cells very rapidly, but more are required so the daughter cells themselves need to divide many times to raise the cell count. Every cell produced can trace its line of descent back to the meristem.

Secondly, the new cells need to grow quite a bit. Meristematic cells are so tiny they make normal plant cells seem huge by comparison. So, growth is about producing more cells that themselves grow larger. However, producing more cells that themselves grow will not by itself produce new tissue types by which we can recognize plant tissue.

This is where the third process takes place where these cells turn into the cell types we recognize as normal plant cells such as epidermal cells or parenchyma cells or any of the other cell types we find. This is the process of cellular differentiation. This is where the plant comes into being so to speak.

Apical meristems, a blob of undifferentiated cells found at the tip of stems and branches and roots produce daughter cells that themselves divide, grow and differentiate and are the source of all new growth both upwards and downwards. Meristems and new growth go together. Without meristems there is no new growth. This is the importance of meristems and most of the action takes place in the first few millimeters of the root or shoot apex.

THE PROBLEM OF GIRTH

Trees grow up and down and they also grow outwards. Trunks and branches increase in girth and each year a new tree ring develops which represents a years' worth of girth. This constitutes new growth and wherever there is new growth there are meristems involved. So, we expect the presence of totipotent, undifferentiated meristematic cells that in turn produce cells that divide, grow and differentiate and ultimately give us new plant growth.

This outwards growth or as it is termed the lateral growth of tree trunks, roots and branches has one major difference to the meristems that occur at growing tips of shoots and roots and that difference is in the structure of the meristem itself. Apical meristems are blob-like, a bit like a slightly flattened sphere sitting at the tippy top of stems, roots and branches. Lateral meristems that produce girth have a different structure completely.

When you look at a tree trunk you can't actually see anything that is alive. The bark is obviously dead and hard up against the trunk is another layer, the "cork" that is made of large air-filled cells that are also dead. The living part of a tree trunk is below all this.

If you carefully pick away at the bark and cork of a tree you will find below an outer layer of living cells. This layer envelops the entire tree. If you had the inclination and a steady hand you could skin the tree and end up with a huge but very thin layer of cells. This is the cork cambium, and it contains a layer of undifferentiated meristematic cells. It is a meristem. A thirty-meter-high tree trunk with a diameter of one meter would have a lateral meristem with an area over 90 square meters and a thickness of tissue paper. Large trees could have hundreds of square meters of this type of meristem. If that isn't enough, you would need to double the area of meristem because there are in fact two layers of meristems under the bark. They are the cork cambium meristem and the second is the vascular cambium meristem.

The cork cambium is fairly specialized and, in most trees, mainly produces the outermost protective layer of cork and bark. The vascular cambium is responsible for producing the vascular tissue consisting of the phloem vessels that move sugars from the leaves to the rest of the plant and the xylem vessels that take water and dissolved minerals up from the roots to the rest of the plant. In both cases the meristematic tissue of undifferentiated cells in that thin layer of cells produces cells that in turn divide, grow and differentiate into different cell types and produce new growth.

The annul cycle of the seasons results in tree trunks growing faster or slower depending on conditions and produces the growth rings that can be seen when trees are cut down. The size of the ring indicates exactly the amount of new trunk growth that has occurred in any year. The initial cause of this are meristem cells existing in two thin layers. The life of a tree trunk is a surface phenomenon where layers of living cells are laid over a core of dead xylem vessels that make up the heartwood of the tree. It is the xylem vessels that are mostly responsible for the formation of wood which gives a tree its structural rigidity which allows some tree species to get so large to become some of the largest organisms that have ever existed.

MERISTEMS IN ACTION

Around us we can see meristems in action. Most plant people are aware that some plants just keep on growing along the main stem with little or no branching at all. Sometimes this can be good, but maybe a bit of branching would be nice. To get that effect the simplest way is to pinch out the growing tip. This may seem to be rather extreme since all that needs to be removed is the apical meristem as we remember is a blob of cells about the diameter of a human hair, but human fingers are not exactly a precision scientific instrument but mostly is all we have at hand.

This is an example of apical dominance where the apical meristem inhibits side branch growth and when the inhibition is removed the side-branches grow. This is new growth so where are the meristems that produce this? The base of a leaf stalk, where it meets the stem is called the leaf axil where buds and shoots can develop. The meristems found at the leaf axil are suppressed by the dominance of the top apical meristem. When that is removed the axillary meristem springs into action and new growth happens and side-branches grow and the entire structure of the plant changes.

Striking roots on plant cuttings is another example of where meristems are of importance. A cut through a leaf node and a dip in a rooting hormone gel is often all that is needed. However, sometimes there are plant species that will never set roots. It could be that there isn't a meristem in the first place, and this is probably the reason that new root growth will never happen.

There is an interesting phenomenon in some of the difficult to root species where they can be given a "hormonal sledgehammer" which turns normal everyday plant cells back into a meristematic state. This is the reverse of everything we have been talking about. We can de-differentiate cells and make them meristematic again and with this comes the power of totipotency and they can be differentiated or changed into the root cells we wanted in the first place. This sort of thing is worth exploring because it can help solve problems with difficult-to-propagate species.

Mericloning of orchids is a specialized type of tissue culture where growing tips of plants are cultured and manipulated using plant growth regulators which act on the actual apical meristem to make it produce the mericlones, exact copies of the mother plant.

CONCLUSION

Next time you stand before a mighty tree you might consider the actual number of meristems there are on that one plant. There are meristems at the tips of every branch and branchlet, every place there is new growth there are meristems in action. Thousands and thousands of them. Think of how many root tips are out of sight but still there pumping out the initial step in root growth, and the simple fact that there are not one but two separate meristems that wrap almost the whole of the tree. And they all have one job and that is to produce new plant growth. Meristems are where tall trees come from.

Acknowledgement

The author would like to acknowledge the kindness and patience of Dr Ranjith Pathirana, Honorary Research Fellow at the University of Adelaide for his assistance in editing the paper.

Combined Proceedings IPPS 74:87-92. 2024.

Cultivating Success: My Journey into the Nursery Industry

Samantha Birkwood

Owner & Director, Bamboo World Production Nursery Northern Rivers, NSW, Australia

Keywords: Bamboo, sustainability, innovation, community engagement, adaptability

Summary

Samantha shares her journey into the nursery industry, highlighting key strategies that have shaped Bamboo World Production Nursery to what it is today – working hard to create a business growth of over 250% income in 5 years. Her experiences span various industries and continents, offering insights that may resonate with you or spark new ideas.

Samantha's journey began in the Army, where she learned discipline, teamwork, resilience, and adaptability. After 11 years of service, she transitioned to roles in travel writing, public relations, and project management across Asia, working with brands like Motorola, Apple, and Ikea. These experiences helped her excel in problem-solving and leadership roles in diverse environments.

In 2008, Samantha met her husband, Matt, who has extensive landscaping experience, in Macau. Matt's impressive projects include the Sydney Olympics and the world's biggest interior vertical garden at MGM Macau. In 2019, they moved back to Australia and took over Bamboo World, a 15-acre nursery in the picturesque Northern Rivers NSW.

Their key strategies include innovation, sustainability, community engagement, and adaptability. They updated operations with e-commerce platforms, automated processes, and online ordering. They prioritise sustainability by minimising their environmental footprint and promoting conservation efforts. Community engagement and customer education are essential; they participate in events, host workshops, and offer personalised customer service. They also stay prepared for unexpected challenges, having navigated droughts, fires, a global pandemic, floods, storms, and more in just 5 years!

Samantha's journey has taught her resilience, innovation, and adaptability. As Bamboo World continues to grow, Samantha and Matt are committed to collaboration and industry progress.

INTRODUCTION

Today, I have the privilege of not only talking about my story, but also delving into some key strategies I've learned over the years and applied to our nursery business. These insights stem from my experiences across industries and continents. I hope that what I share resonates with you or sparks new ideas for your own endeavours.

Before we begin, I want to clarify that I'm someone who can never sit still, always having too much on the go, and often saying 'yes' and then figuring it out along the way. Despite this, I managed to fit all that I'm about to tell you into my life - celebrating my 50th birthday last year.

THE WINDING PATH THAT LED ME HERE

Straight out of school, I chose a career path that would profoundly shape my character and equip me with invaluable skills. Instead of a job lined up in a local resort as a receptionist, I joined the Army. Initially, it was a way to escape my small hometown of Port Douglas, but it became so much more. It was transformative and shaped me into who I am today. The Army taught me discipline, teamwork, resilience, and adaptability. I experienced fitness, travel, excitement, new friends, and of course - challenging times - all rich with learning and growth. In fact, you could almost say my first "horticultural" lesson was during training in the bush, learning to choose carefully the right leaf for toilet paper!

After 11 years of service, filled with both rewarding and tough experiences, I left the military to explore new horizons, starting in Singapore. Working as a travel writer and later as an assistant editor for a magazine, I eventually landed my dream job in public relations. At a boutique agency, I learned the power of words and the importance of understanding and speaking the customer's language. I then moved into client-facing roles in a project management company in Asia, at first handling project deliveries but then moving into large-scale retail rollouts, new business opportunities, and project crisis management. My military background equipped me to tackle challenges head-on and excel in problem-solving and leadership roles – and I could quite easily hold my own working in what was a very male-dominated environment.

Over my 17 years living in Asia, I travelled and worked across the region. I spent 5 years in Singapore, 2 years in Hong Kong, and the remainder in the Gambling Mecca of Asia: Macau. I gravitated towards projects needing transformation and growth. Whether it was resolving communication breakdowns or managing complex installations, I thrived in roles demanding adaptability, clear communication, and strategic thinking. I worked with brands like Motorola, Apple, Tesco, Ikea, Levi's, and The Body Shop, gaining diverse and invaluable experience. Through these roles, I also gained insight into how different companies operate and developed a fascination with best practices, observing how some companies excelled at refining these over the years.

I also spent time giving back and contributing to any of the communities I lived in. I volunteered at shelters and helped with any mother nature crisis clean-ups in the communities – and also volunteered my time with local industry events such as the Construction Industry Lighthouse Club or the International Ladies Club of Macau – both of whom raised money and supported those in need. Because of my work, I was asked along to a blind business lunch, which in turn landed me this position: FAB at Lamborghini Global. As a member of Lamborghini's Female Advisory Board, my role was twofold: to raise the brand's profile among female audiences - and the super fun part - to conduct test drives of vehicles and give them my feedback over the dinner table on how to better expand their female market. As you can imagine, it was a horrible job (insert cheeky smile here).

Whilst in Macau fate led me to meet hubby Matt at a poolside party for expatriates. It seemed like a perfect match: he loved working with plants and had a wealth of experience in landscaping, while I admired flowers but had no clue how to keep a plant alive. Matt's apprenticeship in Australia had seen him work on impressive projects, including the Elephant enclosure at Sydney Taronga Zoo, the Sydney Olympics in 2000, and then lead him to managing external works on one of the casinos in Macau. His mantra was to work himself out of a project on budget and in time - given that landscaping was often the final touch before a grand opening. From Crown to MGM, Wynn to Sands, St. Regis to Venetian, he had managed numerous projects, including iconic installations like the world's biggest interior vertical garden at MGM, as well as a very yummy 8m tall chocolate fountain. I can confirm the chocolate was real.

TRANSITIONING INTO THE NURSERY INDUSTRY

Matt's background in landscaping, paired with my diverse experiences in project management and marketing, laid the groundwork for our foray into the nursery industry. When we decided to move back to Australia, we searched for a couple of years for a wholesale nursery that would afford us more time for our family. And there on a small private site, only 3kms from where Matt first started his landscape and tree farm apprenticeship, was this gorgeous 15-acre property - Bamboo World. I was instantly drawn to its location and the potential for growth. And so began the move.

BUT, in 2019 when we took the reins of our nursery business, I felt like I was stepping into a time machine after the quick-paced life of Macau. Our online presence barely existed, limited to a simple single-page website, while orders trickled in through traditional landline phone calls and faxes. No marketing. No events. No chasing business; they just waited for business to come to them. In their defence, they were in retirement wind-down mode and had been wanting to sell for over 2 years. We were their ticket to retirement, and they were our ticket to a new family life.

GETTING INVOLVED IN THE NURSERY INDUSTRY

The handover left me wondering, is this the norm in the nursery industry? Is this the standard for how a nursery operates? It's not what my business plan had in mind. So I embarked on a quest for industry knowledge, seeking out connections with industry bodies and local organizations. Despite scepticism from the previous owners, I dove headfirst into the Nursery and Garden Industry NSW & ACT (NGINA), and was eventually offered a seat on its board as a director overseeing communications. Simultaneously, I immersed myself in our local NGINA Northern Rivers (NR) committee, driven by a desire to learn more and reconnect everyone in the area. A couple of us put our heads together and came up with the "Pulling back the shade cloth" sessions, which has quickly become a hit, offering a glimpse into the inner workings of local nurseries while fostering community and camaraderie over tea and cake. Similar to what you'll be seeing tomorrow on your outings.

Worth the time?

While some may question the value of committee involvement, I view it as an investment in our business and the broader industry. As the saying goes, "A rising tide lifts all boats." I've always believed that serving on a committee within an industry body isn't just about titles or how much time out of your job it takes; it's about effecting change and driving progress. It's an opportunity to shape the industry's trajectory, forge connections with key players, and grow both personally and professionally. So, while the road may be challenging for someone not born with a green thumb, I'm committed to rolling up my sleeves and making a tangible impact. And in terms of Elevating our Business – we've kept four key strategies top of mind the past few years. Not to mention the hard work Matt's done to maximise the space and resources, doubling the stock on the ground and expanding into other pot sizes and varieties.

OUR 5-YEAR STRATEGY

As we navigate the ever-evolving nursery industry landscape, Bamboo World Production Nursery has embraced a forward-thinking approach centered on innovation, sustainability, community engagement, and adaptability.

Innovation, Change, and Technology:

In today's dynamic digital landscape, embracing innovation, technology, and change is crucial for maintaining competitiveness and relevance in any business. We have prioritized updating and integrating technology into our operations and communications whenever feasible. This includes transitioning to user-friendly e-commerce platforms tailored for the nursery industry, like Evergreen Connect and , and implementing automated production processes and online ordering systems. Our goal is to enhance productivity while streamlining operations, essentially optimizing our processes.

Excitingly, we've also ventured into producing short informational videos featuring Matt, aimed at educating and enriching our audience's understanding of bamboo cultivation – across all our social media platforms. And be sure to keep an eye out for our upcoming feature on ABC Gardening Australia, where we are interviewed with Jerry Colby-Williams and share our insights and passion for educating on the best practices for bamboo - a rewarding experience that will allow us to connect with a broader audience once it airs.

Sustainability and Environmental Stewardship:

As stewards of the environment – you know because we grow things for a living, we should all prioritise sustainability in everything we do. But often we are so busy with day to day we can overlook this. From production practices to packaging materials, we try to stay committed to minimising our environmental footprint and promoting biodiversity and conservation efforts. I even now get to grow flowers in our nursery – and because it's called IPM, and good for our environment, it's allowed to stay! By investing in sustainable practices, not only are we doing our part to protect the planet, but as an industry we're also resonating with environmentally-conscious consumers who prioritise sustainability in their purchasing decisions.

Community Engagement, Customer Education, and Service:

Engaging with our local community and fostering a culture of education and awareness is a priority in our business. We actively participate in community events, host workshops – such as the upcoming visit by the Wollongbar Garden Club and collaborate with schools and organisations to share our knowledge and expertise. We offer customers – a service of 'just call us if your customer needs selection help', we provide guides, selection suggestions and after planting help. Our mantra is that we want to make sure that the person has the right plant for the right spot and have often talked ourselves out of a sale. By engaging with our community and customers in meaningful ways, we're not only building trust and loyalty but also helping to engage and educate, and hopefully along the way inspire the next generation to choose horticulture. And often if customers are looking for product and we don't have it, we help to recommend who might. As we believe that it's a great big industry that we can all share a slice of the pie.

Adaptability and Preparedness:

In the nursery industry, as in any other industry, being prepared for unexpected challenges is essential for survival. Since taking over the business, we've had 2019 drought, 2020 fires, 2021 global pandemic, 2022 floods, 2023 storms and even a high voltage electrical lines fire on our property, and this year – well who knows – let's hope it's not recession as some are predicting. The key is we've learned from our own experiences and understand the importance of maintaining flexibility and preparedness. Whether it's navigating these natural disasters, economic downturns, cold winters, or a shift in consumer preferences because the sun isn't out, or the elections are happening, we're thought outside the box and looked at other ways to diversify the business and manage staffing levels to best weather the ups and downs. And in line with this, being a part of an industry body means that we have support when needed for getting coordinated briefings and assistance or information during hard times.

About Bamboo World Production Nursery

Bamboo World Production Nursery (Bambooworld, 2024), established in the mid-1990s by Victor Cusack, is now proudly owned since 2019 by Matt and Sam Birkwood, who share a deep passion for bamboo and ensuring the right plant for the right spot. In the late 1990s, Bamboo World played a significant role in establishing commercial bamboo plantations for the production of edible shoots and timber.

Today, Bamboo World is one of Australia's top three suppliers of quality non-invasive clumping bamboo to garden centres, retail nurseries, landscapers, and the general public. With over 100 varieties of clumping bamboo growing in optimum conditions on a 15-acre nursery, Bamboo World offers a wide range of sizes, from 140mm, 200mm and 300mm pots to 45L bags and larger. The plants are water-fed daily from their spring fed and recycled dam, ensuring their robust health and quality. Additionally, the nursery provides consignment growing for specific needs and quantities. Located within 10 minutes of the Pacific Highway in the picturesque Northern Rivers growing triangle, the nursery boasts excellent freight connections to Brisbane, Sydney, and Melbourne.

Matt brings over 30 years of experience in landscaping, tree farms and external projects to the business. Sam, with a background in the Australian Army and various Director and CEO roles overseas, also contributes her expertise. In her spare time, she serves as a firefighter with Fire + Rescue NSW Alstonville station and sits as a Director on the Board of Nursery & Garden Industry NSW & ACT (NGINA).

Bamboo World's key focus is on growing the industry through collaboration, education, and best practices. The nursery is committed to fostering awareness of bamboo tropical plants and supporting the industry's development. Join Bamboo World Production Nursery in cultivating a greener, more sustainable future.

CONCLUSIONS

As I stand here today, I am so proud of how far we have come with the business. Our nursery is no longer confined to the limitations of the past but is flourishing in the ever-changing landscape of the 21st century. Our commitment has allowed us an increase in our income of over 250% from when we first took over. Of course, that also comes with increased expenses, but it's all headed in the right direction – moving from 'surviving' to 'thriving'.

My journey – from the army to communications, to construction to the nursery industry – has taught me invaluable lessons about resilience, innovation, and the transformative power of adaptation that has led me here today to have an exceptional range of tools in my toolbox to use – and share.

Each of us brings a unique set of skills and experiences to this industry, and it is through collaboration that we can share and grow together. As I'm sure you all have a great story of how you came to be here. And this week, I'd like you to ask the person sitting next to you what their story is that bought them to work in our industry. In closing, I am reminded of the quote, "Bloom where you are planted". Thank you for joining me on this journey and inviting me to be the opening speaker for the IPPS 2024 Ballina Conference. I look forward to showing you around our nursery tomorrow at the Nursery Tours where we'll show you firsthand how to tell the difference between a running and clumping bamboo, and talk through some propagation techniques.

LITERATURE CITED

Bambooworld (2024) Available at: https://bambooworld.com.au/

Combined Proceedings IPPS 74:93-99. 2024.

Brachychiton Breeding: What's Happened So Far and Where to Next

Des Boorman

South Gundurimba, NSW 2480, Australia

biyamitifbb@bigpond.com

Keywords: hybrids, native trees, tree breeding

Summary

This *Brachychiton* Schott & Endl. breeding program started in the mid-1990s with the acquisition of several Cape York species, *Brachychiton velutinosus* Kostermans, *B. grandiflorus* Guymer, *B. garrawayae* (Bailey) Guymer and the naturally occurring hybrid *B. x carneus* Guymer (*Brachychiton grandiflorus* x *B. garrawayae*) to add to *B. bidwillii* Hook. a low growing cold tolerant, versatile and floriferous species with a short juvenile phase.

This small collection of stunning species highlighted the immense potential for the genus and led to the establishment of the breeding program. Plant breeding and improvement has long been an interest having taken plant breeding as a subject in my final year at the university. Once pollination parameters were established it was easy to produce significant quantities of hybrid seed to grow out for assessment.

Hybrid seed is easy to produce; growing resultant progeny out is where the work starts and in the case of trees can be time consuming and expensive, hence the need to develop strict selection criteria before commencing any breeding program. I did start the breeding with an intentional group of traits that I saw as being a part of a good ornamental tree. Traits such as compact growth habit, free and annual flowering, a short juvenile phase, adaptable to a range of climates and less likely to be weedy or suffer from genetic leakage are all factors to consider. Having been involved in regional, state and national strategic weed management, I felt weediness factors are also quite important.

While having had a couple of failed commercial launches that resulted in setbacks, some great insights were derived from them.

1) Protecting IP and structuring any formal agreements professionally will provide clarity for all parties. Such agreements should always be read in the 'event of the worst case' scenario as they are never referred to when everything is going along well.

2) Throughout the commercialisation process observations may influence the selection of additional or alternate traits that were not previously considered.

Results so far have produced many extremely suitable hybrids to assess for commercialisation and also highlighted the opportunity for ploidy manipulation, including double flowers, multiple tepal layers and petaloid filaments.

This paper is an update of my journey with breeding *Brachychiton* Schott & Endl., what I'm currently working on and what future opportunities I see.

INTRODUCTION

Australia has an amazing floristic diversity, covering nearly every climatic range other than extreme altitude and cold, lending itself to both amenity and functional horticultural applications. Much of that diversity remains unexplored from a crop improvement or plant breeding perspective.

There has and continues to be a massive interest in native plants generally, however this comes with several major risks. The two most serious risks are: 1) non-endemic weediness; a classic example being *Schefflera actinophylla* (Endl.) Harms, a high-altitude North Queensland native dispersed by frugivores that is perfectly suited climatically to much of the east coast of Australia where rainfall is adequate and is quite invasive.2) genetic leakage where a non-endemic species can hybridise with rare or other natives in its cultivated range threatening the integrity of a species and possibly causing its specific extinction as more adept fertile hybrids take over, such as *Acacia baileyana* F.Muell. in the Cumberland Plain of the Sydney Basin and the rare *Acacia pubescens* (Vent.) R.Br. (NSW National Parks and Wildlife Service, 2003). This risk is also extremely high with numerous *Syzigium* cultivars on the market and widespread cultivation of *Eucalyptus* and *Corymbia* species.

Brachychiton Schott & Endl. is a widespread genus of over 33 species, within Malvaceae with the most species occurring in Northern Australia with two species occurring in Papua New Guinea: *B. velutinosus* Kostermans that also occurs on Cape York and the endemic *B. carruthersii* F. Muell. Several species are widely utilised in ornamental and commercial applications across Australia and internationally, *B. populneus* is used as a fodder source in its natural distribution and also cultivated as a hardy street tree in inland parts of Australia and internationally, such as in South Africa.

The genus also has potential to yield more new species and natural hybrids such as the recently described *B. guymeri* J.A.Bever., Fensham & P.I. Forst (Fensham et al., 2019) and the yet to be described *B. sp. Ormeau* (L.H. Bird AQ435851) (DoE, 2024), both being critically endangered due to their limited distribution and habitat loss threats. There are several naturally occurring and quite a few spontaneous and intentional hybrids already in cultivation. These hybrids are versatile and used for street trees, parks and specimen tree applications. While the current hybrids being cultivated are quite good, there is immense potential to intentionally hybridise *Brachychiton* to be more versatile and satisfy a greater number of niche applications and importantly hopefully reduce significantly and ultimately remove the risk of weediness and genetic escape through production of sterile cultivars or complex bred cultivars with reduced fertility.

Egyptian research has identified compounds in *B. populneus* (Schott. & Endl.) R.Br. with extremely effective anti-hypoglycaemic effect, with similar or better effect than commonly utilised pharmaceutical preparations (Ragheb et al., 2019).

I chose to work with *Brachychiton* many years ago because the genus contains many beautiful species, but few hybrids are utilised widely with *B.* x *roseus* selections being the main ones and even fewer are strategically produced considering the immense potential for such work in the genus.

MATERIALS AND METHODS

Brachychiton Schott & Endl. is a relatively easy genus to work with as flowers are functionally unisexual by abortion so accidental hybridisation or selfing is easily avoided. Pollination needs to be undertaken on the first day of flower opening within the first few daylight hours. Night flowering species are a different proposition; flowers are generally receptive throughout the night, so pollen needs to be collected off day flowering species and stored to enable pollination between the two groups. Limited success can be obtained by pollinating early in the morning. Flowering is also synchronised between male and female flowering flushes in most species that I worked with, some only produce male flowers some seasons which can impact on breeding programs and obvious biennial, or triennial flowering is also an issue.

Style length also seems to have an influence on successful seed production and short styled species need to be the female parent not the pollen parent.

RESULTS AND DISCUSSION

Initial Hybrid Development

The initial hybridisation undertaken in 1999 consisted of 3 separate crosses: *B. velutinosus* Kostermans and *B. bidwillii* Hook, *B. grandiflorus* Guymer and *B. bidwillii* Hook., and a complex hybrid between *B.* x *carneus* Guymer and *B. bidwillii* Hook.

The first group of hybrids produced consisted of inter-specific hybrids between *B. velutinosus* Kostermans and *B. bidwillii* and as to be expected with inter-specific hybrid ratios, the progeny was somewhat 50% of either parent in leaf, habit and flower form (**Fig. 1**). However, as is also to be expected, there is a significant variance in how each of those traits are expressed in the progeny so as an example flower colour, size and shape was spread quite a bit in the expected bell curve pattern as traits are not dominant or recessive as such as you would expect when undertaking inter-specific hybridisation. It is noted that red is quite a strong colour in *Brachychiton* and can dominate progeny flower colour, however the brilliant magenta of *B. velutinosus* Kostermans also carries through in many of these hybrids.

Figure 1. A group 1 hybrid displaying an 8 petalled flower. This group commonly produces flowers of 6 or more petals.

The 2nd group of hybrids were *B. grandiflorus* Guymer x *B. bidwillii* Hook. and like the previous inter-specific hybrids were somewhat between both parents though flower size tended to be more on the larger side like *B. grandiflorus* Guymer, and the red of *B. bidwillii* Hook. was not expressed as a pure colour **(Fig. 2)** unlike in the *B. bidwillii* Hook. x *B. velutinosus* Kostermans cross where there were definite all red hybrids.

Figure 2. A group 2 hybrid showing *B. bidwillii* Hook. shape but *B. grandiflorus* Guymer size.

The 3rd group consisted of *B. x carneus* (*B. garrawayae* (Bailey) Guymer, *B. grandiflorus Guymer*) x *B. bidwillii* Hook and resulted in some interesting outcomes as there were three species in the mix. Flower size, shape and colour varied greatly in these tri-specific crosses. While some could be identified like the closely related inter-specific crosses of the second group, there was also some recombinations of flower traits that resulted in beautiful varieties that did not resemble any of the parentage singly **(Fig. 3)**. Colours were also quite varied with red, orange, salmon and potentially white in the mix. The red from *B. bidwillii* Hook. Was carried through in many hybrids but was not necessarily attached to that species' flower shape.

Figure 3. Group 3 hybrids show amazing variation, and several are worthy of cultivation.

Interestingly *B. bidwillii* Hook leaf characteristics were quite dominant in all hybrids, particularly with reference to the leaf lobing and density of trichomes, and the small hairs that cover many *Brachychiton* spp.

From these 3 groups of hybrids, roughly 100 of each cross were produced and there were numerous stunning and quite different looking hybrids that warranted commercial exploration. Several of these have also been utilised in further breeding works. A hybrid possibly expressing polyploidy is shown in **Fig. 4** with eight petals instead if the characteristic five petals.

Figure 4. A possible polyploid from a Group 3 hybrid showing extra petals where this branch has eight petalled flowers, not five.

Second Generation Hybrids

In 2008-9 after several years of not producing more hybrids as I was waiting to see the initial results, some of the best hybrids (in my opinion) from the first group were selected to produce another range of hybrids. There were several combinations to choose from but what I didn't want to do was to duplicate my previous crosses, but rather add additional traits from other species that I consider worthy or important when considering the ornamental potential of this genus. To facilitate this, another four hybrid combinations were produced incorporating new species or complex combinations.

The 4th group was an inter-specific cross between *B. velutinosus* Kostermans and *B. populneus* (Schott. & Endl.) R.Br. that has produced a range of hybrids as expected with one being exceptional and is slated for release later in 2024. Many of these hybrids are yet to flower but the selected genotype has great form and growth habit along with prolific and regular flowering which makes it a worthy commercial specimen **(Fig. 5)** regardless of what flowers in future.

Figure 5. A Group 4 hybrid.

The 5th group is a poly-specific cross (four species) between a group 3 hybrid and *B. australis* (Schott. & Endl.) A. Terracc. - the broad leaf bottle tree. This tree has amazing form and dense lush foliage and is worthy of wider cultivation **(Fig. 6).** There are 18 hybrids in this cross remaining, displaying a range of leaf and growth characteristics including dwarf traits identified by short internode length. Unfortunately, none have flowered yet, however in the past year these hybrids have been revived and are showing great promise with the pachycaul trunk of *B. australis* (Schott. & Endl.) A. Terracc. being obvious on all of them. The intention is to breed a floriferous and compact flowering bottle tree. I expect to see some flowering of these hybrids in late 2024.

Figure 6. A Group 5 hybrid showing unique leaf characteristics.

The 6th group is another poly-specific cross (involving five species) which utilised the commonly cultivated cross *B.* x *roseus* Guymer, (*B. acerifolius* (Cunn. ex G. Don) Macarthur *x B. populneus* (Schott. & Endl.) R.Br.) colloquially known as 'Jerilderie Red' a widely cultivated and accepted variety with many suitable traits. For this cross I wanted to get as much in the mix as I could so selected a group 3 tri-specific cross that was worthy of cultivating. This resulted in the production of 35 poly-specific cross progeny with 5 different species' genetics at play. From observations so far, the red of *B. acerifolius* (Cunn. ex G. Don) Macarthur and its flower shape are heavily expressed in the progeny that have flowered to date (**Fig. 7**). Interestingly this poly-specific cross has also produced some quite unusual and unexpected recombinations such as prolific and bushy free branching growth along with quite a few dwarf plants where internodes are significantly compressed even when compared to the shorter growing species in the hybrid. One genotype selected in this group is currently being bulked up for release.

Figure 7. A selection from Group 6 with elongated flower characteristics.

The 7th group is a group 3 tri-specific crossed to *B. populneus* (Schott. & Endl.) R.Br. though none have flowered yet and may do so in the following season.

The Next Steps

Brachychiton is a precocious genus with many naturally occurring hybrids being identified in both wild populations and cultivation (Guymer, 1988). Currently I have two selections of *Brachychiton* x *vinicolor* Guymer (*B. discolor* F. Muell. x *B. acerifolius* (Cunn. ex G. Don) Macarthur) with seed set to a group 3 hybrid and several natural inter-specific hybrids. These natural hybrid combinations are *B. chillagoensis* Guymer x *B. australis* (Schott. & Endl.) A. Terracc. a yet undescribed hybrid collected near Chillagoe in north Queensland, *B.* x *allochrous* Guymer (*B. grandiflorus* Guymer x *B. muellerianus* Guymer), *B* x *turgidulus* Guymernotho subsp. *tugidulus*, (*B. rupestris* (Mitchell)ex Lindley) Schumann x *B. populneus* (Schott. & Endl.) R.Br.), *B. rupestris* (Mitchell ex Lindley) Benth. x *B. acerifolius* (Cunn. ex G. Don) Macarthur and *B. australis* (Schott. & Endl.) A. Terracc. x *B. acerifolius* (Cunn. ex G. Don) Macarthur. What is exciting about these hybrids is that it was possible to select from a range of unique mature specimens so that desirable traits could be viewed and determine whether they have immediate horticultural value and/or potential for further breeding work. These hybrids will reduce generation time (between 4 - 10 years) as they are a known quantity and deliver results at a significantly lower cost because there has been no need to go through the pollination, growth and evaluation stage to pick a suitable desirable plant or even if one does get produced that is distinctly better.

The hybrids involving *B. australis* (Schott. & Endl.) A. Terracc. or *B. rupestris* (Mitchell ex Lindley) Benth. also bring in an important and highly desirable trait of pachycaul trunks which I have been working towards for over a decade. Unfortunately, I am yet to see any hybrids that contain both *B. australis* (Schott. & Endl.) A.Terracc. and *B. rupestris* (Mitchell ex Lindley) Benth. but you just never know what is around the corner. There are several other more compact pachycaul *Brachychiton* species with two of these viz. *B. compactus* Guymer and *B. collinus* Guymer being included in future breeding.

CONCLUSION

Plant breeding, especially tree breeding can be time consuming, take up space and costly, but the rewards of producing new and unique cultivars ready to commercialise is something else and I urge anyone with the interest to just give it a red-hot go.

My only advice is making sure you have a clear plan of what you want to achieve and initially be conservative because sometimes you will get setbacks but that is par for the cause and when you get thru them the satisfaction will be that much greater. If it was easy everyone would be doing it.

LITERATURE CITED

DoE. (2024). *Brachychiton* sp. Ormeau (L.H.Bird AQ435851) in Species Profile and Threats Database, Department of the Environment, Canberra.
Available from:Brachychiton sp. Ormeau (L.H.Bird AQ435851) — Ormeau Bottle Tree. Accessed Wed, 19 Jun 2024

Fensham, R.J, Beveridge, J.A, and Forster P.I. (2019). *Brachychiton guymeri* JA Bever., Fensham & PI Forst. (Sterculiaceae), a new species from north Queensland. Austrobaileya *10*:443-57.

Guymer, G.P. (1988). A taxonomic revision of *Brachychiton* (Sterculiaceae). Aust. Syst. Bot. *1*:199-323

NSW National Parks and Wildlife Service (2003). Downy Wattle (*Acacia pubescens*) Recovery Plan. NSW NPWS, Hurstville, NSW, Australia. ISBN: 0 7313 6504 6

Ragheb, A.Y,, Kassem. M,E., El-Sherei, M.M., Marzouk, M.M., Mosharrafa, S.A., and Saleh, N.A. (2019) Morphological, phytochemical and anti-hyperglycaemic evaluation of *Brachychiton populneus*. Revista Brasileira de Farmacognosia *29*:559-569.

Combined Proceedings IPPS 74:100-108. 2024.

My Horticultural Journey: Rare, Unusual or Reimagined

Des Boorman

South Gundurimba, NSW 2480, Australia

biyamitifbb@bigpond.com

Keywords: Australia, *Banksia*, *Calodendron*, Grafting, *Grevillea*

Summary

I have been extremely fortunate in my professional career to have made so many connections to further my horticultural knowledge and interests and then be able to become a mentor myself. This fortune is not measured financially but rather on how I think I've achieved the goals I set out with and whether they stayed consistent, were modified slightly or completely changed. I can honestly say that it didn't come easily, and many things were just seemingly coincidence but looking back I think that making your own luck is a big part of it. What I have done is constantly strived to adapt and improve everything that I do and the seven deadly words are always in my mind, "we have always done it that way" and the reason why many people and businesses don't keep up or change when needed which ultimately leads to major issues. Having a good general knowledge across several different fields also helps to foster adoption of new ideas, techniques or adaption of equipment to improve efficiency. Being able to weld, build and fix things may not seem like a horticultural skill but it certainly makes you think about a lot of different industries as you apply their tools of the trade.

IPPS Vol. 74 – 2024

INTRODUCTION

I was extremely fortunate to have always had a passion and vocation for all things plant and animal and my earliest memories are of collecting fig seedlings from rock ledges and burning Banksia infructescences to extract seeds. While in the 3rdgrade at the age of nine seeing the bizarre *Calodendron capense* fruits at a primary school fete, buying two of them which ultimately produced nine plants, several of which are still alive more than 40 years later (**Fig.1**). While these trees don't seem like much, I still get immense satisfaction seeing them on my and other farms that I planted or supplied 40 or more years ago, growing and supporting wildlife or just being beautiful flowering specimens.

Figure 1. The best of the *Calodendron capense* that I grew as a 3rd grader.

The Beginnings

My high school work experience in grade 10 was at the local Department of Agriculture, Alstonville Tropical Fruit Research Station, New South Wales where I did some interesting and cool stuff like plant 700 tea seeds which I think were the first planted in the region as a trial to assess whether tea was a viable crop in the region. The interest in tea cultivation in NSW is still there and continuous improvements are happening (Krahe and Krahe, 2022). Another was to assess a range of new citrus varieties for taste and measure sugar content; one of these being Mineola Tangelo and every time I see one in the shops now, I think how cool is that I did some of the assessment on that variety way back as a high school student.

During my school years I was always interested in plants and nature and was fortunate to have an agriculture teacher at my high school Peter Giblin who managed to get a Certificate II in Nursery Production through as a part of the Higher School Certificate courses, as additional units both within and after school hours and was the pilot for what is TVET (Technical and Vocational Education Training) in high school today (TVET, 2000).Along with Peter's accomplishment of getting the course offered was the range of horticultural trainers at the local TAFE (Technical and Further Education) institute; one being Greg McPhee a local nurseryman, grafter and IPPS member.It was Greg who infected me with the grafting bug as while I was interested in it from an early age remembering trying to graft a tomato onto a weedy daisy. I clearly had absolutely no idea what I was doing as my background in dairy farming and beef cattle gave me no exposure to it. Greg showed our small group of six students how to propagate seeds and cuttings commercially and then how to bud and graft stone fruit that we germinated ourselves.

As a part of that course, we attended a seminar Prof Keith D. Cairncross of Macquarie University delivered at the local Wollongbar Department of Agriculture about his endeavours to graft Western Australian (WA) Banksias onto hardy east coast rootstocks for the 1988 Australian Bicentenary. Having never seen a WA *Banksia* before I was immediately hooked as they are quite showy and different to the east coast species I was familiar with. This began my absolute passion for Proteaceae which has diversified into other genera over time as well.

The University Days

So empowered by a small amount of knowledge and an excessive amount of enthusiasm, I embarked on getting myself a higher education, the first in my entire extended family. I had decided that while I hated school and all things study related, I realised that a degree would grant me a lot more options than a Certificate II or III in Nursery Production. I attended several agricultural campus open days to inspect their facilities and see if I could acquire any of their growing facilities for my Banksia obsession.

I decide that University of Queensland Gatton, formerly Queensland Agricultural College (QAC) had the best nursery and opportunity to borrow some facility space for my passion/obsession. It was a good decision as the nursery was a commercial operation supplying propagated plants for the cut flower industry, also focussing on difficult to propagate species and varieties. I spent every spare moment I could in the nursery learning about cutting and seed propagation and undertaking my own cultivation of *Banksia*, *Hakea*, *Grevillea* and *Dryandra*. Some of that work is shown in **Fig. 2.** These are a few of my earlier grafts conducted in 1989 and were done as cotyledon grafts using double sided razor blades.

Figure 2. Some of my early work during my undergraduate study at the University of Queensland, Gatton. A) *Banksia pilotstylis* grafted 14 April 1989, B) *Banksia ashbyi* seedling tubestock and C) *Dryandra kippistiana* grafted onto *Banksia integrifolia*. Grafts were made using double-sided razor blades.

I met members of the local Society for Growing Australian Plants and this introduced me to legends like Norm McCarthy, a local collector and enthusiast and Merv Hodge, an amazing collector and grafter of *Grevillea* spp. and where I first learnt the mummy graft technique around 1989 or so. I then through Merv met Peter Olde and Neil Marriot who are both massive *Grevillea aficionados* and have written brilliant refence books on the genus (Olde and Marriot, 1994-1995), revised the taxonomy (see for example, Olde and Marriott, 1993; Olde and Marriott, 2009) and contributed greatly to their conservation (see for example Olde and Marriott, 2021) and broader awareness.

As I entered my 3rd year, (1990), I focussed on *Grevillea* and after a 6-month stint on a flower farm in Western Australia which I had arranged on a trip a year or so earlier, I was totally consumed by the immense diversity in the genus and the number of people cultivating them.

While living there I went to Zanthorrea Nursery and collected several *Grevillea* species one of which was simply called 'Black Magic' (**Fig.3**). The next year this plant was identified as the extremely rare *Grevillea calliantha* from Cataby area in south-west of WA (Makinson and Olde, 1991).

Figure 3. The *Grevillea calliantha* plant produced from the original material I collected from Zanthorrea Nursery.

I was also producing a broad range of grafted grevilleas that were sold thru the QAC Plant Nursery as a part of their plant range (**Fig. 4**).

Figure 4. A) *Grevillea thyrsoides* in 200mm pots ready for sale around 1991, B) some of the range of grafted grevilleas in 200mm containers in the Queensland Agricultural College plant nursery ready for sale.

Along with commercial production, I was also involved in grafting recently collected material notably from Peter Olde and Neil Marriot on some of their trips into Western Australia while researching for their Grevillea books, which included some amazing discoveries of extremely rare Grevilleas such as *Grevillea batrachioides* (ANPSA, 2024) and *Grevillea flexuosa* (Environment, 2008) both of which I was successful in grafting and bringing into cultivation in Southeast Queensland (**Fig. 5**).

Figure 5. A) *Grevillea batrachioides*(left) compared with its close relative *Grevillea asparagoides*(right). B) The stunning leaves of *Grevillea flexuosa*

I insisted on doing my year paper on grafting *Grevillea* and while lecturers were reluctant, I still managed to push it through. I was paying for my course so I wanted to get as much out of it for me as I could. My nursery Lecturer at the time was Ian Gordon, a long time IPPS member and it was he who suggested I apply for the Rod Tallis Award which I did, and in 1991 I was awarded this in Canberra. I was a country lad so speaking at a conference was daunting and I was extremely nervous but managed to pull it off. This was before the days of Microsoft® PowerPoint presentations, so I didn't have any cool props - just me and a few slides.

I also was extremely interested in plant breeding. At this time Merv Hodge had developed *Grevillea* 'Superb' and it was a beauty, so I took crop improvement as a 4th year subject with Robert Fletcher, an experienced wheat breeder and within a year was already hybridising *Grevillea* spp. (**Fig.** 6) and then ultimately *Brachychiton* spp. which has become a 26-year odyssey producing many stunning hybrids and more in the pipeline. I won't go into detail as that is the subject of another paper in this Issue and in several previous IPPS Proceedings.

Figure 6. The first hybrid that I produced that survived to maturity was this *Grevillea batrachioides* x *G. bipinnatifida* and while stunning it was as prickly as razor wire so not ideal for cultivation.

The Experienced Horticulturist

Throughout the times I was cultivating *Grevillea* spp. that culminated in over 136 species around 1992, I saw the potential for what plants could look like. This early experience has taught me to look at plants in different ways, particularly how to show them off in the best possible light whether that be a standard *Ixora*, *Gardenia*, *Callistemon* or *Melaleuca*. I have always used my grafting skills to re-imagine what plants could be to make them more desirable or presentable for horticultural uses. Some of these ideas are depicted in **Fig. 7 to 9.** Plant breeding takes that to another level by blending the desirable traits of several or many species into interesting and novel new hybrids.

Figure 7. Standard *Callistemon* 'Little John (A) and standard *Melaleuca quinquenervia* red (B)

Figure 9. Standard *Gardenia* 'Radicans' onto *Gardenia magnifica* (A) and *Ixora* 'Pygmy Pink' onto *Ixora coccinea* (B) .

Figure 8. *Slumbergera* onto. *Hylocereus* (Not my idea - I saw this style of grafting at Bundaberg markets so did my own)

This is far from my entire horticultural journey, but it does spell out some of the earlier highlights and while it seems that luck has played a big part in my career and 'success' as I see it isn't necessarily so. Luck did to some extent play out with the people I have met and learnt from, but it was also my determination to succeed in what I wanted to that drove me to get where I was.Whether it was my grade 7 maths teacher who said in my first maths class at high school "oh you're a Boorman you'll never go anywhere" obviously not knowing this Boorman, or lecturers at university saying oh you can't pick your own 4th year study project theme, but I did and that got me a Rod Tallis Award which still hangs proudly in my loungeroom. I'm not saying that this path I've taken didn't cause a lot of friction because it did, but I'm used to that, being who I am.

CONCLUSION

Find your passion and your vocation, academia is but a tiny part of the real world and the trick is to find your niche and love what you do. If you're good enough or passionate enough you will make an amazing career out of it and be extremely satisfied. Financial reward can be tricky in some fields but being truly satisfied with what you do is also extremely rewarding.

LITERATURE CITED

ANPSA (2024). *Grevillea batrachioides*. Available at: https://anpsa.org.au/plant_profiles/grevillea-batrachioides/ Accessed 20 August 2024.

Environment. (2008). Approved Conservation Advice for *Grevillea flexuosa* (Zig Zag Grevillea). Available at: Approved Conservation Advice for Grevillea flexuosa (Zig Zag Grevillea) (environment.gov.au) Accessed 20 August 2024

Krahe, J., and Krahe, M.A. (2022). Optimizing the quality and commercial value of gyokuro-styled green tea grown in Australia. Beverages *8*:22.

Makinson, R.O. and Olde, P.M. (1991). A new species of *Grevillea* (Proteaceae: Grevilleoideae) from south-west Western Australia. Telopea *4*:351-355.

Olde, P., and Marriott, N. (1994-1995). The Grevillea Book [in three volumes], Kangaroo Press, Kenthurst, NSW, Australia.

Olde, P., and Marriott, N. (1993) A taxonomic revision of *Grevillea angulata* (Proteaceae: Grevilleoideae) and closely related species from the Northern Territory and Western Australia. Telopea *5*:399-417.

Olde, P. M.and Marriott, N. R. (2009). *Grevillea tetragonoloba* (Proteaceae: Grevilleoideae) recircumscribed, with notes on its typification and a new segregate species, *Grevillea nivea*, described. Nuytsia *19*: 229-243.

Olde, P.M., and Marriott, N.R. (2021). *Grevillea merceri* Olde & Marriott (Proteaceae: Grevilleoideae: Hakeinae) an uncommon, geographically isolated species in the Triloba Group from subcoastal areas of southern Western Australia. Telopea *24*:241-245.

Robinson, C. (2000) Developments in Australia's vocational education and training system. Workshop presented to the Central Institute of Vocational and Technical Education Beijing, PR China, August 2000. Available at: https://www.ncver.edu.au/__data/assets/file/0020/2765/china.pdf Accessed 13 August 2024.

Combined Proceedings IPPS 74:109-117. 2024.

Plants in the Classroom Can Improve Student Performance

John Daly[1], Margaret Burchett[2] and Fraser Torpy[2]

[1] Eco-Environment – Unit 3 /5 Deakin Street, Brendale. Q, 4500, Australia.

[2]Plants and Indoor Environmental Quality Group, Centre for Environmental Sustainability, Faculty of Science, University of Technology, Sydney (UTS) Australia.

john@ecoenvironment.com.au Margaret.Burchett@uts.edu.au

Keywords: indoor plants, specialist growing media, schools, children, Queensland, environment

Summary

Many studies in the past have shown that plants and Growing media (as a Biofilter) maintained indoors improve air quality, ambiance and mood of workers resulting in improved staff productivity, performance, job satisfaction and reduced sick leave absence, stress, depression and negative mood states. However, only few studies on classroom performance of school children have been conducted so far. To understand the performance of students in classrooms with and without Potted plants, we conducted the first Trial study involving 360 students in grades six and seven in 16 classes in three schools in Queensland, Australia and student performance was tested across three curriculum course streams: Numeracy, Literacy and Science. The results indicated that the presence of plants and long-term specialist growing media in the classroom consistently led to improved performance in spelling, mathematics and science – i.e., across the curriculum (by removing VOCs from the air). The results were statistically significant with 10 to 14% improvement in all but one of the five sets of scores in two schools, whereas in the third school where results were not significant between groups with and without plants' presence, the students were already involved in an active gardening program, involving both ornamental and vegetable species. Therefore, it can be concluded that the presence of plants in the classroom environment improves student performance.

INTRODUCTION

Numerous studies have now shown conclusively that indoor plants improve many aspects of indoor environmental quality, including cleaner indoor air quality, increases in staff productivity, performance and job satisfaction, and reductions in sick leave absences, and feelings of stress, depression and other negative mood states (for review, see, e.g.., Burchett et al., 2010). However, there has been almost no research conducted on the potential benefits to school student wellbeing of indoor plants in their classrooms. In fact, we have been able to find only two reports of any such studies. The first study, conducted in Sweden (Fjeld, 2002) found that potted-plants reduced sick-leave absences among primary school children. The second report was from a Taiwanese study (Han, 2009), which found that both class marks and behaviour in junior high-school students were improved when plants were installed in the classroom. However, the second study involved only two classes (one with, one without plants) and the researchers conceded that a variety of other factors (e.g. a more engaging teacher?) might account for the differences reported. Plants in the room, however, have been found to improve performance in university students (Shibata and Suzuki, 2004), and lower their feelings of physical discomfort (Lohr and Pearson-Mims, 2000). Also, in one other study the performance of tertiary students was compared in classrooms with and without plants (Doxey and Waliezek, 2009). In this case, the authors reported that, although grades were not significantly affected by plant and potted media presence, there were significant differences in student satisfaction ratings. Those with planted classrooms rated their lecturers more highly on organisation and enthusiasm, than those in the group without media and plants, indicating perhaps that both staff and students were happier with plants in the workspace.

The aim of the current study was to investigate the effects of indoor plants and potting media on classroom performance in composite classes of Year 6 and 7 (i.e. Mid School and Senior Primary) students in three independent schools in the Brisbane region, Australia with a total of over 360 students in 13 classes.

Note some 2 months were needed to process and formulate the test structure of different school education systems, School Curriculum's across three varied school cultures, to maintain a standard Testing method across 3 curriculum course streams, (for trial credibility and consistency) in Numeracy , Literacy and Science (SOZ).

The three schools include:

A) **All Saints School** – Principal contact Steven Mongomery, Albany Creek, Brisbane North, Queensland, (3 classes each of Grades 6 and 7 with an average 25 pupils (150 total).

B) **All Saints Middle School, Merrimac** – Principal contact Sue Daly, Gold Coast, Queensland. (Middle school – 4 classes120 pupils, Grades 6and 7).

C) **St Joseph's School** – Main contact Dianne Pennings Beenleigh, Loganlea, Queensland (3x grades 6and7- 90 pupils).

METHODS

Human Ethics approval for the project was first obtained for the schools concerned. Then half of the participating classes each received a total of 6 plants in Specialist Growing Media in 200 mm pots, while the remaining classes received no plants.

Plant species were supplied by- Sharon Prater and Nick Holt at Advance Plant Services and also Stockade Nursery. Plant containers were supplied by Trevor Murphy at Container Connections and Sid Dyer at A2Z Planter Technology. Planting media was supplied by eCo-Environment as Bioganic Earth Podium indoor blend (Specialist Long-term Growing Media).

Each Classroom was supplied with the same species of three plants to maintain consistency. The plant species and numbers used were as follows: One each of 300mm Staked – *Rhaphidophora aurea* (*Scindapsus*, golden pothos), 250mm *Spathiphyllum* spp. 300mm and *Dracaena fragans* ('Janet Craig').

Students were tested with standard tests before plant placements and re-tested after about six weeks of plant presence (or absence). Test measures included spelling (South Australian Spelling Test, SAST) and mathematics in all three schools, while in one school tests of benchmark reading, and in another school tests in science, were also included.

RESULTS

Differences in student responses were found among the three schools. Students in two schools showed marked improvements in scores in spelling and maths in classrooms with plants present. However, in the third school no differences were found, in spelling, maths or reading, among classes with plants and those without. A comparison of results for the two schools showing improvements with plants present is presented in **Table 1**. For all but one of the five sets of scores influenced by plant presence, the improvements ranged from 10 to 14%.

In comparing planted to unplanted classrooms, two schools showed positive responses in scores with plant presence (**Figs. 1 to 3**). In School A, baseline scores were similar across the classrooms before plants were installed. At mid-term, classes with plants showed higher scores than those without (**Fig. 1**). Then by the end of term, though both sets of classes had further progressed, as would be expected, the classes with plants retained their lead over those without. The end of term results for spelling showed a parallel difference between the two groups of classes (**Fig. 3**).

Table 1. Summary of percentage differences, in each of two schools, in scores on three standard tests, in classrooms with and without plants (Means± Standard Error); 3 classes per school per treatment; totals 70 to 80 students per treatment.

Tests/Differences in scores		% increase in scores with plants present
Mathematics		
	School A	14 (±1.2)*
	School B	5 (±0.8)
Spelling		
	School A	10 (±1.5)*
	School B	12 (±1.7)*
Science		
	School B	11 (±1.2)*

*Signifies difference is statistically significant ($p \leq 0.05$).

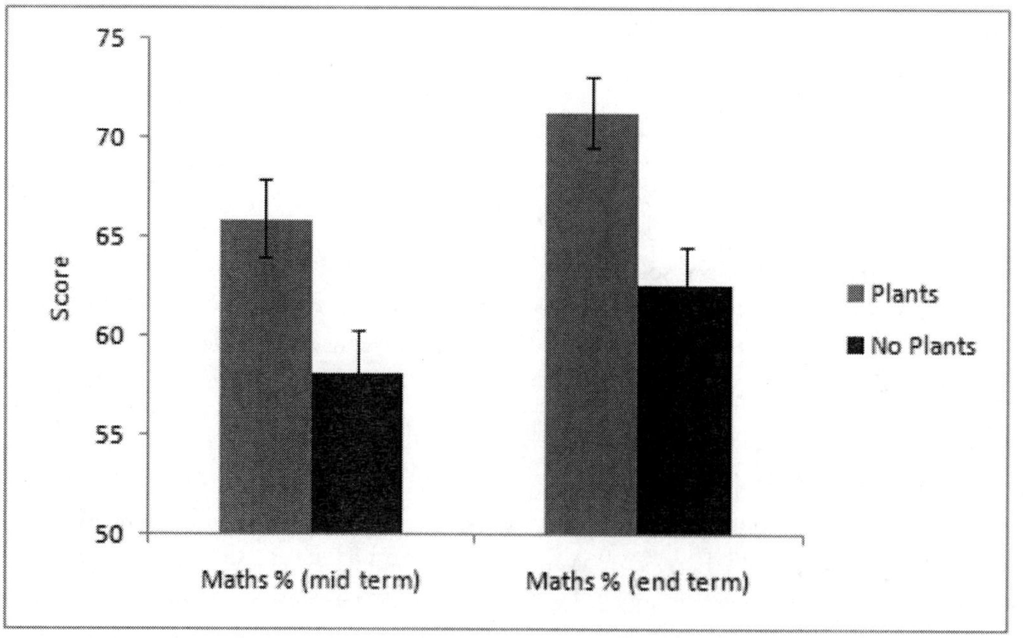

Figure 1. School A (All Saints Albany Creek): Comparison of changes in mid-term and end of term maths scores, in classes with and without plants. (Means and SE; n = 69–72.)

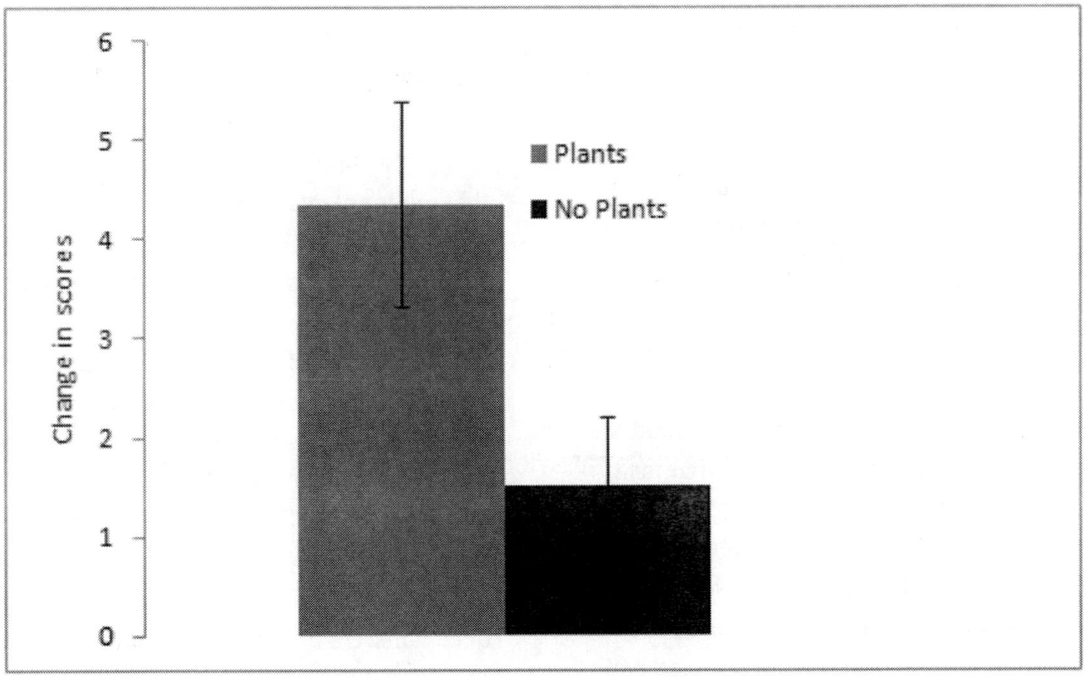

Figure 2. School A (All Saints Albany Creek): Comparison of end of term spelling scores, in classes with and without plants. (Means and SE; n = 69–72.)

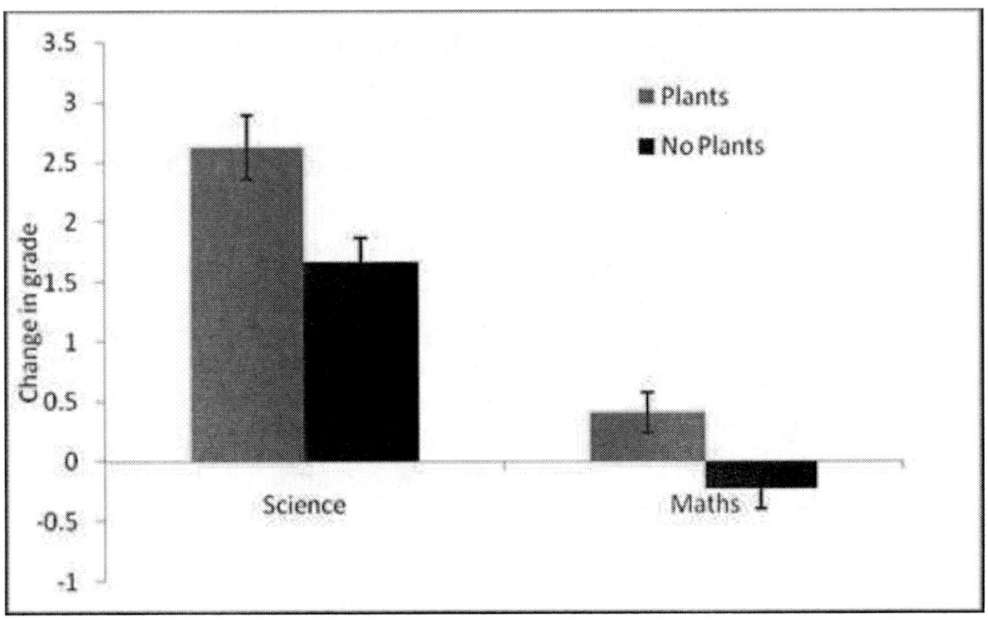

Figure 3. School B (All Saints Anglican Gold Coast): Comparison of end of term Science and Maths grades, in classes with and without plants.(Means and SE; n = 149.)

The results for School B for the end of term science and maths tests showed an improvement in both subjects with plants in the classrooms (**Fig. 3**). The slightly lowered score for maths for the non-plant group is not in itself significant, but the difference between that group of classrooms and those with plants is a statistically significant difference.

DISCUSSION

Because of the number of classes and students involved, it would appear that the differences can be accepted as real improvements in classroom performance resulting from plant and Specialist growing Media presence. Improvements in performance on the fundamental tasks of spelling and mathematics of 10% and more, are generally regarded by educationists as significant in students' progress in school.

How Might Such an Influence of Plants in the Classroom Come About?

First, research has shown that plants can significantly improve indoor air quality in office buildings (with or without air-conditioning). Research at University of Technology, Sydney (UTS) has shown that two or three plants in an office can significantly reduce levels of CO_2 and air-borne volatile organic compounds (VOCs) that are continually outgassing from plastic/synthetic surfaces (furnishings, fittings, equipment eg computers, copiers etc). These are the two major types of contaminants always found in higher concentrations indoors than outside, even in the CBD. However, the participating teachers indicated that, in this case, doors and windows of the classrooms were very often all open, so that this health benefit of plants might not have had much effect on the results obtained. However, the effects may well be found to be significant in winter, in closed classrooms with flueless gas heaters, since raised CO_2 levels causes loss of concentration and drowsiness.

Secondly, other studies have shown that indoor plants improve performance and productivity in adult workers (Lohr et al., 1996,). Also, a UTS study with 55 participants (university staff) showed that office plant presence had strong psychological benefits in reducing stress, anxiety and low spirits in adults. Other research indicates that nearby greenery resets our 'calm' button (Kaplan and Kaplan, 1990), and that indoor plants are also directly attractive, evoking positive responses among building occupants. In this study we found, on visiting the schools, that teachers and students showed great interest in having plants in their classrooms. Students of one class had even named their plants – 'Luigi', 'Mojo', 'Napoleon' and so on.

Urban living involves what has been described as a "disengagement with the natural environment". Re-establishing 'better links with nature' has become an important international public health concern (Maller, *et al.,* 2005; Frumkin, 2001; Kellert and Wilson, 1995; Kaplan, 1995; Wilson, 1984). Evidence shows that, for city dwellers, time spent in city parks and nature reserves is beneficial to health and wellbeing, with improvements in such physiological measures as blood pressure, and psychological measures as 'mood states' (Velarde et al., 2007; Hartig, *et al*., 2003; Herzog et al., 2002). 'Park time' has also been shown to improve concentration performance in children with attention deficit disorders (Taylor and Kuo, 2009).

In the last two decades of the 20th century the 'Biophilia' hypothesis was introduced into environmental psychology (Wilson, 1984; Kellert and Wilson, 1995. This is the proposition that "humans have an inherent inclination to affiliate with nature" (Grinde and Patil, 2009). In line with this hypothesis, that a love of greenery and pets has very deep roots in our being, it seems to us that it is no random chance that three of the top favourite family websites include gardening, weekend get-aways, and fishing, all 'back-to-nature' pursuits. A possible reason, therefore, for the finding that School C showed no differences in performance in classes with or without plants, is that this school has an active gardening program, involving both ornamental and vegetable species. Indeed, these classes have on occasion sold their vegetables to parents and friends of the school, the money raised being spent on excursions or new materials and activities for the classrooms. It is possible, then, that in this school a continuing contact with nature is already being satisfied, and the classroom plants are just a pleasant extra. One teacher here, however, reported that, when children were asked to sit and read quietly or form small groups to discuss some topic, they tended to cluster on the floor around each of the plants.

In summary, the results indicate that, for possibly a variety of interlinked reasons, classroom plants consistently led to improved performance in spelling, mathematics and science – i.e., apparently across the curriculum. This was a preliminary study – the first of its kind in attempting to compare the performance of class populations of school students in classrooms with and without indoor plants, and follow-up studies would be needed for formal confirmation of the changes found here. However, taking the other relevant research evidence into account, and since in our informal discussions at the three schools there seemed to be unanimous agreement among teachers and students that plants in the classroom improved its appearance and 'ambience', a recommendation for indoor plants to be a standard installation of school classrooms appears justifiable and timely.

Plants in the classroom could also be used as a teaching tool in biological science (observations on growth, and flowering, e.g. in *Spathiphyllum*; consideration of the requirements of maintenance and growth; caring for a living organism; comparison of high-light vs low-light plants; geography of origins of various species; environmental principles for vegetation conservation; etc.). A recent Japanese article on the issue of changes in school curricula in that country since the second world war, deplored the reduction of any studies developing 'nurturing' or 'fostering' concepts, with practical demonstrations, e.g. in caring of small animals or plants. The same trends may also have occurred in Australia. The schools garden programs that are growing in this country, could be augmented by the inclusion of indoor plants.

Acknowledgements

Our thanks go to the principals, teachers and students of the participating schools and classes including principal contacts Steven Mongomery, Sue Daly and Dianne Pennings for the considerable amount of extra organisation involved in dealing with the plants and administering the tests.

Thank you to Dianne Pennings for the Help in standardising this Testing Format. (Prior to Naplan testing).

Plant materials were supplied by Sharon Prater and Nick Holt @ Advance Plant Services and also Stockade Nursery. Plant containers were supplied by Trevor Murphy @ Container Connections and Sid Dyer @ A2Z Planter Technology. Specialist Planting Media (Bioganic Earth- Podium Blend) supplied by eCo-Environment.

Author Contributions

John Daly- Initiated the Trials with Prof. M. Burchett, facilitated the negotiation process of different Education systems, School Curricula across three varied school cultures, to maintain a standard testing method across three curriculum course streams, (for trial credibility and Consistency) in Numeracy, Literacy and Science (SOZ).

LITERATURE CITED

Burchett, M. D., Torpy, F., Brennan, J., and Craig, A. (2010). Greening the Great Indoors for Human Health and Wellbeing. Final Report to Horticulture Australia Ltd., Sydney, Aust.

Doxey, J. S and Waliezek T. M. (2009). The impact of interior plants in university classrooms. on student course performance and on student perceptions of the course and instructor. HortScience, *44*:384-391.

Fjeld, T., (2002). The effects of plants and artificial daylight on the well-being and health of office workers, school children and health-care personnel, Proceedings of International Plants for People Symposium, Floriade, Amsterdam, NL.

Frumkin, H. (2001). Beyond toxicity: human health and the natural environment. American Journal of Preventive Medicine *20*:234-240.

Grinde, B. and Patil, G.G. (2009). Biophilia: Does Visual Contact with Nature Impact on Health and Well-Being? International Journal of Environmental Research and Public Health *6*:2332–2343. https://doi.org/10.3390/ijerph6092332

Han, K-T, (2009). Influence of limitedly visible leafy indoor plants on the psychology, behavior, and health of students at a junior high school in Taiwan. Environment and Behavior *41*:658-69 21.

Hartig, T., Evans, G. W., Jammer, L. D., Davis, D. S., and Garling, T. (2003). Restorative environments. Journal of Environmental Psychology *23*: 109-123.

Herzog, T. R., Black, A. M., Fountaine, K., and Knotts, D. (2002). Reflections and attentional recovery as distinct benefits of restorative environments. Journal of Environmental Psychology *17*:165-170.

Kaplan, S. (1995). The restorative benefits of nature: towards an integrative framework. Journal of Environmental Psychology *15*:169-182.

Kaplan, R., and Kaplan, S. (1990). Restorative experience: the healing power of nearby nature. In: M. Francis and R. T. Hester Jr. (eds). The Meaning of Gardens: Idea, Place and Action. MIT Press, Camb. MA, pp 238-243.

Kellert, S. R., and Wilson, E. O. (eds.). (1995). The Biophilia Hypothesis, Island Press, Wash. DC.

Lohr, V. I. and Pearson-Mims, C. H. (2000). Physical discomfort may be reduced in the presence of interior plants., HortTechnology *10*: 53-58.

Lohr, V. I., Pearson-Mims, C. H., and Goodwin, G. K. (1996,). Interior plants may improve worker productivity and reduce stress in a windowless environment. Environmental Horticulture *14*: 97-100.

Maller, C., Townsend, M., Pryor, A., Brown, P., and St Leger, L. (2005). Healthy nature healthy people: 'Contact with nature' as an upstream health promotion intervention for populations. Health Promotion International *21*:45-51.

Shibata, S. and Suzuki, N. (2004). Effects of an indoor plant on creative task performance and mood. Scandinavian Journal of Psychology *45*:378-381.

Taylor, A. F., and Kuo, F.E. (2009). Children with attention deficits concentrate better after walk in the park. Journal of Attention Disorders *12*:402.

Velarde, M. D., Fry, G., and Tveit, M. (2007). Health effects of viewing landscapes – Landscape types in environmental psychology. Urban Forestry and Urban Greening *6*:199-212.

Wilson, E.O. (1984). Biophilia: The Human Bond with Other Species. Harvard Univ. Press.

Combined Proceedings IPPS 74:118-123. 2024.

Rainforest Seed Propagation

Mark Dunphy

387 Dorroughby Rd, Dorroughby NSW 2480, Australia

info@firewheelnursery.com.au

Keywords: Australia, germination, reforestation, X-ray radiography, seed viability, recalcitrance, inbreeding

Summary

Restoring the Sub Tropical rainforests (STRF) decimated in the 1800 and in the early part of 1900s has gathered momentum since the 1970s. Ecological restoration is possible in small remnants, however planting is the only solution in the vast paddocks as just 1% of STRF is left in eastern Australia. Annual plantings grew from mere hundreds in the 1980's to over half a million by 2020's. Over the years it was realised that not only the quantity of plants but also its diversity is important. Hence seedlings have become the planting material for rainforest restoration. As a result seed collection, storage, propagation and growing techniques of seeds of over 450 rainforest species became the cornerstone of research and a multimillion-dollar industry in recent years. Species composition and functional trait representation in these forests is of utmost importance. This paper describes the planning and methods of collection, processing, germinating and establishing seedlings in rainforest restoration and factors to be considered when restoring degraded forest ecosystems.

INTRODUCTION

The goal of rainforest seed propagation is to produce high diversity, high quality stock to grow rainforest. STRF across northern NSW and SE Queensland had mostly been cleared by the late 1800's. The remaining remnants and fragments that survived into the 1900's were largely ignored and unvalued. The interest in rainforest for values other than timber, and the land beneath the trees emerged in the 1970's. By the 1980's rainforest remnants were beginning to be restored by removing weeds and cattle and allowing the natural regeneration processes to repair the ecological function of the forest.

In the Big Scrub, the largest area of STRF in Australia, only 1% of the 78,000 hectares of rainforest remained to regenerate. With 99% ex rainforest country converted to paddocks over a hundred years ago, planting was clearly the only option to restore rainforest.

The interest in growing rainforest from a bare paddock has been increasing over the decades. Some farmers developed an interest; however, an increasing number of people were buying land specifically to plant rainforest trees and restore the rainforest. In the 1980's the individual plantings could be counted in the hundreds of trees, by the 1990's most rainforest plantings were in the thousands. By the 2000's some of the plantings were into the tens of thousands. Today in 2020's the number of rainforest trees planted in Northern NSW and SE Queensland is well over half a million trees per year.

There is clearly an increasing demand to plant rainforest trees. This demand for quantity is coupled with a demand for diversity. STRF are highly diverse and the desire to restore rainforest with broad range of species resembling the original forest is critical to the restoration attempts by the government agencies, landowners, ecologists and restoration practitioners.

GROWING RAINFOREST PLANTS

The vast majority of rainforest restoration plants are grown from seed. This is because it is far cheaper to produce plants by seed than cuttings. Also as the vast majority of plants grown are trees, the bushy form of cutting grown trees is a disadvantage to the management and growth of an emerging rainforest.

The seed collection, propagation and growing techniques for the over 450 rainforest tree and understorey species needed for restoration has developed from a small backyard novelty in the 1980's to a professional multimillion dollar industry. The techniques developed to produce the hundreds of thousands of tube stock needed for restoration have been developed at Firewheel Rainforest Nursery and other smaller nurseries over the last 40 years. All the information is now printed in a CSIRO published Book (Dunphy et al., 2020), (**Fig.** 1).

Figure 1. Australian Rainforest Seeds. A Guide to Collecting, Processing and Propagation published in 2020 (Dunphy et al., 2020).

RAINFOREST SEED COLLECTION

The collection equipment used for seed collection is very 'low-tech', that is only basic equipment like secateurs, pole-pruners, rakes, buckets and bags are needed **(Fig. 2)**. The skill comes with the ability to identify over 400 species and locate these trees in the landscape. The seed collector then needs to know what time of year they fruit, when to collect the fruit, how to collect the fruit and how to test for viability. This knowledge can take many years to acquire and many more years to become proficient.

Compounding this complexity, the seed collector faces the masting of many rainforest species. Masting mostly occurs with mature phase species and some secondary species and not with the faster growing short lived pioneer species.

This means there are hundreds of species that fruit only every two and up to every six years. However, masting may not be regular or predictable. A species may fruit two years in a row and then not again for 3 years. To add more complexity masting isn't always an all or nothing affair, a species may fruit very lightly, or a smaller number of trees may fruit and the other trees of that species not at all.

Figure 2. Basic tools used for rainforest seed collection.

The seed collector needs to understand genetics, which is difficult as the goalposts seem to be constantly moving as more is understood about this increasingly important science. Without going into too much detail, it is safe to say modern genetic testing has confirmed inbreeding is more of a problem than outbreeding. That is, the strongly held belief of the importance of 'genetic provenance' (i.e. collecting seed locally and planting those plants locally) can cause or increase potential inbreeding depression. Therefore, the seed collector needs to collect widely and not from the same mother trees every year and ideally not from the progeny of those same mother trees.

PROCESSING RAINFOREST SEED

The fruit and seed brought into the nursery needs to be processed for a number of reasons. These are primarily to;

1. Increase germination rates
2. Reduce germination time
3. Allow the seed to be stored easily and effectively

At Firewheel Nursery there has been decades of research, observation and experimentation carried out to determine the most effective, efficient way to produce large numbers of germinated seedlings. This has been shared with other nurseries as other nurseries have shared their insights with us. There is always more work to be done and with the help of the great people at Mt Annan Botanic Gardens in Sydney we are always trying new ways to improve our germination rates, times and storage. The use of X-ray imaging to understand seed viability of *Acronychia littoralis* is shown in **Fig. 3.**

The overriding guiding question that constantly runs through the process is "what happens in nature?" How does nature process the seed and how can we imitate and speed that process.

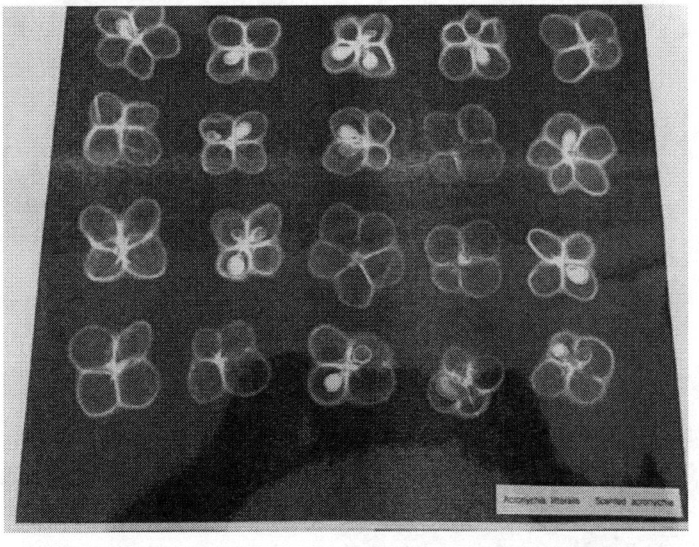

Figure 3. Seed viability testing of *Acronychia littoralis* using X-ray radiography

There are many techniques used to process the fruits and seeds; these are listed below:

- Blending
- Boiling
- Crushing
- De-winging
- Drying
- Fermenting
- Floating
- Leaching
- Macerating
- Manual extraction
- Manual scarification
- Shaking
- Sieving
- Soaking
- Splitting
- Wet and Dry Composting
- Winnowing

These techniques are detailed in the 'Australian Rainforest Seed' Book (Dunphy et al., 2020) and this list illustrates the number of techniques used to unravel the mysteries of rainforest seed propagation. Of course the research, innovation and learning goes on, continuing to improve rainforest seed propagation.

The majority of rainforest seeds are recalcitrant which means they cannot be dried and stored at room temperature or under refrigerated conditions. In fact many species such as White Booyong (*Argydendron trifoliatum*) have very short viability measures in days. **(Fig. 4)** This is in contrast to the majority of species that are orthodox and can be dried and stored often for many years.

Figure 4. Many rainforest species have short viability as in White Booyong (*Argydendron trifoliatum*) and many are recalcitrant.

The answer for many rainforest species to this problem of recalcitrant seed is to sow and store the seedlings rather than the seed. This is done by using a nutrient poor medium and germinating the seedlings at high density in a tightly spaced tray to slow and even stop their growth for up to a number of years. These seedlings forced into suspended animation can be potted at any time and grown into healthy trees without any ill effects. **(Fig. 5)**

Figure 5. Recalcitrant seeds are sown in high density in a nutrient poor medium in tightly spaced trays to slow and even stop their growth for up to a number of years. These seedlings can be potted and grown later as needed.

SOWING AND GERMINATING RAINFOREST SEED

Sowing seeds seems like a simple task for any horticulturally trained person, however tens of thousands of rainforest seeds have been lost to incorrect sowing technique. The most common problem is sowing seeds too deep. To compound this, trays are often kept too wet. It would be logical to assume rainforest seeds can handle very wet conditions, however on the bottom of a rainforest floor most seeds germinate on or in the leaf litter rather than in the soil. The necessary water is supplied by the surrounding moisture and humidity.

It is clear that many rainforest seeds can be sown on the surface and kept relatively dry to achieve maximum germination. In fact, the ability to alternate between a few days of the trays being wet and then a few days of the trays being dry seems to break dormancy in many species. This we assume is imitating what happens in nature.

CONCLUSIONS

Collecting rainforest seed, propagating it, growing it on, then planting it and maintaining the trees to form a young rainforest is a relatively new science. We have been moving ahead in leaps and bounds since the 1980's. We are still learning and discovering ways to improve and speed germination, growing on and establishing rainforest.

Native forest and rainforest specifically is being cleared, logged and burnt at an increasing rate around the world. The science and practise of rainforest restoration in northern NSW is moving against the trend and increasing significantly in diversity of species, number of trees planted and area of rainforest established. And, of course, a crucial component to making this process work is the propagation of rainforest seed.

LITERATURE CITED

Dunphy, M., McAlpin, S., Nelson, P., Chapman, M., Nicholson, H. (2020). Australian Rainforest Seeds: A Guide to Collecting, Processing and Propagation. CSIRO Publishing, Clayton, VIC, Australia.

Combined Proceedings IPPS 74:124-132. 2024.

New Nursery Built on the Back of IPPS Seeking and Sharing

David Hancock

Natural Area Nursery, Natural Area Consulting Management Services, 57 Boulder Road, Malaga, WA 6090, Australia

david@naturalarea.com.au

Keywords: tubestock, planning, logistics, designing, irrigation, efficiency

Summary

Natural Area Nursery in Western Australia started from humble beginnings in 2005 and grew into an 800,000 annual turnover of tubestock from 80,000. Over the 15 years, the Government leased land of the nursery quadrupled in area as well. At the beginning of Covid pandemic, in January 2020, the Government of Western Australia asked for the return of the land for a new train station complex. This paper describes the identification of land, logistics, designing, building and relocation of the massive operation within the two-year timeframe allocated by the Government. In addition to all other factors, the dedication and hard work of the staff of the Natural Area Nursery made this relocation possible.

IPPS Vol. 74 – 2024

INTRODUCTION

This paper underlines the international nature of the IPPS. The Natural Area Nursery (Naturalarea, 2024) is a family run business that had operated from a long-term Government leased site at Whiteman Park in Western Australia commencing in 2005. The nursery area quadrupled in size over 15 years to production of tube stock from 80,000 to 800,000 (**Fig. 1**). The piecemeal expansion led to fragmented logistics, with three separate nursery areas across a tight and disjointed area.

Figure 1. Over 15 years, the old nursery developed into a large operation with three different disjointed area with fragmented logistics. **A**) Open benches **B**) Covered benches and **C**) on ground tubestock. **D**) At the time of moving out to the new nursery.

At the same time that Covid pandemic hit in January 2020, The State Government of Western Australia advised that they wanted the entire site for a new train station complex for a rail line. We were being forced to relocate and needed to exit the site by the end of 2021, giving us less than 2 years. Almost a year went by in searching for a new site, negotiating for relocation, rebuild compensation, appointment of engineer and designs and finding building companies for the works.

A new and level site 4 km away was finally agreed upon in early 2021 and the huge task of managing the existing nursery and at the same time project managing the new nursery construction was underway with a 12-month deadline. The works were undertaken at a time of extreme shortages of labour and material supply delays due to Covid impacts. Despite Government driving the requirement to move, we faced the frustration of dealing with multiple agencies for planning, heritage, environmental, transport and services. The new site had no services and new installations for power, water and communication were necessary. In the very early stages, we had initially hoped to time our relocation to late winter/spring of 2021, but the time taken to deal with all the preliminaries meant that we were facing a stock relocation in late 2021 early 2022 and at high stock levels, in the heat of Perth's summer.

THE NEW NURSERY

The new nursery design and specifications achieved (**Fig. 2**) was a game of endless brinkmanship as the Government was only committed to replace like for like but the Government required that we project manage the entire build and relocation. An enormous amount of management time was involved effectively taking one Full Time Equivalent staff time for 1.5 years. The demands on the nursery team were extreme but the prospect of a much better work environment and nursery facilities drove them on.

Figure 2 shows that we went for bitumen hardstand around all the traffic areas to allow for heavy vehicles and deep compressed bluemetal for the growing areas. The buildings were spaced to allow machinery and trailers adequate room to allow access by vehicles and forklifts. The growing program remained essentially the same as the old nursery but with enhanced space. It was a good result, well above our expectations.

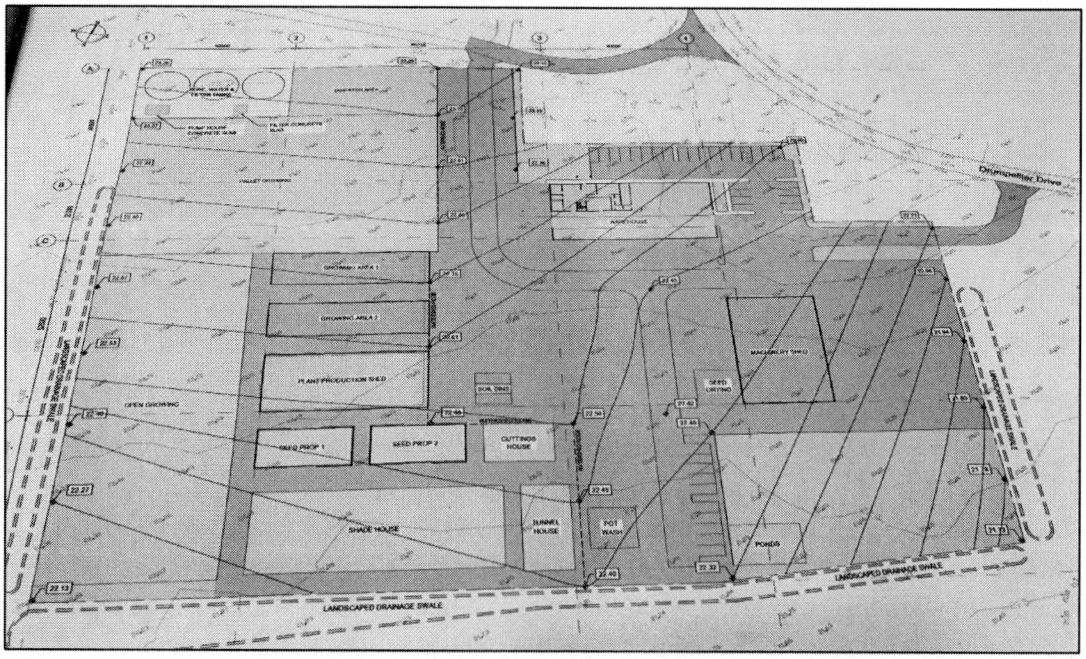

Figure 2. The new nursery plans involved bitumen hardstand around all the traffic areas to allow for heavy vehicles and deep compressed bluemetal for the growing areas. The buildings were spaced to allow machinery and trailers with adequate room to allow access by vehicles and forklifts.

The stock transition in January 2022 was difficult in 40 plus degree Celsius heat and losses in stock and production time were high and for which no compensation was payable. The dedicated staff did an amazing job. In total it took four trucks and four forklifts 10 days and almost the entire company workforce to get it done. The new nursery has superior water quality and systems (**Fig. 3** and **4**) and enhanced production facilities for staff and stock.

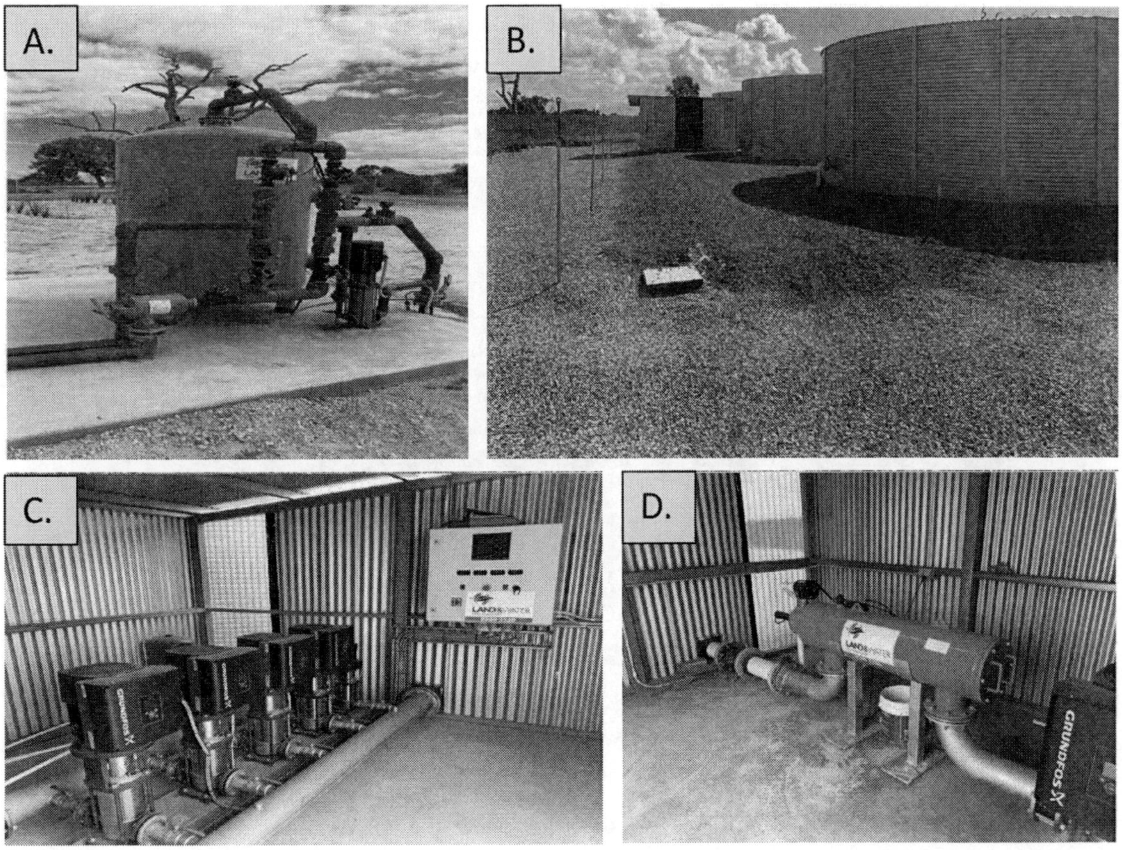

Figure 3. The new nursery has superior water quality thanks to the installation of modern **A**) Bore water filter. **B**) Large water storage facilities. **C**) Irrigation pumps with digital control and **D**) Screen filter.

Water Supply in the New Nursery

There is an onsite groundwater bore with automated supply to storage tanks (**Fig. 3B**). Bore water is filtered via Netafim F600 with clay-based media to remove solids (Netafim, 2024) and to reduce iron levels to below 1 part per million (**Fig. 3A**). Automatic back wash water supplies a seed production area. Water storage capacity is 960 KL in 3 x 320 KL steel tanks independently filled (**Fig. 3B**). Storage is for cover in event of bore pump failure. The irrigation system is installed with 4 x Grundfos CRE 20-4 variable pressure pumps (Grundfos, 2024) each on consecutive demand delivering up to 80 KL per hour (**Fig. 3C**). This is sufficient to water the entire outside nursery area at one time if required.

Nursery is watered by station selection based upon plant needs and to minimise power draw, watering in daylight hours to suit 30 kw solar power system output. Service water is supplied via Grundfos constant pressure jacking pump to 25 watering service points across the nursery. Additionally, the water system is equipped with a post pump filtration unit - Netafim screen guard at full pressure with 120 μm mesh (~1.5 human hair width) (**Fig. 3D**) and automatic back wash water used to supply seed production area.

Irrigation control is managed by a Signal SDS 50 system operated by remote cloud app to adjust station start and run times and monitor pump pressure (Signal, 2024) (**Fig. 4**). Willowbank Frost Watch Environment System® is set to operate short run irrigation cycles at 1 degree and lower. For the emergency power system backup, we have a hard-wired generator to bore and irrigation system. For irrigation outlet controls Bermad Solenoids® and Netafim Orkal® disc filters have been installed to fine spray overhead irrigation areas.

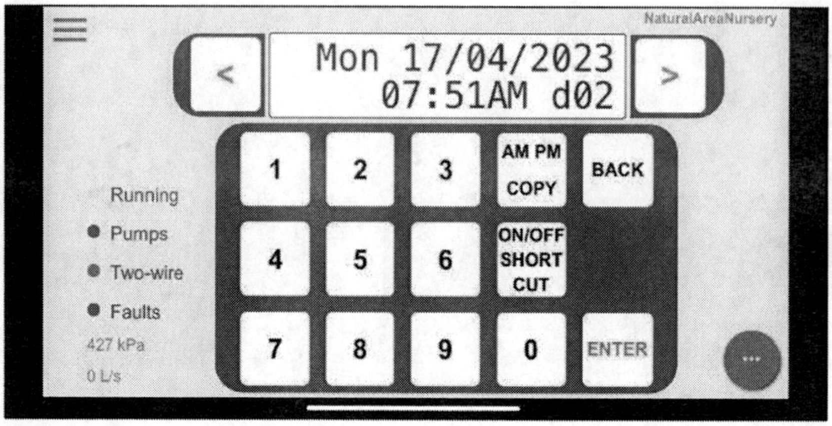

Figure 4. The digital control system of the water pump.

As a result of these improved facilities and practices, the outlook for the 2023/24 season is much improved on what was achieved in 2022/23 and has overcome the stock and sales shortfalls due to the move. The current year (2023/24) production target of 1.4 million plants will be achieved.

Logistics in the Nursery

The site design has allowed the use of imported Combi-trac all terrain forklift and small 4 x 4 ride on vehicles and trailers for improved plant movement (**Fig. 5**) not available to us on the old site and the new covered growing areas provide a wide range of propagation conditions. The production shed is insulated to walls, doors and roof with modern in-built conveniences that provide a quality work environment.

Specifications of the New Nursery

Outside growing areas are installed with 14 irrigation stations supplying 300 Netafim Gyronet turbo rated 160 L/h at 5.5 m rests at 3 bars. The current growing capacity is approximately 1.4 million tubes (**Fig. 6A**). The 1000 m^2 shadehouse is equipped with powered retractable top and side screens (**Fig. 6B**).

Figure 5. A) Improved logistics and better planning of the new nursery means that there is space for large trucks. **B)** and **C)** operation of imported Combi-Trac® all terrain forklift and **D)** small 4 x 4 ride on vehicles.

Figure 6. The modern new nursery. **A)** Outside growing areas. **B)** Shade house, **C)** Seed propagation area and **D)** Palletised growing area.

It has overhead irrigation with Netafim Spinnet 90 L/h at 4 m rests operating at 4 bar. The seed propagation area consists of 2 x 200 m^2 enclosed houses with differing polycarbonate roof finish for light and heat control variation (**Fig. 6C**). One house is set at 48% light transfer and 32 % heat transfer and the other at 26% light and 50 % heat. The benches are overhead irrigated with Netafim Spinnet at 90 L/h at 2 m rests at 4 bar. Palletised growing areas have auto seeding in 2 x 350 m^2 with differing polycarbonate coverings. One at 48% light transfer and 32 % heat transfer and the other at 26% light and 50 % heat. Overhead irrigation is provided by Netafim Spinnet at 90 L/h at 2 m rests at 4 bar (**Fig. 6D**). Our design is custom-made for Perth conditions and is working well for our requirements.

Cuttings propagation house is a double skin polycarbonate structure of 200 M^2 with automatically controlled light, temperature and humidity. It is fitted with both conventional and LED lighting. It is equipped with EnviroSTEP™ control system and overhead Netafim Coolnets with 7.5 L/h at 1.5 m rests, 4 bar (**Fig. 7A**). It was supplied by Argosee Greenhouse Technologies, WA, Australia (Argosee 2024). The tunnel house is a 250 m^2 poly film with powered retractable overhead and side shading (**Fig. 7B**). It has a Netafim Gyronet turbo at 160 L/h at 3 metre spacings. This has a capability of higher water output for hotter conditions and was also Supplied by Argosee Greenhouse Technologies.

Plant production shed has an 800 M^2 single-span insulated roof with Bondor double insulated panels to doors and walls (**Fig. 7C**). It also accommodates the air-conditioned office and break-out kitchen/lunch area and wet areas. The sliding doors are 7 x 4 m to facilitate drive through machine movements.

Figure 7. Cuttings propagation house **A**). Tunnel house **B**). Plant production shed **C**) and KW auto seeder and conveyor **D**).

Equipment and other Facilities of the Nursery

The major pieces of equipment include Kanga mini loader, Combi Trac all terrain forklift, Heli HD forklift, 2 x Kubota 4 x 4 ride on with twin 4-wheel steer trailers, KW auto seeder and conveyors (Kwautomation, 2024) (**Fig. 7D**). Also included are Urbinati tray filler (Urbinati, 2024) and a weather station monitored by phone.

The nursery also has three soil bins each with an area of 30 m^2 (**Fig. 8A**). A drone view of the new facility is shown in **Fig. 8B**.

In summary, the vastly improved efficiency, productivity and quality of facilities are being appreciated by the staff. The microclimate and growing conditions within 4 km distance of the old and new nurseries are surprisingly different and it took a full season to understand the site differences. Such a large relocation doesn't come without some shortfalls. For example, the auto retractable shade cloth should have been in separate sections. We brought weeds with stock transfer, and it took a year to clean up, now effectively under control. Additionally, we find that the soil bins are too small for the scale of the new operation.

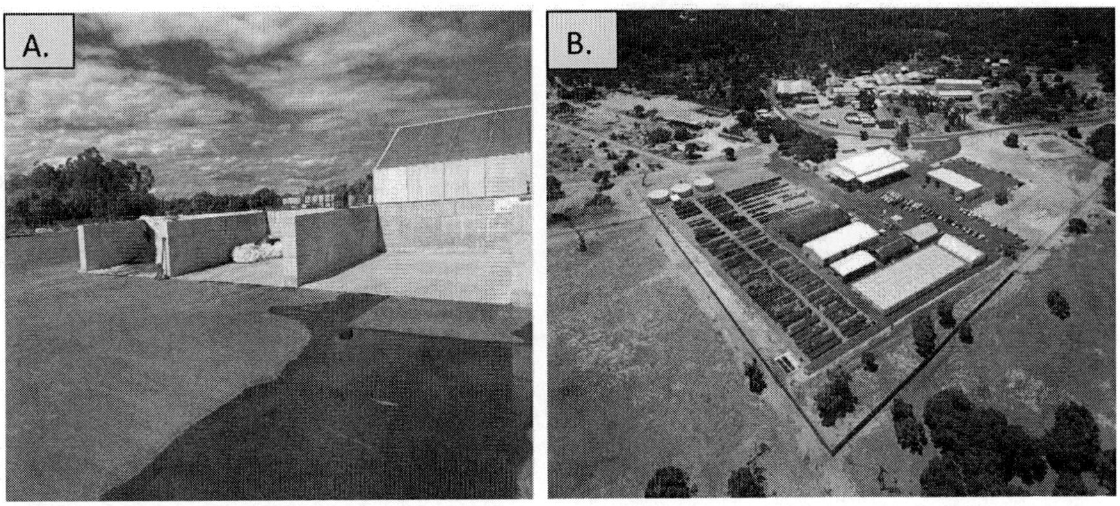

Figure 8. Three soil bins **A**) and drone view **B**) of the nursery.

CONCLUSIONS

Water quality and irrigations standards should be given number 1 priority when planning a nursery. The message from our experience is "Do not underinvest in water systems". Where possible, ease of access, level grounds and hardstand need to be prioritised. Both staff and company will be rewarded when the staff facilities are of high quality. Labour effort can be significantly reduced, and operational efficiency maximised through proper site design. Finally, design space for expansion is important as one never knows how fast operations can expand when properly planned. Without the learnings from IPPS Seeking and Sharing, Natural Area Nursery Team would not have achieved this success.

ACKNOWLEDGEMENTS

The author wishes to thank all the staff of Natural Area Nursery for their dedication during the stressful period of relocation, without which this would have been impossible. Thanks also to Dr Ranjith Pathirana (Editor, IPPS Australia) for editorial assistance.

LITERATURE CITED

Argosee. (2024). Available at: https://argosee.com.au/product/greenhouse-automation-controls/

Grundfos. (2024). Available at: https://product-selection.grundfos.com/au/applications/commercial-buildings/commercial-water-pressure-boosting?tab=products

Kwautomation. (2024). Available at: https://www.kwautomation.com.au/HorticultureNursery/Seeding-460/AutomaticSeeder-538/

Naturalarea. (2024). Available at: https://naturalarea.com.au/

Netafim. (2024) Available at: f600-gravel-filter-july17-v1-0.pdf

Signal. (2024). Available at: https://signal.com.au/product/signal-sds-series-controller/

Urbinati. (2024). Available at: https://www.urbinati.com/en/products/tray-fillers/

Combined Proceedings IPPS 74:133-143. 2024.

Tissue Culture of Red Bayberry, A New Industry for Australia

Jayeni Chathurika Amarathunga Hiti-Bandaralage [1,2,3]* and Shaun O'Brien [1,4]

[1] J&S Plant Biotech Pty. Ltd; 41, Tembler Place, Dulong QLD 4060, Australia.
[2] STC Link Pty Ltd.; 22, Castlecor Street, Ferny Grove, QLD 4055, Australia.
[3] STC Biotech Pvt Ltd.;170C, Janasavi Mawatha, Gurulana, Bope, Padukka, Sri Lanka.
[4] Microplants Pty Ltd.; 41, Tembler Place, Dulong QLD 4060, Australia.

jchiti@jsplantbiotech.com.au

Keywords: in vitro culture; *Myrica rubra,* woody plant

Summary

The successful establishment and propagation of red bayberry (*Myrica rubra*) in vitro cultures require precise optimization of initiation, multiplication, and rooting stages. This study aimed at developing a high-throughput clonal propagation system for red bayberry. Here, the influence of stem growth stage, seasonal variation for clean culture initiation success and tissue culture media composition for bud induction, multiplication and rooting were evaluated.

Un-hardened red stems consistently outperformed semi-hardened stems, achieving a maximum clean culture establishment rate of 90% compared to a 10% maximum for semi-hardened stems across four rootstocks trialled in the study. Seasonal analysis revealed summer and autumn as the optimal material collection periods, with overall success rates of 47.25% and 45%, respectively. Media optimization during the initiation phase enhanced axillary bud vigour, addressing initial poor growth observed on

IPPS Vol. 74 – 2024

BB01 media. Multiplication challenges, including chlorosis and shoot die-back, were mitigated using I09.1 media, which supported robust shoot quality and a multiplication rate of 2,244-fold, over seven subculture cycles for the tested genotype. Rooting experiments demonstrated an efficient auxin treatment protocol, achieving 100% rooting success within six weeks without adverse effects on shoot quality. Rooted plants acclimatized successfully under misting conditions, reached 100% survival and vigorous growth post-acclimatization.

These findings establish a comprehensive commercial tissue culture protocol for red bayberry, from initiation to nursery transfer, demonstrating potential for large-scale propagation. This research provides critical insights into optimizing in vitro systems for woody perennials, enhancing their application in horticultural biotechnology.

INTRODUCTION

Myrica rubra, commonly known as Chinese bayberry or yangmei, belongs to the Myricaceae family. The genus *Myrica*, to which *M. rubra* belongs, comprises approximately 50 species globally, many of which are found in temperate and subtropical regions. *M. rubra* is the most economically significant species within this genus (Zhang et al., 2015). This species holds significant horticultural, economic and cultural importance in East Asia, particularly its native region of China (Mo et al., 2024). *M. rubra* is a dioecious, evergreen tree characterized by its vibrant bright red to dark purple sweet-tart fruit that has a unique textural delight, which is highly valued for its unique flavor and nutritional benefits. The taxonomy of this species has been well-studied, revealing its close genetic relationships within the Myricaceae family and its adaptation to various climatic conditions. It has a long history of cultivation and usage in traditional medicine, making it a subject of growing interest for both agricultural and scientific communities (Zhang et al., 2022).

Myrica rubra is a dioecious species, meaning that individual trees are either male or female. This characteristic necessitates the presence of both male and female plants for successful pollination and fruit production. The tree typically reaches a height of 10 to 20 meters, with a broad, dense canopy. The leaves are leathery, lanceolate, and dark green, measuring 5 to 12 cm in length. The flowers are small, inconspicuous, and appear in early spring. The fruit is a drupe, 1.5 to 2.5 cm in diameter, with a rough, waxy surface and a vivid red to dark purple color when ripe (He et al., 2016; Jia et al., 2019; Liu et al., 2014). The tree thrives in well-drained, acidic soils and prefers a subtropical climate with moderate rainfall and mild temperatures. Traditional cultivation methods have been employed for centuries, but recent advancements in agricultural practices are being adopted to enhance yield and fruit quality (Wang et al., 2017).

Propagation of red bayberry (RB) is commonly done through grafting, which ensures the consistency of desirable traits such as fruit size, taste, and color. Seed propagation is less common due to the genetic variability and the extended time required for trees to reach fruit-bearing age (Chen et al., 2008; Fang-Yong and Ji-Hong, 2014). RB is a difficult to root woody species with no commercial success in rooted cuttings as a propagation method. In China propagation is achieved through grafting. It takes 1.5 years to raise rootstock seedlings and also suffers with inefficiencies due to low/variable seed germination rates (50-60% maximum). Graft is also problematic; success depends on optimum budwood condition making the propagation process highly inefficient (Perkins, 2014).

The primary challenge for the progression of the RB industry in Australia is the "lack of efficient propagation technology" (Joyce and Sanewski, 2010; Perkins and Joyce, 2018). RB is an outcrossing woody species, meaning that genetically divergent seeds from parent plants are unsuitable for commercial cultivation. Vegetative propagation, which replicates exact genetics, is crucial to maintaining consistent fruit characteristics such as colour, flavour, and post-harvest qualities that are vital for market success. (Perkins, 2014).

Micropropagation presents a promising alternative for the mass production of RB plants with desirable characteristics. By exploiting the inherent totipotency of plant cells, micropropagation enables the regeneration of numerous identical plantlets from a small explant, such as a shoot tip or leaf segment. Moreover, the use of tissue culture techniques allows for the production of disease-free plant material, thus reducing the risk of pathogen transmission and ensuring the establishment of healthy orchards and landscapes (Hiti-Bandaralage, 2019; Hiti-Bandaralage et al., 2017).

The successful micropropagation of RB relies on the optimization of several key factors, including the selection of suitable explant sources, the development of appropriate culture media formulations, and the manipulation of growth regulators to induce shoot proliferation and root formation. Additionally, the establishment of proper environmental conditions, such as temperature, light intensity, and humidity, is crucial for the in vitro growth and development of the plantlets. To date commercial tissue culture propagation of this recalcitrant species is not found anywhere in the world due to the less efficient protocols for it to be viable with respect to consistency of results as well as economic feasibility.

This manuscript presents the latest advancements in developing a commercially viable tissue culture platform for the mature rootstock MR06. The developed protocols and some specific details are withheld due to high commercial value and project funders' agreements. However, the results are presented in detail where possible to benefit the scientific community by highlighting the availability of this advanced tissue culture platform and to attract prospective future collaborations. The developed tissue culture platform holds significant potential not only for the propagation of *Myrica rubra* but also for its conservation and genetic improvement, underscoring its economic and ecological importance.

MATERIALS AND METHODS

Explant Sterilisation

Elite RB rootstock mother plants were maintained in grow bags with irrigation at a private property in Brisbane, Queensland, Australia. Three days prior to stem section collection, the mother plants were treated with 1 g/L Mancozeb, systemic fungicide. Initial efforts to establish an effective sterilization process began in mid-summer (January), using 10 cm softwood stem sections with all leaves removed. A series of experiments were conducted to optimize the sterilization process, involving various pre-treatments (e.g., 1 g/L fungicide-Mancozeb soak for 1-24 hours, hot water treatments, and soap water soak), followed by standard 70% ethanol washes and bleach treatments at different concentrations and durations.

However, the material proved to be highly sensitive to ethanol and bleach, often resulting in successful sterilization at the cost of losing bud viability. After numerous trials, the following procedure was identified as the most effective, yielding at least 20% clean cultures with viable axillary buds.

Optimized Sterilization Procedure

a) Remove all leaves by snipping and collect 5-7 stem sections into a 500 ml tall container filled with 200 ml of a 1 g/L solution of systemic fungicide Mancozeb.

b) Incubate the stem sections in the fungicide solution for 24 hours.

c) Decant the fungicide and add 2 ml of antibacterial hand soap to the container. Add approximately 300 ml of 40 °C water and gently shake for 10 minutes. Decant the hot soapy water solution and repeat the hot soapy water washing step one more time.

d) Decant the hot soapy water solution and wash the stem sections under running tap water for 30 minutes.

e) After the tap water wash, transfer the container with washed material into a laminar hood. Under sterile conditions in a laminar hood, add 400 ml of 70% ethanol to the container and gently shake for 3 minutes.

f) Decant the ethanol and wash three times with sterile distilled water.

g) Add 2 drops of Tween 20 solution onto the stem sections and add 300 ml of a 4% commercial bleach solution (Concentrate containing 42 g/L sodium hypochlorite, and 9 g/l NaOH; active chorine 4.0% m/V). Shake gently for 3 minutes.

h) Decant the bleach solution and wash 5 times with sterile distilled water to remove all traces of bleach.

After sterilization, stem sections were cut into 1-1.5 cm nodal sections, ensuring each section contains at least one axillary bud. The nodal sections were inoculated into BB01 media (Commercial IP not disclosed) contained in test tubes (one nodal section per tube). Inoculated cultures were then incubated for 7 days and meticulously inspected them for any signs of fungal or bacterial growth. Bacterial or fungal contaminations were discarded and clean cultures were continued to maintain under the optimum incubation conditions for up to 4 weeks.

Initiated clean cultures were incubated under fluorescent lights with light intensity of PAR 575 uW/cm^2 with 16 h light and 8 h dark regime. The temperature of the growth room was maintained at 25 $^0C \pm 1$ ^0C.

Mother Plant Quality Assessment

Initially four different rootstocks were subjected to the assessment: MR06, MR07, MR09 and MR24. Two different types of material; semi hardened green stems and red soft stems were collected from each rootstock to identify the type of material suitable for tissue culture initiations. To assess the quality of plant material in different growth cycles, initiations were carried out in all four seasons to understand any seasonal effect on establishment of clean cultures and viability of axillary buds in culture. The optimized sterilization procedure explained above was used for all experiments.

Optimisation of Initiation Media

A series of experiments were conducted with MR06 rootstock to optimize initiation media that can support fast axillary bud growth. Due to the high commercial value of the protocols developed, the media composition is not presented in detail for this manuscript. The factors included the type of cytokinin used, concentration of cytokinin, combination of cytokinin with one-two types of auxins, sugar type and concentration, other additives such as coconut water, banana pulp and potato extract tested in independent experiments. Effect of different light spectrums (red/far red/blue) on bud initiation and growth were also tested (data not presented).

Optimisation of Shoot Multiplication

Clean cultures established were used to excise 0.1 – 0.5 cm shoot tips for multiplication. A series of independent experiments were conducted to test individual factors; basal media, type of cytokinin, concentration of cytokinin and combination of cytokinin and auxin to achieve commercially viable multiplication rates. Ten tubs from each multiplication cycle were randomly collected to calculate the multiplication rate (number of explants produced/number of explants used) for each subculture cycle to identify the best media formulation and plant growth regulator (PGR) combination.

Optimisation of Root Induction

MR06 shoots cultured and elongated on I09.1 media with at least 3 open leaves and about 1 – 1.5 cm length were subjected to rooting treatment. Rooting trials were carried out as in vitro agar based rooting trials with the use of different concentrations of naphthaleneacetic acid (NAA), indole-3-butric acid (IBA) or indole-3-acetic acid (IAA) as rooting hormones. Once the effective type and the concentration of auxin were identified in individual experiments the effect of charcoal at 1 g/L concentration was also trialed to improve the root number and growth vigour of the plantlets.

Acclimatisation of Rooted Plantlets

MR06 rooted plants were removed from agar medium and directly planted into 50 mm net pots filled with seed raising potting mix added with perlite at 3:1 ratio. Plants were maintained in a misting chamber at 100% humidity for 7 days and at 65-70% humidity thereafter for another 3 weeks.

RESULTS AND DISCUSSION

Mother Plant Quality Assessment

The comparison between green semi-hardened stems and red un-hardened stems revealed that the higher success rate in establishing clean cultures is associated with newer young stems rather than older stems. Un-hardened red stems achieved over a 50% success rate in clean culture establishment, while semi-hardened stems from all four rootstocks recorded a maximum of only 10% success (**Table 1, Fig. 1**).

Figure 1. Optimal growth stage of un-hardened red bayberry stems for in vitro culture success.

Table 1. Effect of growth stage on culture initiation success.

Rootstock	Percentage clean culture, semi-hardened stem	Percentage clean culture, un-hardened stem
MR06	10%	52%
MR07	9%	78%
MR09	3%	90%
MR24	0%	53%

The higher success rate observed with un-hardened red stems aligns with findings from other woody perennials, where newer growth often exhibits superior culture establishment due to reduced microbial contamination and higher physiological activity (Leelavathy and Sankar, 2015). Similarly, studies on avocado and grapevine have demonstrated the pivotal role of growth stage in tissue culture success, underscoring the advantage of juvenile/young tissues for optimal in vitro outcomes (Cyndi, 2023; Hiti-Bandaralage et al., 2022). These results emphasize the critical need for precise selection of donor plant material to maximize clean culture establishment rates and propagation efficiency.

The season in which the material was collected had a significant impact on the percentage of clean culture establishment for red bayberry (**Fig. 2**). The highest percentages of clean cultures were achieved in the summer and autumn seasons, with overall success rates of 47.25% and 45%, respectively, as a cumulative average across all varieties and multiple initiation trials.

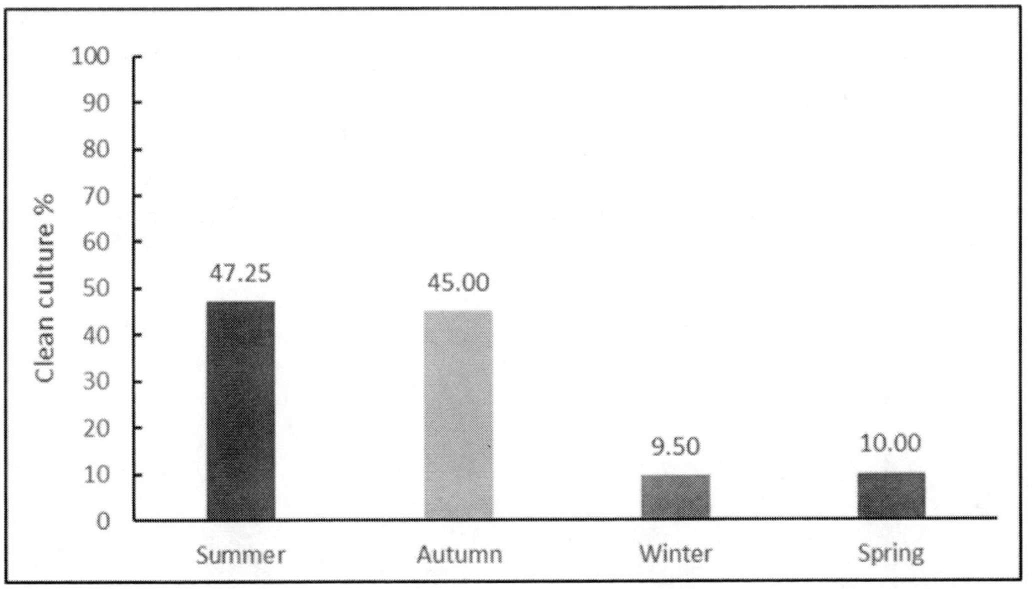

Figure 2: Percentage of clean culture establishment in different growth seasons for red bayberry.

The significant seasonal impact on culture establishment aligns with findings in woody perennials, where summer and autumn promoted active growth and reduced microbial contamination. Seasonal variations affect endogenous hormone levels and microbial load, enhancing tissue responsiveness and clean culture rates during these periods, as observed in mango and fig tissue culture studies (Ray and Savage, 2021).

Red bayberry initiation was successful using BB01 media, but the growth of the axillary buds was notably slow and exhibited poor vigor. The series of optimisations could improve the growth vigour of the axillary buds (**Figure 3**). Similar obasevations were made by Asghari et al. (2013) highlighting the importance of fine optimisation of sterilisation process and initiation media for success with red bayberry.

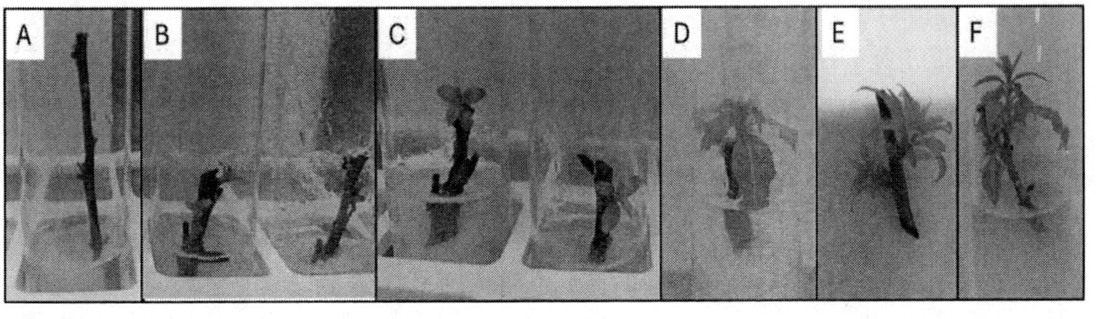

Figure 3. Bud breaking and shoot quality improvement by fine optimization of initiation media composition. (**A** to **F**) are shoot development at 2 weeks in different initiation media from inoculation.

Several problems were encountered in the multiplication stage of red bayberry. Chlorosis, shoot die-back and stunted growth were common to many of the formulations trialed in multiplication experiments. The best shoot quality and multiplication was achieved with media formulation I09.1 which resulted in best quality shoots with highest multiplication rates over the 8 subculture cycles (**Fig. 4**). The multiplication rate achieved over the 7 subculture cycles starting from a single shoot tip was almost close to 2500 times (**Table 2**). Asghari et al. (2013) in their study recorded 5.6 – 6.8 maximum rate of shoot emergence, however, did not report the total multiplication factors for long term culture required for commercial tissue culture process. Woody plants in general are difficult to be maintained in continuous subculture with high multiplication rates and good health (Hiti-Bandaralage, 2019).

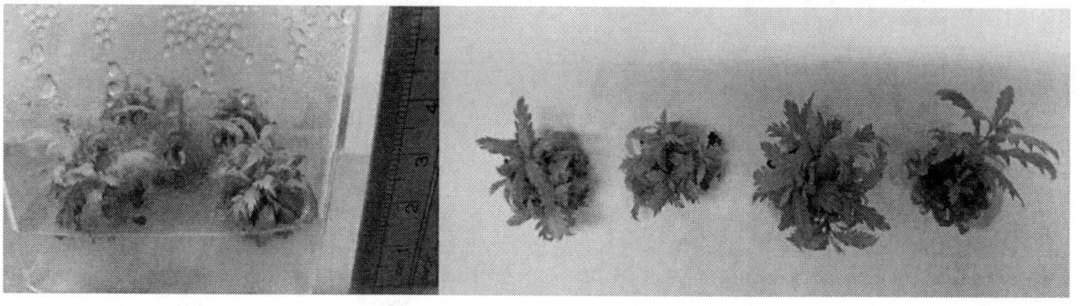

Figure 4. Successful shoot multiplication of MR06 on I09.1 media.

Table 2. Multiplication rate for MR06 with I09.1 media from initiation to 7th subculture cycle.

Initiation (I_0)	I_0 - T1	T1-T2	T2-T3	T3-T4	T4-T5	T5-T6	T6-T7	T7-T8	Total multiplication per shoot tip initiated
1	1	2	4.5	2	4.5	2.6	3.2	3.33	2,244

In general, woody plants display leaf defoliation, tip die back or necrosis when subjected to rooting treatments with auxins (Hiti-Bandaralage, 2019; Xue et al., 2023). Treatments applied on MR06 shoots did not have any adverse effect on shoot quality such as leaf defoliation, tip die back or necrosis. At 3 weeks from inoculation into rooting media, root initiation was visible (**Fig. 5**). The best treatment recorded 80% rooting at 3 weeks and 100% rooting at 6 weeks.

Figure 5. Rooting of red bayberry genotype MR06. **A1**) Microcuttings inoculation to rooting media and **A2**) Rooted plantlets.

Red bayberry plants developed a good root system within 6 weeks of rooting treatment. These plants acclimated with 100% survival under misting conditions. After 4 weeks, roots emerged from the pot and shoot growth was visible with new leaves (**Fig. 6**). The only significant work published on *Myrica rubra* propagation do date is by Asghari et al. (2013), who focused on two commercial cultivars, "Biji" and "Dongkui". For red bayberry tissue culture, successful rooting and acclimatization have only been reported by them and achieved a 95% rooting rate, demonstrating the importance of high-quality shoots to ensure successful multiplication and commercially viable rooting rates.

Figure 6. Fully acclimatized MR06 plantlet ready to move to nursery for repotting.

CONCLUSION

This study emphasizes the importance of optimizing plant material selection, seasonal timing, and media composition for successful red bayberry micropropagation. Un-hardened red stems and summer/autumn collections demonstrated superior clean culture establishment rates, supported by physiological activity and reduced microbial contamination. Improved initiation and multiplication media significantly enhanced shoot vigor, multiplication rates, and rooting success, achieving nearly 2500-fold multiplication. Rooting treatments produced high-quality plantlets with 100% acclimatization success. This research introduces the first successful commercial tissue culture protocol for the challenging red bayberry species. These findings provide crucial insights into overcoming culture establishment challenges, advancing the development of efficient propagation methods, and enabling large-scale production for horticultural purposes. The protocol will facilitate improved growth and mass propagation of red bayberry for commercial applications within and beyond Australia.

Acknowledgements

The authors would like to express their gratitude to Agrifutures Australia for co-funding the project through the Emerging Industries Business Grant 2023. We also acknowledge our project partners: Microplants Pty Ltd, Australian Nurserymen's Fruit Improvement Company (ANFIC), Wide Bay Seedlings, and Australian Horticultural Traders. Special thanks to Dr. Dion Harrison for his invaluable assistance in maintaining the mother plants throughout the project.

LITERATURE CITED

Asghari, Sayed, J.A., Ligeng, C., Xinhua, H., and Yonghua, Q. (2013). Micropropagation of *Myrica rubra* Sieb. and Zucc. using shoot tips and nodal explant. Afr. J. Agric. Res. *8*: 1731–1737. DOI:10.5897/AJAR10.1160

Chen, S.Y., Kuo, S-R., and Chien, C.T. (2008). Roles of gibberellins and abscisic acid in dormancy and germination of red bayberry (*Myrica rubra*) seeds. Tree Physiology *28*:1431–1439. DOI: 10.1093/treephys/28.9.1431

Cyndi, C. (2023). The application of tissue culture for grapevine disease elimination. The Grapevine Magazine. https://thegrapevinemagazine.net/2023/06/the-application-of-tissue-culture-for-grapevine-disease-elimination/ (accessed 12.1.24).

Joyce, D. and Sanewski, G. (2010). The Commercial Potential of Red Bayberry in Australia. RIRDC Publication No. 10/200 RIRDC Project No. PRJ-000527. https://agrifutures.com.au/wp-content/uploads/publications/10-200.pdf

Fang-Yong, C., and Ji-Hong, L. (2014). Germplasm genetic diversity of *Myrica rubra* in Zhejiang Province studied using inter-primer binding site and start codon-targeted polymorphism markers. Scientia Horticulturae. *170*: 169–175. DOI: 10.1016/j.scienta.2014.03.010

He, K., Li, X., Xiao, Y., Yong, Y., Zhang, Z., Li, S., Zhou, T., Yang, D., Gao, P., and Xin, X. (2016). Hypolipidemic effects of *Myrica rubra* extracts and main compounds in C57BL/6j mice. Food Funct. *7*: 3505–3515. DOI: 10.1039/c6fo00623j

Hiti-Bandaralage, J., Hayward, A., and Mitter, N. (2022). Structural disparity of avocado rootstocks in vitro for rooting and acclimation success. International Journal of Plant Biology *13*: 426–442. DOI: 10.3390/ijpb13040035

Hiti-Bandaralage, J.C.A. (2019). Micropropagation as an alternative for avocado clonal propagation (PhD Thesis). The University of Queensland. DOI:10.14264/uql.2019.531

Hiti-Bandaralage, J.C.A., Hayward, A., and Mitter, N. (2017). Micropropagation of avocado (*Persea Americana* Mill.). American Journal of Plant Sciences *08*: 2898. DOI:10.4236/ajps.2017.811197

Jia, H.M., Jia, H.J., Cai, Q.L., Wang, Y., Zhao, H.B., Yang, W.F., Wang, G.Y., Li, Y.H., Zhan, D.L., Shen, Y.T., Niu, Q.F., Chang, L., Qiu, J., Zhao, L., Xie, H.B., Fu, W.Y., Jin, J., Li, X.W., Jiao, Y., Zhou, C.C., Tu, T., Chai, C.Y., Gao, J.L., Fan, L.J., van de Weg, E., Wang, J.-Y., and Gao, Z.S. (2019). The red bayberry genome and genetic basis of sex determination. Plant Biotechnology Journal *17*:397–409. DOI: 10.1111/pbi.12985

Leelavathy, S. and Sankar, P.D. (2015). Curbing the menace of contamination in plant tissue culture. Journal of pure and applied microbiology *10*:2145-2152.

Liu, H., Qi, X., Cao, S., and Li, P. (2014). Protective effect of flavonoid extract from Chinese bayberry (*Myrica rubra* Sieb. et Zucc.) fruit on alcoholic liver oxidative injury in mice. J Nat Med *68*:521–529.
DOI: 10.1007/s11418-014-0829-9

Mo, J., Rashwan, A.K., Osman, A.I., Eletmany, M.R., and Chen, W. (2024). Potential of Chinese bayberry (*Myrica rubra* Sieb. et Zucc.). Food Bioprocess Technol *17*:3506–3524.
DOI: 10.1007/s11947-023-03313-9

Perkins, M. (2014). Pilot production and sales of red bayberry in Australia. RIRDC, Barton, A.C.T.Perkins, M.L., and Joyce, D.C. (2018). Red bayberry – the journey to its commercialisation in Australia. Acta Hortic. *1205*: 281–288.
DOI:10.17660/ActaHortic.2018.1205.1

Ray, D.M., and Savage, J.A. (2021). Seasonal changes in temperate woody plant phloem anatomy and physiology: implications for long-distance transport. AoB PLANTS *13*, plab028.
DOI: 10.1093/aobpla/plab028

Wang, J., Tian, C., Zhang, C., Shi, B., Cao, X., Zhang, T.Q., Zhao, Z., Wang, J.W., and Jiao, Y. (2017). Cytokinin signaling activates WUSCHEL expression during axillary meristem initiation. Plant Cell *29*:1373–1387. DOI: 10.1105/tpc.16.00579

Xue, Y., Hiti-Bandaralage, J.C.A., and Mitter, N. (2023). Micropropagation of *Duboisia* species: A review on current status. Agronomy *13*:797.
DOI: 10.3390/agronomy13030797

Zhang, S., Yu, Z., Sun, L., Ren, H., Zheng, X., Liang, S., and Qi, X. (2022). An overview of the nutritional value, health properties, and future challenges of Chinese bayberry. PeerJ 10, e13070.
DOI: 10.7717/peerj.13070

Zhang, T.Q., Lian, H., Tang, H., Dolezal, K., Zhou, C.M., Yu, S., Chen, J.H., Chen, Q., Liu, H., Ljung, K., and Wang, J.W. (2015). An intrinsic MicroRNA timer regulates progressive decline in shoot regenerative capacity in plants. The Plant Cell tpc.*114*.135186.
DOI:10.1105/tpc.114.135186

Combined Proceedings IPPS 74:144-151. 2024.

Micropropagation and Applications of In Vitro Systems for Grapevine (*Vitis* spp.)

Puthiyaparambil Josekutty

Research Manager, Skybury Farms, 136 Ivicevic Rd, Paddy's Green, QLD 4879 Australia

josekutty964@gmail.com

Keywords: tissue culture, organogenesis, shoot proliferation, rooting, acclimation, high-health plants

Summary

Grapevines (*Vitis vinifera* L.) are of great significance to the Australian economy as they comprise vine grapes and table grapes. Grapevine rootstocks belong to several *Vitis* species such as *Vitis rupestris*, *V. riparia*, *V. berlandieri* and *V. champini* etc. from America. In vitro propagation systems for grapevine like micropropagation, organogenesis, somatic embryogenesis, protoplast culture and cryopreservation are important for various reasons such as rapid and reliable cloning, international transfer of germplasm, virus elimination, rapid crop improvement through cell-level selection, genetic engineering, gene editing as well as in vitro conservation of valuable germplasm. Micropropagation of three open varieties of table grapes undertaken for rapid reliable cloning of virus-free stock material for orchard establishment is described. Also, various in vitro methods applied to grapes and their applications are discussed.

INTRODUCTION

Skybury farms, 136 Ivicevic Rd, QLD 4880 in Australia (www.skybury.com.au) is a diversified agriculture business commercially producing Papaya, Coffee and a variety of value-added products from our farm produce, for example, the papaya vodka that won international award in the London spirit show. Papaya cream is another example of its premier products. Skybury Farms is also a proud supporter of horticulture development in Australia offering cost-effective, results oriented research and development services.

Grapevine industry is a major player in the Australian horticulture sector valued at about 50 billion AUD in 2022. Wine grape industry is the major player with a cultivated area of 146,224 million ha., annual production of 1.48 billion litres wine of which 48% is exported, and employing 136,790 persons. Although much smaller (AUD 1.5 Billion), the table grape industry is significant as 70% of the Australian table grapes are exported according to the Australian Grape and Wine website (https://www.agw.org.au/).

Although there are several reports on micropropagation of grapes as recently reviewed, our objective was to develop a rapid and reliable cloning technology that works at the Skybury lab which is specialising in horticulture research and development. This technology could be used to assist wine growers and wine variety importers in Australia to rapidly clone new varieties of grapes they introduce to Australia. A mid to longer term goal of Skybury lab is to provide crop improvement research service (improved variety development) to wine growers using the in vitro breeding technique we successfully applied to the rapid development of the world's best papaya, the Skybury, Sweet, Red variety (Puthiyaparambil et al. 2023).

MATERIALS AND METHODS

We used healthy (disease-free) grafted plants of three most popular table grape varieties with distinct fruit characteristics {'Menindee seedless' an erly, seedless, green grape; 'Çrimson seedless' a late season seedless, red variety, and 'Autumn royal' also a late variety but with seedless, black berries} obtained from a reputed plant nursery. We maintained the mother stocks in the insect- proof greenhouse in Skybury nursery for 60 days and regularly observed for visual symptoms of common diseases of grapevine. We applied insecticides and fungicides as normally applied to our papaya production nursery which maintains over 50,000 tissue culture papayas for our farm use.

Shoot tips and tender nodal segment were collected and used as explants for micropropagation. The explants were surface sterilised with a 10-15 min wash with Johnson's$^{(T)}$ baby shampoo followed by a one-minute rinse with 70% (v/v) ethanol, then rinsed once with sterile (autoclaved) RO water. Further, treated the explants for 3-5 min with 5% (v/v) household bleach diluted with sterilised RO water and finally rinsed four times with sterile RO water. Surface sterilised explants were trimmed to 2.0 – 3.0 cm shoot apieces and nodal segments with single nodes. The cleaned explants were inoculated on to autoclaved media in a laminar flow hood. All the cultures were incubated in a growth room maintained at 25± 2°C, lighted with red + blue LED lights at approximately 80 μmol/s/m^2.

RESULTS AND DISCUSSION

Culture Initiation

Initiation was easy from shoot tip and nodal explants in MS medium (Murashige and Skoog, 1962) modified with benzyl adenine (1.0-2.0 mg/l) or kinetin (1.0-2.0 mg/l) + naphthaleneacetic acid (NAA; 0.1-0.5 mg/l) + 25 g/l sucrose, with / without 500 mg/l charcoal, pH adjusted to 6.25 before adding 2.8 g/l Gelzan and sterilisation (**Fig. 1**). On average, shoot culture initiation required 25-30 days from the day of culture establishment. Fungal contamination (10-12%) was more than bacterial contamination (3-5%) observed at initiation stage. The tropical climate in which the plants were maintained explains the high rate of contamination at the initiation stage.

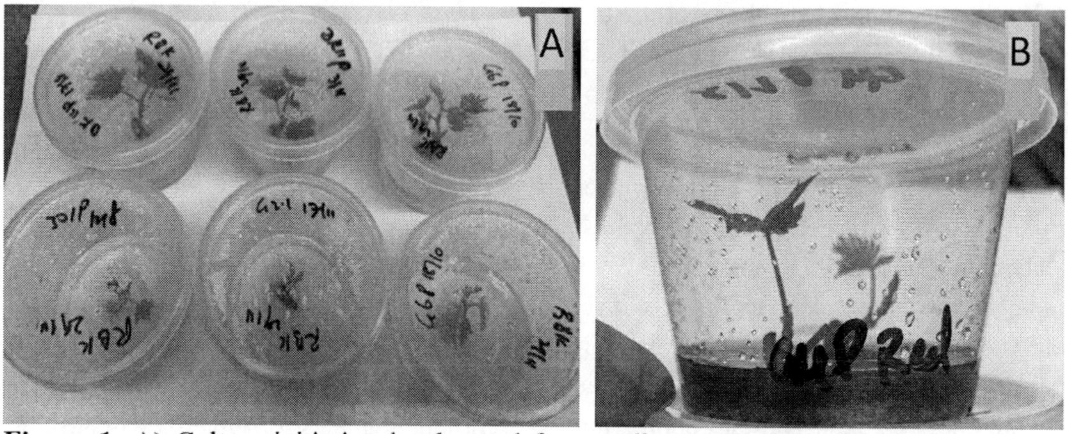

Figure 1. A) Culture initiation in charcoal-free medium and **B)** in charcoal supplemented medium.

Shoot Proliferation

According to recent reviews, (Yancheva et al. 2018; Carimi et al. 2019; XiuMing et al. 2021) there is a significant genotype effect on micropropagation of grapevine varieties. However, in this experiment, we didn't notice much differences between the varieties at the initiation and proliferation stages.

Best medium for shoot proliferation was MS medium supplemented with 25 g/l sucrose, 100 mg/l Polyvinylpyrrolidone (PVP 40), 2.0 mg/L Kinetin + 0.2 mg/L NAA, 2.8 g/l Gelzan and pH adjusted to 6.25 before autoclaving. Shoot proliferation started from day 10-15 of culture and 3-5 shoots developed in 30 days on average from cultured shoots. The shoot multiplication rate was also similar (6-8 shoots / culture) between the three varieties. From the third monthly subculture, a 600 ml jar with 10 shoots easily generated 50-60 shoots and filled up the jar in 3 weeks (**Fig. 2**). Due to the high rate of proliferation and growth, I found that transferring the cultures on to hormone free MS medium containing 20 g/l sucrose for alternate monthly culture cycle kept the shoots with little or no vitrification. Specific media requirements for obtaining high frequency shoot proliferation from different grape varieties is on record as per the recent reviews (Yancheva et al. 2018; Carimi et al. 2019; XiuMing etal. 2021). However, all three varieties I studied responded very well in the same medium.

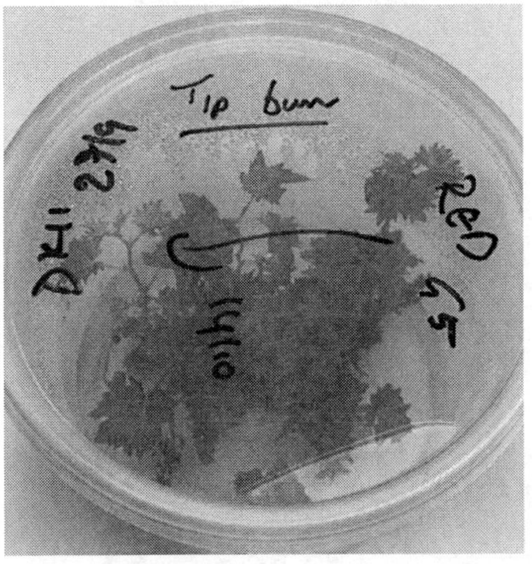

Figure 2. Shoot multiplication

Rooting

All three varieties of grapevine rooted easily on the same i. e. ½ strength MS medium supplemented with 15 g/l sucrose, 0.5 mg/l indole-3-butyric acid, 3 mg/l Thiamine and 6.0 g/l agar, pH adjusted to 6.0 before autoclaving (Fig. 3A & B). It has been reported that different grapevine varieties require their own unique basal media supplemented with different plant growth regulators (Yancheva et al. 2018, Carimi et al. 2019; Xiu Ming et al. 2021). However, all the three grapevine varieties in our experiment rooted efficiently, and roots proliferated in the same medium.

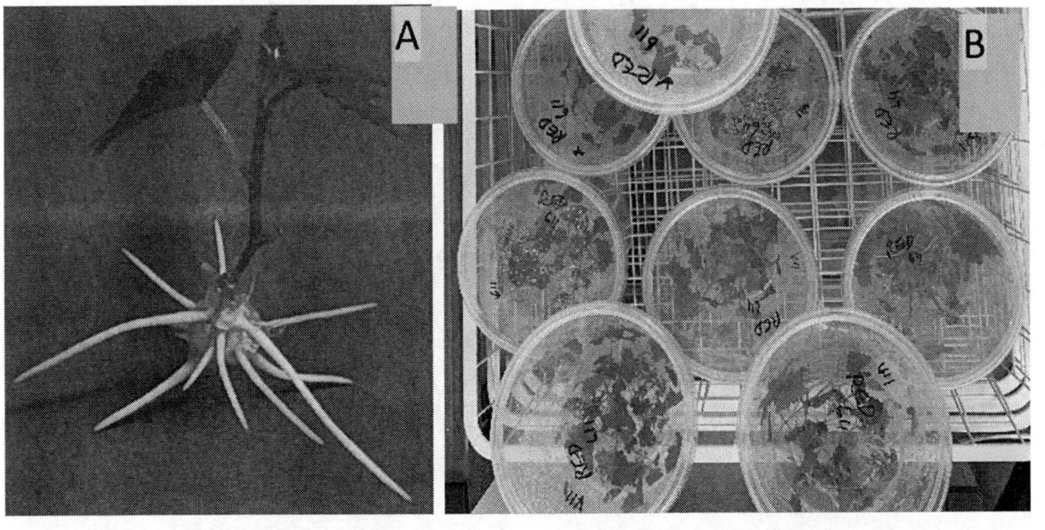

Figure 3. Rooting of microshoots of grapevine. **A**) after 2 weeks, **B**) after 4 weeks.

Acclimation

Well rooted grapevine plantlets when transplanted to porous potting mix (Searles) and maintained in a climate-controlled greenhouse acclimated at high frequency (90-95% survival) and the primary hardening only required 3 weeks (**Fig. 4**). The acclimated plants established and developed further in the net house with 50% shade (**Fig. 5**).

Figure 4. Primary hardening in the climate-controlled greenhouse

Figure 5. Secondary hardening of tissue cultured grapevine in the net house with 50% shade.

They sun-hardened efficiently in a fully open net house and survived field planting. Once planted in the farm with a supporting shade, 100% of them established and produced fruit as normal grafted grapevine (**Fig. 6**). Acclimation of tissue culture plants is complicated by the change in environment, poor development of root-shoot juncture, delicate leaves with poorly developed cuticle and stomata. Therefore, ambient high humidity and moderate temperature are required for acclimation of micropropagated plants. The high humidity and low temperature provided at primary hardening in the climate-controlled greenhouse may have helped to achieve high rates of acclimation of tissue cultured grapes.

Figure 6. Field testing of micropropagated grapevine. **A**) on planting day, **B**) after 30 days of planting, **C**) three months after planting and **D**) maturing red grapes.

Applications of tissue culture techniques for grapes

Somatic embryogenesis can be used for rapid cloning of grapevine varieties and, for selecting non-genetically modified grapevine with resistance to fungal pathogens (Li et al. 2014). Somatic embryogenesis was also used for regenerating genetically modified grapevine with improved characteristics (Dhekney et al. 2016). Cryotherapy and thermotherapy were applied to in vitro shoots of infected plants for virus elimination and rapid cloning of virus-free plant material (Pathirana et al. 2015). Long term storage of several accessions of grapevine germplasm in vitro has been achieved and maintained under cryopreservation (Bettoni et al. 2021). Accelerated mutation breeding of grapevine was also achieved using tissue culture technology (Pathirana and Carimi, 2023). Gene editing was successfully used to make disease resistant grapevine (Wang et al. 2018) and grapevine with Muscat flavour (Yang et al. 2024).

CONCLUSION

Micropropagation of the three grapevine varieties studied was easy and behaved similarly in this experiment. All the three varieties initiated, multiplied and rooted in the same culture medium and with similar efficiency.

Acknowledgements

I acknowledge the support (facilities and finance) for the research by Skybury Farms, 136 Ivicevic Rd, QLD 4880 and the able technical assistance of Mrs Jay Pui Lam.

LITERATURE CITED

Bettoni, J.C. et al. (2021). Grapevine shoot tip cryopreservation and cryotherapy: Secure storage of disease-free plants. Plants *10*:2190. https://doi.org/10.3390/plants10102190

Dhekney, S.A. Li, Z.T., Grant, T.N.L. and Gray, D.J. (2016). Somatic embryogenesis and genetic modification of *Vitis*. In: Germana, M., Lambardi, M. (eds) In Vitro Embryogenesis in Higher Plants. Methods in Molecular Biology *1359*. Humana Press, New York. https://doi.org/10.1007/978-1-4939-3061-6_11

Li, Z.T., Kim, K., Dhekney, S.A., Jasinski, J.R. Creech, M.R. and Gray, D.J.. (2014). An optimized procedure for plant recovery from somatic embryos significantly facilitates the genetic improvement of *Vitis*. Horticulture Research. *24*:1. https://doi.org/10.1038/hortres.2014.27

Murashige, T. and Skoog, F. (1962) A revised medium for rapid growth and bioassays with tobacco tissue cultures. Physiol. Plant. *15*:473-497. http://dx.doi.org/10.1111/j.1399-3054.1962.tb08052.x

Pathirana, McLachlan, A., Hedderley, D., Carra, A., Carimi, F. and Panis, B. (2015). Removal of leafroll viruses from infected grapevine plants by droplet vitrification. Acta Hort. *1083*, ISHS 2015. https://doi.org/10.17660/Acta-Hortic.2015.1083.64

Pathirana, R and Carimi, F. (2023). Studies on improving the efficiency of somatic embryogenesis in grapevine (*Vitis vinifera* L.) and optimising ethyl methanesulfonate treatment for mutation induction. Plants *12*: 4126. DOI: 10.3390/plants12244126

Puthiyaparambi, J.C., Pagie, M., Teressita, S., Jay, P.M., Bongani, J.N., Paul, F., Candy, M., Mark, M., Farion, M., Ian, M, Sanskruti, L., and Nitin, M. (2023) Rapid papaya crop improvement through accelerated in vitro breeding and molecular diagnostics. Acta Hort. *1362*:499-506. https://doi.org/10.17660/Acta-Hortic.2023.1362.67

Xiu-ming, Z, Wu, Y., Li, Z., Song, C., and Wang, X. (2021). Advancements in plant regeneration and genetic transformation of grapevine (*Vitis* spp.). Journal of Integrative Agriculture *20*:1407–1434.

Wang, X., Tu, Ml, Wang, D., Liu, J., Li, Z., Wang, Y., and Wang, X. (2018). CRISPR/Cas9-mediated efficient targeted mutagenesis in grape in the first generation. Plant Biotechnol J. *16*: 844-85524. https://doi.org/10.111/pbi.12832

Yancheva, S., Marchev, P., Yaneva, V., Roichev, V., and Tsvetkov, I. (2018). In vitro propagation of grape cultivars and rootstocks for production of pre-basic planting material. Bulgarian Journal of Agricultural Science *24*:801–806.

Yang, Y., Wheatley, M., Meakem, V., Galarneau, E., Gutierrez, B., and Zhong, G. (2024). Editing VvDXS1 for the creation of muscat flavour in *Vitis vinifera* cv. Scarlet Royal. Plant Biotechnol J. 22(6):1610-1621. DOI: 10.1111/pbi.14290

Combined Proceedings IPPS 74:152-161. 2024.

Understanding Evolutionary Biology of Lavender for Successful Nursery Production

Clive Larkman

Larkman Nurseries, 7 Jurat Rd, Lilydale VIC 3140, Australia

clive@larkmannurseries.com.au

Keywords: *Lavandula*, physiology, origin, adaptation

Summary

For success, operators of plant production nurseries need to understand not only the morphology, but also the physiology, evolutionary biology and genetics of the plants they grow. This aspect is discussed in relation to flowering and production of lavender flowers and oil. The main cultivated species are *Lavandula angustifolia*, *Lavandula latifolia* and their naturally occurring hybrid *Lavandula x intermedia*. These species originated in the northern Mediterranean in southern France under a dry temperate climate with drought conditions in the summer. Soils are low in nutrients except for high levels of calcium. How the understanding of evolutionary biology of lavender and its adaptation can help increase lavender flower production in nurseries is discussed with examples from author's own experience as well as from a field experiment.

IPPS Vol. 74 – 2024

INTRODUCTION

I have been involved in a wide range of industries since my teenage years. The majority have been micro to small enterprises and from entry level employee to owner and managing director. It is well accepted that to make anything a success takes hard work but what is often overlooked is that it also takes understanding, wisdom and clear thinking.

There have been millions of reviews as to why some organisations fail and others succeed. The key ingredient is the ability to analyse, understand and interpret the basic principles of the main component of the business. This is independent of industry, industry sector, product or service. It is most important for small, new and emerging industries – like lavender.

Lavender produces three major products – fresh flowers, stripped and dried flowers, and oil. The demand for oil and stripped and dried flowers is building within Australia, and shortages are regular occurrences. There is a worldwide demand for lavender oil and stripped lavender flowers which can be counter seasonal to Australian production. Australia currently imports large quantities of both products creating a market opportunity for both export and import replacement.

As mentioned above the path to success in small business is the base level of understanding. In horticulture this means that a solid understanding of the biology of the key plants is critical but often overlooked. Not just the morphology but the physiology, the genetics and the evolutionary history. All add up to give a connection to the plant.

TAXONOMY, BOTANY AND PHYSIOLOGY OF *LAVANDULA*

The genus *Lavandula* belongs to the Lamiaceae (or Labiatae) family commonly referred to as the mint family. Lamiaceae is a member of the Division Magnoliophyta, Class Magnoliopsida and the Order Lamiales. The family consists of more than 236 genera and has been stated to contain 6900 to 7200 species (Xu and Chang 2017). These are from all parts of the world with the greatest number occuring in the Mediterranean.

Lamiaceae includes plants that are of some economic importance around the world as sources of aromatic oils and for their culinary uses. The majority are aromatic shrubs. The stems are quadrangular with simple, opposite or occasionally whorled leaves. The main genera are; *Teucrium, Salvia, Mentha, Origanum, Thymus, Rosmarinus* and in Australia *Westringia* (the native Rosemary) and *Prostanthera* (the native Mint Bush) (Upson and Andrews, 2004). Some species of Lamiaceae are shown in **Fig 1**.

As an evolutionary botanist I have a passion for why plants grow where they do. I often say that 'plants grow in the area of greatest adversity'. This means that many plants grow in the harshest conditions they can tolerate better than other plants. The other key concept for plant growth is that we should treat all plant species as a smart human with one purpose; to produce strong viable offspring and this is best done if they can control when and where their seed germinates. The aim of each plant is to produce viable seed that germinates when the local environment is most conducive to successful growth.

Figure 1. Some species of Lamiaceae. **A)** *Thymus longicaulis,* **B)** *Salvia officinalis* and **C)** *Salvia rosmarinus* [syn: *Rosmarinus officinalis*].

Taking these two concepts on board, it is possible to understand the best conditions for plant growth and seed production. Even asexual propagation is influenced by these parameters. It involves timing that is directly related to the movement of nutrients, water and hormones around the plant. Also timing so that cuttings are not taken when a plant really wants to flower. It is this solid understanding of plant physiology and plant evolution that is critical to developing the highest yielding plants. Indeed it is the best option to ensure plant survival in less than ideal conditions. Like humans, plants need a program of good conditions, good food and the right inputs to perform at their best. When well fed they are better able to grow through poor conditions and less susceptible to pests and disease.

The genus *Lavandula* is divided in to three subgenera: *Lavandula, Fabricia* and *Sabaudia*. These are in turn divided into eight sections. Some are large and well known and some are small and rarely seen. Subgenus *Lavandula* is of significant economic value to the horticultural industries. It includes the three sections that produce all the commercial oil forms, all the cut flower forms and most of the ornamental and pot varieties i.e. Sections Lavandula, Dentatae and Stoechas.

The section with real commerical concern is *Lavandula* and it is by far the most important. All the forms have narrow, lanceolate leaves with entire margins. The flowerheads are borne atop long slender peduncles. Most of them have small florets arranged in whorls in long flower spikes with oil sacks at the base of each floret. These oil sacks are why lavender is grown in such large quantities and it is the composition of that oil that determines the value and use of the plant.

EVOLUTIONARY BIOLOGY OF *LAVANDULA*

The main cultivated species are *Lavandula angustifolia*, *Lavandula latifolia* and their naturally occurring hybrid *Lavandula x intermedia* (**Fig. 2**). With thousands of years of cultivation it is surprising there is so little known about the biology of these species. They come from a small geographical range on the northern side of the Mediterranean in southern France. The only geographic variance between the three is the altitude they prefer to grow at and thus their preferred temperature range. The natural growing region is the area called Provence on the Italian/French border along the Mediterranean sea (**Fig. 3**). The region runs from the sea to the Haute Alps in the north along the Rhine river in the west and the Italian border in the east. High points are Mt Ventoux at 1900m in the Sault region and the Maritime Alps at 3500m (Demasi et al. 2021).

Figure 2. Three common Australian cultivars of the three *Lavandula* species: **A**) *Lavandula latifolia,* **B**) and **C**) *Lavandula angustifolia*, and **D**) and **E**) their naturally occurring hybrid *Lavandula x intermedia*.

Figure 3. Geography of origin of cultivated lavender, showing Provence, the historical province and geographical region in southeastern France that is almost the same as the modern administrative region of Provence-Alpes-Côte d'Azur (PACA). PACA is also known as Région Sud, which means "Southern Region".

The climate here is dry temperate with drought conditions over the summer period. The winters are generally mild and there is lots of intense sun – especially during summer and autumn. In a rough generalistion *L. angustifolia* comes from above 1000 m, *L.latifolia* comes from below 1000m and *L. x intermedia* from a band around the 1000 m contour. Although the lines are quite distinct in Provence the plants will all grow under similar climatic conditions elsewhere.

The environment is referred to as a Mediterranean Climate and can be summarised as hot dry summer days, cool dry summer nights, wet winters in low altitudes and heavy winter snows in higher altitudes. The summer dry goes from mid spring until late autumn and although there maybe some heavy rains over the period it would never be called wet at anytime. The summer day length is long – around 15 hours in mid summer. The winter low temperatures are well below zero in the upper valleys but milder in the lower valleys and along the coast. The mean temperature and precipitation in this region is given in **Fig. 4**.

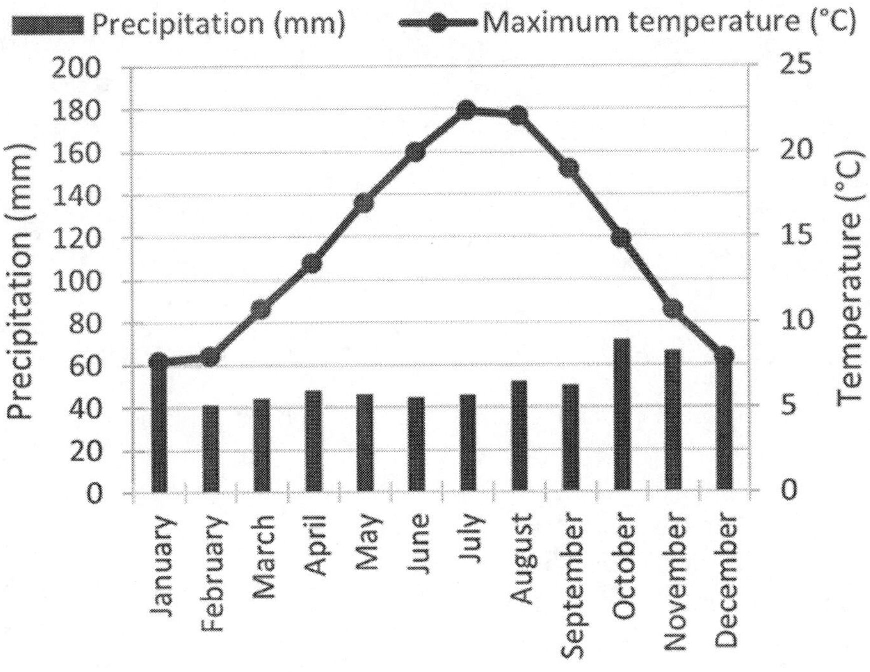

Figure 4. The mean temperature and precipitation in the in Provence region in southern France where cultivated lavender originated,.

The high mountains adjacent to the sea with large interior plains to the north create the perfect conditions for regular, strong dry winds. In this part of France they are called the Mistral and basically blow cold dry air from north to the south. They help to keep the humidity low and the sky clear. They are at their strongest in late winter and early spring which is when most temperate regions have the highest humidity. The winds can clear a cloudy, overcast sky in a matter of hours and leave clear sunny skies. It is thought that this region can have up to around 3000 hours of intense sun per annum. This added sunshine and dry air brought by the Mistral have an important effect on increasing the dryness. The light volumes and dry air have created a unique environment that has given lavender, rosemary and similar plants a world of their own.

As light and rainfall are critical for plant growth so is the substrate in which they set roots. The geology of Provence is unique and sometimes it is hard to understand how anything grows there. As mentioned above the evolution of plants is an interesting science and can be quite convoluted. The physiology of lavender (and other Provencal plants) has its origins in the Jurassic Period when the super continent Pangea started to split up. Over 150 million years the earth went from one super continent to the seven continents we know today **(Fig. 5**). The mountains of

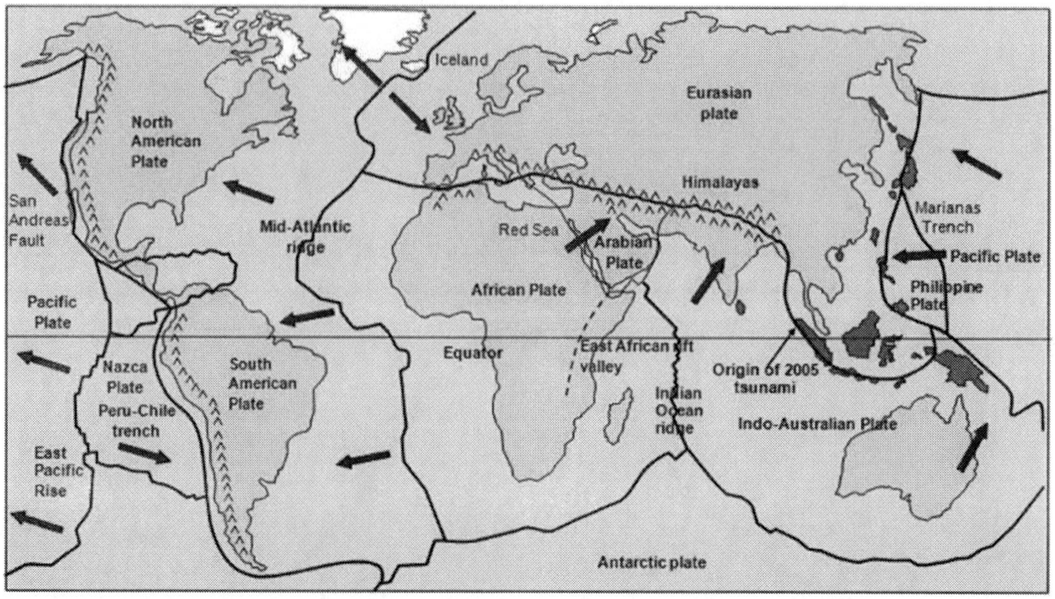

Figure 5. Formation of continents and forces that drifted continental plates.

Provence were created from several 'crashes' of land from Africa into Southern Europe. They forced up land that had been underwater for tens of millions of years and was rich in calcaerous fossils. This created the calcium rich mountains of the region.

The geology of the hills and valleys of Provence is the complete antithesis of what we feel is good growing soil. It is basically straight limestone with an overlay of more limestone. The substrate is solid rock and the topsoil is very gravelly. The type of gravel varies across the region and for some plants the variations are important (e.g. grapevine) whereas for others they have little effect. The gravels range across sandstone, shale, gypsum, lime and some clay – none of which are high in nutrients. All except the clay are very well draining and add to the dry growing conditions. The ground lavender grows in is basically the white calcareous gravel which reflects the intense light back up on the plants. This situation increases the dryness and produces even higher light volumes meaning even more stress on the local plants.

Understanding the geography and geology of lavender details the world this specialised plant has adapted to. In summary, it is an environment of very high light levels, dry air and soil, defined seasons, low soil nutrition and extremely high Calcium levels. As stated above plants evolve to live in the area of greatest adversity. This means they don't prefer these conditions in which they exist but can deal with them better than other plants.

HOW CAN WE APPLY THE KNOWLEDGE TO GROW A SUCCESSFUL LAVENDAR CROP?

To develop the best lavender farm the grower needs to address each of the issues discussed above in a balanced approach. Some need to be replicated (e.g. sun, dryness) and others need to be adjusted for (e.g. Calcium, low nutrients) so as to get optimum growth – and yield. These two approaches are not the same and in some cases are actually contradictive. For best yields per acre it is imperative to understand all the inputs for good growth and how they affect yield.

For decades it was assumed that lavender needed high pH to grow. In reality it needs very high levels of Calcium with a period of wet soils and dry air. Without these it will have low Calcium levels in the plants meaning weak cell walls and thus subject to fungal diseases.Also, calcium exhibits a dual function, both as a structural component of cell walls and membranes and as intracellular messenger. Once mature, lavender needs minimal levels of fertiliser. If oil is the end product and all the discharge from the still is put back on the farm; the levels are in the 100s of grams per acre.

Another key input for healthy and productive plants is light. Unlike humans that need light for vitamin production and psychological health, plants need it for their very existence. There are very few plants that have been able to adapt to very low or zero light. Both plants and animals respire or breathe and expel CO_2 and water, but only plants photosynthesise to absorb CO_2 and produce O_2. It is this balance between the production of CO_2 through respiration and the absorption by photosynthesis that keeps the earth in sync.

Light has the greatest influence on when and how a plant grows. Too little and a plant will "stretch" to get more, too much and the plant will "burn". Unlike every other input into plant growth, light is not simply there or not there. Granted there is the blunt observation that a plant needs light and without it there is no growth. However, there is more to a plant than just growing or not growing. There are a range of growth styles and these have a range of controlling inputs. To start with there is light and then there is light strength, light periods and light volume. Each of these have a major effect on the plant growth cycle depending on the plant and where it comes from. Before looking at how light controls lavender growth we need to understand the parameters that define how light determines plant behaviour. In summary there are three major criteria, the length of daylight, the volume of light received by a plant each day and finally the intensity or strength of that light.

The length of daylight or photoperiod has its own set of criteria that are complex and affect various components of the plant's growth cycle. These are actual daylength – i.e. the number of hours in the day that the sun is 'shining' compared with the number of hours of 'darkness'. The ratio of day to night, termed as short or long daylength and finally the changes in daylength. This can be referred to as increasing or decreasing daylength. Some plants respond to the relative daylength or the changes in daylength. The options are complex and varied.

Then there is the strength and volume of the light. For some plants a certain intensity of light is required for growth or flowering whereas others require a minimum number of light units per day. This can occur in two hours or ten - independent of the daylength. Two hours of clear skies may yield the same volume of light as ten hours of partial cloud cover. This is what controls flowering in lavender. Plants flower earlier and better when the spring skies are clear compared with an overcast spring. Lavender grows in a region with distinct seasons with definite changes in daylength and high light levels – when the sun is shining.

There are also distinct humidity variations between seasons and regular temperature ranges. As mentioned above the mountains and valleys of Provence are subject to the strong early spring winds. They do two things; they create clear blue skies in early spring and also blow away any residual humidity from the wet winter and the melting snows.

TESTING OF OUR UNDERTANDING OF *LAVANDULA* EVOLUTIONARY BIOLOGY IN A FIELD TRIAL

In 2019-2021 we undertook some research in conjunction with Latrobe University. The aim was to analyse the application of different fertilisers in conjunction with the changing seasons. This was done at a farm with poor soils northwest of Melbourne, Victoria, Australia. We were also collecting and growing plants at the University to trial some of the fertilisers in a controlled environment. There was a range of plants across the three groups mentioned above. Half in a polyhouse with 100% environmental control and half in a shade house with natural humidity levels. There were very similar light levels in the two spaces.

We observed the plants in the high humidity were just entering bud whilst those in the low humidity shade house were in full flower. This was strange as the expectation would have been for the plants in the poly house, with a small amount of extra light would flower first. What was more surprising was that as soon as the plants were moved into the shade house they came into flower. The question is why? The assumption by most people is that lavender flowering is initiated by daylength or photoperiod. Problem is that if this was the case the flowering period for a lavender farm would be almost identical every year and independent of cloud level. For a given latitude/longitude the day length cannot vary. The light intensity may vary causing the plant to have small changes in flowering time. Also, we often observe that sometimes lavender flowers at completely odd times after potting. We have some plants in quarantine that are budding up now in May – probably due to them coming from the northern hemisphere in March. However, as mentioned earlier the key to plant success is to think about the plant and where it originates. As a species these plants know that the best time for the young seedlings to start life is early autumn when the summer heat breaks and the autumn rains come. The seed takes three months to ripen after the flowers being on the plant for one month. The flowers need three months to bud and grow. This means that the young buds need to initiate seven months prior to the rains which is the end of winter. At that time the air is suddenly becoming dry and the sky is clear. The plant knows that when the air dries out it is time for the flowering organs to start forming then when sufficient light is accumulated the buds can form and the flowering stems can grow. This results in seeds falling in late summer and germinating in early autumn.

This is conjecture and the laboratory experiments need to be conducted to prove or disprove it. However, assuming it has some substance it shows how important it is to understand where a plant comes from, what the environment does in that region and how you can replicate it to control flowering (or not to prevent flowering) and plant growth. The concept is understood and managed with some crops like poinsettias and is part of the breeding for other plants like strawberries but not used in the production of most plants.

The message here is that this understanding of plant evolution can have a significant impact on the bottom line and should be part of our thinking when planning plant production.

LITERATURE CITED

Demasi, S., Caser, M., Lonati, M., Gaino, W. and Scariot, V. (2021). Ornamental traits of *Lavandula angustifolia* Mill. are affected by geographical origin and cultivation substrate composition. Acta Hortic. 1331: 49-56. doi: 10.17660/ActaHortic.2021.1331.6

Upson, T., and Andrews, S. (2004). The genus *Lavendula.* Botanical Magazine Monograph. Royal Botanic Gardens, Kew, UK.

Xu, Z. and Chang, L. (2017). Lamiaceae. In: Xu,. Z. and Chang, L. (eds) Identification and Control of Common Weeds: Volume 3. Springer Singapore, Singapore, pp 181-265. doi: 10.1007/978-981-10-5403-7

Combined Proceedings IPPS 74:162-170. 2024.

Red Imported Fire Ant – The Multi-Million Dollar Impact on Nursery Production

John McDonald

Director RDE and Biosecurity, Greenlife Industry Australia, Unit 1/250 Orange Grove Rd., SALISBURY QLD 4107, Australia.

john.mcdonald@greenlifeindustry.com.au

Keywords: Insect control, eradication, invasive species

Summary

Red Imported Fire Ant (RIFA) was first detected in South East Queensland (SEQ), Australia, in February 2001 at a property in the suburb of Richlands, Brisbane. Subsequent tracing and investigation identified further outbreaks at the Port of Brisbane and Verrierdale in the Sunshine Coast Hinterland in 2001. The Queensland Government, along with Australian State, Territory and Commonwealth Government, agreed to a national RIFA eradication program to be cost shared between government Parties; however, a cohesive structured funding program was not fully agreed to until 2017.

Over the first 16 years of the program the Queensland Government was forced to secure annually a commitment to cost share funding, from each government Party, totaling $367 million over that period. In 2017 a 10-year plan was agreed by Parties, from 2018 onwards valued at $411.4 million, however in 2023 a revised total of $592.8 million over 5 years to 2027 was agreed (Scott-Orr et al. 2021).

As one of the most invasive species known, it is critical that Australia eradicates RIFA to ensure our way of life for future generations is unaltered and that our environment is protected. RIFA is a pest of the environment, more so than horticulture, where if established it dominates the invertebrate world out competing or killing native insects and small animals. The ant has been known to 'farm' plant pests such as aphids within cropping systems which add to growers' pest pressures and further enhances worker exposure and subsequent health impacts from stings. Estimates put the annual heath impacts to the Australian community at a third being stung of which 25% will develop an allergic reaction, and between 43,000 to 174,000 people in Australia will require medical assistance, annually.

The production nurseries in SEQ have been battling RIFA since 2001 with many having been in the fight for 24 years, others are only now being impacted as the incursion expands in size. The costs associated with the pest are high, however, vary due to market access, treatment, site management, lost markets, etc., costing the sector in SEQ more than $20 million annually. These are costs that are unlikely to be recovered due to the highly competitive plant production sector and are absorbed by growers as a 'cost of doing business'. When governments fail to protect our borders and manage pest incursions it is industry that must bear the cost, often in perpetuity.

The author of this paper has been extensively involved in the Australian RIFA response since inception, first as the Queensland Industry Development Manager (1996 to 2015) and subsequently as the Greenlife Industry Australia National Biosecurity Manager/Director RDE and Biosecurity (2016 – to date).

INTRODUCTION

Red Imported Fire Ant (RIFA) (*Solenopsis invicta*) is a highly invasive species of ant native to South America (Argentina, Brazil, Paraguay, Uruguay) having spread to the United States of America, Taiwan, China, Japan, Philippines and Australia (NFAEP, 2024). The ant is highly adaptable, omnivorous (eats plants and animals), aggressive possessing a stinger, similar to a bee, swarms out of nests, and has serious impacts on economies, environments and human health (Scott-Orr et al., 2021). RIFA is not a specific 'pest' of horticulture, in that it does not attack crops like usual plant pest insects, however the biology of the ant leads to having significant impacts on infrastructure such as vehicles, equipment, electrical systems (equipment and transmission), staff due to stings/hospitalisation, and market access to prevent the spread. The pest has significant impacts on the ecology of any area it infests due to the above characteristics and will decimate native vertebrates and invertebrates as they either feed on them, their young or out compete for food sources (Invasive Species Council 2024).

RIFA was first detected in Southeast Queensland (SEQ), Australia, in February 2001 at a property in the suburb of Richlands, Brisbane. However, it is estimated the incursion originated around 1992 (Scott-Orr et al. 2021), with some suggesting an even earlier incursion in the 1980's. Subsequent tracing and investigation identified further outbreaks at the Port of Brisbane and Verrierdale in the Sunshine Coast Hinterland later in 2001. The Queensland nursery industry, through the Nursery & Garden Industry Queensland (NGIQ) and the national peak industry body (Nursery & Garden Industry Australia now Greenlife Industry Australia (GIA)), have been heavily invested in eradication from the outset of the detection, response and planning from 2001 onwards. These bodies supported the deploying of resources into the response with staff sitting on various committees, forums and working groups to contribute to the eradication of RIFA. These industry resources also worked with government to establish risk mitigation measures growers could apply, market access protocols allowing growers to trade, developed and conducted RIFA information workshops and coordinated industry research and information.

The RIFA infested zone in SEQ **(Fig. 1)** has increased from 40,000 ha in 2005 to more than 800,000 ha in 2024 with further RIFA detections, in 2024, at two locations in Northern New South Wales (South Murwillumbah and East Wardell) and in South West Queensland (Oakey and Meringandan West).

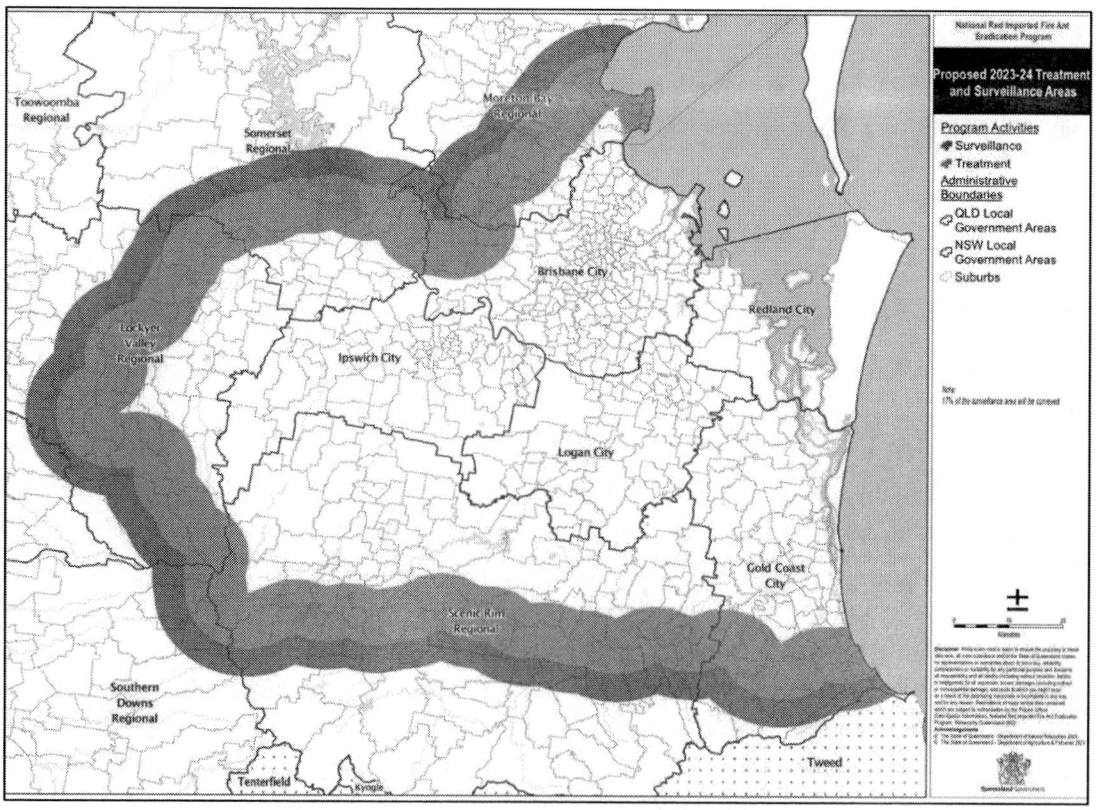

Figure 1. SEQ eradication surveillance and treatment zones for RIFA 23/24. Source: National Fire Ant Eradication Program (2024).

An independent review of the eradication program in 2021 estimated an annual eradication cost between $200 to $300 million per year is required to achieve success by 2032 ($2 to $3 billion). The Australian Institute advising in 2024 the actual cost to the economy without eradication is likely to be more than $22 billion, a public cost benefit of between $3 and $9 per dollar spent on eradication.

In 2006 the Nursery & Garden Industry Queensland surveyed growers in SEQ and established annual RIFA costs to industry (lost markets, regulatory compliance, risk mitigation, etc.) at that time exceeded $9 million per year however with the significant increase in area, more growers impacted, this cost per annum has grown exponentially now estimated at more than $20 million per annum in 2024.

Human health impacts of RIFA infestation are also severe. Approximately a third of the population in RIFA-infested areas are stung each year with about 20% causing a large local reaction and another 0.5% and 2% stings causing systemic allergic reactions which can range from skin symptoms to life-threatening anaphylaxis (Lopez et al., 2024).

THE PEST

There are two genotypes, monogyne and polygyne, with this referencing the cohabitation, or not, of RIFA colonies. A monogyne RIFA Queen and her colony will defend their territory against all comers, including other RIFA, whereas a polygyne RIFA Queen and her offspring will join with other RIFA Queens in super colonies of tens of millions of ants and hundreds of Queens. Disturbing these 'super' colonies can see small animals (e.g., sheep, lambs/calves) killed as they are swarmed over by thousands of RIFA. Queens typically fly within 5 km, some have been reported at 30 km in favourable wind conditions, from launch point however more commonly they appear to land at around 300m - 500m (NFAEP 2024).

RIFA are generally between 2 to 6mm in length (sterile female workers) with Queens approximately 10mm with a nest containing a great diversity of sizes and hundreds of thousands of ants. The ant is a coppery/reddish-brown colour with a black-dark brown abdomen (NFAEP 2024).

Nests can be a small 'bump' in the ground **(Fig. 2)** or a larger dome with both having no obvious entry/exit point as nests can have many dispersed entry/exit holes meters from the 'bump'. The size of nests is often associated with the age of the colony. It has been reported that when left unchecked a population of monogyne RIFA can have a nest density of 500/ha and polygyne populations at 5,000/ha (Scott-Orr et al. 2021). The dominant genotype in SEQ is the monogyne nest which are known to be smaller, less populated, yet travel further in mating flights (NFAEP 2024).

The Queen is the only reproductive member of her colony, with workers being sterile females, while production of alates (fertile winged males/females) are controlled by the Queen and periodically launch from nests (flying ants) to mate in the air from members of other colonies. New Queens land and can form a new colony as no further fertilisation is required across her 7-year life expectancy, male alates die after mating. The new Queen drops her wings upon landing and must find a suitable nesting site within hours otherwise she will die. It is believed that between 90% and 99% of these mated alates do not go on to establish nests/colonies due to high predation rates, missed matings, unsuitable nesting sites, etc., (NFAEP, 2024).

The accepted norm for RIFA nesting sites are those areas that are least travelled, near water and in a more open setting such as agriculture fields, along fence lines, on the verges of roadways, water drains, dams, creeks/rivers/streams, etc. The ant does not enjoy a closed environment such as a forest or dense bushland however will likely establish in a clearing. Nests have been found under concrete slabs such as paths, but also large parking slabs where the ant accesses underneath via cracks, etc., on machinery with large soil build up and in electrical equipment such as ground level junction boxes, switch boxes, runway lights, etc., where they carry soil into these units (Scott-Orr et al. 2021).

Figure 2. (A) Red Imported Fire Ant (RIFA) nest above ground, (B) an alate (circled) with workers, and (C) a worker RIFA.

THE INDUSTRY

The nursery industry was the first horticultural industry significantly impacted due to the nature of the pest, past experience in other countries and the location of growers within the RIFA biosecurity zones established by Biosecurity Queensland. Many growers in the early days of the response to RIFA lost key customers in southern states, afraid they may get RIFA in consignments from growers in SEQ. Over the past 24 years there has not been a recorded movement of a RIFA nest in nursery stock from a professional production nursery, a testimony to the professionalism and diligence growers in SEQ have applied to meeting their biosecurity obligations.

The industry impact however has been significant with 20 plus years of a failed eradication program noting that the NFAEP claim great success in 'preventing the spread' of RIFA due to the ant not having spread as far as north of Mackay, south of Sydney and west of Cunnamulla **(Fig. 3)**. In fact, the program has been a significant failure, as noted above, expanding from 40,000 ha in 2005 to more than 800,000 ha in 2024, a 1,900% increase in infested area. During this period (2005 to 2023) the production nurseries in SEQ have had to manage this pest as the infested zone grew in diameter, collecting more and more growers, plus the increasing ant densities elevating the risk of infestation due to the failed eradication.

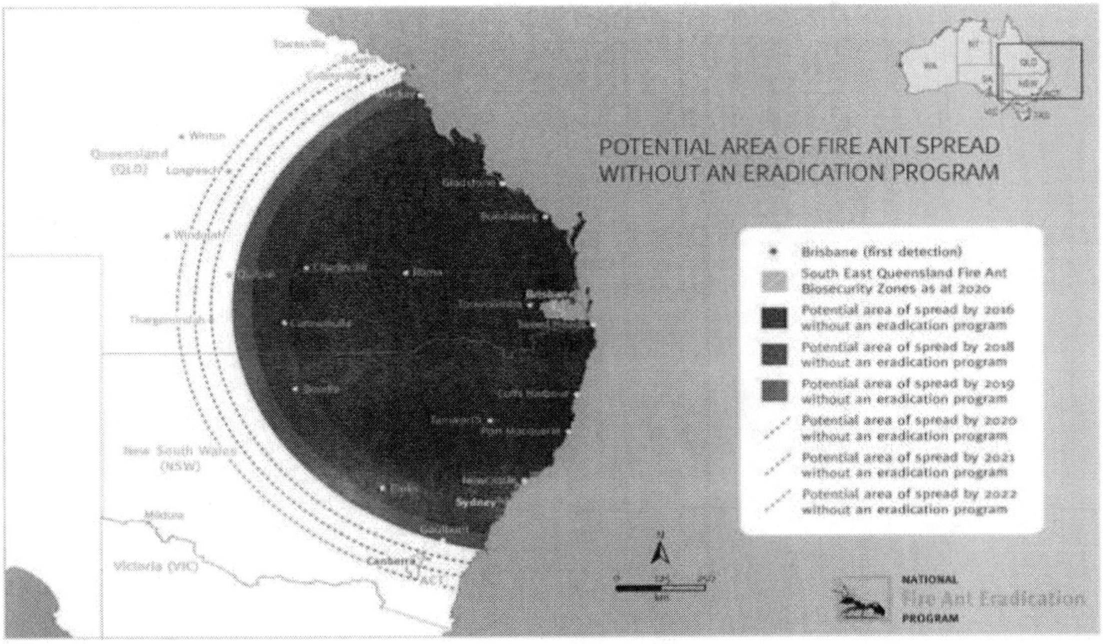

Figure 3. Potential RIFA spread based on NFAEP modelling without program. Source: National Fire Ant Eradication Program (2024).

It is important to note that the fundamental driver for the failure of the eradication program has been due to all state and commonwealth governments lack of action, priority setting, commitment and recognition of the seriousness of risk to Australia's way of life and our environment **(Fig. 4)**. To not have an agreed comprehensive national eradication plan in place until 18 years into the detection of the RIFA incursion is telling of the massive failure in leadership. Of concern is that these same entities have been given clear advice on what is required to eradicate RIFA from SEQ, and the same response is delivered, with a \$500+ million 5-year funding plan when the experts have said it will require this on an annual basis for at least 5years.

The production nursery sector in SEQ have had to bear the brunt of these poor funding and eradication decisions made by government and importantly, as this is a pest of the environment, it does not sit under the cost sharing Emergency Plant Pest Response Deed, nor the National Environmental Biosecurity Response Agreement (NEBRA), which allow governments to fall short of their obligations. The RIFA response is outside of any of the formal Deed arrangements therefore industry has very little say, that is no say at all, in how the response is funded and managed yet the grower impacts are severe and costly.

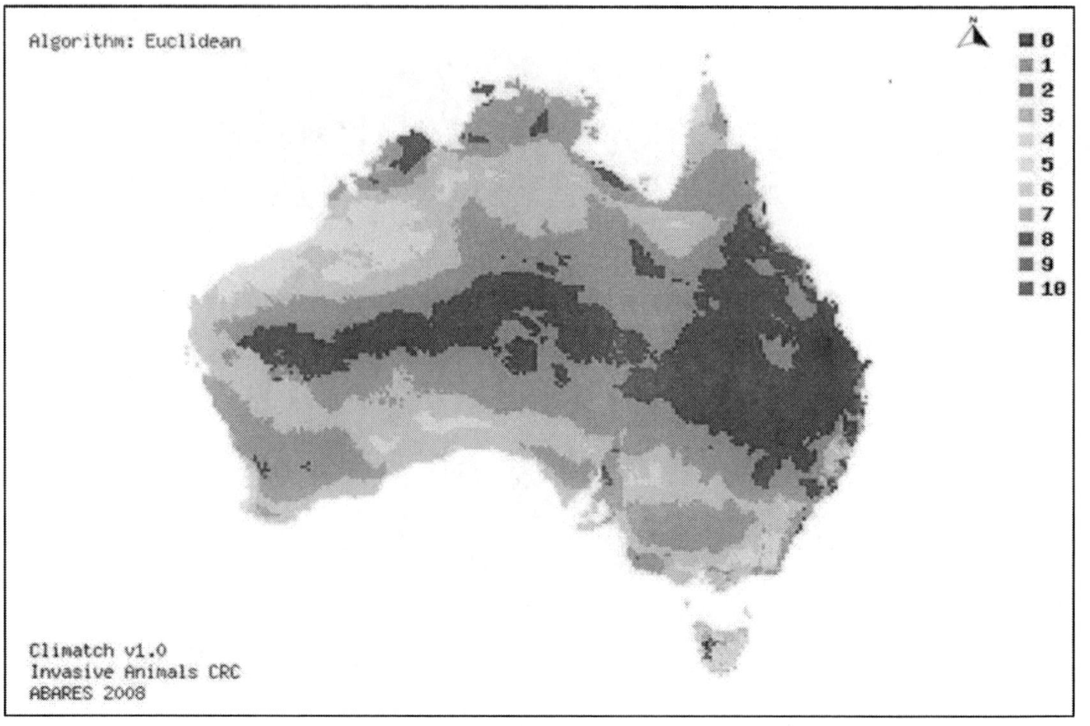

Figure 4. Suitable Red Imported Fire Ant (RIFA) habitat across Australia. Increasing scores represent increased habitat suitability for RIFA: Scores 0-4 (below threshold, or not suitable for RIFA habitat) and scores 5-10 (above threshold, or suitable for RIFA).

The RIFA incursion has, over the past 3 to 5 years expanded at a rapid rate both north and south of the Brisbane CBD. The northern boundary, at this point in time, is just short of Caboolture, with new detections coming almost weekly, and south to Currumbin Waters, 5km from the New South Wales border. This has added hundreds of new production nurseries to the list of Affected businesses that now must meet intra and/or interstate movement conditions. This impact on businesses is exacerbated by the lack of effort put towards suppressing the RIFA infestation within the RIFA Zone.

As the RIFA population within the zone increases in density, noting the eradication program only manages the outer edges of the infested area (**Fig. 1**). Blue and orange zones), the greater the risk to production nurseries of RIFA establishing onsite or nearby which enhances the chance of RIFA infestations. This risk is also exposing the role all businesses in SEQ have in mitigating the risk of moving a RIFA Queen on a consignment of any good that may have been stored in the open and loaded onto interstate/intrastate transport. Unfortunately, those overseeing the eradication program have allocated very little funding to attend to this risk rendering it fundamentally ineffective in reducing overall risk, and industry is paying for this failure.

In 2006 the Nursery & Garden Industry Queensland surveyed growers in SEQ and established that annual RIFA costs to industry (lost markets, regulatory compliance, risk mitigation, etc.) exceeded $9 million per year. With the significant increase in area since then, more growers impacted, this cost per annum has grown exponentially now estimated at more than $20 million per annum in 2024. A significant factor in this increase, aside from more growers impacted, is the changed risk mitigation measure implemented under the Queensland Biosecurity Regulation 2016. The introduced mandatory cover or treat elements of the Regulation have seen growers in SEQ incorporate a granular formulation of the insecticide bifenthrin (e.g., MaxGuard 2G, Superway Bifenthrin, etc.) into growing media to give a maximum protection period of 2 years. The additional cost of incorporating bifenthrin, per m^3 of growing media, is approx. $20 which adds millions of dollars per year to the industry costs of managing RIFA.

MANAGEMENT

Currently in Australia we have the most disconnected RIFA interstate movement conditions since the detection 23 years ago due to regulators adopting poor science, outdated information, and cherry-picking information to suit off-loading regulatory administrative actions, etc. The NFAEP has been a poor source of data and at almost every independent program review has been criticized for the lack of research and data gathered over the almost quarter century of this incursion. Local data can better inform our decision making, assess risk, evaluate mitigation measures, etc., however very little of value exists in Queensland hence we see opinion dominate decision making which is costing the industry millions each year and a failed eradication program.

We have plant biosecurity regulatory institutions that are risk averse, while claiming they meet Australia's 'very low risk' Appropriate Level of Protection (ALOP), however it is obvious they are seeking 'zero risk' and relying on chemicals to achieve it. The 20 years (2001 – 2020) where we applied a 'Systems' Approach' to mitigating the risk of moving RIFA, through Approved Risk Management Plans, focused on property freedom, crop monitoring and dispatch inspections proved a most effective strategy in preventing the movement of RIFA in nursery stock. We can make this claim because there were no infestations/movement in consigned nursery stock recorded across Australia. Importantly, in support of the above claim, industry has produced approximately 5.5 billion plants in SEQ (Australian Horticultural Statistics Handbook 2022) over the past 20 years, moving an estimated 2 billion (35%) interstate (NGIQ unpublished survey 2020) presenting a significant sample size from which to draw, and have confidence in, our above statements.

CONCLUSION

As RIFA continues to spread, and governments cannot lead, we expect to see RIFA move into other Australian jurisdictions as we cannot see how the current program can succeed at a 10th of the funding the experts say is required. This means that we will likely witness RIFA move at a similar rate to that we have seen in Queensland and continue to move further north further along the Queensland coast and south into New South Wales and onwards.

For production nurseries the message is clear, keep RIFA off your site and trade with businesses that are RIFA aware and have their own risk mitigation programs in place. Most production nurseries in SEQ are incorporating bifenthrin granular into the growing media however this does not prevent RIFA Queen's flying onto a palletised consignment such as we saw in

March 2023. Inspecting consignments is an important part of mitigating the risk of moving any plant pest and should be a standard operating procedure for any plant producer.

Property surveillance, crop monitoring and dispatch inspections have been more effective than a simple chemical treatment, particularly over the first 20 years of this incursion, as these were the mitigation measures that saw no RIFA movement in nursery stock from a production nursery. The NFAEP changed the RIFA regulation in 2016 to mandatory chemical treatment to save them an administrative activity (cancelled Approved Risk Management Plans) and placed all their eggs in chemical treatment. Since the removal of the inspection system a plant consignment in 2023 moved a Queen, not a nest, on pallet wrapping of a plant consignment from SEQ detected through the verification inspection upon arrival in Melbourne, Victoria.

There are a significant number of pest management resources available to industry including RIFA specific technical information plus a plant protection program (BioSecure HACCP) that provide all the guidance on surveillance, inspections and crop monitoring. Resources can be found here: https://nurseryproductionfms.com.au/

LITERATURE CITED

Invasive Species Council. (2024). Available at https://invasives.org.au/publications/red-fire-ants/ Accessed 10 July 2024.

Lopez, D., Winkle, K., Wanandy, T., van Nunen, S., Perrett, K. and Lowe, A. (2024). The human health impacts of the Red Imported Fire Ant in the Western Pacific Region context: A narrative review. Tropical Medicine and Infectious Disease *9*:.69. https://doi.org/10.3390/tropicalmed9040069

NFEAP. (2024). National Fire Ant Eradication Program. Available at: https://www.fireants.org.au/ Accessed 10 July 2024.

Ngoc Le, M. and Campbell, R. (2024). Red imported fire ants - the benefits of avoiding a national disaster. The Australian Institute. Available at: https://australiainstitute.org.au/report/red-imported-fire-ants-the-benefits-of-avoiding-a-national-disaster/ Accessed 10 July 2024.

Scott-Orr, H., Gruber, M. and Zacharin, W. (2021). National Red Imported Fire Ant Eradication Program Strategic Review August 2021. Report for the Steering Committee. Department of Agriculture, Fisheries and forestry, Government of Australia. 96 Pp. Available at: https://www.agriculture.gov.au/biosecurity-trade/policy/partnerships/rifa-eradication/strategic-program-review. Downloaded 21 July 2024.

Combined Proceedings IPPS 74:171-175. 2024.

Cleaning Irrigation Systems and Preventing Blockage Problems Coming Back

Gary Murdoch-Brown

Advanced Nutrients Pty Ltd., 13 Hinkler Ct, Brendale QLD 4500, Australia

gmb@advancednutrients.com.au

Keywords: biofilm; aquamate, extracellular polymeric substances, water use efficiency

Summary

Sediments, scale and biofilms cause inefficient function of irrigation pipes, drippers and sprinklers increasing operation costs and lowering production and water use efficiency. Algae, fungi, protozoans or their combinations are responsible for biofilms and are problematic in all production regions. Biofilm fouling is traditionally carried out using oxidising agents, chlorine dioxide, hypochlorite, hydrogen peroxide etc. Most of these products have very high environmental impacts as well as toxicity to humans, water systems and soil and are inefficient in that the biofilms will recur. These are being replaced by modern, environmentally friendly 'green' alternatives that use either physical forces such as hydrodynamics (flushing through high water flows) or the use of substances that are capable of interfering with the matrix structure of biofilms. The former is limited for systems where water is plentiful and inexpensive. The latter group includes biocatalysts (enzymes, phages) and organocatalysts (organic non-enzymes). AquaMate® is a patented organocatalyst that causes destruction of the biofilm matrix by breaking it into simple sugars and flushing out from the system whose repeated use prevents reoccurring of biofilm and is low-cost, nontoxic and non-hazardous.

INTRODUCTION

It is well documented, that irrigation lines, drippers and sprinklers may operate inefficiently when fouling occurs due to scale, sediment and/or biofilms (Kanarek 2023). Inefficient irrigation leads to lower production, higher operating costs and lower Water Use Efficiency (WUE).

Sediment removal is a function of flocculation/filtration and not covered in this paper. Scale is caused when metallic ions are oxidised to inorganic salts and accumulate in-situ. Iron and Calcium are the most prevalent though can be removed using strong acids. This will have significant Occupational Health, Safety and Environment (OHSE) consequences due to the hazardous and dangerous nature of strong acids. Avoidance can be purchased through water treatment; pH correction, EC reduction and/or chelation of ions.

Biofilms are the key focus of this paper. Extracellular polymeric substances (EPS); glycocalyx (exopolysaccharide); biofilms; slime. These may be green, red, brown or black depending on the microorganisms forming the biofilm.

Biofilms In Irrigation Systems

Biofilms may be formed by bacteria, algae, fungi, protozoans or combinations thereof. Common phyla in irrigation systems include *Proteobacteria, Actinobacteria, Chloroflexi, Mycobacterium* etc. (Kanarek 2023). Biofilm producing organisms are endemic to all production regions, they will typically proliferate in water with higher temperature, organic matter, nitrogen and phosphorus.

Because of its universality the biofilm concept impacts virtually all the subdivisions of microbiology (including Medical, Dental, Agricultural, Industrial and Environmental). Data indicate that the highly structured biofilm mode of growth provides bacteria with a measure of homeostasis, a primitive circulatory system, a framework for the development of cooperative and specialized cell functions, and a large measure of protection from antibacterial agents (Costerton, 1995).

The complex matrices that form biofilms restrict water flow causing poor distribution of water and increase pumping energy costs. Kanarek et al. (2024) confirm that biofilms provide a possible haven for pathogenic organisms and increase their resistance to biocide control. De Carvalho (2007) discusses the bioaccumulation of heavy metals and toxic compounds that can have serious effects on food production quality and the safety of field workers. Biofilms increase long term maintenance costs as they are generally corrosive to metal and concrete pipes. it is difficult to develop a single model for the qualitative composition of biofilms in irrigation systems due to factors such as geographic location (climatic conditions), type of irrigation (furrow, drip, sprinkler), materials used in the construction of the irrigation system, age of the system, intensity of operation, origin of water, fertilizer addition, and system disinfection techniques (**Fig. 1**). Each of the factors listed can promote or degrade the presence of particular microorganisms in the biofilm community. (Kanarek, 2023).

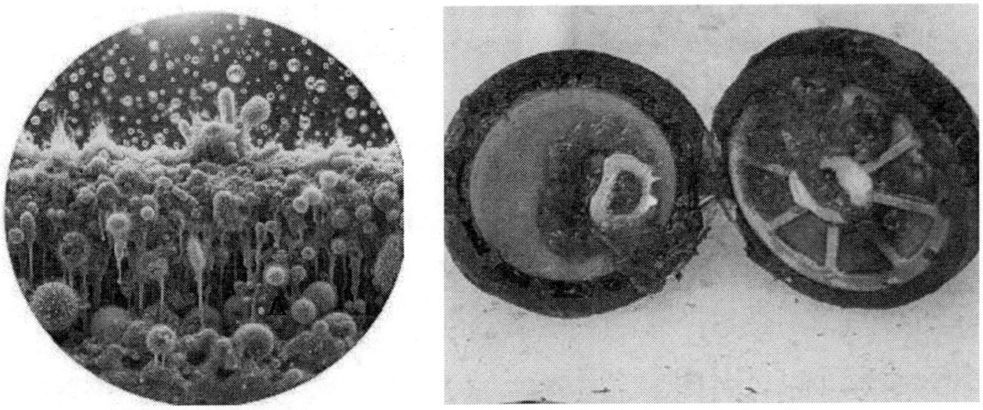

Figure 1. It is difficult to develop a single model for the qualitative composition of biofilms in irrigation systems due to factors such as geographic location, climatic conditions, type of irrigation (furrow, drip, sprinkler etc.), materials used in the construction of the irrigation system, age of the system, intensity of operation, origin of water, fertilizer addition, and system disinfection techniques. Contrasting biofilms are shown here.

Treatments for Biofilms in Irrigation Systems

Chemical applications

The treatment of biofilm fouling is popular via very dangerous and hazardous chemical applications. Oxidising agents are typical and can provide some effect where the microorganisms responsible for biofilm are able to be oxidised. This is often not the case, as biofilms are produced specifically to protect from oxidation. This leads to incomplete control or the requirement of very high rates of chemicals to try and lyse cellular walls.

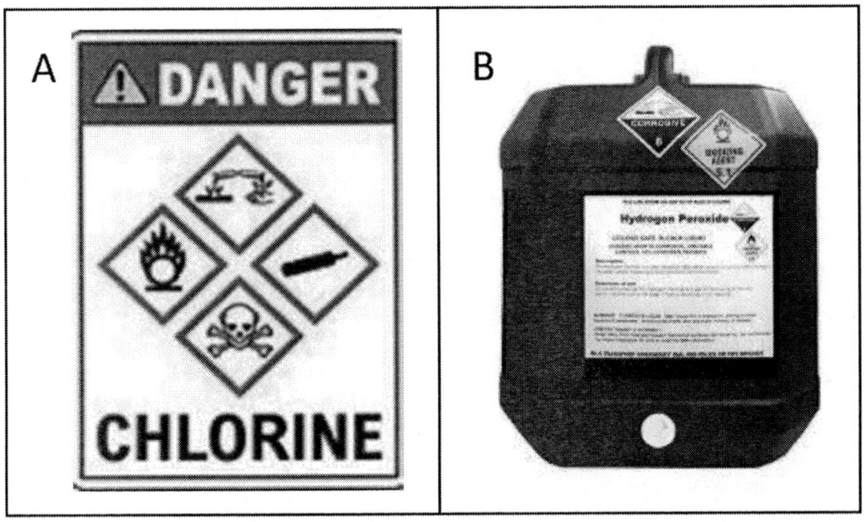

Figure 2. Products used in traditional methods of biofilm control are hazardous and often less effective than modern 'green' technologies. For example, Chlorine dioxide (ClO_2) is an effective disinfectant for oxidisable microbes, however, it is highly hazardous and explosive in air at concentration above 10% and toxic to humans at >0.8 ppm in water (A). Hydrogen peroxide is less severe in environmental impacts but has many limitations for effective use (B).

Chlorine dioxide (ClO_2) is an effective disinfectant for oxidisable microbes, however, it is highly hazardous and explosive in air at concentration above 10% (De Carvalho, 2007) and toxic to humans at >0.8 ppm in water (**Fig. 2A**). Hypochorite solutions offer some biocidal control though not effective for many anoxic biofilms. It carries with it a caveat of being an invasive measure with significant environmental impacts (Kanarek et al., 2024).

Hydrogen peroxide (H_2O_2) is a common treatment offering some biocidal control and some biofilm control. Often very high rates are required for more complete control which comes with significant OHSE and cost impacts. Whilst less severe in environmental impacts than hypochlorites (**Fig. 2B**), effectiveness of hydrogen peroxide can be limited by UV, temperature, organic matter and heavy metals (Thomas, 2021). A small amount of oxygen may permeate soils after treatment; however this is very short lived (hours), and more likely, the persistence of OH radicals will cause greater environmental issues (Watts et al., 1999).

Alternate non-hazardous modern technologies

In modern systems, the use of "green" alternatives is becoming more in vogue. Hydrodynamics uses high water flows to flush systems. This is an alternative where water is plentiful and inexpensive, though consideration must be taken in flushing the problems further downstream.

De Carvalho (2007) advocates the use of substances capable of destroying the physical integrity of the matrix, interfere with bacterial adhesion or initiate cell detachment from surfaces, and are good alternatives to biocides and/or disinfectants. The substances include biocatalysts (enzymes, phages) and organocatalysts (organic non-enzymes). Biocatalysts are generally very specific to a target and environment, whereas organocatalysts are broader spectrum and better used in dynamic systems. Kurtural (2020) advocates a novel organocatalyst, AquaMate®, that has shown to increase gas transfer in liquids (oxygen), decrease surface tension of water, and cleave hydrogen bonds. Whilst not biocidal, this causes the destruction of the biofilm matrix, breaking it into simple sugars and flushing from the system (**Fig. 3**). With repeated use, the biofilms cannot reform as the new environment does not allow adhesion and growth of the matrix. AquaMate® is a patented, novel, organocatalyst that is low cost, non-dangerous, non-hazardous and nontoxic.

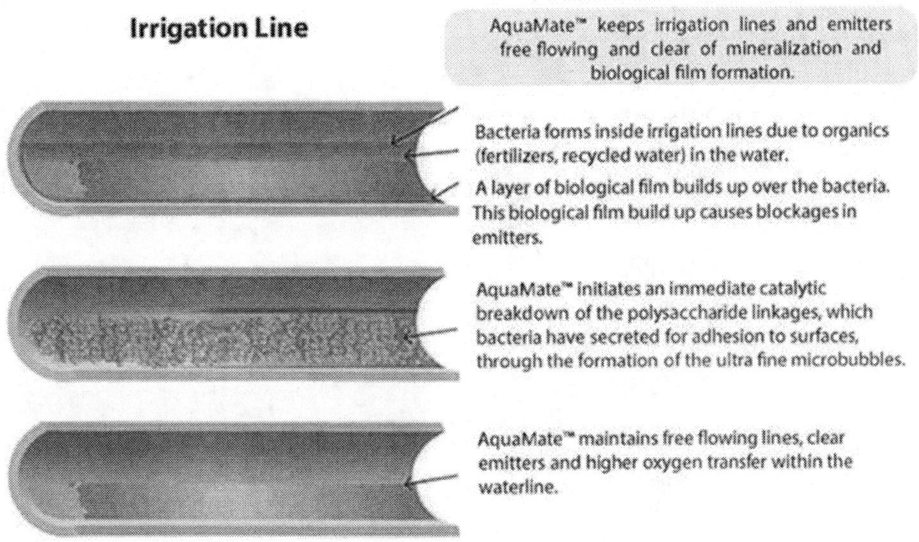

Figure 3. Benefits of AquaMate®, a patented low cost, non-dangerous, non-hazardous and nontoxic organocatalyst.

CONCLUSION

Biofilms are ubiquitous to all irrigation systems, though the level of issue is environmentally dependant. Whilst past control measures rely on highly hazardous and dangerous chemicals, modern techniques have emerged that are lower cost and safe to humans, plants, soils and the environment.

LITERATURE CITED

Costerton, J.W. (1995). Overview of microbial biofilms. Journal of Industrial Microbiology. *15*:137–140.
DOI: 10.1007/BF01569816

De Carvalho, C.C.C.R. (2007). Biofilms: recent developments on an old battle. In: Recent Patents on Biotechnology *1*: 49-57. Bentham Science Publishers.
DOI: 10.2174/187220807779813965

Dale, P.D. (2015). Process for treating organic material. Australian Patent No: 2013267982, Neozyme International.

Kanarek, P., Breza-Boruta, B. and Rolbiecki, R. (2024). Microbial composition and formation of biofilms in agricultural irrigation systems- a review. Ecohydrology and Hydrobiology *24*:583-590.
https://doi.org/10.1016/j.ecohyd.2023.10.004.

Kurtural, S.K. (2020). Application of fractions of phyto-cat affects carbon partitioning of grapevine differentially in a hot climate. Research Report: Dept of Viticulture and Enology, University of California Davis.

Thomas, P. (2021). Optimisation of stabilised hydrogen peroxide use for drip irrigation. CQ University. Thesis.
DOI: 10.25946/20288844.v1

Watts, R.J., Foget, M.K. and Kong, S.-H. and Teel, A.L. (1999). Hydrogen Peroxide Decomposition in Model Subsurface Systems. Journal of Hazardous Materials *69*:229-243.
https://doi.org/10.1016/S0304-3894(99)00114-4

Combined Proceedings IPPS 74:176-186. 2024.

South African Region Student Exchange

Nosipho Phiwokuhle Ndlovu

1-9 Church St, Wynberg, Cape Town, South Africa 7800

Nosiphophiwokuhle97@gmail.com/

Keywords: propagation, nursery, native plants, Ballina, conference

Summary

In May 2024, I was selected as an exchange student, which marked a watershed moment in my career. IPPS's generous support not only made this trip possible but also allowed me to learn more about the field of Horticulture from experienced individuals who are all about knowledge sharing. My trip to Australia lasted two weeks, it was my first time leaving South Africa, and I was extremely nervous, however.

When I arrived, I was greeted by friendly people who made this experience enjoyable and educational. My first trip was to Perth, where I spent time with Mr. David Hancock. The experience in Perth was a highlight of my career; I visited many places with Mr. Hancock and learned much about different nurseries and plant diversity in Australia. We went to Grasstrees Australia, Plantrite, Natural Area Nursery, Caversham Wildlife Park, where I saw a kangaroo for the first time, and the Biodiversity Conservation Centre, which provided me with insights into my MSc project because they do similar work. I also had the opportunity to meet incredible people from Natural Area Nursery, who introduced me to different kinds of seafood. Perth was a beautiful place, rich with history and beautiful native plants of Australia.

After my trip to Perth, I traveled to the beautiful city of Brisbane, where I was hosted and welcomed by Pam Berryman. During the three-day pre-conference tour, I had the opportunity to meet and interact with people who were both knowledgeable and welcoming. We visited several nurseries, and it was fascinating to see the differences and similarities between our nurseries in South Africa and Australia. I couldn't wait to bring back the knowledge I'd gained to South Africa. During my time in Australia, I was fascinated by the machinery in nurseries, as well as the diversity of plants and their similarities to those in South Africa. I also had the opportunity to identify some South African plant species that are invasive in Australia.

The conference began after the tour. This opportunity was eye-opening because I heard from highly skilled and knowledgeable professionals in my field. Being a member of the six packers introduced me to new friends and allowed me to interact and assist at the conference; I felt honored to be a part of the group. Following the conference, I spent time with the incredible Pam and Linda. I had the opportunity to visit the Australia Zoo, and Movie World, and explore the Sunshine Coast.

I would like to thank the IPPS Australia and South African region for supporting me throughout my trip, The funding and support I received from IPPS was instrumental in making my trip to Australia a reality. It provided the financial backing needed for travel, accommodation, and daily expenses, while also opening doors to academic, cultural, and personal growth opportunities. The experience was transformative, broadening my horizons and leaving a lasting impact on my personal and professional life. I would also like to thank my hosts for welcoming me to their homes, introducing me to their families, and taking up time from their schedules to allow me to explore my time in Australia.

INTRODUCTION

In May 2024 I had the privilege of participating in a fully funded horticulture trip to Australia, sponsored by the International Plant Propagator Society (IPPS). This opportunity began by being in an inspiring competition with two other skilled individuals in Stellenbosch, South Africa (**Fig. 1 and 2**). We presented our experiences as young professionals in the Horticulture industry and why the opportunity to visit Australia would be beneficial in our careers. I had the honor of being selected to visit Australia and in May 2024, I began my visit to this beautiful country. The purpose of the trip was to help me learn Horticultural practices outside of my country of residence focusing on various production and propagation techniques and engage with a variety of professionals in the field. My visit to Australia began in Perth and ended in Brisbane where I attended the 52nd IPPS (Australia) conference in Ballina.

Figure 1. IPPS delegates in Stellenbosch.

Figure 2. N. Ndlovu with student exchange nominees.

Day 1-3: Arrival and Introduction to Western Australia

My trip began with a connecting flight from South Africa to Dubai then finally landed in Western Australia. I was warmly welcomed by Mr David Hancock and his beautiful wife who was my host, upon my arrival, we had a special dinner engaging in beautiful conversations that made me get to know their family more (**Fig. 3**). In preparation for the following day, Mr Hancock encouraged me to take as much rest as I only had one day in Perth and plenty to see and learn.

Figure 3. Mr. David Hancock with his lovely wife and family friend Bree welcome dinner.

Our first nursery visit was in Grass Trees Australia where we were given a tour by the lovely Fiona Reynolds. It was a pleasure visiting this leading supplier of grass trees and it was the first time seeing the species *Xanthorrhoea preissi* in person (**Fig. 4 - 6**). Fiona gave us a history of the company and I learned that since 1995, Grasstrees Australia's skilled staff have provided quality grasstrees and zamia palms to the general public and government agencies showing me how conservation of these species is done as well as the extent done for restoration purposes of Australia's precious grasstree. I further learned that the company has been involved in restoration, they have a license to rescue grass trees in the wild. Viable trees are selected on site, taken to the nursery, and treated for 12 months depending on how well roots have developed. Watering is done twice a week.

Figure 4. A) Nosipho & Fiona standing next to a 250-year-old grass tree. **B**) Grass tree plant. **C**) New shoot formation on a grass tree.

Our second nursery visit was to Plantrite Nursery. The drive was really interesting as we spotted several native species along the way. Upon seeing the nursery, I was amazed by how big and automated it is (**Fig. 5**). This is a wholesale native nursery that gave me an overview of native Australian flora and I was able to spot some similarities and differences to South African indigenous flora. I got to learn about the role and importance of precision irrigation systems and efficient water use as Perth experiences issues with drought. The technologies I observed in this large-scale nursery showed me significant potential for application in South Africa's dry regions like the Karoo and Northern Cape. I also got to learn about the services they provide for multi-acre reforestation and environmental planting projects.

Figure 5. **A**) Large scale propagation unit at Plantrite nursery. **B**) Nursery machinery at Plantrite.

Having a host like Mr. David Hancock was an extreme honor for me, and our last nursery visit was a privilege as I got to see Natural Area Nursery.I was given a tour of the nursery and a history of it. Recently moved to a new area, Natural Area Nursery has a reputation for producing quality plants for restoration, revegetation, and landscape around Perth and Western Australia (**Figs. 6 and 7**). I got to learn about all the fascinating restoration projects Mr. Hancock has been involved in and it was inspiring to see the relationship he has with the staff. I got to see a variety of native Australian plants and learned about water efficiency and some challenges presented during drought seasons.

Figure 6. Pellet trays at Natural Area Nursery.

Figure 7. **A**) Seed storage area at Natural Area Nursery. **B**) Seed smoking facility was similar to the smoker for Protea seeds in South Africa. **C**) Checking seedlings in the nursery.

Our next destination with Mr. Hancock was Caversham Wildlife Park, where I got to see and feed a kangaroo for the first time **(Fig. 8)**.

Figure 8. Feeding kangaroos.

We then took a drive to Kings Park, which was one of the most beautiful places I have seen **(Fig. 9)**.

Figure 9. King's Park memorial.

During our visits and chats with Mr. Hancock, I expressed an interest in biodiversity conservation and my current MSc projects. He then arranged for a brief meeting with

Jason at the Biodiversity Conservation Center in Kings Park. I got a tour and got to see their seed science program (**Fig. 10**), genetics, plant biotechnology (their tissue culture laboratories), storage science component which gave me ideas on storing recalcitrant seeds, and their work on restoration ecology.

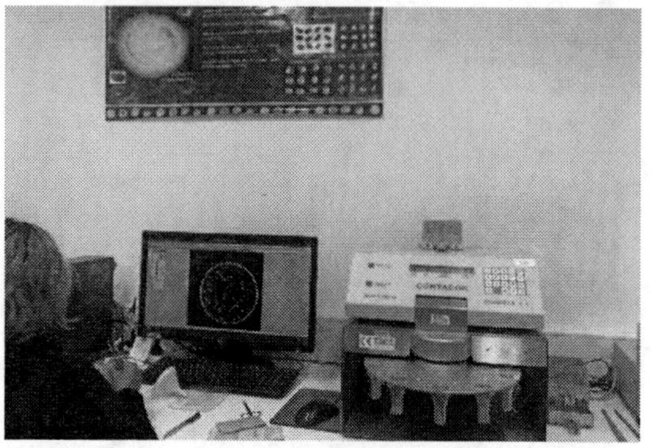

Figure 10. Automatic seed counter at the Biodiversity Conservation Centre.

Following our tour of the conservation center, I got to see the beauty of Kings Park and the history seeing the names of servicemen and women from Western Australia who served in the war was emotional and made me see how rich in history Western Australia is.

After a long day of fascinating visits and tours, we took a drive to the north beach where I went to my accommodation in preparation for a night out with some of the young people from Natural Area Nursery. Driving there we stopped at a few spots for pictures, and I got to see some of the restoration projects Mr. Hancock has done around the North Beach. The night in Perth ended beautifully with local seafood, music, drinks, and good company (**Fig. 11**).

Figure 11. Visiting North beach and having a seafood buffet with Natural Area Nursery staff. Exploring the beauty at the coast and tasting indigenous foods.

Day 4-6: Arrival in Brisbane and attend Pre-conference tour

I then flew from Perth to Brisbane where I was hosted by the lovely IPPS Secretary Pam Berryman. While in Brisbane, I saw a stark contrast with its humid, subtropical climate as compared to Perth. I also got to learn that Brisbane is known for its lush landscapes and thriving horticulture and offers an ideal setting to explore tropical and subtropical crop production. On the first day of the pre-conference tour, I meet up with fellow delegates who share the same enthusiasm for Horticulture. On the first day of the pre-conference tour, we engaged in insightful conversations meeting with all delegates and I got to see Mount Tambourine and explore it with the lovely Pam and Linda (**Fig. 12**).

Figure 12. Exploring Mt Tambourine with Pam and Linda.

The second day of the conference tour began with our first nursery visit to Tambourine Mountain Nursery. We had a tour of the nursery and engaged in a talk with the owner Alicia Atkinson. Our second nursery visit was Mt Nathan Propagation which was a 20 km drive. We saw a wide range of native and exotic trees and shrubs. We then took a short drive to Crystal Waters Nursery and had some refreshments while touring the beautiful nursery. Following that visit, we explored two more nurseries that grow plants for landscape trade and had a delightful lunch paired with wine tasting and a cheese board (**Fig. 13**). The day ended with a lovely Mexican dish and engaging in conversations with fellow delegates.

Figures 13. Day one of the pre-conference tour. Visiting nurseries and ending the day with wine tasting and cheeseboard.

Day three of the pre-conference tour was filled with interesting nursery visits, good conversations, and exploring native flora in Australia. Our day began with a one-hour drive to Limpenwood Botanic Gardens. Having a background in working at a botanical garden in South Africa, I was fascinated by the deep history and beauty of Limpenwood Botanic Gardens. Being around since the mid-seventies, this stunning garden exposed me to a variety of rare and unusual tropical and sub-tropical plants all of which have potential applications for South Africa's tropical regions, like KwaZulu-Natal South Africa. We also got to explore local food in Mt Warning Hotel engaging in mind-stimulating conversations. Following our lunch, we went to Gondwana Nursery, a beautiful horticultural space that is regarded as Australia's leading native propagation nursery (**Fig. 14**). I also observed that the nursery also employed environmentally friendly practices, such as integrated pest management (IPM), which reduces chemical inputs while maintaining high yields. These methods could be replicated in South Africa's nurseries. Our day ended with a buffet of prawns, fresh oysters wine, and good conversations exposing me to the beauty of the coastal town.

Figure 14. **A**) Nursery visit. **B**) Using different colour pots in a production nursery. This could also be done in South Africa for ease in plant identification. **C**) *Grevillea* cuttings stuck in perlite.

52nd IPPS Conference – A breath of Fresh Air

Following the pre-conference tour, I was extremely excited to meet other delegates at the conference. It was an honor to be part of the six packers and assist Pam in preparing gift bags for fellow delegates. Meeting the six-packers was very insightful, it was very inspiring to share our knowledge and get to see fellow young people inspired by Horticulture as I am (**Fig. 15**). Dinner conversations before the conference proceeded got me to understand the role of the six-packers and what an amazing opportunity to be a part of the group. We were introduced to fellow delegates, and I was privileged to represent South Africa in that moment.

The first day of the conference began, and we were welcomed by enriching talks that provided valuable insights into the future of horticulture, a "breath of fresh air" indeed. Keynote speakers at the conference were experienced in multiple industries and lifelong interest in plants. I was fascinated and inspired by all the talks, and I could not wait to share the knowledge back home, particularly relating to micropropagation, plant breeding, and propagation of rainforest plants. I got the opportunity to network with a variety of knowledgeable individuals who are skilled in running businesses, research, and innovation. The conference also included a variety of nursery visits in between. The knowledge I gained from this conference can be applied to improve South Africa's horticultural sector, making it more sustainable, productive, and resilient in the face of our changing environment. The global connections made at the event also offer opportunities for ongoing collaboration and exchange of best practices and I had the opportunity to present about my visit to Australia which was an amazing opportunity for me (**Fig. 30**).

Figure 15. **A**) Meeting fellow six-packers. **B**) Plant identification during a nursery visit. **C**) Presenting at the IPPS conference in Ballina, New South Wales.

The conference closed off with a beautiful gala dinner with exquisite food and an auction. It was an honor to assist in the auction with fellow six-packers and I received a certificate for presenting and being part of the six-packers. The following days were the last days of my visit to Australia.

My last few days of my visit was spent with Pam and Linda who allowed me to see the Gold Coast. We went to places such as Cape Byron State Conservation Area which reminded me so much of Cape Point in Western Cape, the ginger factory and I got to taste a variety of ginger delicacies (**Fig. 16A**), and one more nursery visit where we were given an interesting tour by the inspiring Zoe (**Fig. 16B**). One of my highlights included visiting two iconic attractions: Australia Zoo and Warner Bros. Movie World (**Fig. 16C**). The visits provided a mix of wildlife conservation education and entertainment, offering unique experiences that highlighted both Australia's natural heritage and its thriving entertainment industry.

Figure 16. A) The Ginger Factory. **B**) Nursery visit. **C**) Australia Zoo and D) Warner Bros. Movie World.

Located in Beerwah on Queensland's Sunshine Coast, Australia Zoo is world-renowned for its conservation efforts and its association with the late Steve Irwin, the "Crocodile Hunter." I have always been a big fan watch him and his family on television back in South Africa and the zoo offered me a chance to interact with some of Australia's most iconic wildlife. I got to see and even hand-feed kangaroos, hold a koala and observe wombats, Tasmanian devils, and dingoes up close.

The experience deepened my understanding of the unique fauna of Australia. The zoo gave me an unforgettable wildlife experience with a strong emphasis on conservation. It was inspiring to see the work being done to protect endangered species and learn how human actions can make a positive impact on wildlife preservation.

The following day was an exciting visit to Movie World. Located on the Gold Coast, this popular theme park gave me "the magic of Hollywood" to life through its thrilling rides, live shows, and character meet-and-greets. I got the opportunity to ride a range of exciting rides, from adrenaline-pumping roller coasters, that were scary yet enjoyable experiences for me.

Throughout the day, I got to see and interact with characters from popular franchises such as Batman, Superman, Wonder Woman, and the Looney Tunes gang. The live-action performances provided a playful and immersive element to the experience. I truly enjoyed the experience and am grateful to Pam for taking me there, a beautiful way to close my 2-week trip to Australia (**Fig. 16 C**).

CONCLUSION

IPPS Australian region provided me with valuable insights into the future of horticulture. The trip provided me with a wealth of knowledge and inspiration, much of which applies to South Africa's horticulture industry. The highlights I took from attending the conference and visiting nurseries include:

1. **Water management**: South Africa's water conservation issues may be resolved by adopting Australia's cutting-edge irrigation methods and regulations, such as drip irrigation and rainwater harvesting.
2. **Indigenous plant use and conservation**: Using South Africa's indigenous flora more and having nurseries dedicated to producing these plants for a variety of uses such as restoration.
3. **Technology integration**: The application of automation, sensors, and drones in Australian horticulture demonstrates how technology can increase output while reducing its negative effects on the environment.

Overall, I had an amazing opportunity that allowed me to learn and interact with a variety of skilled individuals in the Horticulture industry.

Combined Proceedings IPPS 74:187-219. 2024.

Tissue Culture Technologies and Their Applications

Ranjith Pathirana

The School of Agriculture, Food and Wine, University of Adelaide, Waite Campus, Urrbrae, South Australia, 5064

Ranjith.Pathirana@adelaide.edu.au

Keywords: micropropagation, doubled haploids, polyploidy, conservation, interspecific hybridization, somatic embryogenesis, microtubers

Summary

Over the last century, plant tissue culture (PTC, in vitro technology) has evolved into a highly sophisticated biotechnological tool not only for understanding biochemical processes within the plant, but also in applied plant breeding, conservation and propagation. It is now the foundation in advanced biotechnologies. The most widespread application is micropropagation for producing clonal plant stocks with the market expected to grow to US$ 2.1 billion by 2030. Micropropagation is widely used in clonal propagation in horticulture, floriculture and forestry. PTC is also used in eradicating viral and bacterial diseases infecting clonal plant material to produce high-health planting material for agriculture, horticulture and forestry. In vitro technologies allow combining different methods such as meristem culture, thermotherapy, chemotherapy, electrotherapy and cryotherapy within one experiment to eradicate multiple diseases and/or to eradicate aggressive pathogens that cannot be eradicated using a single therapy. Also, in vitro technologies have a central role in the development and deployment of new cultivars to the industry much faster and efficiently than traditional field-based plant breeding methods. This application encompasses an array of technologies to produce crop cultivars or even new man-made species with traits of interest. The approaches to cultivar improvement using PTC techniques include the induction of mutations and selection of desirable mutants with improved traits and in developing genetically modified crops using traditional transformation methods as well as gene editing techniques. PTC is the preferred pathway for developing interspecific and intergeneric hybrids that cannot be produced by hybridization in vivo due to various incompatibility issues. Ploidy manipulation, production of doubled haploids for hybrid development, increasing the proportion of hybrid seeds in apomictic species are some other applications in crop improvement. Because our plant genetic resources are not safe in the field due to climate change resulting in vagaries of weather as well as pests and diseases, ex situ conservation is becoming increasingly important. Again, PTC takes a central role in ex situ conservation, whether as tissue culture repositories or as cryopreserved collections. Finally, cell cultures are used in producing biopharmaceuticals, food ingredients, cosmetics, flavours, dietary supplements, fragrances, and biostimulants.

INTRODUCTION

Plant tissue culture (PTC) also called in vitro culture constitutes a critical component in plant biotechnology and refers to the culture of plant cells, tissues or organs on a nutrient medium under aseptic (sterile) conditions. Although the theoretical foundation of this technology was laid in early 1900s, considerable progress in practice was achieved after the discovery of the hormonal control of cell proliferation and organogenesis in vitro in the 1950s (Carra et al. 2024). Plant cell technology is now seen as the breakthrough technology that can help to meet the challenge in the next phase of plant breeding after the yield increases in Green Revolution have plateaued in the last two decades. Micropropagation – efficient mass propagation of clonal plants through tissue culture is the largest commercial application. Other plant biotechnologies that aid in developing new varieties and individual traits within existing plant varieties include cell and tissue manipulation, marker-assisted selection, transgenic technologies, genomics, and molecular breeding. Cell and tissue culture technologies provide a range of applications in the creation, conservation, and utilization of the genetic variability in crops, such as in vitro pollination and embryo rescue for distant hybridization, the production of haploids and doubled haploids, polyploid breeding, in vitro mutagenesis and selection of somaclones, in vitro selection, germplasm preservation (in vitro for medium-term and cryopreservation for long-term), protoplast fusion for producing somatic hybrids, and gene manipulation for producing transgenic crops or the newly emerging techniques that allow for the generation of gene-edited plants (Pathirana and Carimi, 2024).

Additionally, plant-based drugs have been used in all ancient civilisations and continue to be used in modern times. The market has been growing steadily for plant-based drugs and may reach USD 50 billion during the next five years (Bapat et al. 2023). As only about 1 % of the plant species are used in drugs, they are overexploited and threatened in their natural environments. Furthermore, extraction from plants is inefficient and laborious. Plant cell cultures present an alternate sustainable pathway to produce drugs. This approach has the advantage of homogeneity of cell suspensions, scalability, fast growth with no quiescence, less space requirement, and ease of handling and hence offers attractive options for the production of not only drugs but other bioactive secondary metabolites in the food and cosmetic industries.

During the growth in vitro, we control temperature, humidity, light intensity, light cycle and manipulate media components. The cultures are either maintained in highly controlled growth chambers or culture rooms in liquid or solid media. In this review, covering different applications of this technology, examples from author's own work will be presented.

Aseptic Culture

We use both physical and chemical methods to achieve sterile conditions. Equipment that uses physical methods include autoclaves (uses heat and pressure), laminar flow cabinets (filters including High Efficiency Particulate Air - HEPA filters), bead sterilisers or Bunsen burners (heat) and filter sterilising units used to sterilise media containing compounds that can get destroyed by heat. Chemical methods include surface sterilization of laminar hood, gloves and explants (part of plants introduced into culture) using ethanol or other sterilants, antibiotics used to prevent growth of bacteria in media etc.

In tissue culture we grow plant cells or tissues in artificial culture media, whose proportion and concentration of elements vary depending on the objective and plant species. We supply major and minor elements required for growth and development in the form of inorganic salts, for example, magnesium sulphate to supply magnesium and sulphur. There are many different media available commercially for different species and we need to test different media for less studied species.

Plants or cells growing in artificial media have none or very little photosynthesis. Even if leaves are present, their stomata are always open and non-functional. So, we need to supply carbon through the media as the plants in culture can't use CO_2 from the atmosphere of the culture vessel to fix carbon. The main carbon source is sucrose, but often we use other sugars such as maltose, fructose, glucose etc. Plant growth regulators are an important component in culture media, and we can drive the growth and development to our requirements by changing the proportions of auxins, cytokinins, abscisic acid, gibberellic acid etc. Additionally, we have to give support to plants by making media solid using gelling agents such as agar or gellan gum, unless we use liquid cultures.

Why Are Plant Cells and Tissues Amenable to Tissue Culture?

The plants grow and develop throughout their life and for this they have dividing meristematic cells that we can always use to initiate cultures (**Fig. 1A**). In addition to the meristematic tissue in apices and axillary buds, vascular cambium (a layer of cells between primary xylem and phloem) and cork cambium (a major portion of the bark of woody plants) also have meristematic tissue.

Dividing meristematic cells are easier to manipulate than differentiated cells. However, unlike most of animal cells, plant cells have totipotency and pluripotency. Totipotency is the ability of any living plant cell to revert to a meristematic (dividing) state, given the right environmental cues and then produce a complete plant. Pluripotency is the ability to regenerate an organ or tissue from a living cell. Thus, a differentiated plant cell can return to a meristematic state and then differentiate to produce tissues and organs that have specialised cells. This latter process is called de-differentiation.

For an example, we can get an anther to produce a haploid plant (plant with a single set of chromosomes) directly using low auxin and high cytokinin in media (**Fig. 1B**) and this is an example of direct regeneration. When we produce plants from cells through an intervening callus phase by manipulating plant growth regulators (**Fig, 1C and D**), the process is called indirect organogenesis This is generally a two-step process. For clonal propagation direct organogenesis is preferred because when going through a callus phase, there are chances of mutation induction, resulting in somaclonal variation.

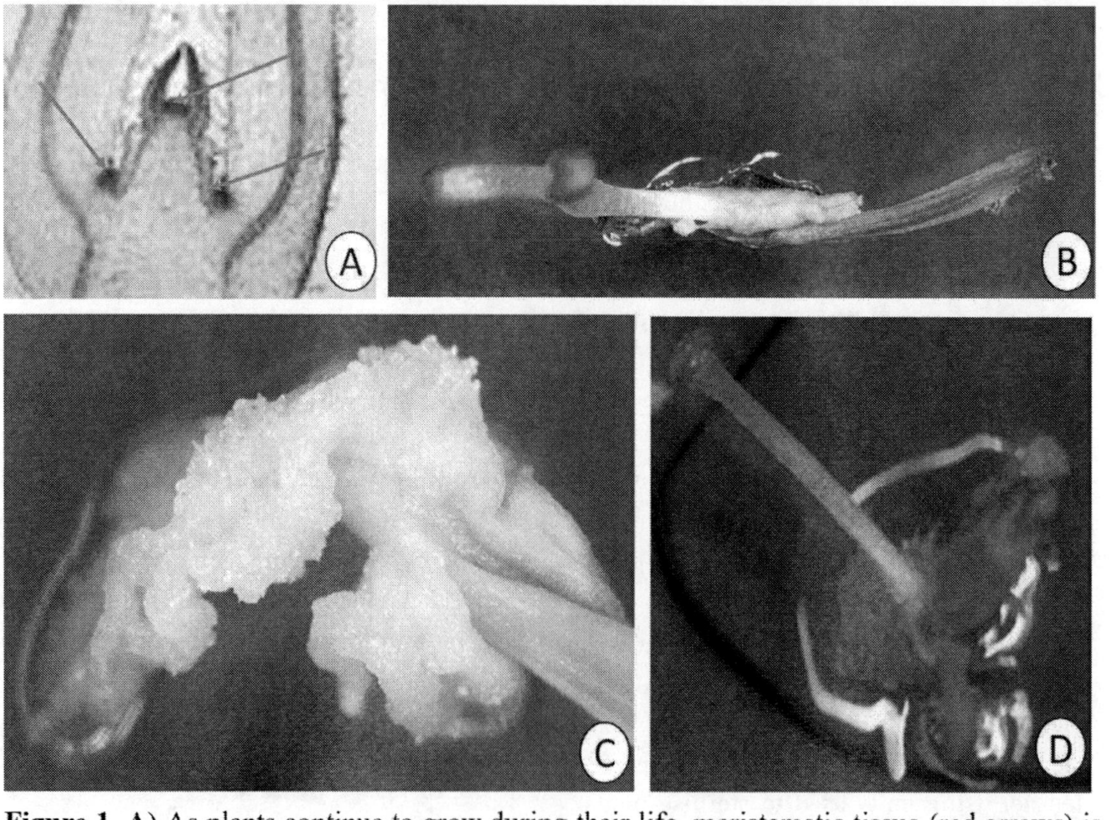

Figure 1. **A**) As plants continue to grow during their life, meristematic tissue (red arrows) is present for easy manipulation – meristems of shoot tip and axillary nodes in *Hibiscus rosa-sinensis;* **B**) Direct regeneration of a haploid plant from an anther of *Gentiana triflora;* **D**) Indirect organogenesis in the same species through an intervening callus phase **C**). (From Pathirana et al. 2011. Plant Cell Reports, 30, 1055–1065).

The ratio of auxins to cytokinins invariably determines structural organisation in vitro, a concept published as early as 1957 (Skoog and Miller, 1957). Since then, we have learnt much more about organogenesis in plants. The differentiation of cells into organs or undifferentiated callus is guided by this ratio as well as concentrations and types of growth regulators used in media. This is illustrated in **Fig. 2.**

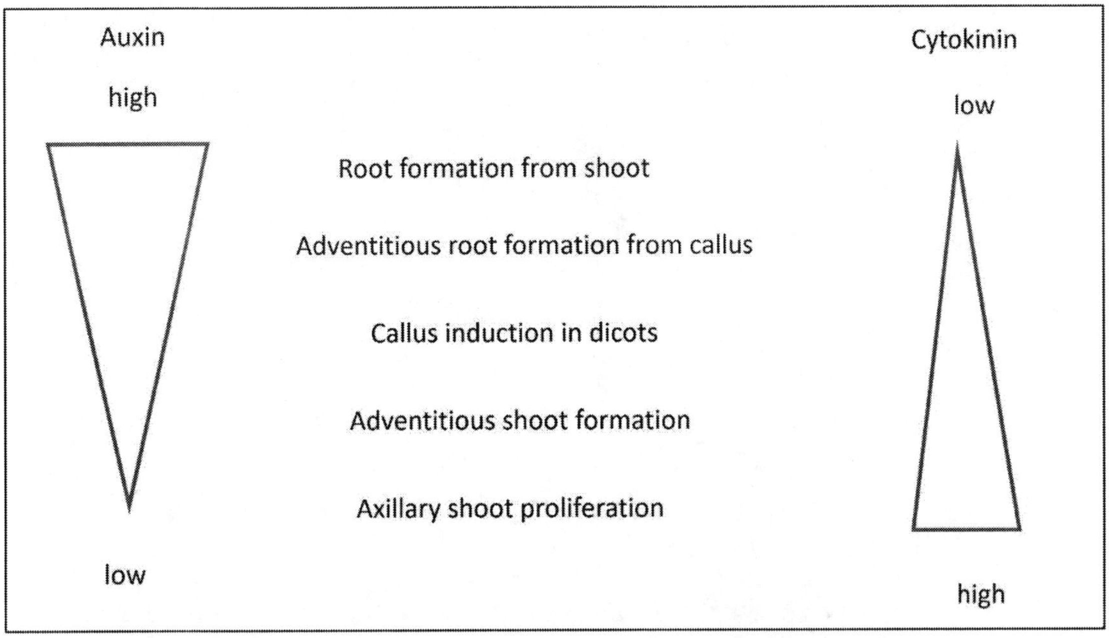

Figure 2. Determination of structural organisation of plants in vitro as affected by auxin/cytokinin ratio.

Clonal Propagation Through Tissue Culture – Micropropagation

The market

Clonal propagation in tissue culture, technically called micropropagation is the most widely used and widely known application among the many in vitro technologies. According to P&S Intelligence (2024), the global micropropagation market stood at US$ 1.28 billion in 2022 and is expected to grow to US$ 2.1 billion by 2030. Due to the growing interest in indoor plants, the largest driver of this growth will be the increased demand for orchids and other indoor plants. Additionally, the wide use of ornamental plants in decoration of commercial facilities such as hotels, airports, restaurants and office spaces add to this demand. The largest micropropagation facilities can be found in China, India, Thailand and South American countries where labor is relatively inexpensive as this is a labor-intensive process and automation is difficult and costly. Micropropagation is also used in large-scale propagation of banana, pineapples, potato, *Pinus radiata* (and many forest species) and many floricultural and horticultural crops, as well as for medicinal plants including medicinal cannabis (with high tetrahydrocannabinol content, which is the psychoactive chemical), with many countries legalizing its medicinal use.

The growth of the micropropagation industry is boosted by the demand exceeding the current production capacity. It is estimated that the global demand for healthy, clean and uniform planting material is about 16 trillion equating to US$ 4 trillion, whereas only 1.5 - 2 billion plants are produced through micropropagation (Wei et al. 1990). Undoubtedly micropropagation is the leading technology that can deliver quality plants round the year. Despite the tissue culture production facilities are located mainly in developing countries, major companies have already installed acclimation facilities for tissue cultured plants they buy, thus reducing costs.

Approaches to Micropropagation

Clonal multiplication in tissue culture is achieved through different pathways, and the main approaches are direct organogenesis as seen in our recent experiments with *Corymbia* spp. (**Fig. 3 A and B**) and kiwifruit (Saeiahagh et al. 2019), indirect organogenesis via a callus phase (**Fig, 1 C and D**), through somatic embryogenesis (SE) (**Fig. 3 C and D**) or using microtubers **(Fig. 4 and 5)**.

Figure 3. *Corymbia* micropropagation through direct organogenesis **A)** Multiple shoot formation from a shoot tip, **B)** rooting of separated microshoots. Production of somatic embryos of kiwifruit **C)** and their germination **D)**. (C and D from Pathirana et al. 2016. Acta Hortic. 1127:17-222).

Organogenesis

Direct organogenesis involves the use of plant growth regulators, particularly cytokinins, to induce growth of existing shoot primordia in the shoot apex or in axillary buds, resulting in multiple shoots that can be separated and further multiplied. For indirect organogenesis, we first produce a callus tissue with undifferentiated cells from which large numbers of plantlets can be generated.

Micropropagation through somatic embryogenesis

Somatic embryogenesis involves producing bipolar structures resembling zygotic embryos from non-reproductive, somatic tissue such as leaves (**Fig. 3 C**), petioles, cotyledons or even roots, without a vascular connection with original tissue. These then go through a maturation process and can be germinated, usually by increasing the concentration of auxins in growth media or by culturing in hormone-free media (Carra et al. 2019) (**Fig. 3 B, 3 D**).

Somatic embryos differ from sexually produced zygotic embryos by the absence of a seed coat, although they have the embryonic axis with radicle and plumule including the cotyledons. The somatic embryos, unlike zygotic embryos, do not have tolerance to desiccation. Therefore, they cannot be handled like dry seeds. Zygotic seeds of many species can be dried to 6-8 % moisture content for storage and used for planting, retaining their viability at that low moisture content. However, somatic embryos cannot be dried. Studying the process of acquisition of desiccation tolerance in fertilized ovules of alfalfa (*Medicago sativa*), scientists at Guelph University in Canada managed to mimic this process in somatic embryos. Adding abscisic acid in correct concentration to media at the cotyledonary stage of embryo development was crucial (Senaratna et al. 1989). Thus, the possibility of imparting desiccation tolerance to somatic embryos exists, but the challenge is synchronizing the process of embryo development (to apply abscisic acid at cotyledonary stage) in tissue culture systems. Therefore, most of the somatic embryo-based systems of micropropagation rely on using alginate beads to encapsulate the embryos for short-term storge or directly germinating the embryos to produce plants. Synchronized production of somatic embryos in suspension cultures, maturing in solid media followed by drying and encapsulating in an inert substance such as clay with incorporation of more nutrients has the potential to revolutionize the clonal propagation industry.

Microtubers for propagation

The other popular method used in micropropagation is microtuber production in the case of tuber crops like potato (**Fig. 4**), ulluco (**Fig. 5**), yams etc. In major production areas of potato, microtubers are the preferred option for 'seed potato' production.

Microtubers can be produced from single nodal cuttings in media with high sucrose (6 – 9 %) supplemented with cytokinins, mainly kinetin. The cultures are maintained in the dark for tuber production. Potato and ulluco (*Ullucus tuberoses* - a South American tuber crop) microtuber production in our work in New Zealand is presented in **Figs. 4 and 5**, respectively. In potato, we wanted to understand if the microtubers can be used to screen potato germplasm for cold-induced sweetening (CIS), a problem encountered in processing potato (Pathirana et al. 2008). A rise in hexose sugar levels during cold storage of potato tubers results in a brown, bitter tasting and unmarketable product. This is caused by invertase enzyme activity in cold storage and this activity and hexose sugar production is different in different potato cultivars. Comparison of field-grown potatoes and microtubers showed that there is a good correlation in CIS and hence microtubers are a good model for selecting potato genotypes for processing (Pathirana et al. 2008). In ulluco, our objective was to produce diverse mutants to adapt this new crop to New Zealand condition (Pathirana et al. 2011b). In vitro mutagenesis is an efficient way to produce large mutant populations and the ability of ulluco to produce microtubers is an advantage for easy transfer to field conditions for screening the mutant populations. Our work resulted in mutants with traits such as early maturity, altered tuber morphology and colour, less geosmin (a component in ulluco tubers that imparts an 'earthy' taste) and altered leaf colour.

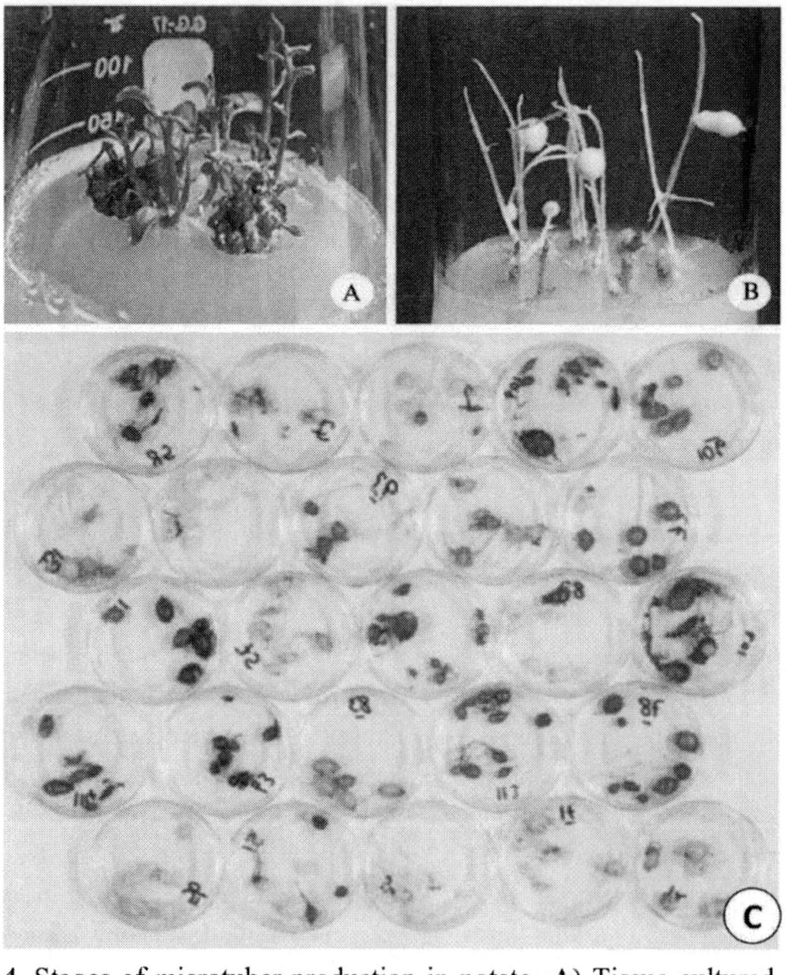

Figure 4. Stages of microtuber production in potato. **A**) Tissue cultured plantlets in culture media for microtuber production. **B**) Microtubers produced within 3-5 months. **C**) Morphological diversity of microtubers from different cultivars - they resemble their counterpart field-grown tubers not only in colour and shape, but also in cold-induced sweetening (From Pathirana et al. 2008. Post Harvest Biol Technol, 49; 180-184).

Figure 5. Production of microtubers of ulluco (*Ullucus tuberoses*) for a mutation breeding program. **A**) Microshoots in culture media with high sucrose and cytokinin for microtuber production, **B**) and **C**) Microtubers produced within three months.

Stages of micropropagation

Growing mother plants in a disease- and stress-free environment is important to initiate cultures for all the methods. Optimally they should be sourced from a greenhouse.

The stages of micropropagation through direct organogenesis include initiation of cultures on proliferation media, multiplication (Fig. 3 A), rooting of microshoots (Fig. 3 B), exflasking and acclimation. For micropropagation through direct organogenesis, only those parts of a plant that have meristematic tissue with shoot initials are used and cultured in media supplemented with higher proportion of cytokinins than auxins (Fig. 2). This proportion is reversed for rooting of microshoots. Sometimes the cytokinin used has an influence over subsequent rooting. Recently we have shown that the use of *meta*-Topolin (*m*T - a natural cytokinin first found in poplar – *topola* is the Polish word for poplar) in place of benzylamino purine (BAP) or zeatin helps rooting of a 'difficult-to-root' red kiwifruit (Saeiahagh et al. 2019). Lower hyperhydricity (an anomaly often found in tissue cultures) and low residual effect of *m*T are some reasons for this success. To enhance rooting, supplementation of media with activated carbon, pulse treatment of microshoots with high concentration of auxins (1-50 g/L; 2 – 30 min), incubation for several days in the dark, reduced mineral and sucrose content in media as well as photoautotrophic micropropagation systems (systems mimicking natural growth conditions with higher light intensities, no sucrose supplementation and often using liquid cultures) have been used, and for each species, the methods need to be optimized. Some species like blueberry can be rooted in the greenhouse, without a rooting phase in tissue culture.

After the rooting phase, the plantlets are taken out from agar media, washed and transferred to specialized potting mixes. These potting mixes are often autoclaved to reduce the infection of tender and vulnerable plantlets in the early stages of acclimation.

Micropropagation through somatic embryogenesis has the potential for scaling up and automation of production of clonal plants. It is amenable to suspension cultures as shown in our work with kiwifruit (Pathirana et al. 2016) and grapevine (Pathirana and Carimi 2023). If the process can be well tested and established, mass production for nursery industry is a possibility as in the case of *Pinus radiata,* with scientists introducing machine learning algorithms to predict and select somatic embryos with high rate of germination success (Scion 2024). Our recent attempt for scaling up grapevine somatic embryo production (in this case for mutation induction for breeding purposes) is given in **Fig. 6** showing different stages and steps, starting with the establishment of cell cultures with proembryogenic masses. The proembryos then go through globular, heart, torpedo and cotyledonary stages in dicots. Abscisic acid is generally used for maturation, and they can be germinated on solid media (**Fig. 6**). Another advantage of somatic embryos is that they are easy to cryopreserve for long-term storage, a huge advantage for in vitro breeding and conservation as we have shown for kiwifruit (Pathirana et al. 2016) and for an endangered grapevine species (Carimi et al. 2016).

In conclusion, tissue culture has immense potential for mass production of clonal species in horticulture, floriculture and forestry and for industrial and medicinal crops species as well as for conservation.

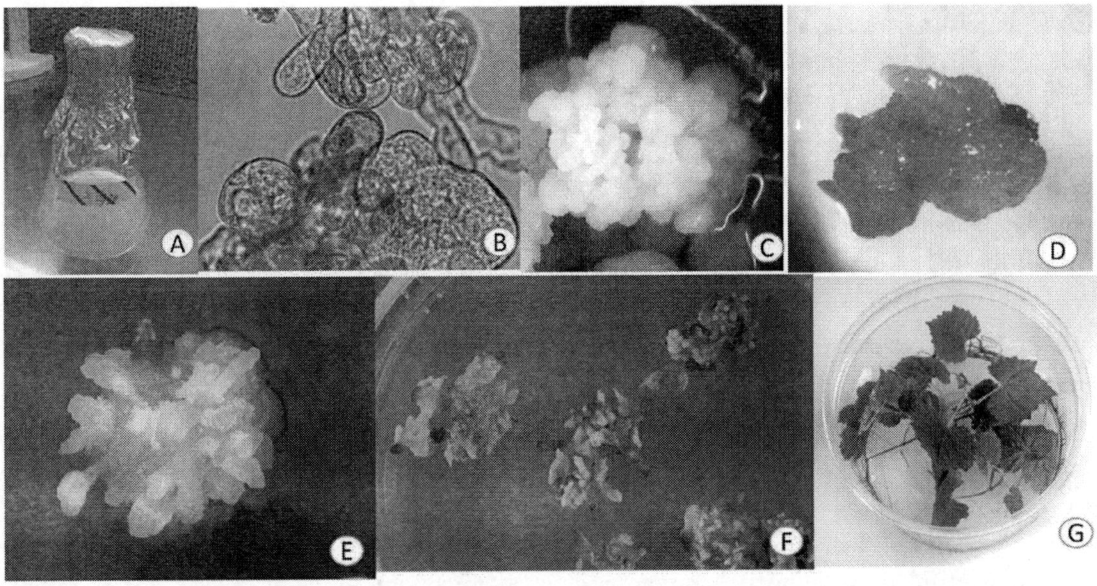

Figure 6. Somatic embryogenesis in grapevine. **A**) Embryogenic cultures in liquid media. **B**) Cells of embryogenic cultures under microscope (x20). **C**) Globular embryo formation. **D**) and **E**) Torpedo shaped embryos. **F**) Fully formed somatic embryos. **G**) Germinated somatic embryos.

Producing High-Health Plants Through Tissue Culture

Another application of PTC is the development and deployment of disease-free high-health plants for agriculture, horticulture and forestry. Often, we deal with clonally propagated crops and most of these are perennials. The vegetative propagation and exchange of budwood among regions and countries contribute to the spread of pathogens. This is true not only for fruit crops but also for root and tuber crops such as potato, cassava, yam and sweetpotato and for many ornamental crops such as orchids, anthuriums as well. The practice of vegetative propagation contributes to the spread of pathogens. Plant pathogens cause significant economic losses, for example in grapevine, fanleaf disease caused by a nepovirus (Grapevine fanleaf virus - GVFV) causes economic losses amounting to US$ 16,600 per ha, and in France where about 2/3rd of the vineyards is affected, it has an economic impact of at least US $1.5 billion per year. Another main viral disease affecting vineyards around the world is Grapevine leafroll-associated virus 3 (GLRaV 3) causing leafroll disease. It is estimated to cause losses from US $25,000 to US $ 226,000 per ha over a 25-year vineyard lifespan depending on the location and cultivar (Fuchs and Lemaire 2017).

In Australia, there are many virus diseases causing millions of dollar losses to the respective industries such as *Banana bunchy top virus* in banana – its exclusion will avoid AUD 15.9 - 27.0 million in annual losses for the banana industry (Cook et al. 2012), leafroll disease in grapevine – even more than 50% yield loss including reduced quality (Nicholas 2006), *Potato virus Y* in potato, *Strawberry mild yellow edge virus* in strawberry, *Cassava mosaic virus* in cassava, sweet potato virus disease in sweet potato, *Blueberry scorch virus* in blueberry etc. In addition to transmission by the vectors, these diseases are transmitted through infected planting materials and cuttings. Furthermore, several undetected viruses and viroids keep reducing yield and quality of Australian crops such as *Potato virus X*, earlier thought to be due to 'degeneration' of cultivars. Therefore, establishment of orchards of horticultural crops and seed production of potato free of damaging diseases is a key control measure. Towards this many countries have sanitary selection programmes and certification of clonal stock. In Australia, potato seed is certified free from *Potato virus Y* by the Australian Seed Potato Certification Authority and grapevine clonal stock needs to be certified free from several viruses. However, once the stock is infected, it is important to have robust methods to eliminate the infecting viruses.

In general, the term eradication is used to describe the process of destroying all infected plant material after an incursion of a new disease. An excellent example comes from Australia in eradicating citrus canker, a disease caused by *Xanthomonas* bacteria. There have been several outbreaks; 1912, 1991, 1993 (in NT), 2018 (NT + WA), 1984 and 2004 (QLD) that were successfully eradicated. While strict quarantine, sourcing plant material from disease-free fields/orchards, inspection etc. are important, often we encounter situations where the need for eradicating infecting microorganisms from planting stock for reuse.

Traditional Practices Used to Eradicate Pathogens from Clonal Stock

Heat therapy is a traditional method often used for this and is successful in eradicating some pathogens from planting material. It consists of keeping plants, or a part of them, at temperatures between 35°C and 54°C, within the physiological tolerance limits of each plant species, for a predetermined period. The selected temperature should represent the best compromise between virus degradation and plant survival. Advantage is that the threshold of thermal sensitivity of some viruses is lower than that of plant cells and that the damage caused to plant tissues by the thermal stress can more easily be reversed than viral damage. Some examples of heat therapy for disease eradication from planting material include mint (*Mentha* sp.) rhizomes infected with mint rust (*Puccinia menthae*). The fungus can be eradicated by immersing in water at 44°C for 10 minutes and then transferring to cold water. This is effective for mint rust existing in the form of urediniospores. Hot water treatment (50°C for 2 h) is also used to control ratoon stunt bacterium (*Leifsonia xyli* subsp. *xyli*) in sugarcane planting material (setts) as well as to control nematode infections in bulb crops. Some seed borne diseases can also be controlled by hot water treatment (e.g. some fungal pathogens in wheat seeds and leaf spot of brassicas caused by *Alternaria brassicae* and black spot caused by *A. brassicicola*). Traditionally heat therapy has been used to reduce the viral load, but some viruses are heat stable. Although heat therapy is useful in reducing the incidence of virus diseases, when used alone it is often inadequate for clean stock certification programs.

Tissue Culture-Based Methods for Eradicating Diseases Infecting Vegetative Plant Material

PTC plays a significant role in eradication of microbial infections in clonal crops as different therapies can be applied alone or in combination when a single therapy is not effective. Also, the methods can be applied anytime of the year and under highly controlled laboratory conditions, making them easily reproducible once protocols are established.

a) Meristem Culture

The simplest method is to use meristem tip culture because if extracted accurately, meristem is devoid of vascular tissue and consists only of actively dividing meristematic cells. Therefore, plants can be regenerated from the meristems without phloem limited bacteria and viruses such as Ampeloviruses causing leafroll disease in grapevine. The meristem is microscopic and is often less than 0.5 mm in most species. Shoot apical meristem (SAM) consists of an apical dome and one or two leaf primordia (Fig. 1A). Often meristem culture alone does not eradicate even phloem limited viruses. Moreover, there are many pathogens that are not limited to vascular tissue such as Nepoviruses (e.g. GFLV in grapevine), infecting the meristem and then therapies need to be combined to successfully eradicate the pathogens.

b) Combining Thermotherapy with Meristem Culture

Thermotherapy, when combined with meristem culture, is much more effective compared to the application of the two methods separately. Thermotherapy, as discussed above, either completely deactivates or partially kills the infecting virus. Thus, the chances of regenerating virus-free plantlets through meristem culture are greater when combined with thermotherapy. There are many examples such as eradicating GLRaV 1 and GLRaV 3 from infected 'Chancellor' grapevine by applying thermotherapy to in vitro plants followed by meristem culture (Dĩaz-Barrita et al. 2008), applying thermotherapy (37–40 °C for 4 weeks under hot air treatment) to shoots harvested from field-grown 'Oregon Spur–II" apple infected with Apple mosaic virus, Apple chlorotic leaf spot virus (ACLSV), Apple stem grooving virus, Apple stem pitting virus and Prunus necrotic ringspot virus followed by in vitro establishment and meristem culture (0.3 – 0.5 mm meristems effective for all viruses and 0.5- 0.6 mm effective for all except ACLSV – only 50% success for ACLSV) (Vivek and Modgil 2018) and applying thermotherapy (35 °C for 3 weeks) to potted nectarine (*Prunus persica* var *nectarina* Max) 'Arm King' infected with P*lum pox virus* (PPV) and *Prunus necrotic ringspot virus* (PNRSV) followed by meristem (1.3 – 2 mm) culture (Manganaris et al. 2003). As thermotherapy inactivates the virus, longer shoot tips that better survive in vitro culture can be used for culture initiation.

Thermotherapy (4 weeks with day/night conditions of 16/8 h and 40/36°C) of two-week-old cassava plants infected with African cassava mosaic virus and East African cassava mosaic Cameroon virus were used by Yéo et al. (2020) to establish tissue cultures from which meristems were isolated and cultured. They reported 88% of the regenerated plants free from both viruses. Combining thermotherapy with meristem culture for improved eradication of Onion yellow dwarf virus and Shallot latent virus from in vitro-cultured shallot shoots co-infected with both viruses (Wang et al. 2021) and *Bean yellow mosaic virus* from infected gladiolus (Sharifi Nezamabad et al. 2015) are among many other reports of successful use of this combined therapy.

c) In Vitro Chemotherapy

Since the first antiviral drug in humans was registered in the 1960s, many antiviral drugs have been developed. Antiviral agents targeting plant viruses are mainly derived from natural products and many products with varied control mechanisms have been developed. However, demonstration of the antiviral activity of a guanosine analogue agent, ribavirin (originally used as a drug for human hepatitis C virus) in tobacco cells infected with *Tomato spotted wilt virus* made this compound the most popular for virus eradication programs in tissue cultured plants. Its broad spectrum of activity against replication of both RNA and DNA viruses makes it the first choice in many virus eradication programs and is often combined with other therapies. Usually, the viricide is added in concentrations of 10 – 100 mg/L in media for growing infected plants and maintained for 2 – 6 weeks during which time thermotherapy regimes can also be applied. It is also used in combination with cryotherapy for viruses that are difficult to eradicate using a single therapy. Therefore, some examples of this method are discussed below under cryotherapy.

d) Cryotherapy

Cryotherapy involves treating the shoot tips of infected plants at ultra-low temperature, often in liquid nitrogen (LN) at -196 °C. The development of novel vitrification-based protocols has enabled cryotherapy to be applied once a protocol is developed. Unlike cryopreservation for conservation purposes, the recovery percentage of treated shoot tips need not be high for virus eradication.

In fact, when optimized protocols are applied with high survival percentages, more cells of the meristem survive the treatment, reducing the chances of virus eradication. For example, we used droplet vitrification to test the suitability of cryotherapy for virus eradication of leafroll disease in grapevine. 'Chardonnay' and 'Lakemont Seedless' were infected with GLRaV-3, Pinot gris' and 'Sauvignon blanc 316' infected with GLRaV-2, and another clone of 'Sauvignon blanc' infected with both GLRaV-1 and GLRaV-3. All the plants regenerated after cryo-treatment (one hour in LN) tested negative for the viruses after six months in the greenhouse. The regeneration percentages were from 13 % (Chardonnay') to 30 % (Sauvignon blanc') (Pathirana et al. 2015a).

In New Zealand the potato germplasm collection of about 950 genotypes was maintained in the field until we could introduce cryopreservation. At the time, the accessions in the collection had many virus infections. Before cryopreserving the germplasm, it is important to clean up the collection. Virus eradication efficiency in potato infected with *Potato virus S, Potato virus A* and *Potato virus M* is variable (20 – 100%) when chemotherapy (2 weeks in 100 mg/L ribavirin) is used alone, whereas combining it with cryotherapy was more effective (80 – 100% efficiency). Cryotherapy and thermotherapy applied alone was also not effective (Bettoni et al. 2022) (**Fig. 7**). Now this method is routinely used to eradicate viruses before cryopreserving the potato germplasm for long-term conservation.

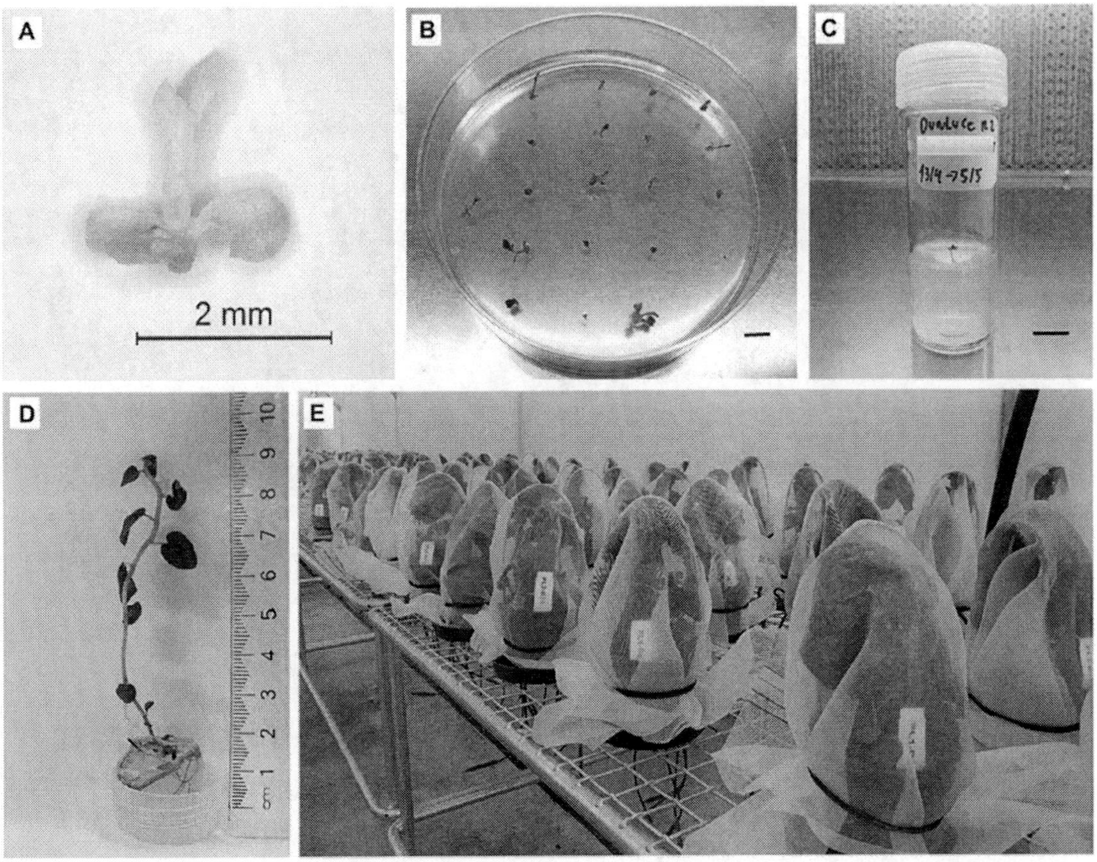

Figure 7. Shoot tip recovery process in potato "Dunluce" infected with *Potato virus S* following a combined chemotherapy + cryotherapy treatment. **A)** Shoot tip 1 week after combined chemotherapy + cryotherapy and **B)** 3 weeks recovery from cryoexposure. **C)** Shoot transferred to vial and **D)** grown for 3 months. **E)** Plants after 3 months of growth in the greenhouse. Bars = B 0.6 cm, C 0.7 cm. Reproduced from Bettoni et al. (2022).

Raspberry bushy dwarf virus (RBDV – a member of the species *Idaeovirus rubi* of the genus *Idaeovirus* in the family *Mayoviridae*) is one of just 17 horizontally transmitted viruses among over 1000 known plant viruses. Horizontally transmitted viruses are pollen-borne and after pollination and fertilization, get into the maternal tissue through the fertilized ovary. Thus, RBDV can spread rapidly within one flowering season and is most damaging and occurs in all raspberry growing regions in the world. Meristem culture, cryotherapy or thermotherapy alone are not effective on their own as it is aggressive and infects parts of the growing meristem as well. Therefore, we tested several combinations of chemotherapy, thermotherapy and cryotherapy. Chemotherapy (30 mg/L ribavirin) combined with thermotherapy (24°C for 8 h in dark and 39 °C for 16 h with light) for two weeks followed by cryotherapy was the most effective with 80 – 100% of plants regenerated after treatment testing virus-free in the greenhouse (Mathew et al. 2021). This work was conducted in New Zealand, and we cleaned up many clones of raspberry in our collection, so that the breeders can now use these genotypes as pollen or female parents in their breeding program without risking virus transmission to the progeny (**Fig. 8**).

Figure 8. Different stages of thermotherapy + chemotherapy followed by cryotherapy of raspberry tissue cultures infected with *Raspberry bushy dwarf virus*. **A**) Cultured meristems on regeneration media after combined therapies. **B**) Regenerating plantlets after 4 weeks in culture, **C**) Shoot transferred to individual tubs after 4 weeks. **D**) Plants acclimatizing in the greenhouse after 6 weeks of exflasking. **E**) Plants in the greenhouse after 1 year from exflasking.

Often, cryotherapy is combined with other therapies for more aggressive viruses as already described for the eradication of three viruses in potato (Bettoni et al. 2022) and for RBDV in raspberry (Mathew et al. 2021). As only the meristematic cells with no vacuoles survive and the differentiated cells with vacuoles get destroyed due to ice crystallization, cryotherapy can be regarded as a very precise meristem culture (**Fig. 9**).

Wide Hybridisation to Create Interspecific And Intergeneric Crosses

Polyploidy (whole genome duplication events) is common in plant evolution, particularly among cultivated plants. Chromosome doubling following hybridisation between two related species produces an allopolyploid with chromosome compliments from both species and can be fertile. There are many examples such as wheat (Durum wheat - *Triticum durum* $2n = 4x = 28$ tetraploid; bread wheat - *T. aestivum* $2n = 6x = 42$ hexaploid), canola - *Brassica napus* ($2n = 4x = 38$, AACC) with genomes from *B. oleraceae* ($2n = 18$, CC genome) and *B. rapa* ($2n = 20$, AA genome) and upland cotton (*Gossypium hirsutum*, AADD, $2n = 52$) (Pathirana and Carimi 2022). While this happens rarely in nature, with the technologies available, it is possible to produce polyploids experimentally.

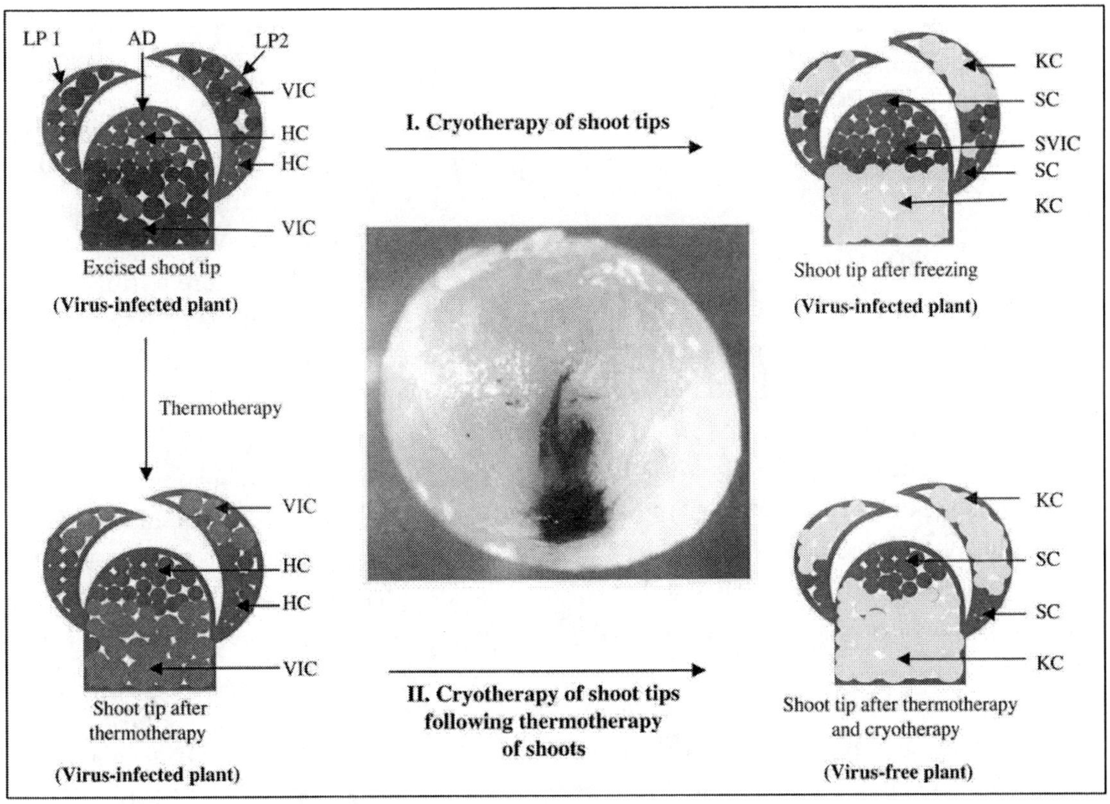

Figure 9. Schematic diagram depicting the effectiveness of the combination of thermotherapy and cryotherapy for enhanced elimination of viruses that can invade the meristematic cells efficiently. I) Most of the differentiated older infected cells are lethally injured whereas the youngest cells in the meristem survive the cryo-treatment. If the virus was not able to enter the meristem, the treatment would result in virus-free plants. However, in cases like here where the virus invades the meristem, shoots regenerated after cryo-treatment will remain infected. II) Additional suppression of virus and increased propensity of infected cells to be injured by cryo-treatment can be achieved by subjecting shoots to thermotherapy before excising shoot tips for treatment in liquid nitrogen. Thermotherapy causes stress and reduces survival of the cells and also accelerates degradation of viral RNA. An encapsulated raspberry shoot tip (1.5mm) used for cryotherapy is illustrated in the middle. AD, apical dome; HC, healthy cells; KC, killed cells; LP1, leaf primordium 1; LP2, leaf primordium 2; SC, surviving cells; SVIC, surviving, virus-infected cells; VIC, virus-infected cells. Reproduced from Wang et al. (2009).

Wide hybridisation is the term used to describe hybridisation between plants belonging to different species and genera. Species, by definition, is a group of plants or animals whose members are able to breed with each other, meaning individuals from different species cannot interbreed. Thus, it is generally difficult to obtain fertile progeny even if we are successful in producing seeds of interspecific hybrids. The reasons for failure can be before fertilization (pre-fertilisation barriers), for example the maternal species may have a longer style than the male species and the pollen tube may fail to reach the ovary even if it germinates on the stigma. An example of post-fertilization barrier is when the fertilized ovary cannot divide by mitosis or endosperm fails to develop. We have solutions to these problems by manipulating floral parts in tissue culture and successfully producing interspecific and intergeneric hybrids.

Man-made Species

The first man made species is *Raphanobrassica*, developed by Karpachenko, a Soviet scientist, back in 1928. His objective was to have a cabbage (*Brassica oleraceae*) on top of a radish (*Raphanus sativa*) root. Although he produced the inter-generic hybrid, the phenotype was not as expected **(Fig. 10 A and B)** (Karpechenko 1932). This research is continuing even today and there are registered Raphanobrassica cultivars used as feed for farm animals in many countries, including Australia and are identified as better suited to lower rainfall livestock systems with the ability to maintain green leaf and withstand dry periods and more consistent production across different growing conditions (Watt et al. 2023).

Another well-known man-made species through intergeneric hybridisation is Triticale, and we must rescue the young embryo from the ovary of the mother plant after hybridisation and grow in tissue culture to produce new fertile hybrid. Triticale has the high yield of wheat (*Triticum*) combined with the resilience (cold, drought resistance) of rye (*Secale*) in one species and is popular mainly in central Europe – Poland, Belarus, Germany, Lithuania, Belarus etc. as well as in China, Australia and New Zealand (**Fig. 10 C and D**).

Interspecific Hybrids of *Vaccinium* (blueberry) – A practical example

In New Zealand, the breeders have been trying to combine the valuable characteristics of highbush, lowbush and rabbiteye blueberries but had difficulty getting fertile hybrids for breeding. We developed methods to successfully culture fertilized ovules and also rescue embryos from successful interspecific crosses and culture them and obtain fertile hybrids. In this research we rescued more than 200 hybrids from 14 different combinations constituting tetraploid, pentaploid and hexaploid blueberries for the breeding programme. As a result of this work, blueberry cultivation could be extended to areas where there were no suitable cultivars and also extend the period of availability of the fruit in the market as a result of release of early and late maturing varieties (Pathirana et al. 2015b). Different stages of the tissue culture of fertilised ovules are given in **Fig. 11**, and the numbers of hybrids produced in different combinations are described in Pathirana et al. (2015b).

Embryo Rescue to Increase Hybrid Seeds in Apomictic Species

Opuntia ficus-indica or prickly pear cactus is an important forage and food source in arid and semiarid ecosystems and is the most important cactus species in cultivation globally. This species is known for its high degree of apomixis, the phenomenon of asexual seed production that occurs without fertilization. While apomixis is useful for clonal propagation, it is a hindrance in plant breeding programs where genetic segregation is sought for the selection of superior genotypes.

In a collaborative project with Italian colleagues, we compared the mature seed-derived seedlings with those regenerated from *in vitro* embryo rescue at different post-anthesis days (PADs) in four Italian cultivars. The zygotic seedlings were discriminated from apomictic seedlings using molecular marker analysis based on inter-sequence single repeat (ISSR) primers. Ovules cultured at 35 PADs resulted in the highest percentage of zygotic seedlings, from 51% to 98% in the four cultivars.

Mature seeds harvested in the field produced much fewer seedlings per seed than in tissue culture-based embryo rescue and a lower percentage of zygotic seedlings (from 14% to 63%) (Carra et al. 2023). Therefore, breeders can now increase the availability of zygotic seedlings in prickly pear breeding programs through *in ovulo* embryo culture. This method may be applicable to breeding in other apomictic species such as apple (Bisognin et al. 2009), citrus (Yadav et al. 2023), mango (Yadav et al. 2023), mangosteen (Baskaware and Deodhar 2023), etc.

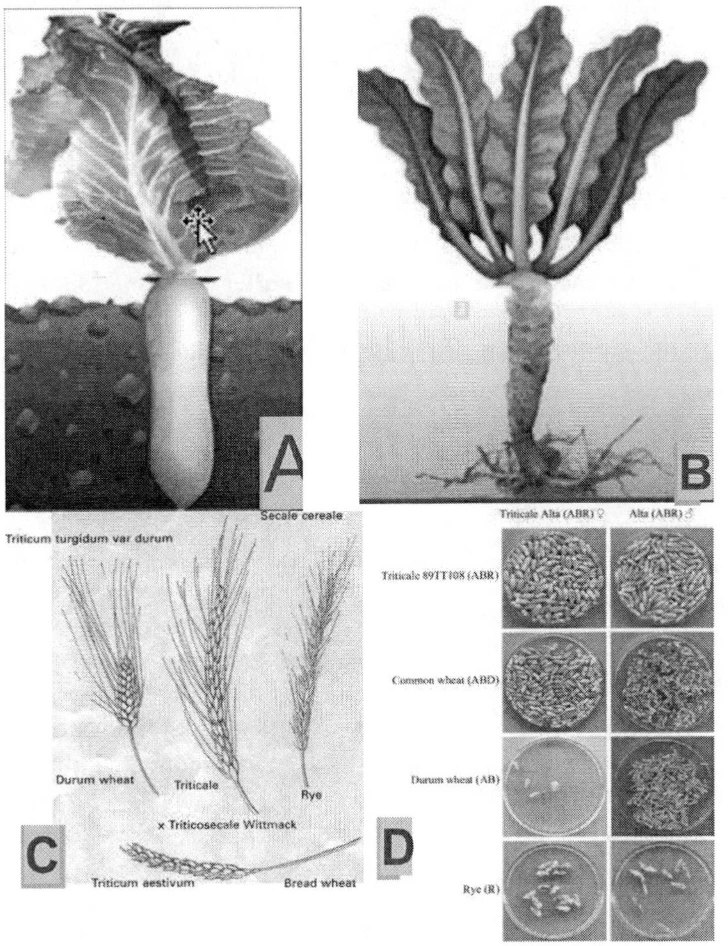

Figure 10. Man-made interspecific hybrids. Karpachenko wanted to produce a hybrid of cabbage with radish that has the root of radish and the top of cabbage (**A**) but what he obtained was something intermediate (**B**). **C**) Comparison of spikes and **D**) seeds of wheat, rye and triticale.

Figure 11. Stages of production of blueberry (*Vaccinium* spp.) interspecific hybrids through culture of fertilised ovules. **A**) Extracted ovules in culture. **B**) Expansion of ovules. **C**) Putative hybrid emerging from fertilised ovule callus, D) shoots growing from hybrid. **E**) rooted shoots of hybrid. **F**) Exflasked hybrids in the greenhouse after confirmation by SSR markers.

In Vitro Mutagenesis for Fast Tracking Crop Breeding

Traditional cross breeding takes about 7-8 years to develop elite material in annual crops and a further 3-4 years for field trials and their release. If back-crossing is involved, many more years will be required. Also, there is a limit to what we can achieve by cross breeding within the genome of the species.

Random mutation and selection

Induced mutagenesis allows to create new recombination through chromosome breakages and can also produce point (gene) mutations. Chromosome breakages that occur when ionising radiation is used as the mutagenic agent has the potential to break linkages of genes, creating novel genotypes. Chemical mutagens on the other hand produce more point mutations. Induced mutations are ideal for perennial crop improvement that have long turnover of generations, and most of the diversity we have in our fruit crops is due to sports or natural point mutations that have been selected and vegetatively propagated.

Traditionally, shoot tips or seeds are used to produce mutant populations, but large mutant populations need to be created for screening. This is expensive and difficult to achieve in the field, particularly for perennial crops. On the other hand, in vitro mutagenesis allows to handle large mutant populations in a small space; metaphorically in a Petri dish. Details of how to create such populations have been described in several reviews (Sharma et al. 2025;

Pathirana 2021; Pathirana 2011). After creating mutant population, we can even screen the population under in vitro conditions for different abiotic stress tolerance traits such as salt tolerance (Patade et al. 2008; Nikam et al. 2015), tolerance to toxic elements (Pathirana et al. 2002), heat tolerance (Das et al. 2000), cold tolerance (McClinchey and Kott 2008) etc. Many crop cultivars with herbicide tolerance have been developed through in vitro mutagenesis combined with in vitro selection, for example in sugarcane (Koch et al. 2012) and pepper (Venkataiah et al. 2005). Disease resistance screening of mutant populations in vitro can be achieved using bacterial suspensions; for example, we screened populations of gold kiwifruit for resistance to *Pseudomonas syringae* pv *actinidiae* – the most devastating pathogen for the New Zealand kiwifruit industry (Pathirana et al. 2016). We have also used a leaf disc method to screen a large mutant population of grapevine for *Botrytis* tolerance (Pathirana 2009).

In grapes, we first developed protocol for inducing embryogenic cultures (**Fig. 12a**) and then optimised the dose of gamma rays or chemical mutagen, ethyl methanesulfonate, and use the dose (dose = concentration of the mutagen x time of treatment) that results in 50% reduction in callus growth to treat a large population of callus cells (**Fig. 12b**) and then regenerate plants (**Fig.12 d & e**) (Carra et al. 2024; Pathirana and Carimi 2023). We can challenge the regenerating plants with the bacterium or its exudates. Alternately, we can have a large, chimera-free mutant population to screen in the field.

Figure 12. A scheme for using grapevine embryogenic cultures for mutation induction and screening. **A**) Embryogenic culture. **B**) Optimizing mutagen dose through growth reduction studies. **C**) Development of somatic embryos after treatment with optimized mutagen dose. **D**) Initial germination. **E**) Screening the germinated embryos for the trait of interest in vitro, **F**) in the greenhouse and **G**) under field conditions. Bars: (A, C, D) = 1 mm; (E) = 5 mm. Figure reproduced from Carra et al. (2024).

A protocol for plant regeneration from caryopses of *Indica* rice via a callus phase was developed and adopted to select regenerants in media simulating iron toxicity. Caryopses of three commercial Sri Lankan rice cultivars (RU 102, LD 355 and AT 353) were induced to produce callus on MS medium supplemented with either 2,4- dichlorophenoxyacetic acid (2,4-D) or chlorophenoxyacetic acid (CPA), and 6-benzylaminopurine (BAP). Plant regeneration was more efficient when 4-week-old calli induced using 9 μM 2,4-D and 0.5 μM BAP were transferred to MS medium supplemented with 2.2 μM BAP and 0.6 μM α-naphthaleneacetic acid (NAA). Three concentrations of Fe, three pH levels and three doses of γ irradiation (0, 100 and 150 Gy) were used in the selection experiment. Varieties LD 355 and AT 353 were more responsive than RU 102 to callus induction and plant regeneration. High iron concentration in combination with low pH significantly decreased callus induction and plant regeneration, but the few plants that were regenerated were more tolerant to iron toxicity and are being used in breeding programmes. The in vitro protocol developed for selecting rice mutants would be useful in supplementing the current breeding efforts to develop rice varieties tolerant to iron toxicity (Pathirana et al. 2002).

Use of In Vitro Culture Techniques in Plant Genetic Modification

Another well-known application of PTC is in plant transformation to produce genetically modified (GM) crops. Genetic transformation uses the natural process of gene transfer by a soil-borne bacterium in the genus *Agrobacterium* that produces crown-gall disease (by *A. tumefaciens*) and hairy root disease (by *A. rhizogenes*). These bacteria have small extrachromosomal circular DNA called plasmids. The tumour inducing plasmids (Ti plasmid) in these bacteria have a small region called transfer DNA (T-DNA) that gets transferred to the host plant and this natural process is used in experimental transfer of desired genes into crop plants. This process is exclusively done using in vitro cultures.

Hibiscus Transformation

In our experiments, we developed an efficient transformation and regeneration system for *Hibiscus rosa-sinensis* **(Fig. 13)** that will enable researchers to improve its quality in a number of ways (Trivellini et al. 2015). For example, by appropriate gene manipulation plants could be produced that have greater tolerance to frost or whose flowers have a much-extended display life.

Improved Selenium Accumulation in Solanaceous Species Through Transformation

Some plant species such as *Astragalus bisulcatus,* a legume, can tolerate growth on soils with high Se content. They are known as Se hyperaccumulators and can convert inorganic Se to methylselenocysteine (MeSeCys) by the enzyme selenocysteine methyltransferase (SMT). MeSeCys has been shown to have anticancer properties in mammals (Dong et al. 2001) and therefore more crop species that can accumulate MeSeCys would help in cancer prevention among populations, particularly those with low selenium intake. We used Agrobacterium transformation of two solanaceous species – tobacco (McKenzie et al. 2008) and tomato (Brummell et al. 2011) and in both species overexpression of *SMT* transgene resulted in increased accumulation of MeSeCys. When transgenic tomato plants were fertilised with selenite or selenate at the stage of fruit development, liquid chromatography mass spectrometry showed that MeSeCys accumulated in the fruit but not in leaves. Also, MeSeCys was produced more effectively from selenite on a percentage conversion basis, but greater accumulation of MeSeCys could be achieved from selenate due to its better translocation from the roots (Brummell et al. 2011).

Figure 13. Plant regeneration from callus induced from axillary buds of *Hibiscus rosa-sinensis* and *Agrobacterium*-mediated transformation. **A**) Callus induced from axillary bud explants on basal medium supplemented with 6-benzylaminopurine and β-naphthoxyacetic acid after 3 weeks. **B**) Callus induced from an axillary bud on selective regeneration medium after 10 days of culture. **C**) A plantlet regenerated on selective regeneration medium after 4 weeks. **D**) A transgenic plantlet after 6 weeks on rooting medium. **E**) Transgenic plants transferred to soil in the containment greenhouse. Figure reproduced from Trivellini et al. (2015).

Golden Rice – Golden Opportunity

It is common knowledge that golden rice, developed through genetic modification allowing the grains to accumulate high amounts of beta carotene (23-fold compared to non-GM rice), can alleviate blindness due to vitamin A deficiency in malnourished children in Africa and Asia (Diaz and Fridovich-Keil 2025). Similar to all other GM crops, Golden Rice also had opposition, mainly due to misinformation. However, rigorous testing for safety for the growers, consumers and the environment resulted in Australia, Canada, United States and several other countries to designate cultivation of this GM crop safe. Philippines is the first country, in 2021, to approve commercial cultivation of golden rice. Unfortunately, in April 2024, in response to a petition filed by Greenpeace and another local group, Court of Appeals in the Philippines issued a cease-and-desist order on the commercial propagation of golden rice, citing a lack of "full scientific certainty" regarding their health and environmental impact. Nevertheless, by 2023 another 12 countries were in the final stages of this approval process including China, Bangladesh, India, South Africa and Vietnam (Diaz and Fridovich-Keil, 2025).

These examples show how transgenic plants produced using in vitro technologies can help not only to understand biological processes but also improve crop species for improved nutrition. Despite the opposition for the technology, many countries have approved growing of GM crops and in the last 15 years, GM crops have given an additional income of US$285 billion and annual incomes are increasing with more acceptance (Wijerathna-Yapa and Pathirana 2022).

In Vitro Culture for Targeted Genome Editing

The advent of targeted genome modification (TGM) technologies, which are based on the application of targeting modules, has helped researchers to work with individual plants rather than large populations when selecting in randomly generated mutant populations. In targeted genome editing, we are able to recognize and modify a particular DNA sequence. The three main targeting modules currently in use are zinc finger (ZF) DNA-binding domains, transcription activator-like effector (TALE) DNA-binding domains and clustered regularly interspaced short palindromic repeat (CRISPR) systems. These techniques offer targeted genome editing capabilities, allowing for precise modifications to the genome of the crop cultivar under modification. Although tissue culture-based systems are traditionally used, researchers are exploring methods to bypass tissue culture as plants of several families are difficult in tissue culture. These include TGM delivery through *de novo* meristem induction, virus mediated delivery and graft-mobile delivery, reviewed by Li et al. (2024). However, these methods are in their infancy and results have so far been achieved only in model plant species.

Development of Homozygous Plants for Heterosis Breeding – Doubled Haploids

Homozygous plants are used in breeding programmes for mining useful genes and for heterosis breeding, to produce high performing hybrids. In cross pollinating species like most of our horticultural and floricultural crops, producing homogeneous lines require backcrossing for 7-8 generations and in perennials such as coconut it can take the lifetime of a researcher. Instead, we can use tissue culture to grow haploid plants from gamete cells found in the flower and diploidise them to produce 100% homozygous plants in one step.

Doubled haploid gentians

Gentians (*Gentiana triflora*) are one of the favourite Japanese flowers. In one of our collaborative projects, we produced doubled haploid lines in Gentians for our client, Hachimantai City in Japan and this program helped them to develop novel hybrids with even pink colour; gentians were known to have only purple flowers until then.

Anthers and ovaries from flowers at mid-late uninucleate stages of microspore development were cultured after treating them at 4°C for 48 h. In this work anthers and ovaries were cultured on a modified Nitsch and Nitsch (1969) medium supplemented with a combination of naphthoxyacetic acid (NOA) and BAP. The explants either regenerated new plantlets directly or produced callus that regenerated into plantlets upon transfer to basal media supplemented with BAP (**Fig. 14**). We used seven genotypes with different ploidy levels. Of the cultured ovaries 0–32.6% regenerated plants among the seven genotypes, whereas only 0–18.4% of cultured anthers produced plants. Nevertheless, all the seven genotypes responded either through ovary or anther culture. We used flow cytometry to assess the ploidy. There were more haploid (4.3 per 100 explants) and diploid plants (3.5 per 100 explants) than those with triploid (0.3 per 100 explants) or tetraploid (0.2 per 100 explants) relative nuclear DNA contents. All the diploid regenerants were shown to be gamete-derived (i.e. doubled haploids) by observing parental band loss using Randomly Amplified Polymorphic DNA (RAPD) markers (Pathirana et al. 2011a). Haploid plants were propagated on a proliferation medium and then treated with oryzalin for 4 weeks before transfer back to proliferation medium. Most of the resulting plants were diploids. Over 150 independently derived diploidised haploid plants were deflasked (Pathirana et al. 2011a).

Anther-derived fast-growing callus as a source for doubled haploid coconut

Coconut is a cross-pollinated perennial tree of immense importance in tropical countries. It can be propagated only by seeds, and crop improvement is through mass selection following mass, controlled or hand pollination. The current methods result in heterogeneous plantations and considering its perennial nature, there is a high demand for uniform, vigorous, elite planting materials for plantation establishment. Therefore, hybrid coconut will be high in demand if produced. Towards this objective, we investigated anther derived fast-growing callus (FGC) as a long-term source for doubled haploid production. After testing different media, sucrose concentrations and plant growth regulators, we found that the highest callogenesis and weight increase (growth) can be achieved with FGC cultured in solid Eeuwen's Y3 medium supplemented with 100 μM 2,4-dichlorophenoxyacetic acid with heat pretreatment at 38 °C for 3 days. Weight increase of FGC was negatively correlated with sucrose concentration. Incorporation of cytokinins facilitated the conversion of FGC to embryogenic callus. Presence of a cambium-like zone, a characteristic feature of embryogenic calli was confirmed by histological studies. Flow cytometry indicated that embryogenic calli derived from FGC originated from reduced microspores (Perera et al. 2021). Our results indicate that anther-derived FGCs are a good source for long-term production of haploid embryogenic callus for developing doubled haploid coconut for heterosis breeding programs.

In Vitro Technologies for Biopharming and for Producing Bioactive Compounds

Biopharming is the use of a living system as a host for the manufacture of biological drugs. In recent years, plants have become increasingly more attractive and acceptable as an expression platform for the production of vaccines and there has been much progress. There are many reasons which make expression in plants a more attractive option.

One main reason is that plants cannot host human pathogens and the production from gene to product is much faster than any other eukaryotic system, very attractive in pandemic situations.

Although plant-based vaccines have been produced for a long time, mainly for vaccines used in animals, recently Health Canada has authorised the use of a plant-produced virus-like particle vaccine against COVID-19 in humans; the vaccine, COVIFENZ®, is made by Medicago Inc.in Québec. However, the association of the company with tobacco industry was a barrier for international recognition of the vaccine (Duong and Vogel 2022).

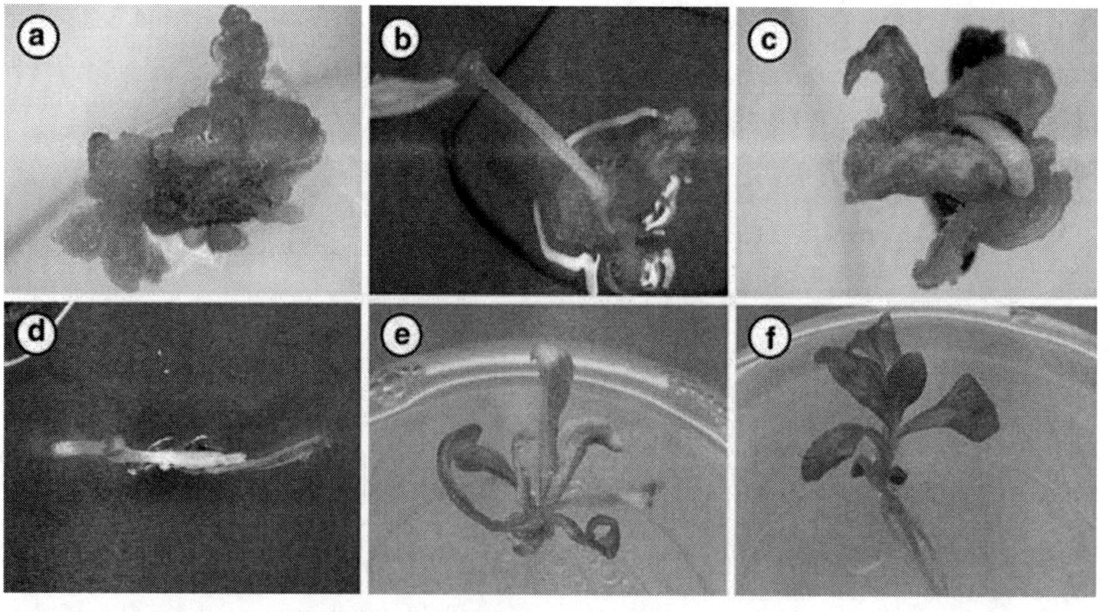

Figure 14. Typical response of *Gentiana triflora* anthers to culture on NOA/BAP supplemented media. **A)** Anther callus producing shoot initials in regeneration medium after 2 weeks under light. **B)** initial stage of plant regeneration from anther callus after 3 weeks on regeneration medium. **C)** Plant regeneration from anther-derived callus induced on NOA (1.4 mg l-1) and BA (1.8 mg l-1) medium after 6 weeks on regeneration medium. **D)** Direct plant regeneration from anther without callus formation in low NOA (0.7 mg l-1) and high BA (2.4 mg l-1) medium for 6 weeks. **E)** Regenerated shoot transferred to rooting medium. **F)** Rooting of shoots after 3 weeks on PGR-free medium.

In addition to the use in the pharmaceutical industry, in vitro plant cell cultures are used for production of natural flavors, agro-chemicals, colors, therapeutic proteins and bioactive compounds. There are many challenges in this area such as retention of the metabolic cues of natural plants in cell culture systems and scaling-up the production to create high yielding cell factories. These aspects have been reviewed in detail by Bapat et al. (2023).

Figure 15. The slow growth facility of kiwifruit at the Palmerston North site of the New Zealand Institute for Plant and Food Research Limited. Each accession is held in eight 35-mL plastic screw cap vials at 5°C after cold acclimation. Figure reproduced from Debenham and Pathirana (2021).

Application of In Vitro Techniques in Plant Conservation

Plant tissue culture repositories are an attractive option to conserve valuable clonal material and endangered species, away from the vagaries of a changing climate. They are much safer in the lab, and especially when duplicated in different labs. The plants from TC repositories are easily available for use, export and research.

Once established, cultures require constant attention and subculture when the plants start to senesce. Subculture not only increases labour costs but also increases the risk of contamination. Low temperature, low concentration of minerals and sucrose, low light intensity are used to reduce the frequency of subculture by slowing down metabolic processes. These repositories should have the backing of good database and IT capabilities so these can be searched for different characters and ordering. This author was involved in the development and methodological improvement of the in vitro repository of kiwifruit (*Actinidia* spp.) in New Zealand, currently holding over 900 accessions, each in six 35 ml screw-cap vials (**Fig. 15**). Before depositing in the cold culture room, kiwifruit tissue cultures are cold acclimated for 10 days at alternating temperatures with a short photoperiod (22°C with 5 μmol m^2 s^1 light for 10 h/2°C dark for 14 h). Thereafter in slow growth media, under low temperature and low light intensity, we can hold some genotypes for over 20 months without subculture (Debenham and Pathirana 2021). A total of 26 species, sub species and interspecific crosses are represented in this collection. Quality control of the collection is conducted regularly by comparing the SSR allele profiles of the parent plant in the field with those in tissue culture (Debenham and Pathirana 2021; Wiedow et al. 2017). The other more attractive and more secure option for conservation of clonal species or even seeds is cryopreservation. This author will elaborate on this method in a subsequent paper in IPPS Proceedings.

CONCLUSIONS

Plant tissue culture has become the foundation in advanced biotechnologies. The most widespread application is micropropagation for producing clonal plant stocks with the market expected to grow to US$ 2.1 billion by 2030. It is widely applied in plant propagation of horticulture, floriculture and forestry. Micropropagation can be achieved through direct and indirect organogenesis, somatic embryogenesis or microtubers. The method is also used in development and deployment of disease-free high-health plants for agriculture, horticulture and forestry. The advantage of tissue culture-based technologies for eradication of pathogens from planting materials is that different methods such as meristem culture, thermotherapy, chemotherapy, electrotherapy and cryotherapy can be combined when a single therapy is not effective for aggressive pathogens.

In vitro technologies have a central role in the deployment of new cultivars to the industry much faster and efficiently than traditional field-based methods. This application encompasses an array of technologies to produce crop cultivars or even new man-made species with traits of interest. Thus, in vitro technologies have a central role in ploidy manipulation, distant hybridization, doubled haploid production, increasing sexual seeds in apomictic species and also understanding biochemical pathways. In vitro technologies are more efficient for the development and deployment of mutants with improved traits and in developing genetically modified crops using traditional transformation methods as well as gene editing techniques. The market share of cell culture technologies is also increasing in the production of plant-based pharmaceuticals, food ingredients, cosmetics, flavours, dietary supplements, fragrances, and biostimulants. For biopharmaceuticals alone the market is expected to reach USD 50 billion during the next five years.

In vitro cultures are used in plant conservation, both in the form of tissue culture repositories for medium term conservation or cryopreserved collections for the long term.

LITERATURE CITED

Bapat, V.A., Kavi Kishor, P.B., Jalaja, N., Jain, S.M. and Penna, S. (2023). Plant Cell Cultures: Biofactories for the Production of Bioactive Compounds. Agronomy *13*:858.

Baskaware, S.V. and Deodhar, M.A. (2023). Apomixis and sexual systems in various species of *Garcinia* with special reference to *Garcinia indica* (Thouars) Choisy. International Journal of Fruit Science *23*:25-33.

Bettoni, J.C., Mathew, L., Pathirana, R., Wiedow, C., Hunter, D.A., McLachlan, A., Khan, S., Tang, J. and Nadarajan, J. (2022). Eradication of *Potato Virus S*, *Potato Virus A*, and *Potato Virus M* From infected in vitro-grown potato shoots using in vitro therapies. Frontiers Plant Science *13*:878733.
DOI: 10.3389/fpls.2022.878733

Bisognin, C., Seemüller, E., Citterio, S., Velasco, R., Grando, M.S. and Jarausch, W. (2009). Use of SSR markers to assess sexual vs. apomictic origin and ploidy level of breeding progeny derived from crosses of apple proliferation-resistant Malus sieboldii and its hybrids with Malus × domestica cultivars. Plant Breeding *128*:507-513.
DOI:10.1111/j.1439-0523.2011.01924.x

Brummell, D.A., Watson, L.M., Pathirana, R., Joyce, N.I., West, P.J., Hunter, D.A. and McKenzie, M.J. (2011). Biofortification of tomato (*Solanum lycopersicum*) fruit with the anticancer compound methyl-selenocysteine using a selenocysteine methyltransferase from a selenium hyperaccumulator. Journal of Agricultural and Food Chemistry *59*:10987-10994.
DOI: 10.1021/jf202583f

Carimi, F., Carra, A., Panis, B. and Pathirana, R. (2016). Strategies for conservation of endangered wild grapevine (*Vitis vinifera* L. subsp. *sylvestris* (C.C. Gmel.) Hegi). Acta Horticulturae *1115*:81-86.
DOI: 10.17660/ActaHortic.2016.1115.13

Carra, A., Carimi, F., Bettoni, J.C. and Pathirana, R. (2019). Progress and challenges in the application of synthetic seed technology for ex situ germplasm conservation in grapevine (*Vitis* spp.). In: Faisal M, Alatar AA (eds) Synthetic Seeds: Germplasm Regeneration, Preservation and Prospects. Springer International Publishing, Cham. pp. 439-467.
DOI:10.1007/978-3-030-24631-0

Carra, A., Catalano, C., Pathirana, R., Sajeva, M., Inglese, P., Motisi, A. and Carimi, F. (2023). Increased zygote-derived plantlet formation through in vitro rescue of immature embryos of highly apomictic *Opuntia ficus-indica* (Cactaceae). Plants *12*:2758.

Carra, A., Wijerathna-Yapa, A., Pathirana, R. and Carimi, F. (2024). Development and applications of somatic embryogenesis in grapevine (*Vitis* spp.). Plants *13*:3131.

Cook, D.C., Liu, S., Edwards, J., Villalta, O.N., Aurambout, J.P., Kriticos, D.J., Drenth, A. and De Barro, P.J. (2012). Predicting the benefits of banana bunchy top virus exclusion from commercial plantations in Australia. Plos One *7*: e42391.
DOI: 10.1371/journal.pone.0042391

Das, A., Gosal, S.S., Sidhu, J.S. and Dhaliwal, H.S. (2000). Induction of mutations for heat tolerance in potato by using in vitro culture and radiation. Euphytica *114*:205-209.
DOI:10.1023/A:1003965724880

Debenham, M. and Pathirana, R. (2021). Establishment and management of an in vitro repository of kiwifruit (*Actinidia* spp.) germplasm In: Ahmad N, Strnad M (eds) Meta-topolin: A Growth Regulator for Plant Biotechnology and Agriculture. Springer Nature, Switzerland pp. 279-291

Dĩaz-Barrita, A.J., Norton, M., Martĩnez-Peniche, R.A., Uchanski, M., Mulwa, R. and Skirvin, R.M. (2008). The use of thermotherapy and in vitro meristem culture to produce virus-free 'Chancellor' grapevines. International Journal of Fruit Science *7*:15-25.
DOI: 10.1300/J492v07n03_03

Diaz, J.M. and Fridovich-Keil, J.L. (2025). Golden Rice - Genetically Modified Grain. Encyclopaedia Britannica. Available at https://www.britannica.com/technology/golden-rice

Dong, Y., Lisk, D., Block, E. and Ip, C. (2001). Characterization of the biological activity of γ-glutamyl-se-methylseleno-cysteine: A novel, naturally occurring anticancer agent from garlic. Cancer Research *61*:2923-2928.

Duong, D. and Vogel, L. (2022). Why is WHO pushing back on a Health Canada-approved Medicago SARS-CoV-2 vaccine? CMAJ *194*:E504-E505. DOI: 10.1503/cmaj.1095992

Fuchs, M. and Lemaire, O. (2017). Novel approaches for viral disease management. In: Meng, B., Martelli, G.P., Golino, D.A., Fuchs, M. (eds). Grapevine Viruses: Molecular Biology, Diagnostics and Management. Springer International Publishers, pp. 599-621. DOI:10.1007/978-3-319-57706-7

Karpechenko, G. (1932). Plant genetics in the USSR (In Russian). Bull Appl Bot Leningrad *4*:101-114.

Koch, A.C., Ramgareeb, S., Rutherford, R.S., Snyman, S.J. and Watt, M.P. (2012). An in vitro mutagenesis protocol for the production of sugarcane tolerant to the herbicide imazapyr. In Vitro Cellular & Developmental Biology - Plant *48*:417-427. DOI:10.1007/s11627-013-9493-0

Li, B., Sun, C., Li, J. and Gao, C. (2024). Targeted genome-modification tools and their advanced applications in crop breeding. Nature Reviews Genetics *25*:603-622. DOI: 10.1038/s41576-024-00720-2

Manganaris, G.A., Economou, A.S., Boubourakas, I.N. and Katis, N.I. (2003). Elimination of PPV and PNRSV through thermotherapy and meristem-tip culture in nectarine. Plant Cell Reports *22*:195-200. DOI: 10.1007/s00299-003-0681-y

Mathew, L., Tiffin, H., Erridge, Z., McLachlan, A., Hunter, D. and Pathirana, R. (2021). Efficiency of eradication of *Raspberry bushy dwarf virus* from infected raspberry (*Rubus idaeus*) by in vitro chemotherapy, thermotherapy and cryotherapy and their combinations. Plant Cell Tissue and Organ Culture *144*:133-141. DOI:10.1007/s11240-020-01829-y

McClinchey, S.L. and Kott, L.S. (2008). Production of mutants with high cold tolerance in spring canola (*Brassica napus*). Euphytica *162*:51-67. DOI:10.1007/s10681-007-9554-8

McKenzie, M.J., Hunter, D.A., Pathirana, R., Watson, L.M., Joyce, N.I., Matich, A.J., Rowan, D.D. and Brummell, D.A. (2008). Accumulation of an organic anticancer selenium compound in a transgenic Solanaceous species shows wider applicability of the selenocysteine methyltransferase transgene from selenium hyperaccumulators. Transgenic Research *18*:407. DOI: 10.1007/s11248-008-9233-0

Nicholas, P. (2006). Grapevine Clones Grown in Australia Part 2. Soth Australian Research and Development Institute.

Nikam, A.A., Devarumath, R.M., Ahuja, A., Babu, H., Shitole, M.G. and Suprasanna, P. (2015). Radiation-induced in vitro mutagenesis system for salt tolerance and other agronomic characters in sugarcane (*Saccharum officinarum* L.). The Crop Journal *3*:46-56. DOI:10.1016/j.cj.2014.09.002

Nitsch, J. and Nitsch, C. (1969). Haploid plants from pollen grains. Science *163*:85–87.

P&S Intelligence. (2024). P&S Intelligence (2024). Micropropagation market. Available at: https://wwwpsmarketresearchcom/market-analysis/micropropagation-market

Patade, V.Y., Suprasanna, P. and Bapat, V.A. (2008). Gamma irradiation of embryogenic callus cultures and in vitro selection for salt tolerance in sugarcane (*Saccharum officinarum* L.). Agricultural Sciences in China *7*:1147-1152.

Pathirana, R. (2009). Production of improved *Botrytis*-tolerant Sauvignon Blanc clones combining *in vitro* and mutagenic technology. Final Report Submitted to MAF/SFF (Ministry of Agriculture and Forestry/Sustainable Farming Fund). Report Number L07/038. Plant and Food Research, Palmerston North, New Zealand. Available at: https://www.researchgate.net/publication/308151012_Production_of_improved_Botrytis_tolerant_Sauvignon_Blanc_clones_combining_in_vitro_and_mutagenic_technology .

Pathirana, R. (2011). Plant mutation breeding in agriculture. CABI Reviews: Perspectives in Agriculture, Veterinary Science, Nutrition and Natural Resources *6*:1-20

Pathirana, R. (2021). Mutations in plant evolution, crop domestication and breeding. Tropical Agricultural Research and Extension 24 (3). DOI:10.4038/tare.v24i3.5551

Pathirana, R. and Carimi,. F. (2022). Management and utilization of plant genetic resources for a sustainable agriculture. Plants *11*:2038.

Pathirana, R., Carimi, F. (2023). Studies on improving the efficiency of somatic embryogenesis in grapevine (*Vitis vinifera* L.) and optimising ethyl methanesulfonate treatment for mutation induction. Plants *12*:4126.

Pathirana, R., and Carimi, F. (2024). Plant Biotechnology—An Indispensable Tool for Crop Improvement. Plants *13*:1133.

Pathirana, R., Deroles, S., Hoeata, K., Montefiori, M., Tyson, J., Wang, T., Datson, P.M. and Hellens, R.P. (2016). Fast-tracking kiwifruit breeding through mutagenesis. Acta Horticulturae *1127*:217-222. DOI: 10.17660/ActaHortic.2016.1127.34

Pathirana, R., Frew, T., Hedderley, D., Timmerman-Vaughan, G. and Morgan, E. (2011a). Haploid and doubled haploid plants from developing male and female gametes of *Gentiana triflora*. Plant Cell Reports *30*:1055-1065. DOI: 10.1007/s00299-011-1012-3

Pathirana, R., Harris, J.C. and McKenzie, M.J. (2008). A comparison of microtubers and field-grown tubers of potato (*Solanum tuberosum* L.) for hexoses, sucrose and their ratios following postharvest cold storage. Postharvest Biology and Technology *49*:180-184.

Pathirana, R., McLachlan, A., Hedderley, D., Carra, A., Carimi, F. and Panis, B. (2015a). Removal of leafroll viruses from infected grapevine plants by droplet vitrification. Acta Horticulturae *1083*:491-498. DOI: 10.17660/ActaHortic.2015.1083.64

Pathirana, R., Sansom, C., Smallfield, B.M. and Heyes, J.A. (2011b). In vitro mutagenesis to adapt ulluco (*Ullucus tuberoses* L.) to New Zealand conditions. Available at: https://www.researchgate.net/publication/268515853 Paper presented at the Australasian Postharvest Conference, Lorne, Australia,

Pathirana, R., Wiedow, C., Pathirana, S., Hedderley, D., Morgan, E., Scalzo, J., Frew, T. and Timmerman-Vaughan, G. (2015b). Ovule culture and embryo rescue facilitate interspecific hybridisation in blueberry (*Vaccinium* spp.). Acta Horticulturae 1083:123-132. DOI: 10.17660/ActaHortic.2015.1083.13

Pathirana., R., Wijithawarna, W.A., Jagoda, K. and Ranawaka, A.L. (2002). Selection of rice for iron toxicity tolerance through irradiated caryopsis culture. Plant Cell Tissue and Organ Culture *70*:83-90.

DOI: 10.1023/a:1016025728740

Perera, P., Pathirana, R. and Vidhanaarachchi, V. (2021). Somatic embryogenesis in anther-derived fast-growing callus as a long-term source for doubled-haploid production of coconut (*Cocos nucifera* L.). J. Natn. Sci. Foundation Sri Lanka *49*:39-49. DOI:10.4038/jnsfsr.v49i1.9834

Saeiahagh, H., Mousavi, M., Wiedow, C., Bassett, H.B. and Pathirana, R. (2019). Effect of cytokinins and sucrose concentration on the efficiency of micropropagation of 'Zes006' *Actinidia chinensis* var. *chinensis*, a red-fleshed kiwifruit cultivar. Plant Cell Tissue and Organ Culture *138*:1-10. DOI:10.1007/s11240-019-01597-4

Scion. (2024). Machine learning whispers tissue culture secrets. Available at: https://wwwscionresearchcom/news-and-events/news/2023-news-and-media-releases/machine-learning-whispers-tissue-culture-secrets Downloaded on 24 January 2025

Senaratna, T., McKersie, B.D. and Bowley, S.R. (1989). Desiccation tolerance of alfalfa (*Medicago sativa* L.) somatic embryos - influence of abscisic-acid, stress pretreatments and drying rates. Plant Science *65*:253-259. DOI:10.1016/0168-9452(89)90072-1

Sharifi Nezamabad, P., Koohi Habibi, M., Dizadji, A. and Kalantari, S. (2015). Elimination of Bean yellow mosaic virus through thermotherapy combined with meristem-tip culture in gladiolus corms. Journal of Crop Protection *4*:533-543.

Sharma, V., Thakur, M., Maan, S.S., Verma, K., Thakur, A. and Penna, S. (2025). In Vitro Mutagenesis: A Non-invasive Technology for Effective Crop Improvement to Assure Food and Nutritional Security—Current Trends, Advancements and Future Perspectives. Journal of Plant Growth Regulation *44*:484-507. DOI:10.1007/s00344-024-11484-8

Skoog, F. and Miller, C.O. (1957). Chemical regulation of growth and organ formation in plant tissue cultured in vitro. In: Symp. Soc. Exp. Biol. pp 118-131.

Trivellini, A., Ferrante, A., Hunter, D.A. and Pathirana, R. (2015). *Agrobacterium tumefaciens*-mediated transformation of axillary bud callus of *Hibiscus rosa-sinensis* L. 'Ruby' and regeneration of transgenic plants. Plant Cell Tissue and Organ Culture *121*:681-692. DOI:10.1007/s11240-015-0738-y

Venkataiah, P., Christopher, T. and Karampuri, S. (2005). Selection of atrazine-resistant plants by in vitro mutagenesis in pepper (*Capsicum annuum*). Plant Cell, Tissue and Organ Culture *83*:75-82. DOI:10.1007/s11240-005-2678-4

Vivek, M. and Modgil, M. (2018). Elimination of viruses through thermotherapy and meristem culture in apple cultivar 'Oregon Spur-II'. VirusDisease *29*:75-82.

DOI: 10.1007/s13337-018-0437-5

Wang, M-R., Hamborg, Z., Blystad, D-R. and Wang, Q-C. (2021). Combining thermotherapy with meristem culture for improved eradication of onion yellow dwarf virus and shallot latent virus from infected in vitro-cultured shallot shoots. Annals of Applied Biology *178*:442-449. DOI: 10.1111/aab.12646

Wang, Q.C., Panis, B., Engelmann, F., Lambardi, M. and Valkonen, J.P.T. (2009). Cryotherapy of shoot tips: a technique for pathogen eradication to produce healthy planting materials and prepare healthy plant genetic resources for cryopreservation. Annals of Applied Biology *154*:351-363.

Watt, L., Bell, L., Stutz, R. and Williams, S. (2023). Managing forage brassicas in Australian mixed farming regions. Available at: mla-forage-brassica-management-guide_v06.pdf

Wei, S.L., Li, P.M., He, Z.P. and Han, B.W. (1990). Rapid propagation of monomerous stem by tissue culture *Dahlia pinnata* Cav. Acta Agriculturae Universitatis Pekinensis *16*:381-385.

Wiedow, C., Debenham, M., Bassett, H.M., Corpe, S., Seelye, J., Pathirana, R. and Gardiner, S.E. (2017). Quality control in a kiwifruit germplasm collection. Paper presented at the Fourth Internaitonal Symposium on Molecular Markers in Horticulture, Napier, New Zealand, 7-10 March, 2017.

Wijerathna-Yapa, A. and Pathiran,a R. (2022). Sustainable agro-food systems for addressing climate change and food security. Agriculture *12*:1554.

Yadav, C.B., Rozen, A., Eshed, R., Ish-Shalom, M., Faigenboim, A., Dillon, N., Bally, I., Webb, M., Kuhn, D., Ophi,r R., Cohen, Y. and Sherman, A. (2023). Promoter insertion leads to polyembryony in mango — a case of convergent evolution with citrus. Horticulture Research *10*. DOI: 10.1093/hr/uhad227

Yéo, E., Kouassi, M., Pita, J., Kouassi, N., Koné, D. and N'guetta, S. (2020). Using thermotherapy and meristem tip culture for producing virus-free cassava planting material from six varieties cultivated in Côte d'ivoire. Int. J. Sci. Technol. Res. *9*:1607-1612.

Combined Proceedings IPPS 74:220-232. 2024.

Plant Quality Control in Action!

Sabine Suess

Natural Area Nursery, Workshop Rd, Whiteman Park, WA 6068, Australia

sabinesuess@hotmail.com.au

Keywords: nursery management, irrigation, best practice, surveillance, scouting, monitoring, natural area, Western Australia

Abbreviations:

AFP: Air Filled Porosity
EC: Electrical Conductivity
IPM: Integrated Pest Management
NACMS: Natural Area Consulting and Management Services
NGIQ: Nursery and Garden Industry Queensland
NIASA: Nursery Industry Accreditation Scheme Australia
WHC: Water Holding Capacity

IPPS Vol. 74 – 2024

Summary

The aim of this paper is to describe my role as Plant Yield Coordinator at Natural Area Nursery, Western Australia in maximising plant yield and optimising water management. The key areas described range from monitoring and identifying areas of concern, water management, determining and implementing actions, documentation, to research and development. The focus is on the role's integration with the nursery team, and practical ways adopted in the overall application. Results achieved include improved plant quality, reduced plant losses and better water management. In addition, the approach allowed for early targeted action, with the ability to be proactive and optimise resources. Consequently, in many areas of nursery operations the efficiencies were able to be improved. At the same time, documentation and research continually is being refined and allows for analysis and improvement of future procedures and management decisions. Finally, the broader benefits of a closely forged and self-driven team are realised.

INTRODUCTION

This paper describes my role as Plant Yield Coordinator at Natural Area Nursery in Western Australia (Natural Area Nursery, 2024), its key aspects, and anticipated outcomes. Natural Area Nursery **(Fig. 1)**, is a part of our company Natural Area Consulting and Management Services, located in Whiteman Park, Western Australia. The company provides environmental rehabilitation services, with teams in consulting, projects, seed, field work, animal management and the nursery.

The nursery is NIASA accredited and produces native plants for our own teams and their projects, and sells directly to clients for revegetation and landscaping.

Plants are produced seasonally, with most plants dispatched within three months in Autumn and Winter. The nursery currently dispatches approximately 1.2 million plants annually. The nursery's capability and specialty are to produce over 600 species and provenance specific stock. Many of the wide range of species are recalcitrant and difficult to propagate. The origin of the plants is traceable, and plants are grown from pure seed, avoiding hybrids.

Water is supplied by a bore, and overhead irrigation. The nursery has open areas using benches and pallet areas, and has several buildings specialised for various growing environments. The large diversity of plants, various pot sizes, growing media, and growing stages present challenges to balance all their care. Many variables for their growing requirements must be considered such as nutrition, growing media, water, sunlight, wind, and temperature exposure.

At the time of writing, the nursery is operated by 12 permanent staff, which increase up to 18 staff during peak production with the addition of a flex team. The area of open irrigation space and buildings amount to approximately 21,000 m^2.

Figure 1: Natural Area Nursery (Photo Matt Wood NACMS).

PLANT QUALITY STORY AND THE INTRODUCTION OF PLANT YIELD COORDINATOR ROLE

Production numbers were a focus previously, however with time, a need to reduce plant losses and improve quality was recognised. In addition, water conservation became crucial. With all this in mind, the Plant Yield Coordinator Role was introduced. The role's principal focus areas were maximising plant yield and optimising irrigation management. The anticipated outcomes included improved nursery efficiency and productivity, solid plant quality and refined water efficiency. It is important to emphasize that Quality Control is a team effort, to which all personnel contribute. However, the role of Plant Yield Coordinator was developed to allow early targeted action, being proactive and optimizing our resources. In essence, being the eyes for the nursery, helping the manager and team to prioritise tasks to avoid undesired outcomes. The Plant Yield Coordinator role involves the key aspects listed in **Table 1**, which are described in further detail.

Table 1. Key aspects of plant yield coordinator's work

- Monitoring of stock, infrastructure, and procedures
- Management of water and reticulation
- Implementation of actions by liaising with teams
- Documentation
- Research and development

Monitoring of Stock, Infrastructure, and Procedures

To begin with, areas of concern need to be detected and identified by performing regular surveillance or scouting. For this to occur, allocation of a regular scheduled time to monitor and assess is crucial.

Detailed Surveillance or Scouting

My approach has been adapted from Greenlife Industry Australia's (GIA) excellent information on the execution of various types of surveillance (GIA, 2024 a, b, c). A wealth of nursery best practice procedures is available on their website under the title technical information: Australian Plant Production Standard (APPS) https://nursery-productionfms.com.au

For surveillance, I walk through the nursery in a pattern to sample areas randomly, including areas at risk or of high value. Some of the scouting walks are shorter and focused on solely assessing water requirements. Other walks are more detailed with surveillance of many areas of the nursery. The water assessment walks are performed more frequently compared to the detailed surveillance.

For detailed surveillance I carry necessary equipment with a toolbelt, as listed in **Figure 2**. Any problem areas noted I record on my clipboard or electronic device for later decision making on required actions. **Table 2** lists examples monitored in detailed scouting procedures regularly.

Table 2. Detailed scouting focus areas.

- Note areas experiencing losses
- Detect hybrids, species corrections, labelling errors
- Check infrastructure problems
- Manage environment of buildings
- Check biosecurity and hygiene (perimeter, disinfestation and hygiene procedures, water and growing media testing)
- Check media moisture and plant stress (short- and long-term watering balance)
- Investigate plant health problems (detection and diagnostics)
- Detect weed focus areas (stock, ground, perimeter)
- Check production technique (potting up, cuttings technique)
- Detect pest animals or wildlife (identify type of animal involved)

Note Areas of Plant Losses

Noting areas of plant losses assists to update and accurately count stock. In addition, this allows actioning prompt removal of empty pots, which in turn results in improved pest and weed control, and optimising space.

Detect Hybrids, Species Corrections, Labelling Errors

Early detection of errors in species purity and accuracy again assists to maintain an accurate count in stock, and reduces grading pressures later in the period of dispatch.

Check Infrastructure Problems

Detecting problems with buildings or surrounds such as erosions, leaks, and damage to structures, allows early addressing of the problem. This aids in reducing costs, improving safety, and staying abreast in maintenance and repairs.

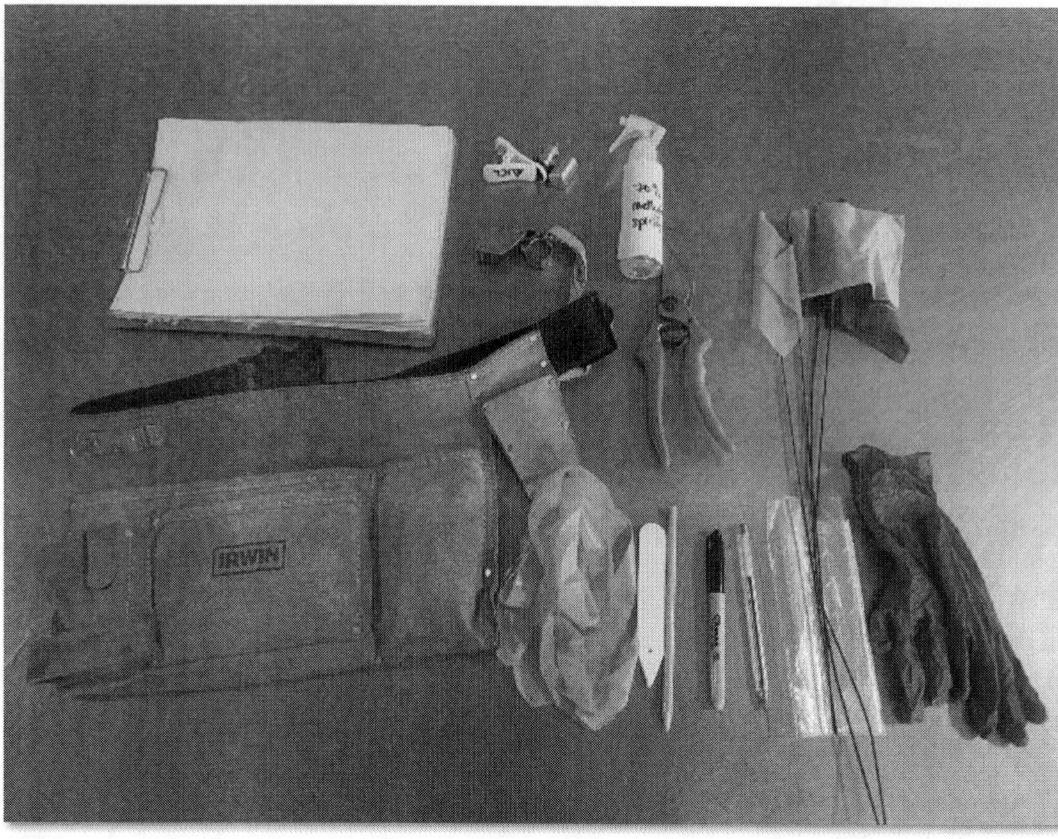

Figure 2. Equipment carried during my detailed scouting walks. (Photo Sabine Suess NACMS). These include: Recording equipment (clip board or electronic device), marker pen, white sheet of paper or plastic, plant tags, coloured markers or flags, disinfectant, gloves, secateurs, sample jar/zip lock bag, magnifying loupe, magnifying lens for camera, mobile phone with camera.

Manage Environment of Buildings

Part of my role is to manage the internal environment of production buildings. Several buildings are equipped with automated or manually regulated features. This may involve operating retractable shade roofs, or sides, and doors. Our cuttings building is highly automated, with light, humidity, and temperature sensors regulating automated fans, a wet-wall and shade curtain. Regular monitoring, maintenance and troubleshooting of building function and environment ensures optimal conditions for the plants.

Check Biosecurity and Hygiene

Monitoring biosecurity and hygiene includes assessing areas such as entrance, exit and perimeter control. In addition, disinfestation, and hygiene procedures, and testing of water quality and growing media quality are all integral in reducing the risk of introduction and spread of disease and weeds.

Check Media Moisture and Plant Stress

As we produce native plants for the purpose of natural area revegetation and landscaping, it is important the plants are hardened to survive in the harsh West Australian environment. Therefore, minimizing overwatering is essential, nonetheless watering according to their need. I visually and physically test random areas and known trouble spots throughout the nursery, including edges exposed to wind or reduced water. Some of the techniques used to assess optimal water applied are summarized in **Table 3** and shown in **Fig. 3.**

Table 3. Visually and physically checking media moisture and plant stress.

- Visually check media surface
- Assess entire growing media moisture by removing plant from pot
- Check for plant wilting or rotting
- Check for chronic overwatering signs (moss, algae, liverwort, fungus gnats)

Figure 3. Assessing media moisture by removing plant from pot. **A.** 140 mm pot, **B.** Forestry Tube (Photos Sabine Suess NACMS).

Investigate Plant Health Problems

Again, my approach has been adapted from the excellent resource GIA (2024 a, b, c), where crop monitoring and site surveillance is described in detail. When I encounter an area of plant health concern, I check for patterns in areas such as listed in **Table 4**. For example, identifying whether factors such as weather condition on the day of production, origin of the seed/cutting, or operator technique, could be a common denominator. Additionally, examining aspects such as positioning relative to wind, water, sun, and proximity to other plants. Some plants may need to be spaced, trimmed, or moved to a more suitable location. Examining media moisture is naturally essential to identify problems in water application.

Table 4. Plant health investigation factors.

- Date Potted, Provenance and Potter ID
- Positioning relative to wind, sun, water, other plants
- Moisture of pot and root health
- Check foliage and stem health
- Check presence of weeds
- Check soil pH and EC, AFP, WHC
- Laboratory sample analysis

A full examination of the entire plant including roots is helpful, and looking with the assistance of magnification, such as referred to in **Table 5** and **Fig. 4**. Sometimes pests in foliage difficult to detect can be dislodged by gently beating the plant against a white sheet, GIA (2024, a, b, c).

Table 5. Magnification options.

- 10 x magnifying loupe
- Macro settings on camera/phone
- Attachable macro lens for phone
- Microscope

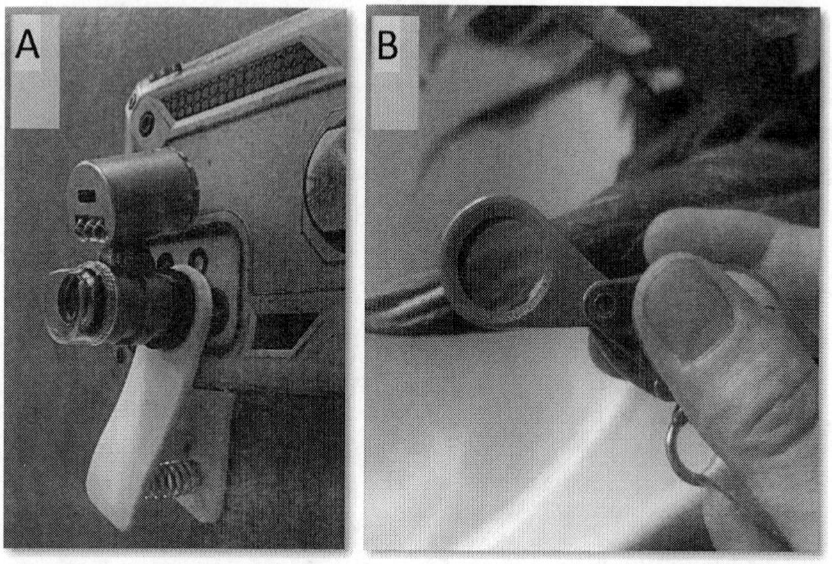

Figure 4. Useful field magnification options **A.** Lens attachment and mobile phone camera **B.** Hand-held loupe 10 x magnification. (Photos Sabine Suess NACMS).

If an unknown pest or disease is encountered, a sample is collected for further analysis under a microscope. Correctly identifying the type of pest or disease allows targeting the problem promptly and develop sound IPM management. In addition, a database of plant species and time of year affected by certain pest or disease pressures can be determined.

In Western Australia, photos and a description of a pest or disease can be submitted to the MyPestGuide Reporter App (https://www.agric.wa.gov.au/apps/mypestguide-reporter or email mypestguide@dpird.wa.gov.au). Another useful resource for pest identification is the NGIQ Pest Identification tool (https://pestid.com.au/).

If the cause of the plant's problem still cannot be found, then I consider further tests such as soil pH, EC, and even AFP and WHC. If needed, a sample could be submitted to an external laboratory for analysis.

Detect Weed Focus Areas

Keeping weeds under control is crucial for efficiencies and plant quality. We found it useful to do what we affectionately call "Power weeding": Rapid hand weeding of the entire nursery, within at least 2 weeks, targeting to remove weeds that go to seed within that time. This is an approach adapted from the excellent management plan for weeds in production nurseries by Conroy and Manners (2020), and article by Barker and Neal (2016).

Any weed focus areas and infestations are then marked with a coloured tag or flag for a team to deal with separately. This strategy ensures the whole nursery is weed controlled in a time-efficient manner and allows a targeted effort where it is needed.

In our strategy, weeds in the ground and perimeter are key areas to control, to prevent spreading into stock. Common rapidly growing and seeding weeds targeted in our nursery include *Cardamine hirsuta*, *Euphorbia maculata* and *Gnaphalium spp.*

We also focus aggressively on *Marchantia polymorpha* and *Sagina procumbens*, as they are difficult to treat once established. These weed examples are displayed in **Fig. 5.**

Any stock plant that begins to develop mature seed is a risk also; it is important seed is trimmed or harvested before contaminating other stock species or the ground. Targeted pre-emergent treatments are used where appropriate to reduce weed pressures. The main points in our weed control are summarized in **Table 6.**

Figure 5. Common weeds of priority in our nursery **A.** *Cardamine hirsuta* **B.** *Gnaphalium spp* **C.** *Chamaesyce maculata* **D.** *Marchantia polymorpha* **E.** *Sagina procumbens* (Photos Sabine Suess NAMCS).

Table 6. Our weed control focus areas.

- Focus aggressively on weeds likely to go to seed within 1-2 weeks, spread rapidly or are difficult to treat
- Use of coloured flags or markers for infestations for targeted control
- Seeds from stock can become invasive and should be harvested
- Ground weed control is crucial
- Targeted pre-emergent herbicide as appropriate

Check Production Technique

As part of investigating plant health problems, I monitor production technique, as per summary in **Table 7** and **Fig. 6**. This allows early action to educate staff or correct the problem where appropriate. Production techniques affect plant growth, survival and growing standard requirements.

Figure 6. Example of production technique problem – "J roots". (Photo Sabine Suess NACMS).

Detect Pests Animals or Wildlife

Pests or wildlife encountered include rats/mice, rabbits, kangaroos, southern bandicoots, and birds. Animals chew on plants, remove seedlings or seeds, pull out tube stock or disturb soil in seed trays and large stock. Determining the animal pest or wildlife culprit as the cause of damage can be difficult. Early detection and identification allow correct and appropriate action.

We use a multimodal approach to investigate the cause of the plant damage, listed in **Table 8**. Again, a database of plant species and time of year for certain animal pressures can be established.

Table 7. Production technique.

- Potting up technique such as J roots, depth of potting, trauma to roots
- Angle of plant stem, branching/growth pattern
- Cutting preparation and sticking technique

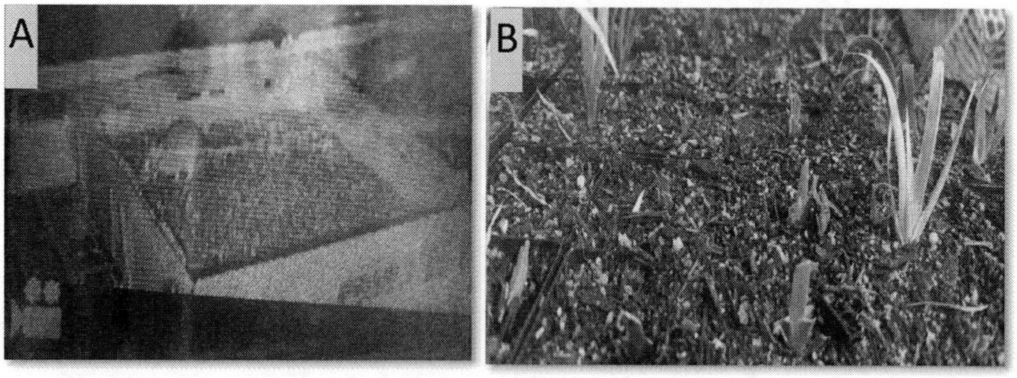

Figure 7. Examples of evidence of animals affecting plants **A.** Motion triggered camera image of a rodent in a seed tray. **B.** Suspected damage from a rabbit to *Machaerina preissii* seedlings. (Photos Sabine Suess NACMS).

Table 8. Methods for detecting and identifying pest animals or wildlife.

- Motion triggered cameras
- Trapping
- Chew/damage pattern, plant species targeted
- Scats
- Nests
- Perimeter walk to check breaches to barrier and tracks

Management of Water and Reticulation

Managing water and reticulation is part of my role. A summary of work related to this role is listed in **Table 9**. Water is becoming an increasingly precious resource, in addition, optimising water application helps to minimise losses and plant quality issues due to weeds, pests and disease. We schedule irrigation based on careful daily assessment, considering forecast data, our weather station records and scout feedback. We can remotely observe our weather station data and monitor and adjust the scheduling using a mobile phone app. To keep our reticulation system working well, I am also responsible for weekly reticulation maintenance and testing, regular water quality and usage monitoring, and responding to system breakdowns and troubleshooting.

Table 9. Management of water and reticulation to optimise water quality and application.

- Irrigation scheduling and system monitoring
- Reticulation maintenance and testing
- Water quality and usage monitoring
- Responding to system breakdowns and troubleshooting

Implementation of Actions by Liaising with Team

After collecting information, the next step is to determine, plan and implement an action. In essence, working out what needs to be done, if it needs doing and how urgent it is, and who will do it. Action decisions are made keeping plant value, efficiency, available resources, and nursery priorities in mind. This means liaising with the team to prioritise and schedule the actions.

Our core members are all assigned coordinator roles, they create the team and determine a time and way to get the action done. For larger work most of the team will be tasked for an allocated time, for example the maintenance team. Other times it might be a specialised small team such as the nutrition team. The coordinators drive the action, and the responsibility is shared by all. **Table 10** lists some of the examples in interaction with coordinators and their teams.

There is a need to remain flexible, as priorities and available resources can change frequently. Succession planning is important, for example, when someone is away on leave, ensuring key staff are aware of current routines.

Table 10. Implementation by liaising with coordinators and teams.

- **Maintenance team:** Grading, weed focus, "power weeding", spacing, trimming, staking, plant movement, covering plants, pre-emergent herbicide application
- **Nutrition team:** Treat for nutrient deficiency, boost growth, water or environmental stress, soil conditioning, beneficial microbes, salt treatment
- **Pest and Disease team:** Treat for insects and disease, ground weeds, control animal pests
- **Production team:** Inform of plant losses and difficult species
- **Seed team:** Inform of seeds ready to collect
- **Plant Movement, and Dispatch team:** Plant placement short and long term
- **Water Management team:** Scheduling, reticulation repairs
- **Management team:** Inform of major issues, trouble shooting

Communication methods are tailored to each coordinator and team and are listed in **Table 11**. Ensuring that the actions are completed, is critical, and is part of my role. Tasks are ticked as they are completed, and I will follow-up with the respective team or manager if a task has become urgent. Informing key staff on outcome and progress for plants of interest is also important.

Table 11. Communication with coordinators and teams.

• Whiteboard (Categorised tasks, and three most urgent marked with asterisk) • Teams posts and groups (including photos) • Prestart meetings • Informal communication

Documentation

Refining record keeping is an ongoing project and part of my and everyone's role. Good record keeping will help in the analysis of data, and therefore in the planning and improvement of procedures. Some examples include irrigation scheduling, fertilising requirements, when to propagate each species, plant spacing or trimming requirements, and plant placement. Some of the documents we are developing and updating are listed in **Table 12.** I also contribute to the creation and updating of Standard Operating Procedures, and to Integrated Pest Management planning.

Table 12. Examples of records being updated and developed.

• Nursery buildings maintenance • Water management • Pest and disease treatment • Plant treatment trials • Plant placement • Nutrition treatment • Plant health • Growing media testing • Losses recording, stock counts • Integrated Pest Management development • Standard Operating Procedures

Research and Development

The final part of my role is to advance our research and development, which is a long-term time investment. I research the current literature to continually improve procedures and foster a proactive and sustainable approach. This involves investigating new and improved products, procedures, equipment, infrastructure, and collating and creating training and information resources. In addition, analysing and solving priority problems presented, such as unique plant production requirements. For example, I have created an electronic resource library, informative and motivational posters, contributed to the development of a nursery training manual and developed product summaries with flow charts. **Table 13** describes the categories of research and development in more detail.

Table 13. Research and development categories.

Plant Production Support	**Infrastructure/Equipment**
• Pest/disease products • Nutrition/tonics • Herbicides • Growing media refinement	• Buildings • Machinery • Tools
Procedures • Efficiency • Innovation • Plant yield • Best practice • Sustainability	**Training/Information** • Library (electronic and paper) • Posters • Training manual • Flow charts • Product summaries

OUTCOMES AND CONCLUSION

In reflecting over our past operations and growth, we can see multiple benefits of targeted nursery quality control. The benefits are both of short and long-term nature, and are summarized in **Table 14**.

One of the most important benefits is perhaps the final point listed in **Table 14**: Pride in all staff, for the combined efforts as a self-driven team to efficiently grow a quality product. Staff morale is held high in a team where every individual can grow, contribute, and drive efficient routines, with a common goal to produce beautiful healthy plants at our Natural Area Nursery (**Fig. 8**), to give back to our precious natural areas in Western Australia.

Table 14. Benefits of targeted nursery quality control

- Reduced plant losses, improved stock quality, increased profitability, happier clients
- Early recognition of problem areas for the production and sales perspective
- Increased efficiency in dispatch grading, weeding and all maintenance tasks
- Optimised water efficiency and quality
- Improved record keeping and documentation
- Motivation for procedure improvement and innovation
- Pride in all staff for the combined efforts as a self-driven team to efficiently grow a quality product

Figure 8. Natural Area Nursery during a beautiful sunrise. (Photo Sabine Suess NACMS).

ACKNOWLEDGEMENTS

The creation of this manuscript would not have been possible without the following people: I thank Andrew Nguyen, Davina Summers, David Hancock, and the nursery team for all their tremendous support. I thank the IPPS Australia Board for the opportunity to present at the 52nd IPPS Conference in Ballina, and for publishing this manuscript. Finally, I thank my editor Dr Ranjith Pathirana for guidance in the completion of this manuscript.

LITERATURE CITED

Conroy, J and Manners, A. (2020). A management plan for weeds in production nurseries. Available at: https://nurseryproductionfms.com.au/wp-content/uploads/download-manager-files/Weed-management-plan-HR.pdf (Accessed on: 08 December 2024).

Barker, A and Neal, J. (2016). Frequent hand weeding saves money. NC State Extension Publications. Department of Horticultural Science. Available at: https://content.ces.ncsu.edu/frequent-hand-weeding-saves-money (Accessed 08 December 2024)

GIA. (2024a). Greenlife Industry Australia. How to implement an IPM management plan. Available at: https://nurseryproductionfms.com.au/elearning/ (Accessed on: 02 June 2024).

GIA. (2024b). Greenlife Industry Australia. BioSecure HACCP – Crop monitoring. Available at: https://nurseryproductionfms.com.au/elearning/ (Accessed on: 02 June 2024).

GIA (2024c). Greenlife Industry Australia *BioSecure HACCP* – Site surveillance. Available at: https://nurseryproductionfms.com.au/elearning/ (Accessed on: 02 June 2024).

Natural Area Nursery (2024) Natural area Perth native nursery. Available at: https://naturalarea.com.au/perth-native-nursery/ Accessed on: 23 October 2024.

Combined Proceedings IPPS 74:233-236. 2024.

South African Exchange

Joshua Taylor

45 Grasslands Rd, Boneo 3939, Victoria, Australia

taylorjosh607@gmail.com

Keywords: load shedding, nursery management, woodfire heating, potting mix, labour cost, Gauteng

Summary

In Australia, a first world nation with generally a gold standard in production Horticulture, we're used to the luxuries of potting machines, tray fillers, seed machines, sticking robots, remote climate control, quality potting mix, fertiliser, pest control and more simply quality and reliable water, gas and electricity supply, whereas South Africa is not. Load shedding (periods of time where power cuts off), similar issues with water, a weak currency, poor potting mix (making out of anything you can find), little nursery automation and the list goes on. Growing plants still isn't easy even with all our modern machines and sterile potting mixes and quality water, imagine how hard growing plants is without all of that! This paper will be going a little bit more in depth into the issues that South African growers experience every day. I was able to experience South African growers' passion and enthusiasm that enables them to overcome these issues daily, truly one of the most inspiring things I've been able to witness.

INTRODUCTION

My name is Joshua Taylor, and I am from Rosebud on the Mornington Peninsula 60 km south of Melbourne. I am currently employed at a propagation nursery, Peninsula Growers as Head Grower. I started my horticultural journey at a young age of 15. I took advantage of the 'work experience' opportunity that my school offered and did a week's work experience at Southern Advanced Plants. After I had completed this week's work experience, I knew that this was the career that I wanted to pursue. I began an apprenticeship at Peninsula Growers shortly after year 10 in 2018. I now have 6 years of industry experience and love my job more and more every day.

Load Shedding

Load shedding occurs when the demand out meets the supply causing power to trip in certain areas for unprecedented periods of time, from my experience this load shedding can last 30 mins to many hours if not days. On the days where temperatures can exceed 45 degrees Celsius you can understand how this could become problematic, disabling (shade screens, extraction fans, air conditioners, etc.).

Many nurseries had coal generators as a reliable source of a backup power supply (**Fig. 1**), diesel generators are too expensive to purchase and too expensive to run for prolonged periods of time

Figure 1. Coal Generator.

Climate Control

Also in Gauteng Johannesburg was a nursery called 'Random Harvest'. This was a retail nursery that specialised in growing native and indigenous plants. In **Fig. 2** you will see concreted propagation benches with assorted trays of freshly propagated plants in a community tray format. These propagation beds were woodfire heated (**Fig. 3**). Gas is a resource that is rather expensive compared to the relatively inexpensive woodfire heater.

Figure 2. Concrete propagation benches.

Figure 3. Woodfired boiler.

Potting Mix

Another common issue in South Africa that growers experience is the lack of quality potting mixes. We are very spoilt in Australia, I can order a fresh load of potting mix and have it delivered less than 24 hours later, sterilised, free of weed seed and any other contaminants or pathogens. This is a rare luxury for most South African growers. This grower was in Gauteng (Johannesburg), they couldn't get good quality potting mix in to their site without the cost being disproportionate. Instead, they source different materials from neighbouring farms including (mulch, grass clippings, soil products, coir, e.g.). After materials have been collected, they are moved into windrows where they are left to decompose and slowly turn into a usable potting mix (**Fig.4**).

Labour

One of the major differences between South African and Australian nurseries is the size of the labour force. In Australia, nurseries face several challenges, including high minimum wages, difficulties in finding institutions to provide effective staff training, and notably high staff turnover rates. These issues often lead Australian nurseries to invest heavily in automation to mitigate labour-related problems and control wage costs whereas in South Africa, labour costs are low and is available to hire anytime of the year (**Fig. 5**).

Figure 4. Decomposing windrows.

Figure 5. Labour for nurseries is available in abundance in South Africa.

CONCLUSION

There are vast differences between South African and Australian Horticultural industries. Despite the many challenges that the South African horticultural industry experiences they have never failed to adapt to the continuously changing conditions and grow exceptional quality plants that are on par with the Australian industry and the rest of the world. During my South African experience, I met many extraordinary people. This experience has enabled me to network and forge relationships with like-minded people that are passionate about this industry.

Acknowledgements

I am very grateful for the opportunity to experience diverse horticultural industries through the IPPS exchange to South Africa 2024 and would encourage other young leaders in this industry to apply for future exchange.

Combined Proceedings IPPS 74:237-250. 2024.

Designing Fresh Air – Future Strategies for Built Environments

Mark Thomson and Anya Rodic

Eco Effective Solutions, Studio 1, Block 1, Witton Barracks, 9 Lambert Road, Indooroopilly, Queensland, Australia. 4068

mark@ecoeffective.com.au

Keywords: biofilter, biophilic design, eco services, indoor air quality, VOCs

Summary

This report explores specific biophilic design strategies and their role in enhancing indoor environment quality through the integration of vegetation and sustainable building practices. As urbanisation increases there is a growing responsibility to address the poor air quality, volatile organic compounds (VOC's), and limited exposure to nature that impact humans physical and psychological health. The assessment of case studies and research evaluate biofiltration techniques and exemplars of green design that highlight the effectiveness of vegetation in improving indoor air quality in commercial office spaces and learning environments.

A key focus of this report is The Revitaliser, a biophilic indoor solution designed by Eco Effective Solutions and Eco Environment, which incorporates vegetation, biofilters and sustainable interior elements to improve indoor air quality and create a productive workspace.

The Revitaliser was built for the International Healthy Buildings Conference and installed in the conference lobby space at the Brisbane Entertainment and Conference Centre, Australia. Findings from case studies such as a German BMW Manufacturing facility and Australian classroom trials, confirm the benefits of integrating vegetation in indoor environments. They include lower indoor temperatures, enhanced cognitive function and improved psychological well-being. The Revitaliser's carefully designed bio filters and materials with specific plants greatly reduce VOC's, stabilise CO2 levels and boost indoor levels. Combined with hygienic space management via commercial cleaning and plant management, effective decision making and learning can be boosted with increased occupant well-being.

This report aims to emphasise the significance of prioritising natural and vegetation solutions in design and planning to transition the built environment towards more sustainable and productive workspaces and built environments. By collaborating with scientific researchers, built environmental consultants and legislators, current knowledge and technologies can be brought on to foster safer and healthy indoor spaces, adopting innovative new ecoservices.

INTRODUCTION

Concerns about indoor air quality and general environmental health are growing as urbanisation increases. Reduced access to nature, high concentrations of volatile organic compounds (VOCs) and poor air quality all have a detrimental impact on human productivity and health. Fresh air is considered good for people as it does not contain dirt, pollutants or dangerous substances. Typically, in urban environments fresh air is confused with Outdoor air, sourced for office and living environments. Breathing in pollutants causes symptoms such as, but not limited to, headaches, fatigue, dizziness, respiratory disease, asthma, runny nose, etc. In the workplace this is vital for the productivity and safety of occupants who spend their working and living hours inside. The integration of vegetation in an indoor environment can aid in improving air quality, noise reduction and occupant well-being. Utilising vegetation to improve indoor spaces can be optimised with biofiltration plant design, often changing the composition of the air. The benefits of biofilters include a reduction of office stress levels via the psychological benefits of plants in the workplace, absorption, diffraction and reflection of noise resulting in improved space acoustics, the reduction of carbon dioxide levels, absorption of electromagnetic radiation, and an improvement of humidity levels through leaf transpiration. The Revitaliser was a built demonstration of indoor environment quality which was designed in collaboration with Eco Environment (1) with specific biophilic principles. The utilisation of natural elements, especially vegetation, to improve indoor spaces is emphasised by biophilic design principles. This paper discussed how vegetation can be strategically used to improve air quality, temperature control, noise reduction, and occupants' well-being in workplaces and built environments.

Exemplars

A comprehensive review of studies and case examples was conducted to evaluate the impact of indoor vegetation on the environment and its occupants.

Real-time testing methods were employed to access information such as temperature, air quality and occupation conditions.

Fig. 1 shows the green spaces created at BMW manufacturing. Indoor air quality improvements based on the absorption of CO2 and VOCs by plants.

Figure 1. Indoor space at BMW Manufacturing created for temperature regulation through plant use in office spaces

Fig. 2 shows the use of green space in Manly Hotel, Brisbane, Australia. The structure has a west-facing wall with a green façade that helped to lower the indoor temperature by up to 5°C.

Figure 2. The structure at Manly Hotel in Brisbane has a west facing wall; the integration of a green façade helped to lower the indoor temperature by up to 5°C.

At the Waste Administration Centre in Logan, Queensland, a greenhouse is strategically placed to provide airflow in to the building through a fernery (**Fig. 3**).

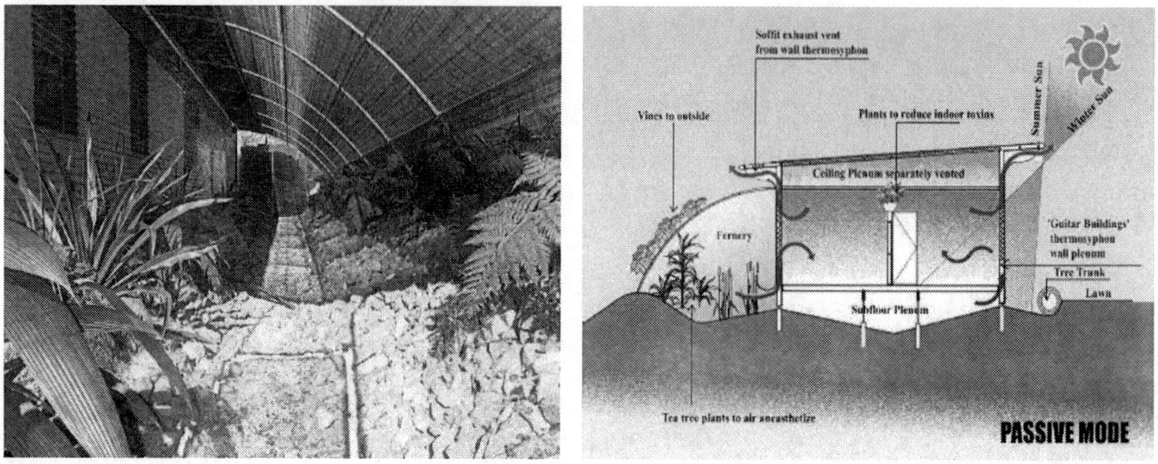

Figure 3. A greenhouse strategically designed to provide air flow through a fernery at the Logan Waste Administration Centre in Queensland.

EVIDENCE-BASED PRACTICES

The Impact of CO2

At Berkeley Lab, USA researchers found that even moderately elevated levels of indoor carbon dioxide resulted in lower scores on six of nine scales of human decision-making performance (The Australian 2012). (**Fig. 3**). When occupants struggle to be productive that will affect work efficiency as well.

Figure 4. Effect of elevated carbon dioxide content in workspace on nine scales of human decision making.

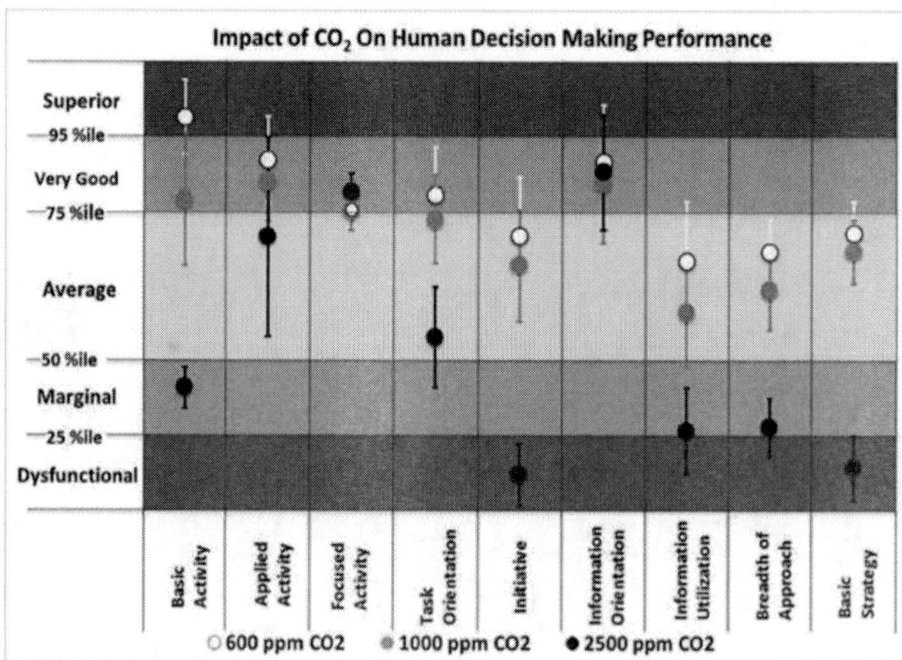

Plants as a treatment

Psychological benefits of plants in the office environment include reduction in stress, anxiety, and fatigue, as measured in studies conducted by the University of Technology Sydney. A 50% reduction in stress and a 65% reduction in negative emotions among occupants were recorded (Burchett et al., 2010).

Indoor Biophilic Trials – Queensland School Trials

A study performed by John Daly and Professor Margret Burchett was aimed at investigating the effects of indoor plants on classroom performance in the composite classes of Year 6 and 7 students in three independent schools in the Brisbane region. Students were tested with standard tests before plant placements and retested every 4 weeks for 6 months of plant presence or absence. The data collected showed that plants in the classroom can improve student performance (**Fig 5 – 7**).

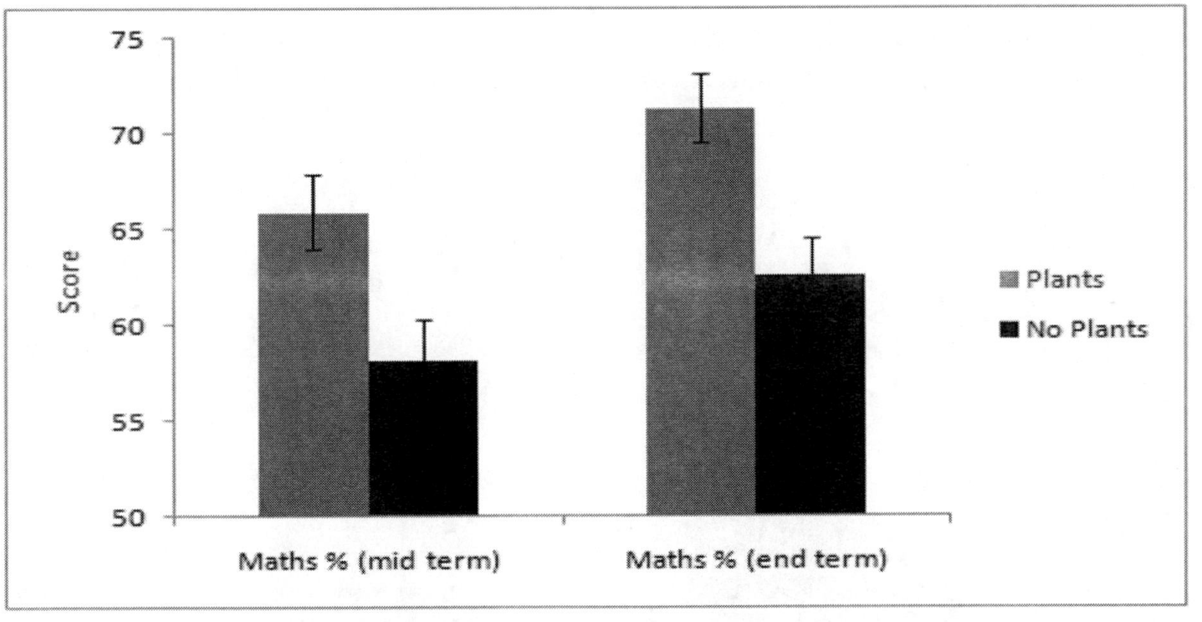

Figure 5. Comparison of changes in mid-term and end of term maths scores, in classes with and without plants in School A (All Saints Albany Creek) (Means and SE; n = 69–72).

Figure 6. Comparison of changes in mid-term and end of term spellings scores in classes with and without plants in School A (All Saints Albany Creek) (Means and SE; n = 69–72).

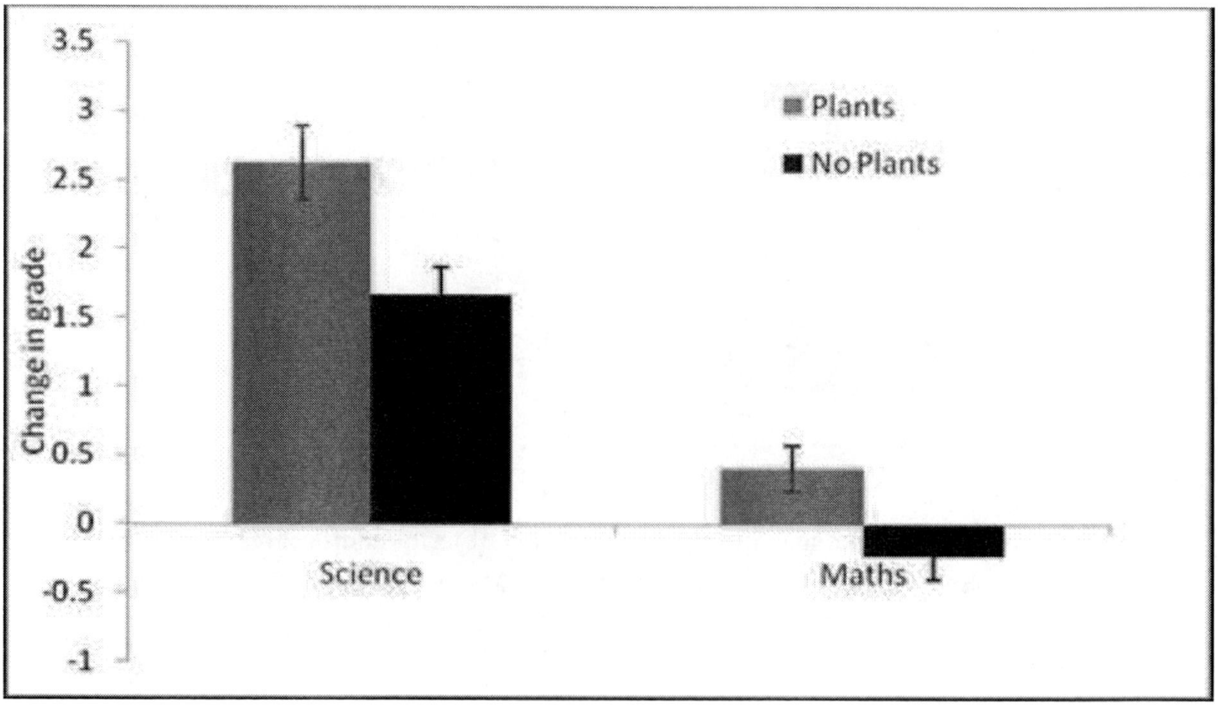

Figure 7. Comparison of end of term science and maths grades in classes with and without plants in School B (All Saints Anglican Gold Coast). (Means and SE, n = 149).

The results indicated that the integration of plants in the classroom led to consistently improved performance. The improvements ranged from 10-14% in every subject. According to this research two or three plants per 100 m^2 in an office can significantly reduce levels of CO2 and VOC's (Daly et al. 2024).

THE REVITALISER

The Revitaliser is an innovation in interior design to improve occupants' decision-making performance and learning environments. Designed by Eco Effective Solutions, under the supervision of Mark Thomson, the Revitaliser combines vegetation, bio filters, and contemporary fit-out elements while incorporating biophilic principles to deliver healthier and more productive workspaces **(Fig. 8)**.

Figure 8. Interior design to improve air quality and staff performance using plants.

The principles are:

1. Healthy vegetation can improve fresh air
2. Construction techniques can reduce VOCs
3. Interior design can improve productivity
4. Particular furniture and fittings can reduce VOCs
5. Acoustics, ergonomics and lighting are critical
6. Workspaces are processes not products

Principle One

Improving air quality through vegetation requires a strategic approach that involves correct soil management, plant selection, positioning and maintenance. Indoor soil mixes need to be designed and managed by specialists to ensure optimal plant health and filtering properties. Maintaining regular rotation and management of plants will aid in their ability to purify airborne pollutants. Selecting plant species with high oxygen generation and VOC absorption will further improve indoor air quality. Additionally, the positioning and maintenance of plants should be designed and managed to ensure adequate airflow around them, maximising circulation and the ability to eliminate pollutants.

Principle Two

In order to reduce VOCs in construction, sustainable maintenance methods, effective building techniques, and cautious material selection are required. Selecting durable materials and suitable repair techniques is crucial to reducing emissions over time. By lowering occupant exposure to dangerous chemicals, the use of low-toxic adhesives and zero-VOC paints enhances indoor environmental quality. Additional advantages of off-site prefabrication offer reduced on-site VOC and indoor air quality. Effective cleaning and waste management, including the use of low VOC cleaning agents and regular air duct cleaning are critical. Additionally, by lowering exposure to pollutants at crucial points, and adhering to a carefully planned construction sequence improves both safety and indoor environmental quality. To maintain sustainability future disassembly and repurposing of materials should also be considered to minimise long-term environmental impact.

Principle Three

Interior design can be crucial in enhancing productivity by creating functional, comfortable and sustainable workspaces. An experienced eco-interior designer should lead the design team to ensure an environmentally responsible and efficient approach. Integrating multiple desktop computer screens has been shown to increase occupant productivity by up to 30%, providing a more efficient workflow. The use of adaptable, modular, and mobile furniture is essential for creating flexible work environments that accommodate changing needs. Sourcing furniture from ISO14001-certified manufacturers ensures sustainability and reduced environmental impact. A holistic design approach that considers ergonomics, aesthetics and sustainability fosters a well-balanced and productive workspace, ultimately improving employee well-being and efficiency.

Principle Four

Selecting the right furniture and fittings aids in reducing VOCs and improving indoor environmental quality. Using third-party environmentally certified furniture ensures that materials meet strict sustainability and health standards. Opting for zero or low-formaldehyde bonding agents in substrates minimises harmful emissions. Additionally, applying zero-VOC paints and powder coated surfaces further enhances air quality by eliminating toxic off gassing. By reducing VOC exposure, workspaces become safer and more comfortable, leading to improved employee productivity and overall well-being.

Principle Five

Acoustics, ergonomics and lighting are critical elements in creating a comfortable and productive workspace. The use of recycled plastic interior linings provides excellent acoustic absorption and durability, helping to reduce noise levels and enhance concentration. Ergonomically approved chairs and furniture improve occupant comfort, reducing strain and increasing overall productivity. A dual lighting concept, incorporating both overhead and task lighting, minimises computer screen glare and enhances visual comfort. By integrating these design elements, workspaces can foster a healthier, more efficient and user-friendly environment.

Principle Six

Workspaces should be viewed as processes rather than products, requiring continuous evaluation and adaptation to maintain optimal environmental quality and occupant well-being. Annual National Australian Built Environment Rating System (NABERS) indoor environmental tests should be conducted to assess air quality, lighting and overall comfort. Regular occupant surveys help monitor trends and identify areas for improvement, ensuring that the workplace evolves to meet user needs.

Collecting data on energy, water, waste management, and indoor environmental quality provides valuable insights for sustainable operations. Effective communication between facilities managers and office management is essential for implementing improvements, while proper induction programs ensure that occupants understand workplace practices and contribute to maintaining a healthy efficient environment.

Revitaliser conceptual diagram by Eco Effective Solutions is given in **Fig. 9.** Revitaliser was installed at the Brisbane Convention and Exhibition Centre for the 10th International Healthy Buildings Conference in 2012 (Ecoeffective, 2025).

Figure 9. Conceptual diagram of Revitaliser by Eco Effective Solutions.

The Revitaliser's specifically selected pants, bio filters and materials dramatically reduce VOCs, stabilise CO2 levels, and boost oxygen to create a space that is safe for occupants and improve their cognitive function in the workspace.

Measurement of Ultrafine Particulates

Ultrafine particulate concentrations inside and outside the Revitaliser over a period were monitored. It was observed that the concentration of ultrafine particles inside the Revitaliser is generally lower than outside, with the outside measurements showing to be higher (**Fig. 10**). This confirms that the Revitaliser reduced particulate levels in air.

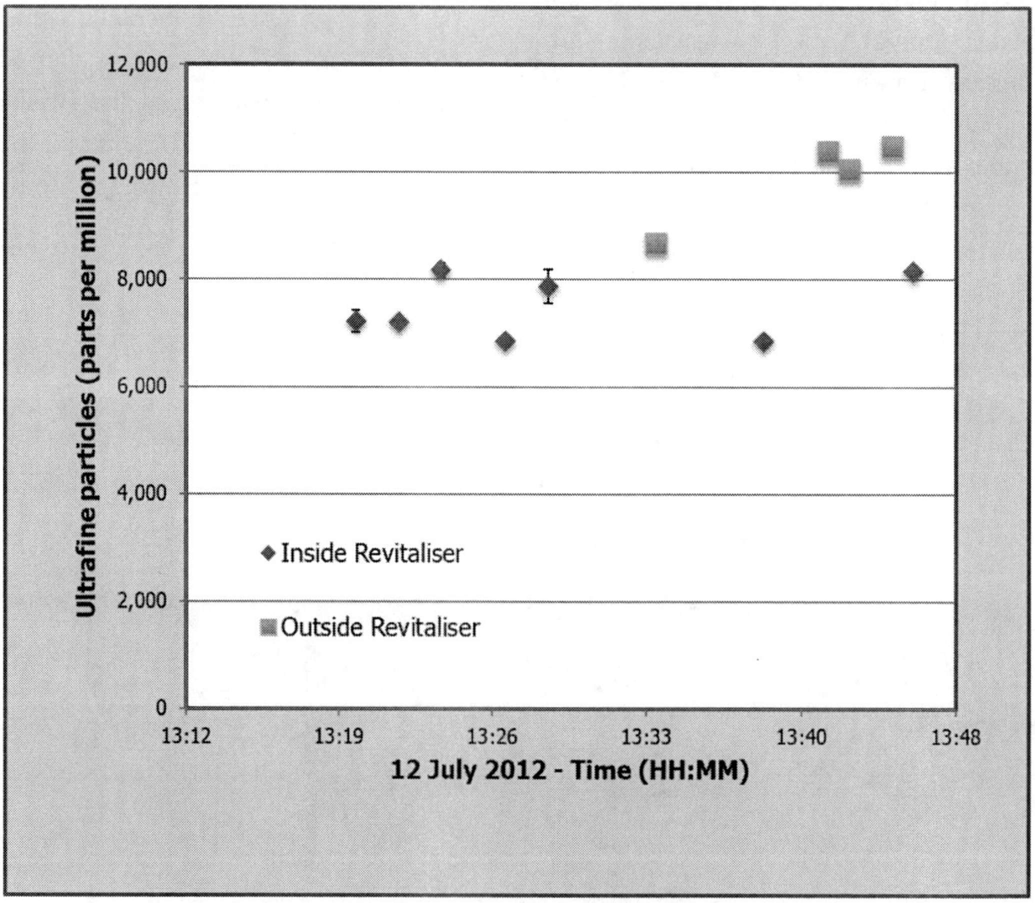

Figure 10. Ultrafine particulate concentration inside and outside Revitaliser over 35 minutes.

Volatile Organic Compounds (VOCs)

VOC concentrations were measured over a period of time with and without plants. Thereafter the Revitaliser was switched on. The level of VOCs that were lower when plants were present, further reduced dramatically when the Revitaliser was switched on (**Fig.11**).

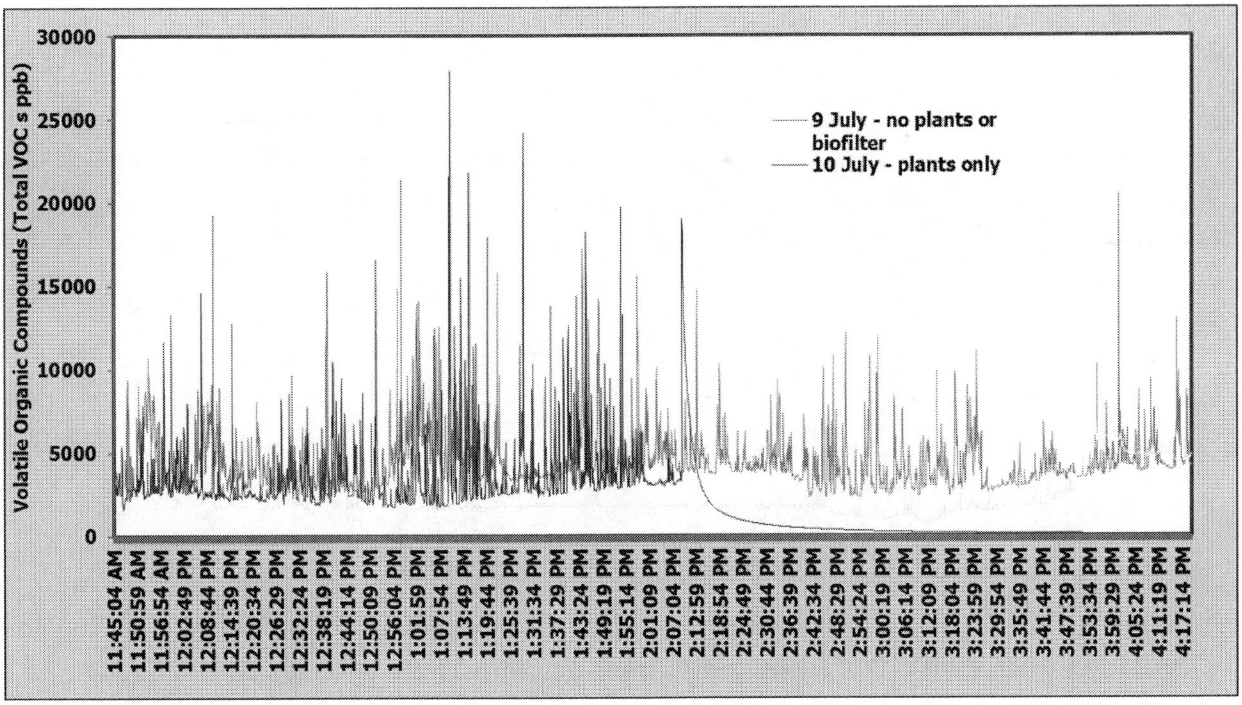

Figure 11. The total VOCs concentration over time on two different days.

Day one – 9th July which is the green line shows the concentration with no plants or biofilter, and the 10th of July which is the blue line showing the concentration with plants only. The VOC levels on day two - 9Th July (no plants) were significantly higher throughout the measuring period, whereas the VOC levels on 10th July (day two with plants) were considerably lower with a further decline in VOC levels once the Revitaliser was engaged at 2.12pm. This indicates that the presence of plants had a strong effect in reducing airborne VOCs.

Indoor Air Quality

Indoor air quality measurements taken within the Revitaliser during the Healthy Buildings Conference in July 2012 are presented in **Fig. 12.**

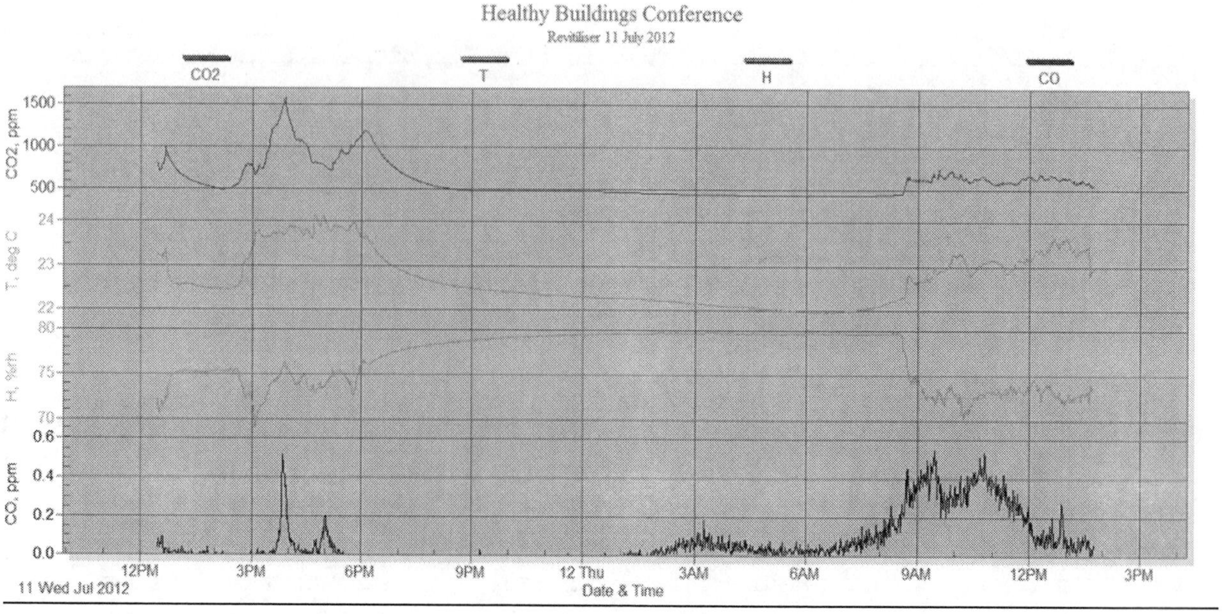

Figure 12. Environmental parameters during the Healthy Buildings Conference on the 11^th^July 2012. CO (ppm, blue), CO2 (ppm, red), Temperature (Celsius, orange), and humidity (%RH, green) were measured within the Revitaliser.

LIVING WITH NATURE

In the global assessment report on biodiversity and ecosystems services key findings were that the health of our environment is decreasing. It is noted that Australian cities are expanding, damaging biodiversity and ecosystems which are impacting our health and wellbeing. Our cities need more green spaces, and due to the significance of this issue The Brisbane Green Factor Tool was introduced into the Brisbane City Council's City Plan (Greenfactor, 2025). The Green Factor tool is a measure that captures the ratio between green areas and its total area. The aim for Brisbane using Green Factor is that it provides urban development practitioners with a method to optimise landscape designs and maximise the ecosystem services by assessing the quality of green infrastructure in new developments. This can be used for residential areas, hotels, retail centres, public spaces, community buildings, industrial and service stations (Greenfactor, 2025).

New Landscape Approval Requirements for Brisbane City Council, Australia

Legislation plays a considerable role in influencing our built environment. Brisbane City Council's new landscape development approval requirements emphasise the importance of sustainable and healthy designed urban development. The use of the Brisbane Green Factor Tool provides practitioners with measurements to optimise green infrastructure and ecosystem services such as deep planting areas, wind studies, and maintenance plans. By enforcing these requirements and regulations, the Council seeks to mitigate urban heat and enhance biodiversity.

Brisbane City Council has established requirements for landscape development to ensure it aligns with city planning and environmental standards. The key requirements include:

- Green Factor Ratio Area to be scheduled, areas calculated and signed off.
- Deep Planting to be 15% of Site Area and a minimum area of 4m x4m.
- Shade and Aspect studies to be conducted for each site.
- Wind report in meters/second for container planting locations.
- Weekly water consumption calculated every month by a certified Designer.
- Details of Structural assessment by RPEQ Engineer for green element planters including material selection and fixing specifications.
- Maintenance Plans for Body Corporate / Community Management adoption
- As built drawings of installations with minimum falls in substrates with 1.20 details plus plans and elevations.

With these requirements there is the opportunity to further urban development within the city and provide green spaces to restore lost biodiversity and address increasing issues such as urban heat and storm water quality. Indoor Air Quality legislation in Australia have proven ineffective during times of extreme heat and external pollution. Indoor Air Guidelines exist, although the design, building and development industries rely on Australian Standards for Air Quality delivery. Australian Standards typically identify the minimum standard to be achieved, whilst healthy built environments require higher standards to achieve productivity and wellbeing benefits. Best Practice Air Quality solutions have received attention since the COVID 19 pandemic, when minimum air quality standards were proven ineffective in curtailing the spread of the COVID 19 virus in Australia.

Healthy Vegetation in Built Environments is now understood to be a best practice ingredient, to deliver productivity benefits, improved occupant well-being and healthier living and working environments. The Revitaliser demonstration created valuable insights to understand the ingredients of healthy working environments.

CONCLUSION

The findings in this report demonstrate the significance of integrating vegetation within the indoor environment and in turn improves air quality, reduce pollutants, and enhances occupant well-being. The Revitaliser and other biophilic design strategies show quantifiable advantages from lowering VOC levels and stabilising CO_2 which improves cognitive function and reduces stress. The research further confirms the significance of strategic implementation of biofiltration and sustainable construction practices and methods to creating healthy workplaces and urban areas. While indoor environments are vital to occupants, the Brisbane City Council's updated landscape development approval requirements support the necessity of incorporating green spaces into proposed urban development. In the future, fostering collaboration between architects, scientists, urban planners and legislators will be crucial in developing built environments into healthy and self-sustaining ecosystems. This collaboration will reinforce human and environmental health while paving the way for a sustainable urban future with proven good indoor air quality and increased indoor wellbeing for building occupants.

Acknowledgements

(1) Eco Environment- specialist consulting firm founded by John Daly, Brisbane, Australia.

(2) BMW manufacturing case study -interview with Architect Dieter Schemp, LOG ID Turbingen Germany, 2005 by Mark Thomson (photos used with Architects permission).

(3) Manly Hotel Brisbane (TVS Architects) Project Architect: Mark Thomson, Landscape Architect: Nial Fraser.

(4) Eco Environment – John Daly Research.

(5) Revitaliser data undertaken by Octief Laboratories under the supervision of Dr. Claire Bird, founder of LITMAS laboratories.

LITERATURE CITED

Australian. (2012). Office air puts you to sleep. A Berkley Study reported in the Australian Magazine. Oct 31st, 2012.

Burchett, M., Torpy, F., Brennan, J., & Craig, A. (2010). Greening the great indoors for human health and wellbeing. Sydney: Plants and Indoor Environmental Quality Group, Centre for Environmental Sustainability (CEnS). Available at: Final Greening-the-great-indoors.pdf

Daly, J., Burchatt, M., and Torpy, F. (2024). Plants in the classroom can improve student performance. Comb. Proc. Plant Prop. Soc. 74.

Ecoeffective. (2025) The Revitaliser at HB 2012. Available at: https://www.ecoeffective.com.au/archives/design/the-revitaliser-at-hb2012/

Greenfactor. (2025). Brisbane Green Factor. Available at: https://brisbanegreenfactor.brisbane.qld.gov.au/

Combined Proceedings IPPS 74:251-257. 2024.

KW Automation: Excellence in Nursery Automation

Luke Weisenberger

158, Old Bowling Green Rd., Palmwoods, QLD 4555, Australia.

luke@kwautomation.com.au

Keywords: pneumatic needle seeder, hoppers, conveyors, elevators, soil mixers, tailored solutions

Summary

Founded in 1979 by Kurt Weisenberger, KW Automation has grown from humble beginnings into a leading provider of nursery automation solutions. Kurt, originally an avocado orchard owner on Queensland's Sunshine Coast, transformed his engineering expertise into innovative machinery for fruit and vegetable post-harvest processing. Recognizing a niche in the nursery industry, he developed the revolutionary KW needle seeder, which set a new standard for efficiency and productivity.

Today, as a third-generation family-owned business, KW Automation continues Kurt's legacy of quality and innovation. Our diverse range of machinery includes soil mixers; hoppers, conveyors, & elevators; pot & bag fillers; needle seeding; tray fillers; potting machines; tray & pot washers; watering tunnels; customizable equipment and more.

Each product is designed with durability, reliability, and customization in mind. Our stainless-steel soil mixing systems, for instance, deliver consistent results with options for high-volume production rates and features like nutrient hoppers and customizable mixing recipes.

We pride ourselves on being a one-stop shop, offering tailored solutions such as boom conveyors and advanced potting machines. Customers like Matt from Hope Valley Nurseries attest to our commitment to quality, recounting decades of flawless operation with minimal maintenance.
Our cleaning solutions, including tray washers, support sustainability by enabling the reuse of trays and pots. With the capacity to clean up to 1,000 trays per hour, they help reduce waste while maintaining sterile conditions.

At KW Automation, we value customer feedback, using it to refine and improve our products. From pioneering designs like the needle seeder to modern innovations, our goal is to help customers build efficient, dependable operations. As automation becomes essential in overcoming industry challenges, KW remains committed to delivering robust, user-friendly machinery backed by dedicated service and support. Trust KW Automation to provide consistent results and a reliable partnership for years to come.

INTRODUCTION

As the third generation in the Company it is an honour to be able to describe what we do here at KW Automation. Our company started as KW Engineering, Founded by Kurt Weisenberger my grandfather in 1979. Originally, Kurt travelled from Germany with his wife and purchased an Avocado orchard in Queensland on the lovely Sunshine coast. With his creative thinking and experience in engineering, he started to concentrate on solutions for post-harvest equipment for avocadoes **(Fig. 1)** and eventually moved into all types of solutions for fruit and vegetables. However, in the 1980s, Kurt recognized an opportunity to support the nursery industry, leading us to develop specialized nursery equipment such as the KW needle seeder which had made a huge impact at the time. This shift allowed us to align our expertise with the specific needs of nursery operators, making a significant impact on the nursery industry's efficiency and productivity. KW has now evolved into a third-generation family-owned and operated business.

Figure 1. With his creative thinking and experience in engineering, Kurt recognized an opportunity to support the nursery industry, leading us to develop specialized nursery equipment such as the KW needle seeder which had made a huge impact at the time.

So How Can You Use Kw Automation to Help You?

With almost five decades of experience in the nursery industry, our expertise encompasses a wide array of units designed to help you create an efficient and successful business. Our offerings include soil mixing systems (**Fig. 2**), storage hoppers, conveyors, elevators (**Fig. 3**), multiple seeding systems, plug popping machines, pot and bag filling machines, tray fillers and potting machines, and tray and pot washing equipment.

Additionally, we are always interested in what is best for our customers so we do offer custom equipment as we know one solution doesn't always work for everyone, this ensures our ability to meet your unique needs with tailored solutions.

Figure 2. Soil mixing systems developed and offered by KW Automation.

Figure 3. From left: 4Cbm Soil Storage Hopper (Left), 4Cbm Soil Storage Hopper (Middle), 2 Cubic Meter Soil Mixer with Soil Bagging Conveyor (Right).

Soil Mixing

If you had to ask any of our customers what words come to mind when thinking about our machines, you would hear "Quality," "Reliability," and "Robust" among others. We pride ourselves on this image, especially in the soil mixing sector where long-lasting quality machines that consistently provide quality results time & time again whilst minimising downtime, to keep you focused on what matters most. This paired with our ability to customise machinery and layouts with our dedicated in-house engineering team can set you and your business up for success. Our focus on quality and robust designs is clearly displayed in our range of standalone soil mixers. With fully constructed stainless steel frames, they are built to provide results for the many years ahead, available in one, two, and three cubic meter options. Additional options such as replaceable body liners and hard-faced spirals increase the longevity of your investment.

KW Automation's inline continuous soil mixing lines (**Fig. 2**) leverage our experience and knowledge, giving our customers reliable and consistent soil mixing systems that require on demand high-volume production with rates starting at 40 cubic metres per hour. Incorporating bale breakers, large stainless-steel peat or coir storage hoppers, nutrient storage hoppers, vermiculite/perlite hoppers and unique mixing heads combined with HMI driven potting mix recipe menus, these all add up to providing the end user with a flexible system with the ability to make changes to your recipe when the time comes. These systems can be setup to automatically fill tray fillers or potting machine hoppers using soil sensing devices to drive the inline soil mixing systems as required.

Hoppers, Conveyors & Elevators

The great thing about KW Automation, is the range of storage hoppers, conveyors & elevators we have. We do our best to set ourselves up as the one stop shop for our customers. If you have a conveying need that doesn't sound like a straightforward solution, run your idea past us and we can provide the support you need to make that idea come to life. A couple of months ago we had a customer who had an idea for a boom conveyor that could balance itself to reach over their greenhouse benches, this conveyor was used to layout their seedling blocks or trays for the growing process and then retrieving them for dispatch. We were able to design and provide a working conveyor for them.

If it's just a straightforward solution we have the standard equipment ready to go for you. Most frames are constructed from aluminium and or stainless steel. We produce all different size hoppers from 40 L nutrient hoppers all the way up to our largest soil hoppers which reach up to ten cubic meters. We can do any size conveyors and elevators where required and we also provide handy little additions to soil mixers like this Bagging conveyor attached to our Soil mixer to fill pots quickly & easily with a foot pedal control.

Pot and Bag Fillers

Our Pot & Bag filler was developed by Grandpa back in the eighties but has recently undergone a revamp to keep our machines at the fore front of automation, it is available in single and dual outlet platforms **(Fig. 4)**. It can be mounted on our standard adjustable fixed legs or have a forklift frame to make transport easy around your nursery, we can even make this on its own trailer style frame with large tyres to be easily moved through your nursery. These units can fill all types of pot & bags up to a 45-litre capacity container. To give you an idea of figures, each outlet is capable of filling 240 28ltr pots, and up to 400 5ltr grow bags per hour.

One great addition that's available with these machines is our fertiliser dispenser. This allows ease of mind with consistent accurate dosing with each fill designed to consistently produce quality products with every pot or bag. Dosages can be changed with the on-board HMI Screen so if you're using larger or smaller containers you can change this to suit whatever vessel you wish to fill. The Hoppers are specially designed to stop bridging as well, so once your hopper is full, you can fill with the knowledge that your mix won't bridge and stop your production halfway through.

Figure 4. Pot and bag fillers are available in single and dual outlet platforms.

Needle Seeding With KW

With one of the longest standing platforms at KW Automation the Needle seeder holds a special place in our hearts. Developed by my grandfather Kurt Weisenberger who came up with the idea after looking at medical needles on one of his visits to the hospital and thought, "I bet I could pick up single seeds with these!"

Sadly, my Pa recently passed away in the last 2 years, so it is great to see his invention still thriving & performing so well to this day. After 3 decades of experience and testing with the needle seeders we have been able to evolve the seeder to deliver a great seeding platform to service all types of operations. The pneumatic seeding options come in 4 different platforms, From semi-automatic, automatic with Vermiculite dispensers, Watering bars to PLC programming on our high range seeder, this has the ability to multi-seed in cells & switch between different tray & seed programs easily. With the versatility of our multiple needle sizes we are able to pick up a large range of seeds used in your industry.

If required, our seeders can be optioned with our wide body solution, so you are able to feed your trays in width ways to speed up tray fill times even more. Just a testament to our reliability and build quality; two weeks ago we had a customer drop off a seeder which was built in 1998, that's when I was born. Probably in better condition than I am. We changed a couple filters, air hoses and fittings and sent this back to our customer ready for more seeding.

Tray Filler's and Potting Machines

I was speaking with one of our long standing customers, Matt from Hope Valley Nurseries yesterday to ask him for his honest opinion on KW, he said "KW's it, he wouldn't go to anyone else, his parents purchased their first potting machine back in 1999 and the unit has been awesome, the only components he's needed to replace is two Augers & two Gearboxes. It's Still in operation and still pumping out pots to this day" since then he has also purchased a tray filler in 2009 which can do 130mm pots in shuttle trays, and 100mm square pots in 10 and 6 packs. Matt said he produces 40,000 units a day with this machine. He purchased another Potting Machine in 2018 for his new facility & finally another potting machine which is currently in production at our factory for a new pot size. We also has another potting machine out at Pohlman's Nursery since 2004, after doing a few calculations, we worked out their unit has filled an estimate of over 50 million pots in its time and is still going strong. It has been the constant narrative of our story. The simple, low maintenance designs that can provide great results and a smile on our customers faces over & over again, it's the reason we do it.

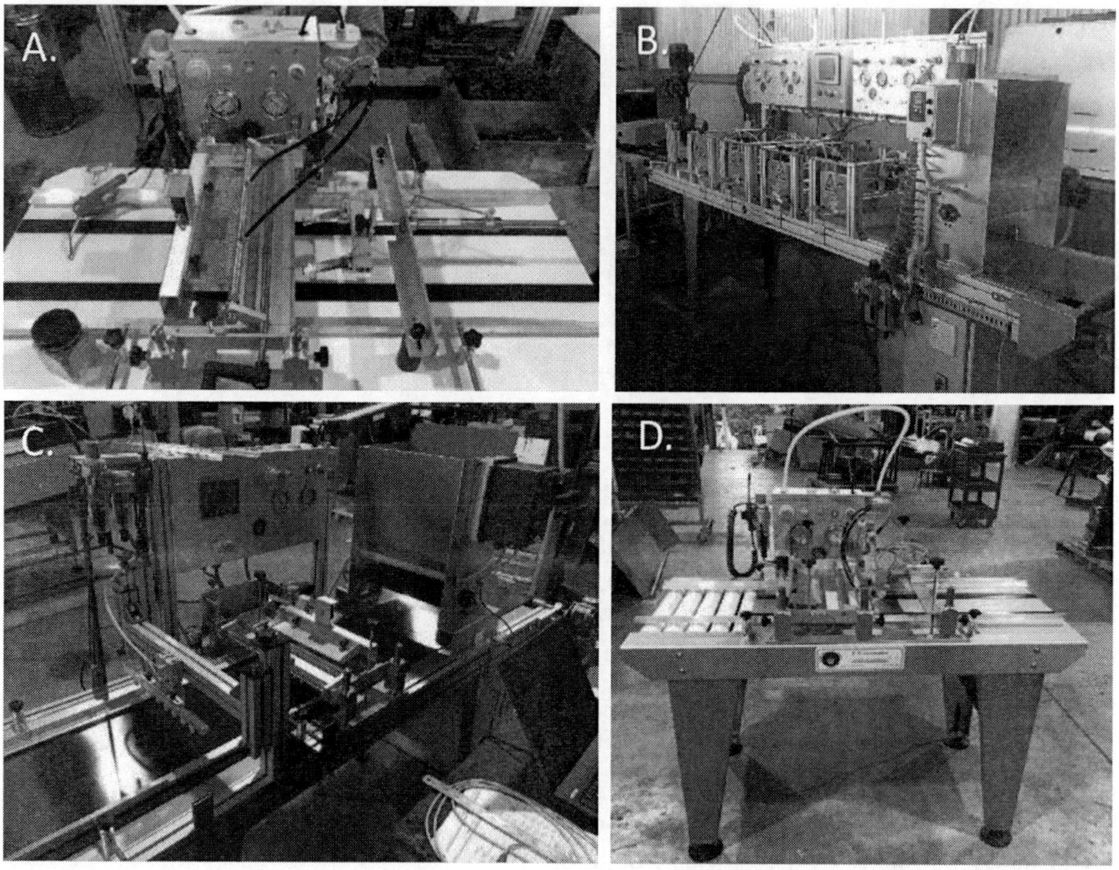

Figure 5. Needle seeders developed at KW automation: **A)** H35 Semi-Automatic hand seeder. **B)** Multi seeding head 100EM automatic seeder. **C)** 100EM Low Drop automatic seeder, **D)** A35 automatic seeder on an aluminium base.

Cleaning Solutions and Watering Tunnels

Living on the beautiful Sunshine coast all my life you can't help but develop an appreciation for creation. Our KW site is situated in Palmwoods which if you've had the pleasure of visiting, is an amazing place. My grandfather loved his garden which he planted around his house and the KW premises, it makes it a special place to be around, once you've seen it you can understand why he had such a fascination for the nursery industry. We would love to preserve this creation we've been blessed with and create a more sustainable future in the nursery industry. KW Automation has a wide range of cleaning solutions for trays & pots. This allows us to reduce on single use plastics and make a viable way for reusable trays & pots to be recycled as much as we can and still have clean sterile products to reuse.

KW has a tray washer at one of our local Nurseries just down the road at Cedar Hill and I was speaking to James and Rose last week and I asked them how their tray washer was going, he responded with, "Everyone loves it, our employees are fighting to be the ones who can use the tray washer."

What once was a daily chore of cleaning trays has now become a set time to clean all their trays once a week. With the ability to clean up to 1000 trays an hour they are able to save time, money and know it's one step closer towards a more sustainable future.

CONCLUSION

Customer success is our commitment, and KW has a dedicated service team so if customers have any issues or a breakdown, we have an experienced team that is able to get the equipment up and running ASAP. Our best advice is to always keep the machines in good condition with regular check-ups from the KW team.

Our ability to be able to look at our customers' history with a full set of detailed drawings and parts list and know every single part on that machine, gives us the ability to quickly supply replacement parts when they need it most. Every customised job is designed in our 3D drawing software, and we are able to show our customers what the machine will look like before it's even produced and go through any changes that might need to be made before manufacture.

Automation is important and it can help you survive the difficult times. We had a larger customer who had done a large automation project in the early 2000's; he stated that if he hadn't made this decision to automate his operations he would most likely not be in business today.

At KW Automation:

- we endeavour to prioritize our customers
- we have been specialising in the nursery automation industry for almost five decades
- we collaborate with our customers to deliver tailored solutions
- we value feedback from our customers no matter good or bad. This is what makes our equipment more user-friendly and long-lasting. We have incorporated many customer ideas into making our equipment into what it is today. Like the IPPS say's, to seek and to share our knowledge, we can grow together.

We are trusted by many and have had the pleasure of being able to provide so many customers with awesome machines across the world. The most important thing for us is to be able to build the best operation for the customers with a manufacturer they can trust; trust in the build quality and trust that we will be here to help whenever you need it. We want you to know when you press that go button your machines are going to produce consistent results for the future to come. If that's what you want in your operations, keep KW at mind.

ACKNOWLEDGEMENT

I Thank IPPS Australia for giving me the opportunity to share our story with its members and a massive thank you to the IPPS for publishing this story.

Combined Proceedings IPPS 74:258-267. 2024.

Paclobutrazol Trials in Commercial Micropropagation of *Grevillea* species

Lisa J. Wightwick

Plant Tissue Culture and Micropropagation, Peninsula Growers, 45 Grasslands Rd, Boneo VIC 3939, Australia

ljwightwick@gmail.com

Keywords: plant tissue culture, acclimatization, abiotic stress

Summary

Grevillea 'Bonnie Prince Charlie' is a popular landscaping plant from the Proteaceae family that is often propagated via plant tissue culture. Despite good growth and multiplication in culture, it possesses morpho-physiological characteristics that decrease its quality and survival during the latter stages of micropropagation. Excessive internode elongation and soft, thin stems exacerbate negative abiotic stress effects that occur once removed from the nurturing culture environment. Trials of the growth inhibitor paclobutrazol were undertaken to test its potential to alleviate the issues encountered during deflasking and acclimatization. Supplementation of the growth media with 2 mg/L proved a success and all problematic traits of the cultured *Grevillea* were counteracted by the application of paclobutrazol. The noted positive effects included: drastically reduced internode elongation, thickened stems capable of supporting their own weight, increased desiccation tolerance and reduced wilting, increased axillary bud growth, broader and deeper green leaves and increased consistency and density of root growth, with 97-100% of plants rooting. Outside of the laboratory environment, deflasking and acclimatization survival rates, quality of sale stock and production efficiency were all greatly improved. The only disadvantage noted was a 1-2 week increase in the holding time in the laboratory. This slight increase in passive storage time was deemed to be a minimal trade-off in return for the multitude of advantages. Large-scale trials proved 2 mg/L paclobutrazol to be the ideal concentration to ameliorate the disadvantageous traits of *Grevillea* 'Bonnie Prince Charlie,' improving its commercial viability. Further testing with paclobutrazol has already begun on other cultured species to ascertain if there are equal improvements to be achieved.

INTRODUCTION

Plant Tissue Culture (PTC) is an asexual method of growing plants *in vitro* - a state in which plant cells, tissues and whole organs are aseptically cultured in a fully controlled environment, on chemically defined media. Plant tissue culture has a wide variety of applications, ranging from use in medicine and pharmaceuticals, scientific, horticultural, and agricultural research, species conservation to micropropagation (**Fig. 1.**).

Micropropagation utilizes plant tissue culture techniques for large scale propagation. There are several stages in micropropagation including: conditioning, multiplication, rooting (rhizogenesis), deflasking and acclimatization (ex vitro). The conditioning phase involves the harvesting and sterilization of *in vivo* plant material and culture onto media containing varying combinations of plant growth regulators to induce juvenile vegetative growth. After the plant material has been successfully initiated, the next stage is multiplication of the tissue culture plants. The last stages in the PTC cycle are the rooting phase; in which the application of auxins induces rhizogenesis and primes the plants for deflasking and acclimatization - the final stage of the cycle where the plants are removed from culture and slowly adapted to the outside environment.

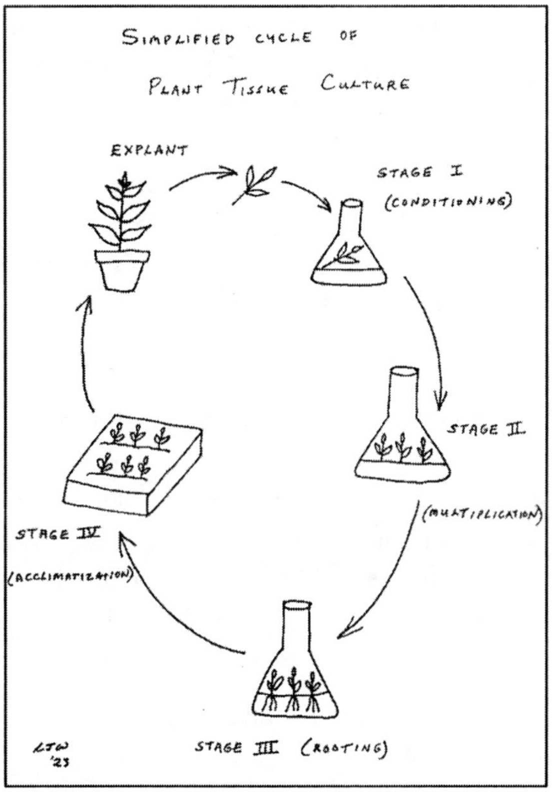

Figure 1. Diagram of a plant tissue culture cycle.

Morpho-Physiological Impediments of in Vitro Plants

The controlled growth environment triggers significant morphological and physiological changes in the cultured plants. Exogenous sources of nutrition, humidity, heat, light, growth regulators and water prompt the plants to revert to a juvenile state, with the average TC plant growing to only a few centimetres tall. Plants grown in culture become heterotrophic and are unlikely to be actively photosynthesising, despite low levels of chlorophyll present in the leaves. The enzymes that play a role in photosynthesis (such as ATP synthase) are either inactive or absent in tissue culture plants. Carbon dioxide uptake and net exchange in cultured plants is either very slow or halted (Conner and Thomas, 1983).

Characteristics and processes in ordinary plants that help regulate water loss are vastly altered in tissue culture plants. Stomata, the pores on leaves that open and close for gaseous exchange and control loss of water vapor (transpiration), are extremely slow to close and often do not close at all. Water is readily able to escape from the vegetation of TC plants through the stomata. A less developed vascular system leads to poor water conduction between the roots and shoots of cultured plants. The leaves of plants *in vitro* do not develop epicuticular waxes in typical amounts or with the same chemical properties as plants outside of culture (Davies et al., 2018). This combination of factors put the plants at considerable risk of desiccation and death unless certain precautions are taken.

It is the combination of these characteristics which necessitate the acclimatization process. The final step of the PTC cycle serves to slowly allow the cultured plants to regain regular function and harden themselves to the outside environment. Acclimatization can be difficult even in ideal conditions, let alone when some plant species have features that make it challenging to micropropagate.

Case Focus: *Grevillea* 'Bonnie Prince Charlie'

Some species carry-over traits that make them difficult to handle in and out of culture. Woodiness, thin stems, etiolation, recumbent and prostrate habits are among many factors that can have a challenging impact on their commercial micropropagation viability. If a certain species is soft and less woody, it follows that its cultured counterpart will also be soft and less woody. *Grevillea* 'Bonnie Prince Charlie' (BPC) is among many popular *Grevillea* varieties sold commercially, but unfortunately, this variety has a fine and delicate form when initiated into culture. Stems average 0.5-1mm in diameter, in contrast to other *Grevillea* species which range from 1-2mm (**Fig. 2**).

Figure 2. A typical *Grevillea* 'Bonnie Prince Charlie'.

'Prince Charlie' cutting in tissue culture. This variety grows faster than other observed cultures at an average rate of 2cm per week. It has been suggested that plants *in vitro* excessively produce gibberellins (Roberts *et al.*, 1992), therefore, it is likely this species secretes high endogenous levels. The speedy growth and thin, fragile stems contribute to a strong wilting predisposition both *in* and *ex vitro*. The plants are also easily damaged when handling with tweezers in culture and during the deflasking process. This form and fragility, coupled with elevated water loss, means these plants are at a significant disadvantage when they leave the culture environment to commence the acclimatization process.

(slower production rates), but most critically, greater losses post deflasking and significant reduction in plant quality. Issues encountered during deflasking included damage to the plants from handling (e.g. snapping of stems), excessive wilting and deformation of plants (unable to support their own weight, stems adopted a sideways 'S' shape) and greater prevalence of fungal rot due to plants drooping over into the media (**Figs. 3 and 4**). Any plants that survive the acclimatization process will likely have an undesirable form to customers and low saleability. Techniques can be employed to help salvage the stock but are labour intensive and involve triple-handling.

Figure 3. A) and **B).** Low quality *Grevillea* 'Bonnie Prince Charlie' cultures at various stages of hardening off - note the warped stem shapes and inability to support their weight.

Attempts to alleviate and/or resolve these problems included:

- Making size adjustments to the cuttings taken for rooting/pre-acclimatization phase: decreasing height of cuttings, harvesting stem cuttings.
- Alteration of PTC media recipes: trialling different types/combinations of plant growth regulators, different basal salts.
- Ventilation of culture flasks prior to deflasking: containers were opened slightly for one week to enable air exchange and start the acclimation process earlier.

- Minimizing plant exposure by speeding up the deflasking process as much as possible.
- Optimizing the process of deflasking: only having one open culture container at a time.
- Conducting deflasking in a greenhouse with high humidity to help minimise shock to the plants.
- Use of anti-transpirants.

Figure 4. A) A tray of deflasked *Grevillea* BPC - sizes are inconsistent, most plants have collapsed under their own weight and there is die-off and botrytis. **B)** An example of a high quality (right) and a poor (left) quality sale stock.

Unfortunately, even in combination, none of these methods yielded significant success, however, did serve to slightly improve the survival rate of the plants *ex vitro*.

MATERIALS AND METHODS

Paclobutrazol (PBZ) is a growth inhibitor which prevents the biosynthesis of gibberellins. It can be used as a pre-treatment in the rooting phase of the PTC cycle to help overcome abiotic stresses such as water loss. It is known to reduce elongation of internodes, increase leaf quantities and thickness of stems (Abdalla et al., 2021) and act as a systemic fungicide (Desta and Amare, 2021). These qualities made PBZ an ideal trial candidate for amelioration of issues surrounding *Grevillea* 'Bonnie Prince Charlie'.

A bottle of 'Trimmit' growth regulator was sourced for the trials, containing the active constituent 250g/L Paclobutrazol. The original product was diluted to a concentration of 0.125g/L with distilled water. Extensive research culminated in a concentration of 2mg/L being chosen as the starting point for testing. As 'Trimmit' is not a product intended for use in plant tissue culture, an initial test on four *Grevillea* varieties ('Bonnie Prince Charlie', 'Sunkissed', 'Gold Rush' and 'Fire Cracker') was undertaken to screen for any potentially detrimental or fatal effects. 2mg/L PBZ was added to a rooting media pre-sterilization and dispensed into small sterile culture containers. Of each variety, twenty plants were placed onto the trial media. No ill effects were observed on any of the trial varieties, and all had improved root growth. Rhizogenesis was faster, consistent, with more dense, thick root balls.

RESULTS AND DISCUSSION

The most notable vegetative improvement was in the *Grevillea* BPC, which exhibited deeper green, more rounded, and dense leaves, thicker, firmer stems which did not elongate further than the height at which they were cut. Axillary bud proliferation was also stimulated. There was an increase in the holding time between the rooting stage and deflasking - trial plants took an average one to two weeks longer to reach ideal form for deflasking. A small amount of trial stock was held to observe the long-term effects of PBZ. A half-life of approximately 35-45 days was discovered and varied based on plant variety and media constituents. After this period, the PBZ would wear off and internodal elongation would resume from the lower region of the plants. Some plant varieties exhibited the physiological disorder hyperhydricity (excessive hydration and malformation of tissues) after long periods of exposure to PBZ. As the trials were designed for temporary use in the rooting phase these issues were deemed to be of no consequence.

The BPC trial was deflasked and it was found that the plants were easier to handle, had better roots and rarely wilted. A second, large scale trial was conducted, in which half a batch of *Grevillea* 'Bonnie Prince Charlie' contained the 2mg/L media addition of Paclobutrazol. Flasks containing 1350 trial plants and 1350 control plants were deflasked. There were very notable differences between the trial and control plants; with the majority of trial plants maintaining a height half that of the control plants (**Figs. 5, 6,** and **7**), with thicker stems, increased quality and survival rate. The trial stock was inspected at the latter stages of acclimatization and progression throughout the nursery and was found to have no adverse effects and remained well-formed and healthy.

Figure 5. Trial versus control: while the control is still healthy, viable stock, it is thinner and more fragile than the PBZ trial which has deeper green, thicker, more compact vegetation and more progressed and better-quality root growth.

Effects of subsequent trials remained consistent, with significant improvement to the strength and resilience of the vegetative tissues, plus faster, more consistent, better quality root growth. As seen in the images left and below, the trial plants have a greater surface area and thickness. The control plants are more than twice the height of the PBZ trial, despite production on the same day. The control plants have warped stems due to excessive vertical growth. Smaller trials were undertaken to test different concentrations of

PBZ at 1mg/L and 1.5mg/L but were found to be too weak for the desired results. Concentrations of more than 2mg/L were not tested due to the risk of adverse effects on the plants.

Figure 6. Side by side comparison of *Grevillea* 'Bonnie Prince Charlie' trial versus control. The control plants are double the height of the trial, have poor form, and damaged vegetation due to abiotic stress.

Figure 7. **A)** comparison of the trial and control plants *in vitro* – there is a clear difference in the plant morphology. **B)** Heights of control and PBZ trial plants side-by-side.

Based on the success of the large-scale trial, all further batches were incorporated with

PBZ and observed closely for quality and consistency. The 2mg/L concentration remained in place as it yielded ideal results. A combined total of 9090 *Grevillea* BPC plants have been successfully produced since the start of the trials, with a substantially higher survival rate and greatly increased quality. There were also increases in productivity throughout the chain of production from the lab all the way to tubing and potting. Time spent grading the stock drastically decreased, and due to the improved resilience and quality of the plants, their progression through the nursery was fast and trouble-free. The trial stock also had a greater holding capacity, transforming *Grevillea* BPC into a low-maintenance, flexible and commercially viable product (**Figs. 8, 9,** and **10**).

Figure 8. A) Uniform trial of *Grevillea* BPC in early stages of acclimatization. **B)** and **C)** Trial of *Grevillea* BPC in early stages of acclimatization.

Figure 9. A) Fully hardened, high quality sale-ready stock. **B)** Ideal form of trial *Grevillea* BPC stock.

CONCLUSIONS

Grevillea 'Bonnie Prince Charlie' can be a challenging plant to propagate. While it may grow quickly and multiply well in culture, it has too many morpho-physiological characteristics that decrease its quality and survival rate when deflasked. Excessive internode elongation and soft, thin stems exacerbate the effects of abiotic stress that these TC plants are exposed to when they are removed from the culture environment. The addition of the growth regulator paclobutrazol into the PTC media to improve survival and quality of *Grevillea* BPC proved a success. All problematic traits of the TC *Grevillea* were counteracted by the PBZ.

Noted benefits included:

- thicker, straight stems, capable of supporting their own weight
- very little wilting
- increased desiccation resistance
- more axillary bud growth
- improved leaf form
- no excessive internode elongation
- more consistent rhizogenesis, ranging from 97-100% consistency
- denser root growth
- significantly higher survival rate ex vitro
- higher quality sale stock
- increased production efficiency in the laboratory and nursery

The only slight disadvantage of the PBZ addition was a 1-2 week increase in the holding time in the laboratory. Ultimately, a slight increase in passive storage time is a very minimal trade-off in return for the multitude of advantages. Further testing with paclobutrazol has already begun on other species in culture to ascertain if there are equal improvements to be achieved. Many of the trials have shown positive results thus far. More research and testing will be undertaken to investigate other potential uses for PBZ within the plant tissue culture and micropropagation fields.

ACKNOWLEDGMENTS:

Many thanks to my wonderful work family at Peninsula Growers for their incredible support, encouragement and for the opportunity to undertake these trials.

IPPS has my deepest gratitude for its amazing opportunities like the Rod Tallis award, which give starting horticulturalists a platform to experience growth and accomplishment.

LITERATURE CITED

Abdalla, N., Taha, N., Bayoumi, Y., El-Ramady, H. and Shalaby, T.A. (2021). Paclobutrazol applications in agriculture, plant tissue cultures and its potential as stress ameliorant: A mini-review. Env. Diversity Soil Security *5*:245 – 257.

Conner, A.J. and Thomas, M.B. (1981). Re-establishing plantlets from tissue culture: a review. Comb. Proc. Intern. Plant Prop. Soc. *31*:342-357.

Davies, Jr, F.T., Geneve, R.L. and Wilson, S.B. (2018). Plant Propagation, Principles and Practices. Hartman and Kester, Ninth Edition.

Desta, B. and Amare, G. (2021). Paclobutrazol as a plant growth regulator. Chem. Biol. Technol. Agric. *8*:1.

Roberts, A.V., Walker, S., Horan, I., Smith, E.F. and Motley, J. (1992). The effects of growth retardants, humidity and lighting at stage III on stage IV of micropropagation in chrysanthemum and rose. Acta Hort. *319*:153-158.

Combined Proceedings IPPS 74:268-271. 2024.

Some Observations on the Plants of the Chatham Islands

Phil Dunn

Ribbonwood Nurseries, 89 Glenelg Street, Bradford, Dunedin, 9011 New Zealand

ribbonwoodnatives@gmail.com

Keywords: Chatham Islands, New Zealand, megaherb, *Myosotidium hortensia*

Summary

This is a report on observations from the New Zealand Region field trip to the Chatham Islands in February 2024. Included are plant observations that provide some insights into the biodiversity of areas preserved by the Department of Conservation or private owners.

INTRODUCTION

The first thing that strikes you on arriving in the Chathams is the lack of forest cover. Most has been burned in Polynesian and European fires. What remains, however, in a few Department of Conservation and private reserves is quite distinctive and well worth checking out. The other noticeable feature is the lack of tall trees in the forest. Dry land only appeared above the ocean 3 million years ago on the Chatham Islands and consequently none of the tall mainland New Zealand species, including all of the podocarps, has managed to cross the 800 km of sea and turn up there. The biggest trees are probably the ubiquitous *Olearia traversiorum* (called akeake by the locals)

(**Fig. 1**) and karaka (*Corynocarpus laevigatus*) which is considered to have been introduced by Maori or Moriori up to 800 years ago, and the nikau (*Rhopalostylis sapida*, Chathams form).

Figure 1. Ancient *Olearia traversiorum* on Pitt Island.

While the biodiversity on this island is not huge, there are about 52 species or subspecies that are either unique to the Chathams or are distinctly different from their mainland counterparts. A number of these make great garden plants and there is potential to see more of them in cultivation. Coastally, some interesting plants have survived the fires and large numbers of feral cattle and pigs. We saw a few surviving forget-me-nots (*Myosotidium hortensia*) growing in pure sand in the dunes, as well as hundreds that have been propagated and planted around some of the beaches. Another striking plant in the dunes was the sow thistle *Sonchus macrophyllus*. Despite the unfortunate name, this is another very cool megaherb with huge chunky leaves and would be great in the garden. Beyond the dunes we saw *Corokia macrocarpa* (**Fig. 2**), *Astelia chathamica, Carex trifida*, a *Pimelea,* and on coastal cliffs, *Geranium traversii*. A local form of the ice plant *Disphyma australe* was common on rock outcrops and had particularly bright pink flowers. We also saw some kowhai (*Sophora chathamica*) growing on steep coastal banks and reaching 3 - 4metres tall. Some researchers believe it may have been introduced by Maori or Moriori a few hundred years ago rather than being indigenous to the Chathams.

Figure 2. *Corokia macrocarpa*

We also encountered all three hebe species that occur on the Islands. *Veronica chathamica* which is a very flat groundcover growing in dune areas and could be used much more widely as a garden plant as it is in very limited cultivation in New Zealand. *Veronica dieffenbachia* is a midsized shrub quite widely distributed on the

Island and *V. barkeri* which can grow to 13 m tall – this must be the tallest hebe on the planet. *Dracophyllum arboreum* is another species that seems to be on steroids. While most New Zealand mainland species are shrubs or even groundcovers, this one can reach a very impressive 18 m making it a canopy species here. *Aciphylla traversii* was encountered with seed heads on female plants and also *A. dieffenbachii* with its unusual blue/green divided leaves which lack the vicious spikey leaf tips of other species. This one is of course widely cultivated as an ornamental garden plant in New Zealand.

Horticulturally, plants with real merit that we saw were *Brachyglottis huntii* (known as the Chatham Island Christmas tree) and still flowering in February (**Fig. 3**), and two stunning olearias; *O. chathamica* and *O. semidentata* whose major drawback seems to be the difficulty in actually keeping them alive when planted out (**Fig. 4**). *O. semidentata* was growing on raised humps in a peat bog and still covered in attractive purple flowers. *Olearia chathamica* also has a purple centre but white petals.

Figure 3. *Brachyglottis huntii.*

Figure 4. *Olearia semidentata.*

The Chatham Island lancewood was widespread and curiously lacks the distinctive juvenile form of mainland species. *Phormium tenax* was also common and often has drooping leaves here. It may well prove to be a different species. *Melicytus chathamicus* was also quite common and variable, but with thick, waxy serrated leaves it is quite attractive and certainly more frost tolerant than *M. ramiflorus* in southern New Zealand **(Fig.5)**.

Figure 4. *Melicytus chathamicus.*

Some other plants that we saw had less obvious differences from their mainland equivalents and included *Myrsine australis, Plagianthus regius* (larger leaves), *Coprosma chathamica*, *Dicksonia* squarrosa, *Myoporum laetum* and *Macropiper excelsum* (kawakawa).

The Chatham/ Pitt Island nikau (*Rhopalostylis sapida*), while not currently regarded as a separate species, is a much better garden plant than the mainland form in my opinion. The leaves are wider, it is a bit quicker growing, fairly frost hardy and a tidier, more compact plant. It is also more stable in strong winds and very rarely blows over – not surprising given its occurrence on a small island hundreds of kilometres from anywhere in the Pacific Ocean.

Combined Proceedings IPPS 74:272-277. 2024.

Monochromatic Red LED Light Supplementation: A Dual Solution for Disease Resistance and Yield Enhancement in Glasshouse Production in New Zealand

Gagandeep Jain

Department of Land Management and Systems, Lincoln University, Lincoln 7647, Christchurch, New Zealand

Gagan.jain@lincoln.ac.nz

Keywords: LED, PAR, plant growth, red light, lettuce

Summary

This paper explores the potential of monochromatic red-light supplementation as a sustainable and innovative solution to challenges in greenhouse production, including low winter yields and high disease pressure. Greenhouse cultivation systems, though efficient, often rely heavily on chemicals for disease control, posing environmental and health risks. Monochromatic red light, delivered through energy-efficient LED systems, has been shown to enhance plant immunity by inducing secondary metabolite production and activating defence pathways, reducing the need for chemical interventions. Drawing on recent studies and ongoing research at Lincoln University, this paper emphasises the need for further investigation into the role of red light in improving crop yield and quality while minimising environmental impact. Our preliminary trials with lettuce plants suggest promising outcomes, laying the groundwork for sustainable disease management strategies that align with New Zealand's goal of promoting environmentally responsible agriculture.

INTRODUCTION

Greenhouse (GH) cultivation systems are rapidly becoming essential components of modern agriculture due to their ability to maximise crop yields in controlled environments. These systems are especially beneficial in regions with challenging climates such as New Zealand, where off-season and year-round production are vital to meet domestic and international market demands. For example, New Zealand exports significant quantities of fresh produce, including capsicum and tomatoes, with 4,394 tonnes of fresh capsicum and 3,304 tonnes of fresh tomatoes exported in 2021 alone (Aitken & Warrington, 2020). Greenhouse cultivation allows for precise control over environmental factors such as light intensity, temperature, CO_2 levels, water, and nutrient supply, which are all essential for optimising plant growth and maximising crop yields.

However, despite the many advantages, greenhouse production faces several challenges, particularly during the colder winter months. The reduced natural light during this time, combined with the need for heating, makes off-season production expensive. Using supplementary LED (Light Emitting Diodes) lighting can help overcome the low crop yield in winter as LEDs are compact, energy saving light source that can provide specific wavelengths for plant photosynthesis as well as specific photoreceptor-mediated reactions, finally aimed at maximizing plant growth and light-driven metabolite accumulation (Landi et al., 2020). Many growers in New Zealand are already using this technique to extend photoperiod in winter months, for example, tomato grower Gourmet Mokai, are using Philips LED lighting to increase the production during winter (Vogrincic, 2018).

Another pressing issue in greenhouse production is pest and disease management. Greenhouses provide a controlled environment that can also be a haven for various pests, including whiteflies (*Trialeurodes vaporariorum*), aphids, and fungal pathogens such as *Botrytis cinerea* and powdery mildew (*Sphaerotheca fulginea*). These pests and pathogens thrive in the warm, humid conditions typical of greenhouses, posing a significant threat to crop health and yield. Integrated pest management (IPM) programmes, timely monitoring and use of biological control agents can help to reduce the insect pest population in the glasshouse, but the disease pressure is still majorly dealt with using chemical sprays. As per Manktelow et al. (2005) New Zealand agriculture uses around 3,400 tonnes of agrichemicals (including herbicides, fungicides and pesticides) active ingredients annually. And a huge amount of these (up to 51%) are presumed carcinogens that can increase the risk of cancer in people working closely with pesticides.

Considering these challenges, there is an increasing need to explore alternative, more sustainable methods for enhancing plant growth and protecting crops from diseases. One such promising solution is the use of monochromatic LED light supplementation. LED lighting allows growers to provide specific wavelengths of light that not only support photosynthesis but also influence plant physiology in ways that can enhance plant health and disease resistance (Landi et al., 2020). Recent research suggests that specific wavelengths of monochromatic light, particularly red and far-red light, can trigger plant responses that increase secondary metabolite production, which can strengthen the plant's natural defence mechanisms against pathogens (Gallé at al., 2021). For instance, studies have shown that exposure to far-red enriched light can protect tomatoes from *Botrytis cinerea* (Mihaly Cozmuta et al., 2016), similarly, red light has been found to reduce the incidence of fungal infections in strawberries (Lauria et al., 2023).

Here in this paper, I briefly discuss the potential of monochromatic red-light supplementation as a promising and sustainable approach to greenhouse disease management and crop production. Drawing on insights from recent studies and ongoing research at Lincoln University, the work underscores the need for further investigation into the role of red light in enhancing plant immunity and reducing reliance on chemical pesticides.

Monochromatic red-light supplementation: A sustainable approach

In greenhouse environments, the primary challenge is often maintaining optimal light conditions for plant growth. Light intensity and photoperiod (the duration of light exposure) are critical factors influencing plant growth, yield, and overall health. During the winter months, when natural light is limited, it becomes necessary to supplement the light supply using artificial lighting. While conventional grow lights, such as high-pressure sodium (HPS) or metal halide lamps, have been used extensively in this role, they are energy-inefficient, produce significant heat, and are not always tailored to the specific light needs of plants (Katarzyna et al., 2020).

In contrast, LED lighting provides a highly energy-efficient alternative. LEDs consume less energy while offering precise control over light intensity, spectrum, and duration, all of which can be optimized for plant growth. The ability to use specific wavelengths is particularly important. For instance, blue light (around 450–495 nm) is known to enhance photosynthesis and regulate plant growth, while red light (around 640–680 nm) has been found to promote flowering, fruiting, and secondary metabolite accumulation (Human, 2023; Landi et al., 2020) (**Fig. 1**).

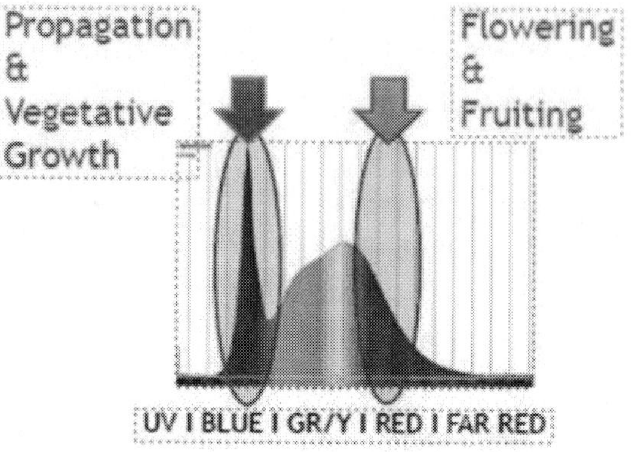

Figure 1. Monochromatic light spectrum and its influence on plant growth. Source: Human, 2023.

Beyond the energy savings, one of the most promising aspects of monochromatic LED light supplementation is its ability to influence plant immunity and disease resistance, particularly red light. Plants have various photoreceptors that perceive light as an informational signal, such as red/far-red light sensing phytochromes, blue light sensing cryptochromes, and phototropins and UV-B receptor UVR8 (Gallé et al., 2021; Landi et al., 2020). When plants are exposed to red light, the cytosol localised inactive phytochrome (Pr) changes to active far-red light absorbing form (Pfr), which can induce physiological responses in plants through the transcriptional regulation of various genes (Gallé et al., 2021). These physiological responses can enhance plant defence by activating signalling molecules such as salicylic acid and induce the production of secondary metabolites like flavonoids and phenolic compounds, which are known for their antioxidant properties (Su et al., 2017).

Several studies have shown that red light supplementation in green house crops can improve disease resistance, for instance, (Hui et al., 2017) showed that *B. cinerea* induced lesion development was significantly reduced in detached tomato leaves when exposed to red light for 4 days as compared to leaves in dark. Additionally, red light exposure significantly decreased the superoxide and H_2O_2 content and enhanced antioxidant enzyme activity of SOD, CAT and POD after 2 days of *B. cinerea* infection in detached tomato leaves (Hui et al., 2017). Similarly, (Meng et al., 2019) showed that strawberry leaves when developed under red light are more resistant to *Botrytis* infection as compared to when developed under white, blue, and blue+red lights. A recent study by (Lauria et al., 2023) showed that strawberry plants exposed to red light supplementation showed higher fruit yield, and lower disease severity in harvested strawberries at 36 hours post inoculation as compared to white, blue, green or no light supplementation treatments. Enhanced disease resistance in post-harvested strawberries was linked to induction of secondary metabolites and the upregulation of defensive genes in plants exposed to red light supplementation (Lauria et al., 2023). Thus, monochromatic light supplementation offers a two-fold advantage: increasing crop yield while simultaneously reducing disease incidence and the need for chemical pesticides.

In response to the growing need for sustainable disease management in New Zealand's greenhouse production, researchers at Lincoln University, including Prof. Rainer Hoffmann and Dr. Gagan Jain, are collaborating with Assoc. Prof. Marco Landi from the University of Pisa. Together, they are investigating the potential of red-light supplementation in greenhouse production systems. This project is supported by the Royal Society Catalyst Grant (Grant reference number: 23-LIU-002-CSG).

The objectives of this project are to:

- Develop a collaborative research and development plan with the University of Pisa to explore the effects of monochromatic light on disease management in greenhouse crops.
- Identify and standardise optimal supplementary lighting techniques for controlling specific pests and diseases across various greenhouse cropping systems in New Zealand.
- Translate these findings into an innovative and sustainable mechanism for managing diseases in greenhouse production.

An initial experimental trial has been established in the nursery glasshouse at Lincoln University. In this study, lettuce plants are subjected to two treatments: supplementary red light and no supplementary light, over a period of 10 weeks. The supplementary red light is provided at an intensity of 250 $\mu mol/m^2/s$ for 5 hours daily (from 11:00 a.m. to 4:00 p.m.).

Preliminary results indicate that lettuce plants exposed to supplementary red light, for four weeks, have accumulated **20% more biomass** compared to those without supplementary light (**Fig. 2**, results unpublished). In the next phase, these plants will be inoculated with a *Botrytis* spore solution to assess disease severity in the red light-treated plants versus those grown under standard conditions without supplementary light.

Figure 2. A) Plants exposed to supplemented red light, in Lincoln University glasshouse. **B**) Plants exposed to no light supplementation, and **C**) red-light supplementation for four weeks.

CONCLUSION

In conclusion, monochromatic LED red light supplementation can present a sustainable and innovative approach to addressing some of the key challenges in greenhouse production. It can offer a viable solution to increase crop yield and quality while reducing disease pressure, minimizing pesticide use, and promoting environmental sustainability, which aligns with New Zealand's broader goal of sustainable agriculture. Further research and practical implementation of this technology could pave the way for transformative advances in greenhouse production systems.

LITERATURE CITED

Aitken, A. G., & Warrington, I. J. (2020). Fresh facts 2020. In New Zealand Institute for Plant and Food Research Ltd. https://www.freshfacts.co.nz/

Gallé, Á., Czékus, Z., Tóth, L., Galgóczy, L., & Poór, P. (2021). Pest and disease management by red light. Plant Cell and Environment *44:* 3197–3210. https://doi.org/10.1111/pce.14142

Xu, H., Fu, Y., Li, T., and Wang, R. (2017). Effects of different LED light wavelengths on the resistance of tomato against *Botrytis cinerea* and the corresponding physiological mechanisms. Journal of Integrative Agriculture *16*: 106–114. https://doi.org/10.1016/S2095-3119(16)61435-1

Human, A. (2023). Horticultural lighting in nurseries from the ground up. Comb. Proc. Int. Plant Prop. Soc. *73*:10-14.

Kowalczyk, K., Olewnicki, D., Mirgos, M., and Gajc-Wolska, J. (2020). Comparison of selected costs in greenhouse cucumber production with LED and HPS supplemental assimilation lighting. Agronomy *10*: 1–14. https://doi.org/doi:10.3390/agronomy10091342

Landi, M., Zivcak, M., Sytar, O., Brestic, M., and Allakhverdiev, S. I. (2020). Plasticity of photosynthetic processes and the accumulation of secondary metabolites in plants in response to monochromatic light environments: A review. Biochimica et Biophysica Acta - Bioenergetics *1861*: 148131. https://doi.org/10.1016/j.bbabio.2019.148131

Lauria, G., Lo Piccolo, E., Ceccanti, C., Guidi, L., Bernardi, R., Araniti, F., Cotrozzi, L., Pellegrini, E., Moriconi, M., Giordani, T., Pugliesi, C., Nali, C., Sanità di Toppi, L., Paoli, L., Malorgio, F., Vernieri, P., Massai, R., Remorini, D., and Landi, M. (2023). Supplemental red LED light promotes plant productivity, "photomodulate" fruit quality and increases Botrytis cinerea tolerance in strawberry. Postharvest Biology and Technology *198* (December 2022). https://doi.org/10.1016/j.postharvbio.2023.112253

Manktelow, D., Stevens, P., Walker, J., Gurnsey, S., Park, N., Zabkiewicz, J., Teulon, D., and Rahman, A. (2005). Trends in pesticide use in New Zealand: 2004. Ministry for the Environment Wellington,, New Zealand.

Meng, L., Höfte, M., & Van Labeke, M. C. (2019). Leaf age and light quality influence the basal resistance against *Botrytis cinerea* in strawberry leaves. Environmental and Experimental Botany *157*:35–45. https://doi.org/10.1016/j.envexpbot.2018.09.025

Mihaly Cozmuta, A., Mihaly Cozmuta, L., Peter, A., Nicula, C., Vosgan, Z., Giurgiulescu, L., Vulpoi, A., and Baia, M. (2016). Effect of monochromatic Far-Red light on physical- nutritional-microbiological attributes of red tomatoes during storage. Scientia Horticulturae *211*:220–230. https://doi.org/10.1016/j.scienta.2016.08.031

Su, J., Liu, B., Liao, J., Yang, Z., Lin, C., and Oka, Y. (2017). Coordination of cryptochrome and phytochrome signals in the regulation of plant light responses. Agronomy *7*: 1–22. https://doi.org/10.3390/agronomy7010025

Vogrincic, S. (2018). Gourmet Mokai – LED installation inspiring. Grower2Grower, August, 2–4. https://www.grower2grower.co.nz/gourmet-mokai-led-installation-inspiring/

Combined Proceedings IPPS 74:278-289. 2024.

Forestry Roles for Propagation

Paul Schroeder

Proseed, 177 Amberley Beach Road, Amberley 7481, New Zealand

paul@proseed.co.nz

Keywords: forestry, *Pinus, Eucalyptus,* somatic embryogenesis, stock plants

Summary

This paper gives a brief outline of Proseed nursery's main propagation programmes, some New Zealand plantation forest research developments, and thoughts on how modern propagation systems might assist native reforestation.

INTRODUCTION

Proseed is a part of the Ngai Tahu Holdings portfolio. It was originally established in 1966 as the NZ Forest Service Seed Store. As part of Rogernomics dissolution and sale of the NZ Forest Service, in 1987 it was created as a separate business by the NZ Forestry Corporation and was eventually sold to Ngai Tahu in 2001.

Proseed now operates a 161 ha specialist forest seed orchard, nursery, seed production and extraction facility near Amberley. It contracts to the NZ Radiata Pine

Breeding Company to run their breeding archive near Waipara.

Proseed is a member of all major NZ tree breeding programmes: the Radiata Pine Breeding Company, the Specialty Wood Programme (non-durable eucalypts, cypresses and Douglas-fir), and NZ Drylands Forest Innovation.

Proseed sells genetic gain! Seed is the vehicle for that. Around 3 tonne of seed (mostly control and open pollinated, improved *Pinus radiata*) is sold annually. Last season it was 4.7 tonne (equivalent to 94,000 ha of planting)! Proseed is the largest forest seed producer in Australasia.

Proseed nursery programmes include grafting *P. radiata* and eucalypts, propagating pine and eucalypt species from cuttings, plus various incidental industry support projects.

Grafting *Pinus radiata*

Most (110 ha) of Proseed's 161 ha is given to production of control pollinated and open pollinated *P. radiata* seed. Production is from grafted elite selections from the NZ Radiata Pine Breeding programme **(Fig. 1)**.

Figure 1. Grafted elite *Pinus radiata* ramets in Proseed's Amberley orchard, bagged ready for controlled pollination.

Proseed completes up to 18,000 grafts annually for orchard establishment and renewal. Planting a recently purchased, neighbouring, 27 ha property will have this continue for a few years yet.

Candle buds are grafted onto regular 1yo container rootstock in winter using a wedge and cleft technique. Grafts are planted out the following winter **(Fig. 2)**.

Figure 2. Newly completed *Pinus radiata* grafts. Candle buds are grafted onto one year old seedling rootstock using the wedge and cleft technique.

Grafting Eucalypts

Proseed has established seed orchards with selections out of national improvement programmes for *Eucalyptus nitens*, *E. fastigata*, *E.regnans*, and *E. saligna*.

As a founding member of NZ Dryland Forest Innovation, a group initiative to develop a hardwood industry based on improved durable eucalypt species, Proseed is supporting the groups programmes through establishment of *E. bosistoana*, *E. globoidea*, and *E. quadrangulata* orchards using clonal selections out of respective breeding programmes.

Grafting eucalypts is a summer activity (**Fig. 3**). Freshly collected scions ("ripened" shoots trimmed of all foliage and drenched with a systemic fungicide) are fitted to rapidly grown rootstock (about 16 weeks from sowing) using a cleft graft technique. Grown on in a greenhouse environment, take (and failure…) is rapid: 4 – 6 weeks will show.

Figure 3. *Eucalyptus bosistoana* scions starting to move 6 weeks after completion. Stem sections are grafted onto vigorously growing seedling rootstock (about 16 weeks from sowing) using the wedge and cleft technique.

Some species (*E. nitens*, *E. bosistoana*) are relatively straight forward while others are more difficult. *E. globoidea* is particularly so and a study this summer is investigating efficacy of grafting clones onto their own, open pollinated (half sib) progeny.

Eucalypt Cuttings

Vegetative propagation of elite genotypes from cuttings affords fastest deployment and greatest genetic gain compared to propagation from seed, that is, growing cuttings produces an exact copy of the donor tree. With seedling offspring, the maternal genome is diluted by the pollinating parent and may not come so true.

Thus, parallel to establishing a breeding population and production of improved seed, the NZDFI wished to establish a programme for cutting propagation of early selections from *E. bosistoana* and *E. quadrangulata* base populations. Proseed accepted a mission to root cuttings for establishment of clonal testing trials and for commercial scale planting. Over 6 years NZDFl tasks were:

- Evaluate coppice as a material source.
- Develop a protocol for cutting propagation of *E. bosistoana* and *E. quadrangulata*.
- Produce ramets of 2000 selections to establish clonal testing trials.
- Produce stools (**Figs. 4 and 5**) of 31 juvenile plus trees (~3 years) selected on growth rate, early form, and wood properties, for commercial scale production.
- From those stools, produce 25,000 rooted cuttings for commercial deployment.
- Evaluate strike of cuttings from 10-year-old ortets.
- Report on best practice for propagating *E. bosistoana* from cuttings.

Figure 4. Select *Eucalyptus bosistoana* stools in a crop cover tunnelhouse, used to produce a bulk supply of rooted cuttings for commercial scale planting.

Eucalypt Cuttings for Clonal Trials

Initial work, modelled on what had been seen during a study tour of Narromine Nursery in Australia (owned and managed by David Cliffe, also an IPPS member) was to produce plants for clonal trials.

Cuttings were set into propagation cell trays, filled with loose perlite and peat media, and placed over bottom heat under mist.

From cuttings taken off 2,000 field selections over two seasons, 13,000 plants representing 1,100 clones were produced for clonal trials.

Figure 5. Pilot stool bed establishment with *Eucalyptus bosistoana* (left) and *Pinus x attenuradiata* (right). *E. bosistoana* stools were started as rooted cuttings potted into PB3/4. Once established and growing strongly, some were set up in the hydroponic dripper installation shown here. PB bases were cut away before they were stood on coir growbags for roots to grow through. *P. x attenuradiata* seedlings on the right were grown on in planter bags, potting into larger grades, up to PB8, as growth necessitated.

Eucalypt Cuttings for Commerce

Twenty-one clones for commercial deployment were drawn from the selections that had been propagated for clonal trials. One year into propagating those 21 clones, all breeding values were updated with new additional information from breeding trials. Consequently, nearly all the original set of clones were replaced. That left Proseed nursery just two years to turn about 300 ramets of 31 clones into 25,000 rooted cuttings.

The adopted strategy was first, to produce as many stools as fast and as big as possible in the time available. Most were grown on as regular potted plants in a shade house while a few were put into a hydroponic set up inside a small tunnel house. Growth of hydroponic stools was spectacular, but nutrition was tricky and plants in the shade house produced more robust material.

Early on, bad root distortion in the propagation cell trays was detected. At the time, work setting pine minicuttings into plugs was working well so that same system was successfully applied to the eucalypts.

Nearer to scheduled sales for planting, potting rooted cuttings switched from PBs into forestry tubes to emulate presentation of standard forestry planting stock. In the end, 15,000 plants were produced. Plants were distributed to privately owned sites through the length of New Zealand: North Canterbury to Northland.

In summary, the average strike improved from just 30% at the beginning to 75% at the end. Huge variation between clones, from nothing to 90%, was consistent throughout the programme. Average strike from older (10 year) material was similar, which was encouraging for bringing in older selected trees from breeding trials after completion of more conclusive assessments.

Supporting government funding for this breeding programme has all but dried up. Because viable commercial production of rooted cuttings was not demonstrated, this propagation programme has been suspended, though stools of the clonal set used have been kept.

Propagation Plugs

At the moment there are 2 plug systems in New Zealand to choose from: Ellepot, a paper tube filled with your own media (**Fig. 6**), and Jiffy Preforma plugs, moulded from peat and binding agents (**Fig. 7**).

Proseed evaluation of strikes in each system were similar. Cost is about the same (~$0.10 ea). For Proseed's purpose, the Jiffy system proved more convenient: no special machinery nor messy filling required, preformed dibble holes and no propensity to drop media from the bottom.

Unlike soft stemmed pine cuttings, eucalypt cuttings don't require a dibble hole, so plugs were simply inverted when traying up. (Plugs without a dibble hole are available.)

Conversely, for other nurseries, the ability to produce a range of plug and tube sizes at scale and filled with specialised media would be more advantageous.

Figure 6. *Eucalyptus quadrangulata* cuttings in Ellepots.

Figure 7. *Eucalyptus bosistoana* cuttings in Jiffy Preforma plugs.

Table 1. Pros and cons for plug systems:

System	Pros	Cons
Ellepot	Biodegradable container	Needs specialised filling machinery
	Fill pots with custom media	Dibbling may be required
	Range of pot sizes	May lose media from base
	Specialised tray systems	Paper may turn roots if not wet
	Easy patching and sorting	
	No transplant shock	
	Automation	
	Easy packing and shipping	
Jiffy	No container to restrict roots	No media options
	Ready to go ex packaging	Negligible size options
	Preformed dibble hole option	
	High root ball integrity	
	Easy patching and sorting	
	No transplant shock	
	Automation	
	Easy packing and shipping	

Pinus x _attenuradiata_ Hybrid Cuttings

Field trials of *P. radiata* x *P. attenuata* hybrids, established by Scion and Proseed in the late 1990's, are showing them to be tolerant of cold and dry conditions and to have good resistance to snow.

The hybrid has rapidly become popular with companies for planting hard sites, particularly in higher altitude, snow prone areas. Popularity is now such that Proseed is unable to meet demand for seed. As practiced for similarly precious control pollinated *P. radiata* seed, Proseed has developed a technique to further bulk plants as rooted cuttings from stool plants.

In New Zealand, radiata pine cuttings are usually collected from open ground grown stools, once, in winter, and set directly into open ground beds under frost cloth.

Proseed has adapted another technique developed by Arbogen's Colac nursery in Australia. Potted stool plants are grown under cover and multiple collections of minicuttings are set year-round in cell trays, on bottom heat, under mist.

Whereas Colac set directly into the final growing container (almost 100% strike!). Proseed wanted to propagate in a smaller cell with a view to forwarding rooted plants to client nurseries for growing on. Small, rooted plants in a small, robust container would be better suited to machine/robotic handling and dispatch, so attention turned to propagation in plugs.

Cutting stools are grown in PBs in a crop cover tunnelhouse. Cuttings are collected to soak for at least 15 minutes in a bucket of IBA + NAA solution. When the bucket is full, the heap of cuttings is turned (so oldest are brought to the top) and cuttings are set into Jiffy Preforma propagation plugs. The plugs are supplied complete with a preformed dibble hole.

Cutting material is soft and requires frequent misting until cuttings harden and can maintain turgor again, usually 1 – 2 weeks. Misting is then dialed right back to irrigation once or twice a day. The Proseed misting system (a Multigrow controller by Autogrow) is driven by solar integral or timer or both.

Rooting is completed within 3 months of setting. A single collection from one mature, open ground grown stool might yield ~40 cuttings but, with this alternative system, a single stool might yield ~120 cuttings in a season.

While cuttings can be rooted year-round, bare root nurseries want plants for lining out October – November only. Storage of rooted plants through interim periods was an issue. Plants can be put out into a shadehouse and, without nutrients in the rooting media, hold reasonably well. Occasional foliar feeding helps keep condition without producing excess growth. However, the small plugs are vulnerable to rapid drying in hot weather.

Alternatively, plants can be held up to 7 months in cool storage without apparent undue effect. Storage in stackable "Lettuce" cell trays from Ellepot has proved both convenient (apart from a need for intermittent bottom watering) and effective (**Figs. 8** and **9**).

Proseed is currently producing up to 150,000 rooted cuttings a season. These are being supplied to a second party container nursery for growing on as soon as they are rooted, so storage has not yet been problematic.

Figure 8. Ellepot lettuce trays have proven useful for cool-storing rooted *Pinus x attenuradiata* cuttings. The trays are designed so that, when aligned the same way stacks nest into each other. But when trays in stacks are alternated end for end, stacks open up enough to accommodate small plants such as rooting pine cuttings in between.

Figure 9. A pallet of lettuce trays packed for cool storage. A shadecloth wrap (fastened with Velcro) proved better than film. The latter kept air inside humid and still, encouraging botrytis issues. Trays do require disassembly and bottom watering about once a week but have stored like this up to seven months with no adverse effect, either during or after storage.

Future Horizons – Tissue Culture and Plugs

As already stated, vegetative propagation of elite genotypes from cuttings affords fastest deployment and greatest genetic gain compared to propagation from seed. Tissue culture offers an opportunity to rapidly proliferate elite genotypes on a spectacular scale. For decades, plantation forestry has worked to realize this. Apart from difficulty, as with standard nursery techniques, of rooting older plant material, two persistent impediments to bringing such a process to scale have been cost of labour-intensive laboratory methods and transitioning soft plant material out of a sterile laboratory into a septic nursery environment. Recent technological advances seem finally to be bringing realization within reach.

Development of techniques to multiply seed embryos and advances in DNA analysis (not GM!) have led to genomics: a revolutionary new field of plant breeding. Instead of determining the breeding value of plus tree selections from the growth performance of their progeny, candidates can now be screened straight away for genetic markers associated with desirable growth traits.

Breeders are now moving to make forward selections even before seeds are fully mature, and to have a pathway that bypasses the adverse effect of physiological aging on propagation success.

For *Pinus radiata*, immature seed from elite pedigree crosses is dissected from green cones and callus is grown from somatic (undifferentiated) tissue. Some callus from selected genotypes (say 1,200 out of 10,000) is placed into cryostorage while other callus is matured to produce entire (shoot with root) plant embryos. The clonal embryos are "germinated" and grown on for further field testing to confirm and refine the genomic selections. Then, selected, still juvenile calluses are brought out of cryostorage and further proliferated for commercial establishment.

Rapid advances in automation and robotics look to be making bringing this technology to scale a real possibility. Examples include:

Bioreactors

In a bioreactor, tissue is placed on a plate in a tagine casserole like vessel that is periodically flooded with nutrient and hormone solutions to direct growth, either for callus or embryo development **(Fig. 10)**.

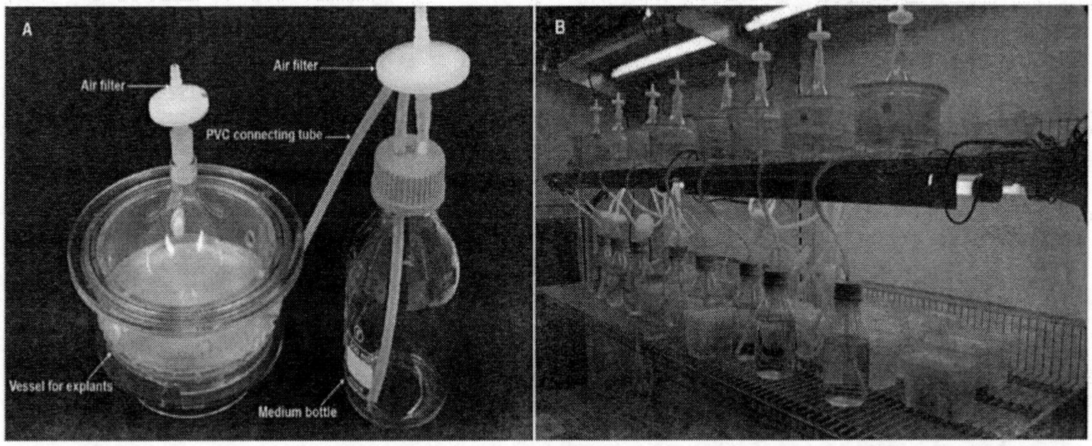

Figure 10. Use of liquid instead of a solid culture medium for the micropropagation of plants offers advantages such as better access to medium components and scalability through automation of the process. Scion Research has partnered with New Zealand Forest Industry to develop a protocol for large scale multiplication of selected *Pinus radiata* using temporary immersion bioreactors as pictured above. Compressed air drives nutrient and hormone solutions between the medium bottle and bioreactor, to bathe tissue cultures in the bioreactor. Bottles of different solutions can easily be changed as tissue growth develops.

Optical sorting of embryos

Embryos are placed into suspension and, through clever fluid dynamics, are sorted to single file past a camera for software to clean out and grade individual embryos through a high-tech drafting gate (**Figs. 11 and 12**).

One process is to drop each minute embryo onto a small filter paper that can be rotated to "right end up" and then gently folded into a mini taco to facilitate easy, automated handling for pricking out/germination.

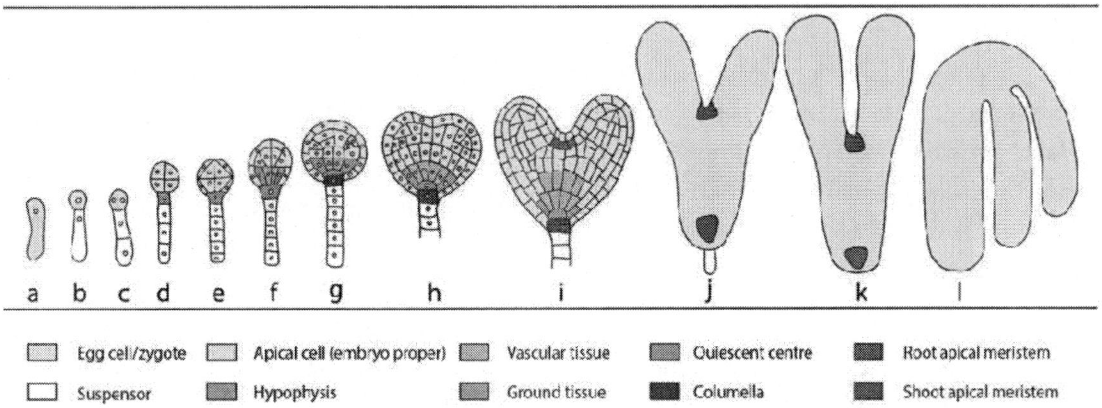

Figure 11. A schematic of seedling embryo development from fertilization of the egg cell through to a whole embryo seedling. Somatic embryogenesis begins with tissue harvested from the zygote at the very beginning of the development. Used with permission, M. Abrahamsson, Department of Plant Biology, Uppsala, Sweden.

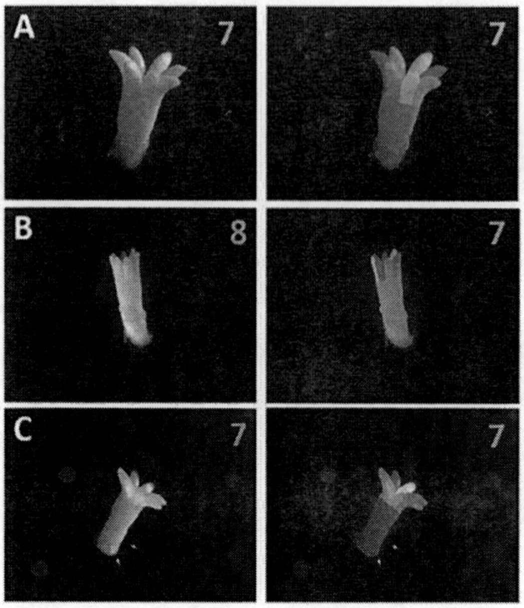

Figure 12. Computer images of embryos with a machine count of cotyledons shown down the left. Portions identified by the computer are colour coded on the right together with a manually corrected count.

Introducing the rightsoil flora

Protocols are being developed to prepare plants for a septic nursery environment by proper hardening and inoculation with healthy fungal flora before release from the laboratory.

Automation

Expectation is that most handling (pricking into plug trays, blanking, potting on) will be automated much as now with bedding plants provided to our box stores. Large forest nurseries already use automated systems for sowing and growing seedlings. Handling somatic embryos raised in plugs won't be too great a stretch for them.

For now, Proseed is watching this research work closely as, when successful, the implications for its business could be profound. Likely a move away from the large control pollination orchards and bulk collection it currently manages to perhaps smaller archives and fewer, carefully chosen controlled crosses from which developing seeds are taken for somatic embryogenesis. As an economic alternative to somatic embryogenesis, seed production from open pollinated seed orchards would likely continue. Meantime, it is prudent for Proseed to develop skills and experience in handling plants in plugs and trays since, like orchids, strawberries, and amenity plant varieties already, that is how the new tree breeding process will finish.

Growing Planting Stock for Native Forest Establishment

Currently there is repeated calling for New Zealand to reestablish native forest. That is canopy trees: not scrub but climax communities that used to be. MacFlora to go please.

Unfortunately, withstanding issues around planting and establishment, plant succession, biodiversity, ecosourcing, or what is a more effective carbon sink, for various reasons, from seed availability and/or dormancy through to specific growing conditions and sheer time to grow, nurserymen are finding many species, particularly some canopy species, difficult to grow. In general, smaller, pioneer species are easier, evidenced by their domination of highway and farm native planting to date.

Difficulty growing preferred, larger grade planting stock of native forest tree species translates to increased cost. However, buyers used to having exotic plantation species that grow to sturdy size in 1-2 seasons and retail around $1 each expect, and indeed need cost to be low. Trickier clonal stock like redwood and Leyland cypress can be had for around $5, yet some large grade native species retail for $20 or more, if they can be had at all.

If large scale planting of native canopy species is to fly, criteria include that cost of planting stock must be kept as low as possible. That means lifting plant percent (high germination/strike rates); shortest growing times; scaling up to spread overhead costs; and minimizing labour inputs through mechanization, automation and efficient handling systems. Plug systems offer remarkable opportunities to achieve all this (**Fig. 13**).

In support of Ngai Tahu interest in recloaking Papatūānuku (Mother Earth), Proseed has experimented with applying techniques described above (rooting mini-cuttings collected from a range of juvenile genotypes) to native forest tree species. Totara and kahikatea performed very well: near 100% and large cuttings quickly developing into large plants. Rimu, red and black beech didn't do nearly as well, though Auckland Regional Authority has reported good success with rimu in the past.

All food for thought re' bypassing tricky seed propagation issues.

In summary, propagation is pivotal in the future of forestry. To meet challenges sustainably and successfully that future will be highly technical and data intensive.

The future is exciting!

Figure 13. Rimu and totara cuttings 14 days after setting 8 June 2023 under mist. Larger cuttings in Transplant Systems TS45 cell trays and smaller cuttings in Jiffy Preforma plugs. The large group at left are on bottom heat while the cell trays at right have that turned off.

Figure 14. Totara tip cuttings from Transplant Systems TS45 in April 2024, 10 months after setting in June 2023 and potting up in January 2024, 8 months after setting.

Combined Proceedings IPPS 74:290-296. 2024.

Acquiring and Marketing "New" Plants

Randy Baldwin

San Marcos Growers, 125 S. San Marcos Road, Santa Barbara, California 93111 USA

rbaldwin@smgrowers.com

Keywords: San Marcos Growers, plant introduction, plant breeders

Summary

San Marcos Growers nursery has come to be known in the industry for the introduction of exciting new plants. This could not have been accomplished without the many people met along the way with similar interests.

INTRODUCTION

San Marcos Growers was founded in 1979 in Eastern Goleta Valley in Santa Barbara, California and ever since that time we have worked to "introduce" new plants into the nursery trade or reintroduce those plants that no longer were available. After 45 years of operation, this year will be our last full year that we will actively be propagating and planting new crops as the nursery will close at the end of 2025 to make room for a housing development. It will be up to others to continue discovering, then growing, and promoting the plants we have come to call "Plants for the California Garden".

Morgan Bill Evans was one person to introduce us to many new and unusual plants. He began his horticultural career working at his father, Hugh Evans' famous Los Angeles Nursery, first called Evans Gardens and later Evans and Reeves. Bill gained fame later working as the landscape architect for Walt Disney when building Southern California's first major theme park. Notably he also helped introduce plants he discovered in his travels, such as the Shogun collection of Asian plants later grown by Monrovia Nursery.

I fondly remember when I would ask him about a plant, I thought new in cultivation and he would point out when it was grown before saying, "It has all been done before my boy." This was a case with *Fabiana imbricata*, which I thought we introduced (both the white and purple form), but only to discover it listed in the 1928 edition of Bailey's Cyclopedia of Horticulture (later volumes renamed Hortus) as common in southern California at the time.

This talk is a story about how an interest in plants helped shape our nursery that has since come to be known in the industry for the introduction of exciting new plants, but it is also about the many people met with similar interests. I have come to the realization that many people have helped shape San Marcos Growers' goal of introducing sustainable and exciting new plants. It really is not what one knows, but who they know and there are way too many people I need to thank for helping us achieve this goal.

One particular person who has helped us out over the years is John Bleck. Dietes 'John's Runner', *Aeoniuma* 'Jolly Green 'Superb', and *Senecio* 'Jolly Gray' are just a few that he gave us and we named and introduced.

A more complete list of the plants John Bleck has given us, but even this is lacking as last week he dropped off cuttings of a new *Puya* hybrid that he had in his garden.

One of the plant meccas for unusual plants was certainly Western Hills Nursery. Many "new" plants into the nursery trade, including some that came from us, were picked up from this small nursery in Occidental, California. Co-owner Marshall Olbrich once noted to me that many such plants came from them, claiming that he and his partner Lester Hawkins introduced from sources in the UK the Sunrose (*Helianthemum*) cultivars that became very popular in California in the 1990s.

Such "introductions" made by larger nurseries from plants acquired from smaller ones was quite common and we experienced this when our 1992 introduction *Anisodontea* 'Tara's Pink', named for our beautiful Rhodesian Ridgeback Tara, who roamed the garden, was picked and displayed at a tradeshow as a new introduction by Hines Nursery.

I was still going to college and working at a Santa Barbara retail nursery when I met Fred Meyer. His interest in a wide diversity of plants opened my eyes wide to an amazing new world. He was most famous for the Meyer's hybrid *Alstroemeria*, but he was also quite interested in and worked with other bulbs, coral trees, kangaroo paws, restios and New Zealand flax. It was Fred's interest in *Phormium* hybrids for cut foliage for the floral market that led to my visit with Margaret Jones in New Zealand and the importation in 1985 of the many New Zealand flax hybrids that we became known for and that became extremely popular in the nursery trade in subsequent years. I even gave a talk to about these first *Phormium* introductions at the 1998 Western Region IPPS annual meeting. While on the topic of *Phormium*, it was a group of Bay Area horticulturists and designers who first gave me the amazing large red flax with weepy foliage that we named 'Wildwood' to honor the southern California nursery where it came from, Ray Walsch's Wildwood Nursery.

Another well-known destination nursery for great plants was Carman's Nursery in Los Gatos, California that was established in by Hugh Carman in 1937 and operated later by his son Ed. Jewell plants were abundant in this nursery. One such plant was a rosemary that Ed had as a hedge across the front of his property. I got a plant of it on a visit in the mid-1980s. Ed swore that this plant was the true 'Tuscan Blue' rosemary cultivar, but it was so different than what ALL other nurseries were growing under this name that we, and Dave Fross of Native Sons Nursery, eventually introduced it into the trade, calling it 'Ed Carman'. Note that rosemary plants are now included in the huge genus *Salvia*. I will not elaborate on this other than note that this trend is what my friend Kathy Musial calls "Namenclutter". Another plant originating at Carman's Nursery was *Tetrapanax papyrifer* 'Steroidal Giant', which he received from Japan in the 1970s and we got later from Sean Hogan at Cistus Nursery. It is a giant of a plant and most importantly it regularly produces viable seed, so we are able to produce seed grown prodigy of it.

One of the most important plants we first purchased from Ed Carman has since become the most popular tree we sell and we owe thanks no only to Ed, but also to Victor and Carla Reiter for planting it in their San Francisco garden in 1944 and also to the Saratoga Horticultural Foundation who introduced this tree and contracted it to be micropropagated by Briggs Nursery in Washington. This tree shown on our website in the Reiter's garden, from which all *Arbutus* 'Marina' were propagated, no longer exists but the tree planted in our own garden last measured over 50 feet tall and 65 feet wide and is the listed California Big Tree and National Champion on Cal Poly San Luis Obispo's "Big Tree" registry.

Our nursery is closely associated with Ganna Walska Lotusland Botanic Garden and one event each fall called The Exceptional Plant Auction brings out all serious plant people and great plants, some that we have later propagated and offered for sale. The 2024 auction on October 5th will coincide with our last San Marcos Growers Field Day the day before. Stay tuned for more information about these plant parties!

Lotusland's original fern garden was designed by talented designer and plantsman Bill Paylen. It used good quantities of famed *Begonia* hybridizer Rudy Ziesenhenne's plants, including one he named 'Lotusland'. Rudy Ziesenhenne created and named many great landscape begonias, naming one for his son and others after locations in Santa Barbara. I believe San Marcos Growers is one of the only nurseries going these Ziesenhenne begonias.

Lotusland's fern garden designer Bill Paylen was memorialized with the naming of a beautiful *Dyckia* by another incredible plantsman, Bill Baker. Many of our *Dyckia* hybrids that we selected and named were Bill Baker seedlings and most importantly he was our original source of *Agave victoriae-reginae* 'Albo-marginata', what has come to be called the White Rhino Agave. Twenty years after acquiring our first plants of this beautiful agave we have built up our propagation stock by coring our 400 stock plants so that we are now the primary supplier of this beautiful plant in California. There is one of these plants on the raffle table.

For all things succulent, be they agaves, aloes or echeveria, Brian Kemble at the Ruth Bancroft Garden, is a primary source of plants, seed and information. Aloe 'Birds & Bees' and Aloe 'Red Bird' are two hybrids were selections that we made and named from open-pollinated seed from *Aloe arborescens* that Brian Kemble sent to us in 2007 and we introduced the plants in 2014. We have been building stock on this hybrid of *Aloe dhufarensis*, which was grown from seed supplied by Brian, that we call 'Dhufar Rose'. We will not have it in any great quantity by the time we close the nursery.

Tom Cole of Cold Spring Aloes has also provided us with seed and plants of Kenyan and Ugandan aloes, including the amazing *Aloe lukeana* that Tom named for his late brother Luke. We are the primary supplier of this plant in the nursery trade and with the 179 taxa of aloe in active production, we supply many of the aloes sold in California.

David Verity was the manager of UCLA's Mildred Mathias Botanic Garden and he made many aloe hybrids, including the one introduced by the Huntington Botanic Garden named after him. We also named *Aloe* 'Dave's Delight' to honor him. Dave Verity was also well known as one of the first people to breed and release native monkeyflowers.

We were also fortunate to work with Rich Persoff with the introduction of both his Kids and Jelly Bean Series of monkeyflowers. One of his last introductions before he passed was the beautiful picotee 'Fiesta Marigold'.

My college schooling focused on native plants, so it was natural that this interest continued at the nursery. I was fortunate in this regard to meet and learn propagation techniques from Dara Emery, the plant breeder and propagator at the Santa Barbara Botanic Garden. His breeding created the great 'Canyon Snow' Pacific Coast Iris and the Quartet Series of Heuchera. Dara tutored Carol Bornstein when she came to the Santa Barbara Botanic Garden in 1981, and we still grow most of her great native plant introductions.

After Carol left the Santa Barbara Botanic Garden, she provided us with additional great plants that we introduced at our nursery, including *Solanum xanti* 'Mountain Pride' in 2011, *Keckiella cordifolia* 'Mountain Flare' in 2016 and *Encelia californica* 'Paleo Yellow' in 2017.

We also introduced several native plant selections we made or that were given to us. *Salvia leucophylla* 'Point Sal Spreader', introduced in 1986 was a plant from John Bleck. *Juncus patens* 'Elk Blue', introduced in 1994, was a plant I collected on private property near Elk, CA and *Ribes viburnifolium* 'Spooner's Mesa' introduced in 1998, was a plant collected by Dylan Hannon. This year we will be introducing a vivid red form of *Eriogonum grande* var. *rubescens* that was found growing in a southern California garden.

I was fortunate to be introduced to the UC Santa Cruz Arboretum in the early years of the nursery, again thanks to Fred Meyer and it was on a trip to get our hands on this wonderful lilac purple form of the blue hibiscus. This led to a long relationship with the garden as a partner in the Koala Blooms Australian Plant Introduction Program for which I could give a whole talk on the 64 plants that were introduced between 2001 and 2018.

Fred Meyer also introduced me to kangaroo paws (*Anigozanthos*) in 1979 and later got us together with Angus Stewart, one of the primary breeders of these plants. Angus Stewart's Landscape Series Kangaroo Paws was exclusively available from San Marcos Growers.

Kangaroo paws and other Australian plants are also a fascination of Kathy Musial at the Huntington Botanical Gardens and so many great plants came our way through association with the Huntington staff.

Kathy and Bart Obrien, then at Rancho Santa Ana Botanic Garden, with Pacific Horticulture introduced the very fine plant *Deppia splendens* 'Cristóbal' that we started growing in 2001.

Through the Huntington Botanical Gardens and its International Succulent Introductions (ISI) program came the many *Aeonium* hybrids of Jack Catlin. This ISI program at the Huntington Botanical Gardens is managed by their Desert Garden curator, John Trager. He has been instrumental in the introduction of many plants and with Pacific Plant Promotions and the Huntington's ISI of *Agave* 'Blue Flame', a hybrid created by David Verity that has become extremely popular.

Greg Starr in Tucson has been an inspiration with his work on *Agave* and after he described and named *Agave ovatifolia* we got our first plants from him in 2004 that we propagated from and began selling three years later.

A few of our own named *Agave* selections that have become very popular. *Agave potatorum* from seed collected near the town of El Camarón in Oaxaca, Mexico was one such selection. 'Mateo' was a selection we made from our *Agave bracteosa* crops and was named for one of our salesman. I named *Agave* 'Stained Glass' because of the showy variegation reminded me of stained glass and also to honor the original Lotusland curator and Cactus and Succulent Journal Editor Charlie Glass, since the plant had originally come from him. This plant and 'Mateo' are now grown by many other nurseries.

We had this very nice small agave from the collection of Alice Waidhofer that we shared with Tony Avent. The next thing I knew we were getting lab-micropropagated plants of it from Hans Hanson, then at Shady Oaks Nursery in , which were from plants that Tony sent to him. This was the beginning of a nice relationship with Hans as he worked on some of our own agaves. We purchased many others that Yucca Do and Plant Delights Nurseries had come up with. Hans has worked on a diverse amount of plants, but is now most famous in succulent circles for his *Mangave* selections. We are fortunate to be licensed to propagate and sell some of the Walters Gardens Mad About Mangave collection, including the beautiful 'Kaleidoscope' which we maintain propagation stock in our greenhouse of. We core these plants to promote pupping and also taken the bulbils from the inflorescences for our crops.

In Jeff Moore's book, "Agaves: species, cultivars and hybrids," there is a picture of our display collection of 35 different *Mangave* cultivars, which are all part of the incredible Mad About Mangave collection from Hans Hansen at Walters Gardens.

The *Clivia* lily (*Clivia miniata*) is beautiful in this mass planting under oaks, but starting in 1993 we were able to offer 1-gallon yellow clivia plants from our own breeding program whose objective was to produce reliably yellow-flowering plants by see propagation.

At a fateful plant sale in 1984, I met the "grassman" and meadow master John Greenlee. We were already growing a few ornamental grasses, but we were suddenly growing many more and also getting known for them. Ornamental grasses and our friendship and collaboration with John Greenlee has been important to our nursery ever since!

John Greenlee gave us a fastigiate form of *Cupressus guadalupensis* in the late 1990s that he had selected from a seed lot. In 2002, we introduced it as 'Greenlee's Blue Rocket'. The picture (not shown) on the left taken in the parking lot at Descanso Garden where a group of this cultivar was planted and on the right the original tree, now gone, that stood as sentinel years after John had moved from his original Pomona Garden.

John also introduced me to Texas plantsman Scott Ogden, who had introduced many fantastic plants, including *Parthenocissus* 'Hacienda Creeper' and *Pittosporum* 'Oakleaf' which were both named by Scott, but we popularized them in California and *Pittosporum* 'Oakleaf' has become one of our signature plants that we are known for.

Jay Kapac is a well-known geranium breeder who partnered with us to release this beautiful series of *Pelargonium* that he named for the song "Lily, Rosemary and the Jack of Hearts" on Bob Dylan's "Blood on the Tracks" album.

I have always had a fascination with 'Pride of Madeira' and over the years introduced quite a few selections including 'San Bruno Pink' in 2006 and 'Rincon Blue' in 2011. We got from Lance Reiners, the Paintbox Plantsman's *Echium candicans* 'Star of Madeira' at the 2003 San Francisco Garden Show and have grown it ever since. We selected and named the wild 'Starburst' after noting this unusual vegetative sport from 'Star of Madeira'.

We began growing *Lomandra* at San Marcos Growers in 1990 and a majority of this first crop went into this planting at Madame Ganna Walska Lotusland in 1992 where the planting remains as an attractive large-scale groundcover under 'Blue Gums' (*Eucalyptus globulus*). These were among the first *Lomandra* in California. In 2019, I spoke to IPPS about what I call the "*Lomandra* Revolution" and these two shown, 'Breeze' and 'Seascape,' were among the first cultivars that started it off, but we currently grow 19 varieties. We have a page on our website with them all listed in case you are interested. There is a plant of the beautiful, variegated cultivar 'Lucky Stripe' on the raffle table. There are now many *Lomandra* in cultivation with 'Platinum Beauty' being the most popular and 'Baby Breeze,' which we exclusively grow, being one of the newest and smallest growing. These are tough plants that are available in a diverse range of form and foliage colors.

The first tissue-cultured, mass produced *Lomandra* was Ozbreed's 'Tanika,' marketed in the US as 'Breeze.' We first received this plant in 2003 and planted it in our own garden in 2004.

Starting with my friend Fred Myer's encouragement, I have long been interested in the *Protea* family. Time limits me from beginning to mention all of the fantastic Australian *Grevillea*, but I do need to touch briefly on the South African pins and *Leucadendron*.

We have participated with plant introductions from other commercial partners and this image shows their logo. These companies with our help have introduced into the California nursery trade many great plants – too many to list all in this presentation but will mention one particular plant introduction.

We are a trial site location for Plant Haven and a plant brought to us in 1999 by a local Santa Barbara gardener. Thinking the plant worthy, we steered her to PlantHaven and *Salvia leucantha* 'Santa Barbara' received a US plant patent and in the year 2000, we introduced the plant into the nursery trade at the Western Nursery and Garden Expo held in Las Vegas that year. We have still done a few trade shows to market plants, but in a couple weeks will do our last one, the NorCal show in northern California.

Over the years, San Marcos Growers has held many open houses that we call Field Day. Later this year, on October 4th, we will hold our fourteenth to celebrate our 45 years in business. Since it will be the last one we hold before closing the nursery next year, we are calling this one "The Last Dance at SMG".

Our best marketing method over the years has been our website with its active server pages about the plants we grow that attracts about 160,000 hits per month. This reminds me that I gave a talk to IPPS in the year 2000 about "Establishing & Maintaining a Nursery Website".

Just a quick story if there is time about a great plant that somewhat got away from us. We saw *Alstroemeria* 'Indian Summer' all over the UK when visiting in 2017, but back stateside the only supplier of this cultivar called 'Tesronto' was a small mail-order nursery in Oregon called Edelweiss Perennials. It was expensive, but I knew I had to go for it and we started building stock.

A few months later, I happened to see patent information on an "improved" selection also marketed as 'Indian Summer' called 'Tesronto' and then started seeing this plant available from plug suppliers so purchased it, as we had to have this great plant. We planted both the original 'Tesronto' and the improved 'Tesronto Imp' in the garden and have never noticed any difference between the two. Both provide nearly year-round color in our garden. Was there really any improvement?

Combined Proceedings IPPS 74:297-299. 2024.

Rex Begonia Propagation: Propagation Tips and Wives' Tales. Keeping Plants True-to-Type for 30-Plus Years

James L. Booman

Booman Floral, 2302 Bautista Ave, Vista, California 92084 USA

jin@boomanfloral.com

Keywords: begonia, propagation, leaf cuttings, production protocol

Summary

This presentation revisits the propagation methods for rex begonia plants, updating knowledge shared 20 years ago. As nursery production has evolved, the challenge of scaling up the propagation of rex begonias, plants that did not fit conventional production protocols, was addressed by reassessing five key factors: heat, water, nutrients, humidity, and light. After experimenting with six propagation systems, a combination was found that allowed for efficient and scalable production of clean, true-to-name rex begonias. Methods discussed include leaf-vein nicks, pizza pie cuts, rectangular leaf cuttings, leaf petiole cuttings, the cookie cutter method, and the cannoli method. Each technique offers unique advantages in producing uniform, vigorous plants, while the choice of method depends on plant variety and production needs. The importance of clean stock, sanitation, and strong genetic material is emphasized to ensure high-quality propagation. There is value and utility in revisiting and refining methods for efficient commercial production of rex begonia plants.

INTRODUCTION

It is a pleasure to be here with you today. I have met many wonderful plant people through IPPS meetings. They have encouraged me and shared their experiences freely. My being here today is my chance to pay back that investment made by others, and share some of my experience.

I was a bit confused as to why I was asked to speak about rex begonia propagation again as I shared most of this information 20 years ago at another IPPS meeting. It dawned on me slowly that most in the audience today were toddlers at that time.

So, it is time to repeat, update, and share again. We had built a large plant propagation business. At peak, we rooted two million plugs a week. Unfortunately, our beautiful Rex begonias we bred and selected did not fit any of our production protocols nor timing. If we cannot book a sale and ship on time, we don't have a propagation business.

So, we examined the five things every propagator does control: heat, water, nutrients, humidity, and light. Then we stepped back and reviewed the basics of clean stock, clean soil, and sanitation taught to us from the University of California system for producing healthy container grown plants.

While any 12-year-old can physically do the steps of making and sticking a cutting, a skilled propagator has to do this on time, with profit, true to name, and true to variety. We tested six propagation systems before landing on a combination that allowed us to scale up production of clean, true-to-name, vigorous Rex begonia liners.

MATERIALS AND METHODS

Leaf-vein nicks. This method is the oldest. A sharp knife cut across the thickest leaf veins will yield a new plant at each nick. We generally limit the leaf cuts to 15 per leaf. We lay the leaf on clean peat/perlite mix and place it under intermittent mist.

Pizza pie. This method involves a sharp knife to shape 2-inch x 2-inch x 2-inch triangles from a leaf. Rex leaves have palmately arranged vein orientation. Each successful cutting includes one or more veins at one point of the triangle, which is planted down into a plug of soilless media. Including a piece of the thick petiole makes the cutting more likely to produce multiple shoots.

Rectangular leaf cuttings. This method yields only one or two cuttings per mother leaf. The idea is to use a razor blade, such as an Exacta knife, to cut a 0.75-inch x 2-inch uniform piece for planting. The lower edge always includes a thick vein near the petiole. The vein runs up the rectangle. Flats planted with this method have uniform appearance, finish, and ship on the same date.

Leaf petiole cuttings. This is perhaps the easiest method to root begonias. It is about the only method that works 100% of the time for rhizomatous types. Simply cut the petiole about 1 inch below the lamina and place the leaf and petiole on clean peat-perlite soilless mix. Roots form rapidly at the cut petiole. Stem and leaf tissue differentiates in about a month.

Cookie cutter. This is our go-to method. The leaf cuttings end up uniform in size, substantially sized, and loaded with energy to root quickly and produce multiple shoots per leaf cutting. We use simple tin or aluminum cookie cutters to cut a 2-to-3-inch circle from a Rex begonia leaf. We place the cookie cutter so that it cuts the petiole 0.5 inch below the lamina. We then insert the trimmed petiole into a preformed hole in the soil plug. The leaf lamina stands up like a peacock's feathers on full display. To permit maximum light to reach the emerging new leaves and shoots, we always stick the leaf cuttings "back-to-back" in rows so that miniature canyons of sunlight are formed.

Cannoli method. This is an odd system that we discovered only recently, which works for rhizomatous begonias that for some reason have low yield when using the leaf-petiole method. We lay the leaf flat and excise the leaf petiole completely. We then roll the leaf lamina into a pointed cannoli shape and plant the narrow end down into a plug of peat-perlite soilless mix in a cell tray. All cut leaf veins root fairly rapidly and then shoots follow. We assume this method works because almost the entire reservoir of sugars and photosynthesis from the entire leaf is available to newly forming roots and shoots.

We always dip cut ends into a dilute IBA solution to make rooting more uniform across the flats, so we can meet predicted shipping dates. Stock plant management is key to any propagation business for Rex begonia plants. Not only do the mother plants need to be tested for viruses and bacteria, but the production facility needs to maintain sanitation protocols to keep the leaves clean.

Perhaps the most important part of any good Rex begonia propagation program is having good genetics. We have developed many great varieties over the past 25 years, which have proven strong and uniform in production, as well as performed well for customers. If a variety cannot be propagated uniformly, it is not a good variety for commercial use, no matter how beautiful it may look in a collector's garden.

Water Use Efficiency and Water Footprint in Ornamental Crops

Raul I. Cabrera

Department of Plant Biology, Rutgers University and Rutgers Agricultural Research & Extension Center, Bridgeton, New Jersey 08302 USA.

cabrera@njaes.rutgers.edu

Keywords: irrigation management, conservation, efficiency, nursery production

Summary

Water is crucial for plants, constituting up to 95% of their fresh weight, and is essential for growth and nutrient transport. However, only a small portion of water absorbed by plants is retained, with most lost through transpiration, which also helps cool leaves and allows nutrient uptake. Water-use efficiency (WUE) is the ratio of plant biomass to water used, and it varies across crops. For instance, roses grown hydroponically have a WUE of 2.3-3.0 g/L. The concept of effective WUE considers water lost to drainage. The water footprint (WF) expands on WUE by including water sources and pollution, divided into blue (irrigation), green (rainwater), and gray (pollution). Greenhouse crops like roses have a WF of 8-26 liters per stem. Reducing water use and recycling effluents can improve WUE and lower WF, supporting sustainability and market preferences for eco-labels.

IPPS Vol. 74 – 2024

INTRODUCTION

Water is essential to life, and in the case of plants, it constitutes 70 to 95% of their herbaceous (non-woody) fresh weight. In addition to its contribution to these herbaceous tissues, water transports minerals and metabolites through cells and tissues, and provides the positive pressure, or turgor, against cell walls, which is the main driver of plant growth through cell expansion. Interestingly, only a small fraction (as low as 1%) of the total water absorbed by a plant through its entire life is retained in this biomass, and the rest is 'lost' through transpiration. This apparent inefficiency in the use of water by plants is a consequence of the leaves opening their stomata to capture CO_2 to photosynthesize. This stomatal opening leads to loss of water (i.e., transpiration), which facilitates uptake and transport of nutrients from the soil, and helps control the plant's temperature by cooling its leaves through transpiration.

Water-use efficiency. The trade-off of CO_2 capture and water loss from the leaves of plants and crops has been defined since the early 1900s by the concept known as "transpiration ratio" or water-use efficiency (WUE). Transpiration ratio or biomass WUE refers to the unit of plant biomass (grams or pounds of fresh or dry weight) produced per unit of water used or evapotranspired by the crop (like liters or gallons). For example, greenhouse roses growing in recirculating hydroponic and open (free drainage) soilless substrate growing systems were reported to have average biomass WUE of 2.3 to 3.0 g of harvested flower dry weight (DW) per liter of water evapotranspired. The woody ornamental, Texas privet (*Ligustrum texanum*), grown in 1-gallon containers in southern California was reported to have biomass WUE of 0.7 to 2.2 g/L, whereas Japanese privet (*Ligustrum japonicum*) growing in northern Florida showed values of 2.8 to 3.6 g/L. In comparison, intensively managed greenhouse-grown vegetable crops have been reported to have maximum biomass WUE of 3 to 6 g of DW per liter of water evapotranspired.

Considering the rather large inputs and large drainage and runoff losses of water to intensively managed greenhouse and nursery crops, some researchers prefer to use the concept of effective WUE, which relates the yield dry weight biomass produced in relation to the total volume of *applied water* from irrigation and precipitation. This concept effectively accounts for the volume of water that is lost to drainage and runoff, in addition to what was actually used by the crop (evapotranspiration). Considering again the example of greenhouse-grown rose crops, the effective WUE reported for these crops range from 0.7 to 2.3 and 2.3 to 2.8 g of harvested DW yields per liter of water applied, respectively, for soilless substrate and recirculating hydroponic growing systems. Compare these values to the range of 0.8 to 2.2 g/L reported for other irrigated agronomic and vegetable crops.

Water footprint. Serious issues with the availability and pollution of water resources suitable for irrigation and pressing competition from urban and industrial uses led to the development of the water footprint (WF) concept. From a sustainability viewpoint, WF is more comprehensive than WUE, as it specifies the volumes of water applied, consumed, and polluted by source to produce a unit of agricultural product. The overall value of WF includes a "blue" component which is the irrigation volumes applied to (including evaporation and other losses) and consumed (or incorporated) by the crop. It also includes a "green" component, which denotes the consumption of the volume of rainwater stored in the soil. Lastly, it's also comprised of a "gray" component, effectively a pollution factor, which is defined as the volume of freshwater that would be required to dilute (or assimilate) the load of agricultural pollutants from the production cycle (in drainage and runoff water) to existing water quality standards. In a nutshell, this "gray" component of WF is what effectively distinguishes this concept from WUE.

The WF of an agricultural product is expressed as the total liters of water (including the green, blue and gray components) used and polluted to produce one unit (one piece) or one unit (gram) of fresh or dry weight of product. According to a WF global database, for example, one average-sized unit of tomato, apple and banana will have a WF of 50, 125 and 160 liters of water, respectively. Expressed in units of fresh weight, the global WF averages estimated for vegetables, fruits, species and nuts are 0.35, 1.0, 7.0 and 9.0 liters per gram, respectively.

Information on the WF of ornamental greenhouse and nursery crops is extremely limited in the literature, and mostly based on modelling exercises. For example, the WF of export cut flower crops grown in Kenya have been modeled to range from 0.3 to 0.4 liters per gram of fresh weight. In the specific case of cut rose flowers, the modeling exercise suggested a WF range of 7 to 13 liters of water per single rose stem, with one-third of it associated to the gray (pollution) component. Using data from a scientific study on the annual water and nitrogen balance of California-grown rose crops, we were able to calculate their actual WF. Across the various irrigation and nitrogen fertilization treatments, the WF values ranged from 8 to 26 liters per stem, or 0.3 to 0.7 liters per gram of fresh weight. These range of experimental values validated, to a large degree, the values previously modelled for the export roses growing in Kenya. Using data from another water and nitrogen balance study in the woody ornamental *Lagerstroemia* x *fauriei*, plants growing in 1-gallon containers and receiving a nutrient solution of 60 mg of nitrogen per liter had a WF value of 47 liters per plant or 0.3 liters per gram of fresh weight. Fertigating this crop with higher nitrogen concentrations produced much higher WF, as the polluting "gray" component of the total WF rose to significantly higher values, thus requiring higher volumes of water to dilute the excessive nitrogen applications.

Any greenhouse and nursery production system that captures drainage and runoff effluents and recycles them back into production will lead to significant increases in water-use efficiency and reductions in their overall water footprint. In addition to meeting environmental pressures and mandates, a reduced use of water and fertilizers, and containment of agricultural tailwaters (runoff volumes rich in fertilizers and other agrichemicals), effectively a reduced total WF, are among the cultural practices that could lead to eco-labels and "green" certifications, which are beginning to be expected or preferred by some consumers and markets.

LITERATURE CITED

Cabrera, R.I. (2021). Irrigation and nutrition management, p. 224-257. *In*: J. Faust and J. Dole (eds.) Cut Flowers and Foliages, CABI, Wallingford, UK.

Cabrera, R.I. (2003). Nitrogen balance for two container-grown woody ornamental plants. Scientia Horticulturae *97*:297-308

Jarrell, W.M., Whaley, S.J. and Miraftabi, B.. (1983). Slow-release fertilizer and water management with container-grown *Ligustrum texanum*. Scientia Horticulturae *19*:177-190.

Jones, H.G. (2004). What is water use efficiency? p. 27-41. *In*: M.A. Bacon (ed.) Water Use Efficiency in Plant Biology, Blackwell Publishing, Oxford, UK.

Mekonnen, M.M., Hoekstra, A.Y. and Becht, R. (2012). Mitigating the water footprint of export cut flowers from the Lake Naivasha, Kenya. Water Resource Management *26*:3725-3742.

Miller, M.N., Gunter, D.L., Melton, B.E., Johnson, C.R. and Midcap, J.T. (1980). Selecting economically optimal levels of fertilization and irrigation for container production of woody ornamentals. Journal of the American Society for Horticultural Sciences *105*:766-768.

Raviv, M. and Blom, T. (2001). The effect of water availability and quality on photosynthesis and productivity of soilless-grown cut roses. Scientia Horticulturae *88*:257–276.

Combined Proceedings IPPS 74:304-306. 2024.

Growing with Less: Substrate Stratification can Improve Crop Productivity and Resource Efficiency

Kristopher S. Criscione and Jeb S. Fields

Hammond Research Station, Louisiana State University, Hammond, Louisiana 70403 USA

kcrisc2@lsu.edu, jfields@agcenter.lsu.edu

Keywords: nursery production, root growth, soilless media, wood bark

Summary

Soilless substrate stratification offers a sustainable solution to address inefficiencies in greenhouse and nursery crop production. By layering finer media, such as peatlite, over coarser substrates like bark, researchers have demonstrated significant reductions in water and fertilizer use, improving resource efficiency. Stratification helps balance water retention and drainage, mitigating perched water tables in containers. Studies show that root growth doubles or triples in both bark- and peat-based systems when using stratified media. Additionally, this method decreases peat inputs by up to 50%, addressing the current peat crisis. Manipulating stratified layer depths has shown that growers can safely reduce peat use without compromising plant quality. Overall, stratified substrate systems yield higher-quality plants with increased root development while using up to 25% less water and 20% less fertilizer than conventional systems, making it an attractive alternative for sustainable crop production.

INTRODUCTION

Decades-long-worth of soilless substrate research has studied how and why our greenhouse and nursery soilless substrates are either (1) inefficient or (2) unsustainable. In recent years, we have taken this information of what we have learned to develop an eloquent solution to improve these inefficiencies and increase production sustainability.

Soilless substrate stratification is a basic and natural concept, serving as an applied solution for the specialty crop industry. It most often involves layering a finer textured media on top of a coarser textured substrate, such as fine bark over coarse bark, or peatlite over bark. Through several studies, researchers continue to see similar results. That is, we are growing equal or superior crops with less resource inputs. The stratified concept helps control the balance between water and air storage, as well as helping with fertilizer management.

Gravity quickly pulls water downward the second we irrigate a container. As a result, scientists can easily measure that the medium dry in the top of the container. However, the bottom stays wet, which leads to a perched water table at the container base). How a stratified substrate works is by placing finer textured media on top, the smaller pores help hold onto water that is quickly lost to either gravity or evaporation at the top surface. When we place a coarser textured substrate that doesn't store water well and has good drainage, we can alleviate the perched-water-table effects.

The results of our research suggest that we can produce a similar or better crop than using conventional media in nursery containers. Using stratified media led to using up to 25% less water and 20% less fertilizer. In addition, a grower can use a single screen to separate fine and coarse bark particle and still receive the stratified substrate benefits.

Working with some Louisiana growers, we found that stratified systems can help address the current peat crisis. Results from our initial peat-based stratified substrate research showed that we can successfully layer an expensive and high-quality peat-based medium, such as peatlite, on top of a less costly medium, such as bark. Doing so reduces peat inputs by 50%, but still leads to the production of high-quality plants.

Another advantage of substrate stratification in nursery and greenhouse production is the enhancement of plant root growth and development in containerized systems. In one study, we found that in nursery systems (bark-based; fine bark over coarse bark), root growth (based off dry weight) doubled in the top half of the container. In addition, root growth tripled throughout the entire container profile. In a greenhouse system (peat-based; peatlite over bark), an opposite pattern emerged. In the earlier stages of root establishment, we found they accumulated in the top half of the container (that is, in the peatlite layer) for a longer period of time than in conventional media. However, when the roots finished growing in the top half of the container, and began growing into the bottom half, root growth dramatically increased along the entire container profile. Again, root growth doubled, but this time in the bottom bark layer, with root growth tripling throughout the whole container.

When we present our research on stratified substrates, we are often asked, "What happens if you change the stratified depth layer?" We conducted a greenhouse study exploring if manipulating the depth layer changes growth and development and allows us to further reduce the amount of peat applied. We found that growers can safely stratify up to 50% volume per volume with no negative impact on plant growth. However, we observed that layering 25% peatlite atop 75% pine bark negatively impacted plant growth. We are still exploring these relationships in nursery production systems.

The overall benefits of growing plants in a stratified substrate system relative to conventional substrate systems include higher-quality plants that are taller with greater root growth with the use of much less peat (often up to 50%), water, and fertilizer.

LITERATURE CITED

Fields, J.S.; Owen, J.S., Jr.; Altland, J.E. (2021). Substrate stratification: layering unique substrates within a container increases resource efficiency without impacting growth of shrub rose. Agronomy *11*:1454. https://doi.org/10.3390/agronomy11081454

Criscione, K. S., Fields, J. S., Owen, J. S., Jr., Fultz, L., and Bush, E. (2022). Evaluating stratified substrates effect on containerized crop growth under varied irrigation strategies. HortScience *57*:400-413. https://doi.org/10.21273/HORTSCI16288-21

Fields, J. S., Criscione, K. S., & Edwards, A. (2022). Single-screen bark particle separation can be used to engineer stratified substrate systems. HortTechnology *32*: 391-397. https://doi.org/10.21273/HORTTECH05018-22

Fields, J.S., & Criscione, K. S. (2023). Stratified substrates can reduce peat use and improve root productivity in container crop production. HortScience *58*: 364-372. https://doi.org/10.21273/HORTSCI17019-22

Fields, J. S., Criscione K. S., and Owen, J.S. Jr. (2024). Conventional sphagnum peatlite substrate can be reduced by 50% in stratified substrate systems while growing equivalent size and quality *Pentas lanceolata*. Scientia Horticulturae *331*. https://doi.org/10.1016/j.scienta.2024.113121

Combined Proceedings IPPS 74:307-308. 2024.

Aloe and Agave Mites

Eric G. Middleton

University of California Cooperative Extension, 9335 Hazard Way, Suite 201, San Diego, California 92123 USA

egmiddleton@ucanr.edu

Keywords: eriophyoid, infestation, pest, succulent

Summary

Aloe and agave mites, both eriophyoid mites, are serious pests of ornamental succulents, with aloe mites being more extensively studied than agave mites. Aloe mites cause tumors, leaf distortions, and dense offsets, requiring either removal of affected plants or use of miticides such as fenpyroximate, spiromesifen, or spirotetramat for management. In contrast, agave mites, which reside at the base of leaves or within the core, remain less understood. Symptoms, including greasy smudges and yellowish areas, appear several months post-infestation. Effective management of agave mites involves vigilant monitoring, removal of symptomatic plants, and application of predatory *Neosiulus californicus* mites or miticides. Miticides should be selected based on ongoing research, with fenpyroximate, spiromesifen, and spirotetramat being potential options. Continuous monitoring and high magnification inspection are crucial for evaluating treatment efficacy and ensuring that mites are eradicated, as symptoms can persist even after treatment.

IPPS Vol. 74 – 2024

INTRODUCTION

Both aloe mites and agave mites can be serious pests of ornamental succulents. Both are types of eriophyoid mites and are functionally invisible to the naked eye. More research has been done on aloe mites, while we are currently conducting research on the less-studied agave mites.

Aloe-mite feeding causes tumors, leaf distortions, and growth of tightly packed offsets. Symptoms will not heal and must be either grown out or removed. To manage aloe mites, removing symptomatic plants and using miticides containing fenpyroximate, spiromesifen, or spirotetramat are good options. Thorough coverage is a must.

For agave mites, things are more difficult and less understood. Agave mites feed on the surface of agave leaves, living hidden at the very base of leaves or inside the core of the agave. Symptoms only appear once the damaged leaves have grown out, around 3-5 months, at a minimum, after the agave mite infestation first begins. The most characteristic sign of agave mite feeding is a greasy smudge or streak appearing on agave leaves. Areas around greasy spots frequently appear yellowish and will lack the powdery blue-white surface color that many agaves have. Mites concentrated at the core of the plant can severely damage the new leaves and cause the core to collapse from their feeding.

The following is a plan for managing agave mites based on what we know so far.

<u>Monitoring</u>. Check if any of your agave plants have symptoms and strongly consider getting rid of any with advanced symptoms. Cover and dispose of these plants downwind of the rest of your agave plants. Continue to monitor your plants for symptoms, especially plants close to previously infested agave.

Prevention. After removing already infested agave plants, you can deploy predatory *Neosiulus californicus* mites or use miticides prophylactically to help prevent infestations. Sachets of *N. californicus* are available to purchase and may provide preventative control for several weeks. If using miticides, products containing fenpyroximate, spiromesifen, or spirotetramat are likely good choices, although research is still ongoing regarding these options.

Curing infestations. If agave mites are established in your plants, applications of miticides labeled for use against eriophyids are the most likely to be effective. Predatory mites will not be able to clean up existing agave mite infestations.

Monitor again. Check your plants multiple times after treatment to evaluate what has worked and what hasn't. Remember that just because symptoms appear later, it does not necessarily mean plants are still infested with agave mites. If possible, cut up one or more plants and check for mites under high magnification (30x or greater at minimum) with a light microscope to see if they are still present.

Combined Proceedings IPPS 74:309-312. 2024.

Nitrogen Management in Nursery Production

Bruno J.L. Pitton

University of California Cooperative Extension, 11477 East Avenue, Auburn, California 95603 USA

bjpitton@ucanr.edu

Keywords: controlled-release fertilizer, groundwater contamination, management plan, nitrogen leaching

Summary

Many years of nitrogen fertilizer use have resulted in nitrate contamination of groundwater in the Central Valley of California, prompting the introduction of the Irrigation and Nitrogen Management Plan (INMP) in 2019. This plan aims to track nitrogen inputs and outputs, estimating potentially leachable nitrogen not removed by the harvested product. Estimating nitrogen output is challenging for nursery crops due to complex production systems. We conducted an experiment to assess nitrogen fate in container-grown *Lagerstroemia indica* plants. Our results showed that 61% of applied nitrogen was retained in the plant or media, while 28% was lost as gaseous emissions and only 6% as runoff. Importantly, just 3% of nitrogen was potentially leachable, far lower than INMP estimates. A separate study on CRF incorporation revealed that mechanical incorporation of Osmocote Plus 15-9-12 into the media resulted in higher nitrogen leaching compared to manual incorporation, likely due to prill coating damage. These findings suggest that INMP calculations overestimate nitrogen leaching from nursery crops, underscoring the need for targeted best management practices. The adoption of CRF incorporation methods that minimize leaching could further reduce nitrogen contamination, making best practices more effective than rigid reporting requirements for nursery operations.

INTRODUCTION

Many years of nitrogen fertilizer application has led to nitrate-contaminated groundwater in large portions of the Central Valley of California. In 2014, this contamination led the Central Valley Regional Water Quality Control Board to require plant producers to submit Nitrogen Management Plans to their water quality coalitions. In 2019, the original Nitrogen Management Plan was replaced by the newer Irrigation and Nitrogen Management Plan (INMP) that also included irrigation management information. The INMP is a balance sheet that reports nitrogen inputs from fertilizer, irrigation water, and container media and output from the harvested product. Total nitrogen output is subtracted from total nitrogen input to estimate total potentially leachable nitrogen.

The idea behind potentially leachable nitrogen is that any applied nitrogen not removed in the harvested product has the possibility to leach into groundwater. Estimating harvested nitrogen is straightforward for a crop like almonds because 136 pounds of nitrogen is removed for every one ton of almonds harvested. As nursery growers will recognize, estimating harvested nitrogen is not as straightforward for nursery crops due to the complex production system and variety of plant taxa and sizes grown at a single location. Possible fates of applied nitrogen to nursery crops include plant uptake, leachate from the container media, remaining in the container media, or possibly emitted as nitrogen gas from denitrification. We initiated an experiment to determine the fate of fertilizer nitrogen and answer the real question the Central Valley Regional Water Quality Control Board was asking, "How much nitrogen leaches from container-grown plant nurseries?"

MATERIALS AND METHODS

We collaborated with a nursery in the Central Valley of California to document nitrogen input and output during production of *Lagerstroemia indica* "Whitt II" plants grown in a Douglas-fir bark media incorporated with Osmocote Plus (15-9-12) and Apex polymer-coated sulfur-coated urea (9-2-0). Plants were transplanted from a #1 container into a #3 container in the beginning of May. On the third day after planting, the growing media was top-dressed with 20-9-9 fertilizer. We measured all nitrogen inputs, including well water applied as irrigation, total nitrogen in the growing media, and surface-applied fertilizer. Nitrogen outputs included shoot uptake; amount remaining in growing media at harvest time; nitrogen gas emitted; and soluble nitrogen in leachate/runoff.

To capture runoff nitrogen, we lined half of the growing beds in the test area with polyethylene sheeting sandwiched between sediment fabric before covering all the beds with gravel. The total nitrogen that infiltrated into the growing bed soil was the difference in the nitrogen in the runoff from the lined and unlined growing beds. After approximately three months, we harvested the plants when the grower was ready to ship them for retail sale. We cut the shoots off at the crown of the harvested plants and measured the total nitrogen in the shoots and growing media separately.

RESULTS

We determined that 61% of the applied nitrogen was in the plant or media when the plants were ready for shipping. Five percent of applied nitrogen was taken up by the plant shoots and 56% remained in the media as controlled-release fertilizer or as organic nitrogen in plant roots or immobilized by microbes. Maintaining a fertilizer nitrogen reserve in the growing media ensures that the plants will remain healthy and attractive while awaiting purchase by home gardeners or landscapers. Six percent of the applied nitrogen was in the growing bed runoff water, predominantly as nitrate. Irrigation runoff water capturing and recycling is common in California nursery production and the nitrogen in runoff water could reduce future fertilizer application costs. In agreement with other nitrogen balance research from container-plant production systems, 28% of applied nitrogen was lost as gaseous nitrogen emissions from denitrification.

DISCUSSION

The question the INMP calculations were supposed to answer is how much nitrogen is leaching into the soil and potentially contaminating groundwater. If harvested nitrogen from a *Lagerstroemia indica* production system was documented in the INMP by a grower, then 61% of applied nitrogen would be used as output in the calculations. This results in 39% of applied nitrogen deemed potentially leachable by the INMP worksheet, when our research results recorded that a mere 3% was potentially leachable. These results indicate that the INMP worksheets and calculations overestimated the amount of potentially leachable nitrogen from container-plant production. Three percent of applied nitrogen is equal to 20 lbs. per acre and coupled with the comparatively small total area of nursery production relative to other crops in the Central Valley, it is unlikely that nursery production is a significant contributor to nitrate contamination of groundwater in the Central Valley. However, areas with a high density of nursery producers could result in localized nitrate contamination of groundwater.

These results are becoming more significant because all Regional Water Quality Control Boards are requiring nitrogen management reporting plans. Some areas have large concentrations of nursery growers. Nurseries would have to conduct studies similar to our research to develop nitrogen input and output values to accurately fill out the INMP worksheet. Due to the large variety of plant taxa and sizes grown and different fertilizer programs, a significant and possibly debilitating cost could be incurred. Therefore, instead of requiring INMPs for nurseries, California's Regional Water Quality Control Boards should facilitate implementation of irrigation and nitrogen best management practices at nurseries to reduce potentially leachable nitrogen. Numerous best management practice guides exist and consultation with University of California Cooperative Extension Advisors could further facilitate implementation.

Uniform incorporation of controlled-release fertilizer (CRF) is a recommended best management practice to reduce nitrogen leaching losses from container-plant production. The potential for damage to CRF prill coating when mechanically incorporated into a soilless substrate was tested. Osmocote Plus 15-9-12 was uniformly incorporated mechanically or manually at the same rate into a soilless substrate and leachate was collected over 76 days. Two experiments were conducted.

One experiment included lavender plants planted into soilless substrate, the other experiment did not. Leachate volume, electrical conductivity (EC), and pH were recorded. Aliquots were later analyzed for inorganic nitrogen content. Electrical conductivity and leachate volume were used to calculate total salt content. Greater total salts, ammonium, and nitrate were leached from mechanically incorporated soilless substrate with and without plants relative to manually incorporated soilless substrate with and without plants. Plants grown in soilless substrate with mechanically incorporated CRF did not have decreased plant shoot biomass even though leachate EC was consistently greater throughout the experiment. Mechanically incorporating CRF in soilless substrate results in greater leaching losses and is likely a result of CRF prill coating damage during incorporation. Researchers should report incorporation method when publishing results on CRF in container-plant production. Container-plant producers should ensure that their mechanical-incorporation equipment does not cause unintended damage to their CRF of choice.

LITERATURE CITED

Pitton, B.J.L., Oki, L.R., Sisneroz, J., and Evans, R.Y. (2022). A nursery system nitrogen balance for production of a containerized woody ornamental plant. Scientia Horticulturae *291*: 110569.

Combined Proceedings IPPS 74:313-316. 2024.

Emerging Pests and Pathogens

Chris Shogren

University of California, Cooperative Extension, Alhambra, California 91801 USA

cjshogren@ucar.edu

Keywords: invasive, introduction, plant disease, spread

Summary

Since the arrival of Europeans in North America, an estimated 50,000 non-native species have been introduced, causing significant ecological impacts. In California alone, more than 17% of plants, 35% of freshwater fish, and over 1,000 invertebrates are considered exotic pests. Among recent invasive pests, *Thrips parvispinus* (pepper thrips) and *Nipaecoccus viridis* (Lebbeck mealybug) pose serious threats. Pepper thrips, detected in Florida in 2020, attacks at least 43 plant species and has spread to multiple states, causing extensive damage to leaves and flowers of affected plant species. Effective control requires careful monitoring and rotation of insecticides. Lebbeck mealybug, native to India, targets over 140 plant species and deforms leaves and fruit, causing wilting and dieback. Biological controls are recommended alongside chemical treatments. Other emerging threats include the *Cydalima perspectalis* (box tree moth), which damages ornamental boxwoods, and diseases like pine 'ghost canker' and fusarium wilt of palms, both of which are exacerbated by drought and warmer temperatures.

IPPS Vol. 74 – 2024

INTRODUCTION

An estimated 50,000 non-native species have been introduced into North America since the arrival of Europeans about 500 years ago. In California, more than 17% of plants are exotic, more than 35% of all freshwater fish are non-native, and more than 600 species of invertebrates have invaded California, with another 400 species deliberately introduced as biological control agents. Records in California for the period 1955–1988 indicate a constant influx of non-native species at an average rate of 6.1 species per year. Between 1989-2010, that rate of acquisition jumped to approximately nine exotic species per year, or one every 40 days. The continuous establishment of non-native species poses many challenges, which include identification of species that are a high risk of becoming serious pests; monitoring for those high-risk species; creation of management strategies to combat invasive species; the implementation of the management strategies; and the costs associated with any potential management program.

Thrips parvispinus (pepper thrips) is an invasive pest native to the Asian tropics. In May 2020, it was found causing damage on *Anthurium* plants in Florida, USA. Since its intitial detection in Florida, it has spread to at least four other states (Colorado, Georgia, North Carolina, and South Carolina). The species has also been detected in Ohio and Pennsylvania, and caused extensive damage to greenhouse crops in Ontario, Canada. Pepper thrips are small (~1mm long), with females brown-black in color and males entirely yellow and smaller in size (~0.6mm). The invasive thrips attacks at least 43 species including; *Gardenia*, *Mandevilla*, pepper, *Anthurium*, *Hoya*, *Ficus*, *Hibiscus*, jasmine, *Chrysanthemum*, *Schefflera*, and strawberry. Pepper thrips feeds on leaves and flowers causing extensive damage. Regular monitoring and establishment of action thresholds for the pest is extremely important for control. The University of Florida has published a recommended pesticide list for *T. parvispinus*. It is important to rotate insecticide groups to reduce the risk of the thrips building resistance.

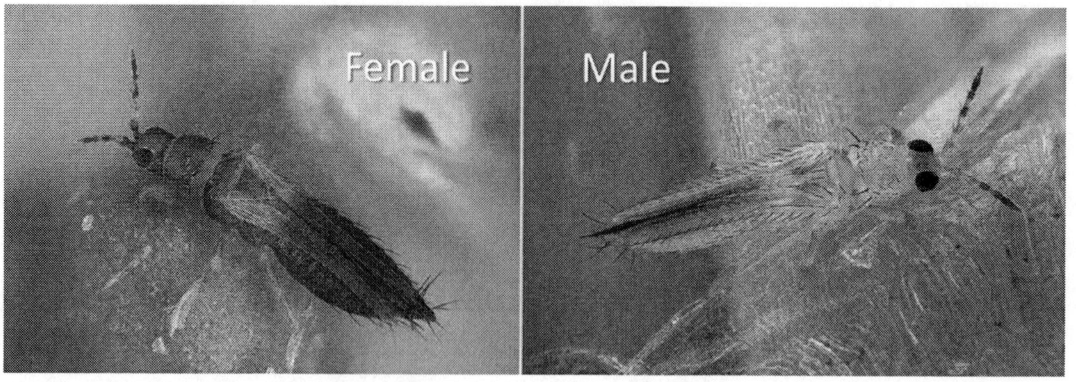

Figure 1. Pepper thrips.

Nipaecoccus viridis (Lebbeck mealybug) is a destructive polyphagous pest native to India (**Fig. 2**). Lebbeck mealybug, also called hibiscus mealybug, has a host range of over 140 species, which includes citrus, gardenia, jasmine, oleander, and hibiscus. *Nipaecoccus viridis* prefers to feed on actively growing tissues, such as new growth, new branches, and fruit. Feeding can cause twisted/distorted fruit and leaves; branch dieback; wilting; and even plant death. Lebbeck mealybug seems to mostly be an issue in absence of natural enemies or the overuse of insecticides. The invasive mealybug is difficult to control with insecticides, and the use of adjuvants to penetrate its waxy covering is critical for gaining control with chemical applications. Biological control options include *Anagyrus aegypticus*, *Anagyrus dactylopii*, *Anagyrus indicus*, *Leptomastix phenacocci*, and *Cryptolaemus montrouzieri*.

Figure 2. Lebbeck mealybug. Photos by Erin Powell, Ph.D., Lance Osborne, Ph.D., and Muhammad Z. Ahmed, Ph.D.

Cydalima perspectalis (box tree moth) is a destructive lepidopteran pest spreading quickly in North America. Boxwoods (*Buxus* spp.) are the preferred host of box tree moth. While there are no native boxwoods in North America, potentially slowing its spread, boxwood is a widely used plant in landscapes for topiaries and hedges. Box Tree Moth is easily recognized by the webbing the caterpillars create throughout the shrub. Caterpillars cause extensive feeding damage to boxwoods, feeding first on the leaves and leaving only the midrib. Once the leaves are gone, caterpillars feed on the bark, leading to girdling and plant death.

Figure 3. Box moth.

Pine 'ghost canker' (*Neofusicoccum* spp.) and Fusarium wilt of palm (*Fusarium oxysporum* f.sp. *palmerum*) are two emerging plant pathogens. Pine 'ghost canker' is a disease recently detected in southern California affecting multiple pine species in urban forests and parks. Multiple *Neofusicoccum* spp. have been isolated from symptomatic trees. Symptoms first appear as lower branch death and if not properly managed will lead to tree death as the canker spreads. Drought and higher temperatures may be predisposing pines to this disease. *Fusarium* wilt of palms primarily attacks queen palms (*Syagrus romanzoffiana*) and Mexican fan palms (*Washingtonia robusta*). Initial symptoms occur on the lower or older leaves in the palm canopy, and exhibit one-sided discoloration or necrosis. Symptoms move from the older leaves to the upper, younger leaves, killing the spear leaf in the top center of the canopy last. Fusarium wilt of palms is fast acting and can kill palms within two to three months.

Combined Proceedings IPPS 74:317-322. 2024.

How Container Color and Root Zone Temperature Affect Plant Growth and Fertilizer

Jacob Shreckhise

USDA-ARS, U.S. National Arboretum, Otis L. Floyd Nursery Research Center, 472 Cadillac Lane, McMinnville, Tennessee 37110 USA

jacob.shreckhise@usda.gov

Keywords: albedo, nursery production, container, root temperature

Summary

Black plastic pots, the nursery industry standard for over 50 years, are poor at managing root-zone heat. They readily absorb solar radiation, which raises substrate temperatures to harmful levels. On hot days, root-zone temperatures (RZT) in black containers can exceed 133°F in the southern U.S., resulting in plant stress and fertilizer inefficiency. In contrast, white containers reflect solar radiation, reducing RZTs and improving plant growth. Recent studies in Tennessee and Ohio show that using white containers, especially with cyclic afternoon irrigation, significantly lowers RZTs and improves plant growth, nutrient retention, and fertilizer efficiency. Plants grown in white containers exhibited less nutrient leaching and, in some cases, doubled in size compared to those in black containers. Although white containers cost 10-20% more, they offer better protection from heat stress, making them a cost-effective solution, especially in hot climates. For nurseries, investing in white containers may improve plant quality and yield long-term benefits.

IPPS Vol. 74 – 2024

INTRODUCTION

Managing root-zone heat. Despite being the industry standard nursery container for over 50 years, black plastic pots are terrible for root-zone heat management. Dark colors absorb solar radiation, effectively converting it to heat. Therefore, on a hot, sunny day, the substrate temperature in a black plastic container can be 30°F higher than ambient air temperature, with peaks as high as 133°F in the southern U.S. Lighter-colored (optimally, white) containers reflect more and absorb less solar radiation, resulting in lower maximum daily root-zone temperatures (RZTs) compared to black containers. Historically, the limited availability or high expense of light-colored containers was a serious barrier to their adoption by commercial nurseries. Nursery container manufactures have since streamlined the production of affordable, white, plastic containers (5 gal or less in size) that have an opaque, black interior wall to block sunlight from reaching the roots. These non-branded, white pots typically cost about 10-20% more than the equivalent black pots, which is a relatively small price to pay for the heat-stress protection they can provide roots.

Excessive heat in the container substrate can also impact the life of controlled-release fertilizer (CRF). All leading polymer- and resin-coated CRFs used for container-based nursery production in the U.S. are highly influenced by temperature, with the nutrient-release rate increasing with increasing temperature. As such, CRF manufacturers include on each label an estimated effective longevity based on the average substrate temperature (e.g., 7 months at 70°F, 6 months at 80°F, 5 months at 90°F). When substrate temperatures exceed ~104°F, a common occurrence in nurseries throughout the U.S., CRFs "dump" nutrients into the substrate. This surplus of available mineral nutrients comes at a time when plant nutrient uptake is impaired due to heat stress. Thus, most of these nutrients will be leached from the container and enjoyed by the algae in the nursery retention reservoir.

Over the past three growing seasons, USDA-ARS researchers, Jake Shreckhise and Jim Owen, have been conducting experiments concurrently replicated in McMinnville, TN and Wooster, OH to explore how container color (black vs. white plastic) and irrigation schedule (once daily at 7:00 am vs. three times daily at 12:00, 3:00, and 6:00 pm; matched total daily irrigation) affect RZT and resulting plant growth; CRF nutrient release rate; and nutrient leaching in two disparate climates. Below are some of the key takeaways from this research.

Tennessee site takeaways. After logging temperature every 10 min in the west-facing quadrant of fully exposed shrub rose (*Rosa* x 'ChewPatout' Oso Easy® Urban Legend®) root balls for 14 weeks, the cumulative time RZTs exceeded the lower threshold for indirect injury (i.e., stunted growth, impaired nutrient uptake, increased susceptibility to diseases) was 332, 234, 152, and 22 hrs, respectively, for the once-daily irrigated black containers, cyclic irrigated black containers, once-daily irrigated white containers, and cyclic irrigated white containers, respectively (**Fig. 1**). As such, in TN, white containers had a stronger heat-mitigating effect than cyclic irrigation. However, the combination of these two practices, by far, provided the greatest protection from temperatures associated with root-zone heat stress.

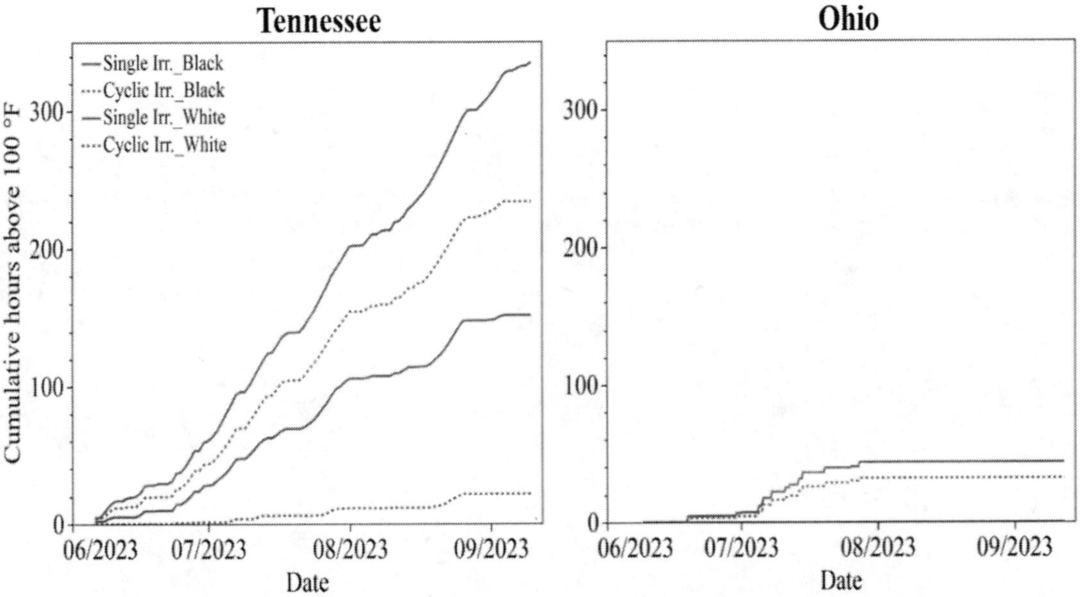

Figure 1. Cumulative time the root-zone temperature of shrub rose grown in McMinnville, TN (left) or Wooster, OH (right) exceeded the lower threshold for indirect injury (100°F) in 2 gal. black or white containers irrigated once-daily at 7 am (single irrigation) or three times daily at 12, 3, and 6 pm (Cyclic irrigation.). Temperature was recorded every 10 min in the western quadrant, 2 in. from the container wall, of three plants per treatment.

The effect of container color on root and shoot growth varied depending on the species. Compared to black containers, white containers had little effect on crapemyrtle (*Lagerstroemia* ×'Natchez') but dramatically improved growth and quality of all other evaluated species, including flowering dogwood (*Cornus florida* 'Appalachian Snow'), red osier dogwood (*C. sericea* 'SMNCSBD' Arctic Fire®; Figure 2), shrub rose (**Fig. 2**), eastern redbud (*Cercis canadensis* 'Forest Pansy'), panicle hydrangea (*Hydrangea paniculata* cultivars), and smooth hydrangea (*H. arborescens* 'NCHA3' Invincibelle® Ruby). In many of these species, including the heat-tolerant shrub rose, plants in white containers were up to twice the size as those in black containers when plants received once-daily overhead irrigation at 7 am. However, when plants were grown using cyclic afternoon irrigation, growth differences between the white and black containers were less dramatic.

Figure 2. Shrub rose (*Rosa* x 'ChewPatout' Oso Easy® Urban Legend®) Red osier dogwood (*Cornus sericea* 'SMNCSBD' Arctic Fire® The south-facing side of the root balls were oriented toward the camera.

To shed light onto container color effects on controlled-release fertilizer longevity, granules of a 6-month (80 °F) CRF, which had been incorporated into a pine bark substrate at the time of planting were picked from the substrate after 14 weeks of outdoor production. Analyzing the partially released CRF for nutrients revealed that the prills from the white containers had 18-35% more nitrogen, 14-18% more phosphorus, and 18-25% more potassium than those from black containers. Cyclic irrigation also conserved fertilizer nutrients, with CRF granules containing 8-15% more nitrogen, phosphorus, and potassium than those collected from once-daily irrigated plants. Periodically analyzing nutrients in the leachate draining from the containers showed consistently higher nitrogen and phosphorus concentrations from the black, once-daily irrigated containers compared to the other treatments.

To summarize, at the Tennessee site (AHS Heat Zone 7), using white containers alone, and especially in combination with cyclic afternoon irrigation, produced larger, higher-quality plants while conserving CRF and limiting the amount of wasted nitrogen and phosphorus leaving the container through the drainage holes.

Ohio site take-aways. Between June 6 and September 10 in Wooster, OH (AHS Heat Zone 4), the cumulative time that RZTs exceeded the lower threshold for indirect injury in the western-facing quadrant of shrub rose root balls was 44 hrs for once-daily irrigated black containers and 32 hrs for cyclic irrigated black containers (**Fig. 3**). In white containers, regardless of irrigation schedule, RZTs never exceeded 100°F. Despite these differences, white containers did not notably improve shoot or root growth in shrub rose, crapemyrtle, flowering dogwood, panicle hydrangea, or smooth hydrangea.

One noteworthy exception was when red-osier dogwood liners were unintentionally exposed to *Botryosphaeria* canker a few days after transplanting into white or black containers. Those in black containers were noticeably more severely infected than those in white containers. Instead of terminating the study, we managed the disease as a grower would. That is, we pruned out infected stems and applied fungicide. We then continued evaluating the plants. At 10 weeks after transplanting, shoots of the red-osier dogwoods in white containers were approximately 50% larger those in black containers, regardless of irrigation schedule (**Fig. 3**). Cyclic afternoon irrigation, compared to once-daily morning irrigation, also improved shoot and root growth, but to a lesser degree than white containers. When the experiment was repeated a year later using a different red-osier dogwood cultivar and the prevention of canker with preemptive fungicide sprays, plant growth and quality was essentially the same across container color and irrigation treatments.

Figure 3. Red-osier dogwood (*Cornus sericea* 'SMNCSBD' Arctic Fire® Yellow) after being grown in Wooster, OH for 10 weeks in black or white plastic 2-gal containers under once-daily overhead irrigation. The south-facing side of the root balls were oriented toward the camera.

Controlled-release fertilizer granules collected from the substrate in white containers 10-14 weeks after transplanting, depending on the year, had 7-12% more nitrogen, phosphorus, and potassium compared to those gathered from black containers, whereas irrigation treatment had no effect on CRF release.

While leachate nutrient concentrations trended higher from black versus white containers, a consistent statistically significant difference was not detected.

A frequently asked question regarding container color choice, particularly in the northern US, is whether the warmer temperatures in black containers in the spring give plants a "head start." To investigate this, black and white 1-gal panicle hydrangeas and 2-gal red-osier dogwoods were removed from an overwintering house in early April, placed on an outdoor gravel pad, and monitored weekly for flushing and stem elongation. No differences were detected between container colors in the aerial portions of the plants in either TN or OH, although early-spring root growth should be compared in future studies before making definitive conclusions.

Are white pots right for you? In the southern US and regions with high solar radiation during the summer months, using a root-zone heat mitigation strategy appears to be a *necessity* for maximizing quality and minimizing finishing time of most container-grown woody landscape plant species. Cyclic afternoon irrigation and using white or light-colored containers are just two options in the toolbox. Overhead shade cloth; jamming plants together until their canopies can provide shade; using porous-walled containers (e.g., air-pruning plastic, fabric, or fiber pots); or adopting a pot-in-pot systems are all improvements, with varying degrees of efficacy, over solid-walled, black plastic in full sun.

Keep in mind that a plant produced in black containers can lose the south-facing half of its root system after less than a day of exposure to full sun. In McMinnville, we found that when plants in white or black containers were faced with this scenario due to trimming or removing the shade-providing neighboring plants on a hot July day, those in white containers had substantially less root death than those in black containers. The same would likely be true when setting plants outside the shade house for a customer pickup. Consequently, relying on shade alone could be risky as plants get shuffled around the nursery.

At higher latitudes, like our Ohio trial site, the use of white containers and other practices for managing root-zone heat are similar to an insurance policy. They may not provide noticeable benefits for every species every year, but when the next "heat dome" or disease outbreak comes around, you'll be glad you had them. To determine whether using white containers would be beneficial under your current production system, consider purchasing a pallet of 1- or 2-gal white containers and doing an on-site, side-by-side comparison with some of your "problem species". There is little to lose and, potentially, much to gain.

Combined Proceedings IPPS 74:323-325. 2024.

Measuring Pressure, Distribution Uniformity, and Improving Irrigation Management in Nurseries and Greenhouses

Gerry Spinelli

University of California, Cooperative Extension, 9335 Hazard Way, Suite 201, San Diego, California 92123 USA

gspinelli@ucdavis.edu

Keywords: irrigation efficiency, pressure regulation, system performance, water-distribution uniformity

Summary

Effective irrigation management in nurseries and greenhouses hinges on maintaining uniform pressure across irrigation systems to ensure consistent water distribution. Variations in pressure, due to factors such as elevation changes, friction in pipes, and improper system design, lead to non-uniform irrigation, causing inefficient water use. Measuring distribution uniformity helps assess system performance, and pressure measurements at sprinklers or driplines can guide adjustments. Distribution uniformity is calculated by comparing the water received in the lowest quarter of the irrigation block to the average, which informs how much additional water is needed to compensate for inefficiencies. Poor uniformity increases water usage, potentially resulting in runoff or infiltration that threatens environmental compliance. Implementing corrective measures like pressure regulators, filters, and proper pipe sizing can enhance system performance, reduce water waste, and improve overall irrigation efficiency.

INTRODUCTION

The main strategies to improve irrigation management in a nursery or greenhouse are measuring pressure, knowledge of the operating pressure of drippers or sprinklers; and understanding how pressure changes as water travels in a pipeline. If drippers or sprinklers in an irrigation block are exposed to different pressures, they will produce different flowrates, and irrigation will be nonuniform. Therefore, much of the irrigation manager's work is to ensure that pressure is close to the value recommended by the manufacturer and that it is uniform across the irrigation system. Poor system performance can also by caused by other factors, such as emitter plugging; wind affecting overhead sprinkler systems; uneven line drainage etc. Measuring distribution uniformity is an excellent way to investigate how these factors may affect the system and to compare different dripper or sprinklers brand/models; system designs; or corrective measures.

Generally, drippers are low-pressure devices, they need 8 to 12 psi of pressure, while impact sprinklers need 45 to 60 psi. Spray stakes need 15 to 25 psi and microsprinklers may vary between 25 to 45 psi. Senninger Wobblers sprinklers are the exception to the rule, since they only need 15 to 25 psi.

Both sprinklers and drippers will produce more flowrate with more pressure, increasing the irrigation system's application rate and the water depth applied per time. Increasing pressure in sprinklers will also increase the throw (or radius) and hence the circular area wetted by the sprinkler.

It is important that irrigators and irrigation managers are provided the tools to measure pressure at the sprinkler or at the dripline. Schrader valves can be installed on PVC pipes or on polyethylene hose fittings for measurement with a hand-held pressure gauge, such as with a car tire. Access these QR codes to view overviews of this process.

Pressure changes in an irrigation system due to two phenomena with additive effects, elevation and friction. The elevation effect on pressure is directly proportional the change in altitude, following the relationship of 1 ft = 0.44 psi or 2.31 ft = 1 psi. Every foot of difference in vertical elevation decreases pressure by 0.44 psi. For example, if a terrace is 10 ft higher in elevation than the next one, the pressure in the higher terrace will be 4.4 psi lower than in the lower terrace. This phenomenon applies whether water moves or not. You may have noticed that your ears hurt under the effect of pressure the deeper you swim deep in the ocean, the same is true in an irrigation system.

Friction affects pressure only when water flows. A certain pressure drop will occur per every foot of pipeline, the longer the pipeline the (linearly) larger the pressure loss. Additionally, more pressure losses occur in a pipe when the diameter of the pipe is smaller. Moreover, more pressure losses occur when the flowrate through the pipe is larger. Both diameter and flowrate affect pressure exponentially. Doubling the length of the pipe will double the pressure loss, but halving the diameter of the pipe or doubling the flowrate trough may increase pressure loss ten times. A common hydraulic misconception is that a reduction in pipe diameter will increase pressure or that a smaller diameter pipe will reduce pressure loss. These statements are simply incorrect. Watch this video if you don't believe me.

Distribution uniformity measurement is performed by setting containers to collect and measure the volume of water distributed by the irrigation system. In overhead systems, we recommend setting containers of the same diameters as the containers where plants are grown. In drip systems and spray-stake systems, care must be taken to collect all the water that the plants would have received. We recommend at least 36 containers per area and 100 in each irrigation block. Various measures of data dispersion can be used, including variance, standard deviation or coefficient of variation. One of the measures traditionally most common in irrigation is the distribution uniformity of the low quarter. It is calculated by dividing the average of the lowest quarter (e.g. the lowest 9 values if the total number of containers was 36) by the average of all containers. If the average of the low quarter was 13 oz and the average of all containers was 16 oz, then distribution uniformity is 13/16=0.81, which is a high value, in the range we find in drip systems. If instead the low quarter average was 11 oz, distribution uniformity is 11/16= 0.69, similar to what we may found in an overhead sprinkler system.

Distribution uniformity is a quantitative measure of irrigation system performance useful for comparison, but it also has another application. It is used to calculate how much more water needs to be applied to the irrigation system to make up for lack of uniformity. This is done by simply dividing the irrigation requirement by distribution uniformity. For example, if the plants need 1 inch of irrigation per week, with the distribution uniformity of the drip irrigation system mentioned above, the grower will need to apply 1/0.81, which equals 1.23 inch. With the sprinkler system the grower would apply 1/0.69, which equals 1.45 inch. Note that 23% more water in the drip system and 45% more water in the sprinkler system than the plants needed was applied to make up for lack of uniformity. This additional water can produce runoff that carries pesticides and fertilizers to surface water or infiltration that can carry nitrate to groundwater. These can be a challenge for compliance with environmental regulations.

Corrective measures to improve distribution uniformity include improving pressure uniformity by correct pipe sizing and irrigation system design, installation of pressure regulators to equalize pressure in irrigation blocks or in pressure-compensating emitters; installation of filters; regular line flushing to minimize emitter plugging; replacement of worn sprinkler nozzles; and installation of check valves to minimize uneven system drainage.

Combined Proceedings IPPS 74:326-333. 2024.

New and Underused Plants for the Western Region of the United States

Nicholas Staddon

Everde Growers, 1045 W. Ketella Ave, Suite 230, Orange, California 92867 USA

nstaddon@everde.com

Keywords: new plants, underused plants

Summary

Both new and underused plants have a place and value in modern landscapes. Incorporation in landscapes of older varieties alongside newer selections should be considered. California's spring bloom, especially after winter rains, was featured along with recommendations for exploring areas like Sand to Snow National Monument in southern California. The plant list includes selections from genera such as *Echinacea*, *Philodendron*, *Callistemon*, and *Cercis*. Recognition was also given to plant breeders, such David Salman and Dennis Werner. These plant introductions provide various benefits, including drought tolerance, pollination services, wildlife habitat, and fruit production.

INTRODUCTION

It was indeed a terrific privilege to be invited to speak before such an esteemed group of nursery professionals. My presentation topic was new and underused plants. Part of my professional work with Everde Growers is the pursuit of new plants as well as older varieties that still have a place into today's residential and commercial landscapes. Too often we are focusing on the new, which is all fair and well, but there are many older plants that have a place in today's marketplace.

IPPS Vol. 74 – 2024

The first portion of the presentation was devoted to the incredible spring bloom we can have in California after the winter rains. I suggested places to visit, Sand to Snow National Monument, created by President Obama during his term in office. The Monument is some 157,000 acres in size and located northwest of Palm Springs. I also spoke about the coastal ranges east of Interstate 215 in the Lake Elsinore and Corona areas, and finally the eastern side of the High Sierras in the Lone Pine and Big Pine regions.

I then went on to describe numerous varieties of new and not-so-new plants as well as recognizing the breeders who were involved or responsible for creating the plants. Plantsmen included:

David Salman, High Country Gardens, Santa Fe, NM.

Dennis Werner, North Carolina State University, Raleigh, NC.

Randy Baldwin, San Marcos Growers, Santa Barbara, CA.

Ron Gass, Mountain States Wholesale Nursery, Glendale, AZ.

The following list of plants can all be used in various regions of the United States. Most, if not all, will perform well in California, but some will need winter protection. All offer unique benefits to the landscape depending on the varieties. Such benefits include their drought and disease tolerance; pollination services; habitat for wildlife; and their fruit production.

Echinacea x *hybrida* Double Scoop™

Watermelon Delux PPAF

Alocasia micholitzians 'Frydek Variegata'

Philodendron 'Strawberry Shake'

Philodendron billietiae Variegated

Callistemon viminalis 'CV01' P.P. #24,444

Slim™

Flower Carpet® 'Mini Cherry' PPAF

Groundcover Rose

Rosa persica

The Hollywood Hibiscus® Collection

Lavandula allardi 'Meerlo' P.P.#25276

Laurus nobilis 'Monrik' P.P.#25915 Little

Ragu®

Mulenbergia reverchonii 'Pundui1s' Undaunted®

Punica Angel Red® P.P.#16578

Bouteloua gracilis 'Blonde Ambition' P.P.#22048

Selected additional species include:

Figure 1. *Bergenia cordifolia* 'Peppermint Patty' PPAF CPBRAF

Figure 2. *Callistemon* Light Show®

Figure 3. 'NC2016-2' P.P.#31260 Flame Thrower®

Figure 4. *Cercis* x 'Merlot' P.P.#22,297

Figure 5. *Cercis siliquastrum* 'Bodnant'

Figure 6. Flower Carpet® 'Peach' PPAF Groundcover Rose

Figure 7. *Hesperaloe* Desert Dusk® P.P.#28909

Figure 8. *Grevillea* x Spirit of Anzac'P.P.#29372

Figure 9. *Penstemon* 'Dark Towers P.P.#20013 Beardtongue

Figure 10. *Senecio* 'Angel Wings'

Figure 11. *Vitex* x Summertime Blues Vitex PPAF

Figure 12. *Bouteloua gracilis* 'Blonde Ambition' P.P.#22048

Figure 13. *Fallugia paradoxa*

Combined Proceedings IPPS 74:334-335. 2024.

Biocontrol in Controlled Environment Agriculture

Clint M. Taylor

North Willamette Research and Extension Center, Oregon State University, Aurora, Oregon 97002 USA

clint.taylor@oregonstate.edu

Keywords: biocontrol, pests, predators, parasitoids, insect

Summary

Biocontrol, a fundamental approach within integrated pest management (IPM), utilizes living organisms to manage pest populations while minimizing environmental and economic impacts. This method involves predators, parasitoids, and pathogens, including beneficial organisms like ladybugs, parasitic wasps, and nematodes. Biocontrol agents are released through three main strategies: inoculation, inundation, and conservation. While inoculation and inundation focus on introducing organisms in response to pest levels, conservation creates habitats that sustain beneficial organisms. Effective against pests like aphids, biocontrol provides an alternative to chemical controls. However, challenges such as incomplete pest eradication and the need for precise monitoring demand careful implementation and ongoing adaptation.

IPPS Vol. 74 – 2024

INTRODUCTION

Biocontrol, the use of living organisms to manage pest populations, is a crucial aspect of integrated pest management (IPM). Unlike chemical controls, biocontrol does not strive for complete eradication of pests, but aims to limit or reduce their populations. This approach fits well within the IPM model, balancing environmental, economic, and efficacy considerations.

Biocontrol methods include the use of predators, parasitoids, and pathogens. Predators such as ladybugs, big eye bug, minute pirate bug, lacewings, and predatory mites are effective in controlling various pests. Parasitoids, particularly wasps, play a significant role in managing pest populations. Pathogens like viruses, bacteria, and nematodes are also utilized in biocontrol strategies.

There are three primary strategies for releasing biocontrol agents: inoculation, inundation, and conservation. Inoculation involves introducing a small number of beneficial organisms strategically to establish populations before pest levels become critical. Inundation entails releasing a large number of beneficial organisms quickly to overwhelm the pest population. Conservation focuses on creating refuges within the greenhouse to harbor beneficial organisms during non-pest times.

Aphids are common greenhouse pests that cause significant damage to plants. Parasitic wasps, particularly those in the Aphidiinae subfamily, are natural predators of aphids. These wasps lay eggs inside aphids, and the larvae consume the aphids from the inside, eventually killing them. This method is effective in controlling aphid populations and maintaining healthy crops.

Implementing biocontrol methods comes with challenges, such as the impossibility of complete pest eradication and the need for precise timing and monitoring. Overcoming these obstacles requires careful planning, resource management, and adapting strategies as needed.

Combined Proceedings IPPS 74:336-340. 2024.

Technical Sessions of the International Plant Propagators' Society – Southern Region of North America (SRNA) Annual Meeting – Presidential Address

Cheryl R. Boyer

Department of Horticulture & Natural Resources, Kansas State University, 3601 Throckmorton PSC, 1712 Claflin Road, Manhattan, Kansas 66506 USA

crboyer@ksu.edu

Keywords: IPPS-SR, awards, scholarships

Summary

The 48th Annual Meeting (49th year) of the International Plant Propagators' Society-Southern Region of North America (SRNA) convened at 8:00 am on 28 October 2024 at the Hyatt Regency Tulsa, Oklahoma with President Cheryl R. Boyer presiding.

INTRODUCTION

President Boyer welcomed everyone to Tulsa, Oklahoma for the 48th Annual Meeting of the SRNA. Many of you know me, but for those who do not, I want to take a moment to share my background and why I am here as someone who lives and works in Kansas (part of the IPPS-Eastern Region of North America). First, I was born and raised in Oklahoma. In 2003, I earned a Bachelor of Landscape Architecture from Oklahoma State University, followed by a Master of Science in Horticulture focusing on nursery crop production (2005). My thesis advisor was Dr. Janet Cole and I conducted field research in Park Hill at Greenleaf Nursery where we toured yesterday.

I had the tremendous opportunity to conduct doctoral research in alternative substrates for nursery crop production at Auburn University under the guidance of Drs. Charles Gilliam and Glenn Fain, graduating in 2008. I will never forget that first October being told to pack a bag and get in the van—we were going to an IPPS conference. That year (2005) the meeting was in Gainesville, Florida and I was able to meet growers, tour outstanding nurseries, and watch my fellow graduate students compete in the student competition. The next year, I was on that stage! Since then, I have only missed the IPPS conference during the years my two children were born and the first post-COVID event when I was still unable to travel. I even had the privilege of helping deliver the 2020 IPPS North American Summit to a global audience—what a thrill!

The IPPS student competitions gave me confidence and a connection to people who wanted to learn from and use the research I conducted. I will never forget Charlie Parkerson asking me challenging questions and thinking through implementation at his nursery, Lancaster Farms. I also learned so much from Dr. Mack Thetford, who led the student competition and Dr. Fred Davies who edited my writing for the proceedings. What an unprecedented opportunity to learn from and live life with people who have fundamentally changed what and how plants are produced. Look around you—we are surrounded by Legends!

IPPS is addictive! We care for one another. We share with each other. We change the world together. IPPS is also unique with an industry/academic mix—it doesn't work without the mix. Academics like me are here to serve industry—we need to know how to help and here at IPPS we get to talk shop with you, brainstorm, and fine-tune ideas. I'm excited to hand the baton to 1st Vice President Michael Roe of Windmill Nursery and see him carry forward the projects we have been working on as an Executive Committee. Our partnership over this last year has been remarkable with our strengths alternating to lead our organization well. It's that mix of industry and academic that makes us click, makes us work, and makes us relevant. I'm grateful for the opportunity to serve you.

If you are wondering why I work for Kansas State University after having spent my life and earned an education in the Southern Region, it's very simple—they offered me a job in 2008 just before the economic recession hit! My appointment is 75% Extension, 25% Research in nursery crop production and marketing, but I do lots of other work in the horticulture Extension and research world. I have come to know and love the great state of Kansas and I am proud to work for the nation's first operational land-grant institution, opening in 1863. The work I do is national and international in scope and you all are a major influence in the projects I work on and the Extension products I produce. If it helps IPPS members, it helps Kansas stakeholders, too.

I want to show you a video created Kansas State University about what it means to be a next-generation land-grant university (**Fig. 1**).

Figure 1. The land-grant university system of teaching, research and extension at Kansas State University.

https://youtu.be/jfs9jPXdZZo

This story is similar in every one of our states. This one is about Kansas, but you can insert your state's land-grant university and the story is still true. Those who came before us walked a different road with different challenges, but their goal was the same--a better life.

The land-grant idea inspires me every day.

- **Teaching**: Learning to think critically, collaborate, and focus on research-based choices.
- **Research**: Innovation and increasing what we know to be true. Generating knowledge.
- **Extension**: Taking that knowledge out to the people who need it to make a difference in their world. We do that through meaningful learning experiences that result in behavior change.

Why do I continue to engage with IPPS Southern Region? Because the mission is the same: "To seek and to share". From Florida to California, Minnesota to Texas, and right in the heartland…the mission is to grow our great nation (and by association, the world) by helping our people find community, build lives and livelihoods, and make better choices…TOGETHER.

Every one of us will leave this IPPS conference with an idea and many will implement that idea—we will have changed TOGETHER. We will have made our industry and our lives better TOGETHER. This week we have the FREEDOM and the JOY to seek and to share TOGETHER. Let's get to it!

President Boyer had the dual challenge of serving as both President and Local Site Chair (while located 250-miles from Tulsa). She thanked her Local Site Committee: Paul Havenar, Mark West, Dustin Stoll, Bruce Dunn, Alicia Torres and Todd Lasseigne. She also recognized Holly Dobbs of Oklahoma State University for managing the Silent Auction.

The SRNA remains financially strong, with the due diligence of Sec-Tres, Donna Foster. There are 259 active/paying members, including 30 new members for 2024. For the past 10-years the annual membership has averaged 243 – so there is controlled growth.

Boyer thanked the Executive Committee, and the Sponsorship Committee: Chair, Dr. Anthony Witcher, Boyer, Jeremy Pickens, and Richard May - who raised $42,000, which is outstanding! Boyer encouraged the membership to thank, visit and show their support of our sponsors during the meeting. The SRNA is deeply indebted to our loyal sponsors who make our annual meeting financially possible. She encouraged all members to make new members and first-time attendees feel welcome — share with them and seek from them.

Boyer announced that the SRNA is in its seventh year of the Southern Region Educational Endowment, with a base donation of $20,000 from an anonymous donor. The Education Endowment balance is now at $153,965 – and growing. It will greatly enhance our region's ability to support students and early career professionals – and ensure continued quality of the outstanding educational, out-reach programs our region is known for. All of this year's contributions to the silent and live auction are to go to the Endowment Fund – so please contribute! She thanked Kevin Gantt for leading the endowment effort.

The SRNA currently supports 5-scholarship programs – that cost around $20,000 per year. This includes: Vivian Munday Young Horticultural Research Scholarship Work Program, Charles Student Research Competition, Early-Career Exchange Program, Margie Jenkins Industry Scholarship, and Vince Dooley Scholarship.

Four years ago - the IPPS-SR initiated the **Margie Jenkens Industry Scholarship** to support industry professionals attending our conference for the 1st time. Margie Jenkens had a major impact on the Louisiana Nursery industry with her plant selections. Boyer recognized this year's recipient: Christina Reid of Texas A&M University. She recognized the 2024 Vince Dooley scholarship recipient: Teagan Young, University of Florida. Coach Dooley was a big proponent of landscape horticulture, taking a plant materials course with Dr. Michael Dirr at the University of Georgia – and developing great interest in new landscape plant selections.

For the **Early-Career Exchange Program** between the SRNA and the European Region, Max McKeown, Mobile Botanical Garden, represented the Southern Region at the European Conference and toured the European nursery industry. This is a wonderful exchange program for an early career professional to go to Europe and visit the European Region.

This is the thirteenth year the SRNA is doing the **Vivian Munday Young Horticultural Professional Scholarship Work Program** (formerly Vivian Munday Scholarship). She introduced the four interns: Katelyn Raffler- Moody Gardens; Hayden Lee- Stephen F. Austin State University; Tanner Hamerling - University of Georgia; and Kayla Morrison Oklahoma State University. These young professionals are making a strong contribution to this year's program. Boyer remarked that our future is young people!

At the 2024 mid-year board meeting in Orlando, Florida – the SRNA board had strategic planning session. In 2025, the board will be presenting a 5-year strategic plan for the SRNA – with metrics.

Boyer encouraged the membership to fill out questions for the Question Box – and attend the Tuesday night Question Box/Ice Cream Social. The Question Box was to be moderated by Christopher Brown and Matt Nagel.

Boyer thanked Program Chair and 1st Vice-President, Michael Roe for the excellent program and incredible group of speakers he assembled!

PROGRAM CHAIR, MICHAEL ROE

Program Chair Roe welcomed all members, guests and students. He acknowledged President Boyer for her leadership and very capably serving as both President and Local Site Chair. He thanked the membership for the opportunity to serve them, and then reviewed the scheduled program. There were six outstanding paper submissions for the Charles Student Research Competition. There will be four students competing in the oral competition, and poster presentations for the membership to visit with student presenters during the meeting. The first speaker of the technical session was Dr. Jeb Fields of Louisiana State University Agriculture Center.

Figure 1. President Dr. Cheryl Boyer (right) with Michael Roe (left), Program Chair of the 2024 Tulsa, Oklahoma, 48th annual conference.

Combined Proceedings IPPS 74:341-350. 2024.

Wood Fiber Type and Substrate Temperature Affect Growth of Knockout Rose

Amanda Mizell[a] and Jeb S. Fields

Hammond Research Station, LSU Agricultural Center, 21549 Old Covington Hwy, Hammond, Louisiana 70403 USA

amizel5@lsu.edu

[a]First Place – Charles Student Research Paper Competition

Keywords: soilless media, wood fiber, peat moss, alternative potting media, nursery production

Summary

Peat moss and pine bark are important resources in the horticulture industry but have sustainability concerns. Research efforts have been made to find suitable amendments to reduce reliance on these resources, with the most promising amendment being wood fibers. This study evaluated the effects of two substrates amended with differently engineered wood fibers made from *P. taeda* on rose growth in white and black containers with two fertilizer rates. Results showed that differently engineered wood fibers and container color can influence the performance of crops grown, as well as the physical properties of substrates.

INTRODUCTION

In 2021, the ornamental horticulture industry reported a 10.1% increase in the cost of production (McClellan, 2022), and there continues to be concern around rises in the cost and sustainability of materials essential for container production, such as peat moss and pine bark (Bilderback et al., 2013; Dunn and Freeman, 2011; Fields et al., 2023). Numerous alternatives have been studied to relieve the reliance upon any particular substrate material for soilless production, with the most promising being wood fiber (Bobo and Jackson, 2024; Boyer, 2008; Gruda and Schnitzler, 1999). However, nitrogen immobilization remains a major concern when utilizing wood fibers in production.

Wood is composed of easily available carbon which is consumed by bacteria and fungi within the substrate by utilizing nitrogen, which can cause a decrease in plant health due to nutrient deficiencies (Jackson and Wright, 2008). Many commercial wood fibers are processed using heat and pressure (e.g. disc-refining) to alleviate concerns of microbial activity and neutralize chemical toxins found in wood materials (Bunt, 1988; Dickson and Helms et al., 2022). Low-input processing methods (e.g., hammermilling), do not generate heat and pressure during processing, and therefore must be aged prior to use to avoid nitrogen immobilization (Poleatewich et al., 2022). Another proven method to alleviate the concern of nitrogen immobilization is additional fertilizer applications (Jackson, 2009), yet some studies have shown comparable crop growth in wood-produced plants without additional fertilizer treatments (Fain et al., 2008).

Overall substrate physical properties are influenced by the type of wood fiber (e.g., wood processing method), as more fibrous wood fibers tend to hold more water compared to a coarse wood material. This is important to consider to be able to provide optimal conditions for the specific crop being cultivated and to suit the region in which the crop is being produced. In regions such as the southeastern U.S., temperatures during summer months can reach upwards of 40°C. Therefore, pot color is another consideration for optimal crop health. The standard black plastic nursery container will absorb heat and can cause extensive root damage and impact overall plant health (Ingram et al., 1989).

Considering these factors, this study was developed to compare crop production with two common wood fiber amendments and adjusted container color to evaluate substrate temperature variations, and subsequent effects on nursery crop growth and fertility.

MATERIALS AND METHODS

Substrate preparation. Two unique substrates were developed consisting of 10:30:60 (v/v/v) peat moss (Lambert Peat Moss Inc., Quebec, CA)/wood fiber/aged pine bark (Phillips Bark Processing Co.; Brookhaven, MS). The wood fibers used to amend the substrates were both derived from loblolly pine (*P. taeda*) but were processed utilizing two different methods. One wood fiber was a commercially available disc-refined wood fiber material (HF; HydraFiber Ultra; Profile Products, Buffalo Grove, IL), the second wood fiber was processed via hammermilling (HW) at North Carolina State University using a hammer mill fitted with 6.35 mm screen (Meadows

Mills, Inc., North Wilkesboro, NC, U.S.). Both substrates were separately blended on a clean concrete surface using a shovel. Substrates were limed using pelletized dolomitic lime (6 $lb \cdot yd^{-3}$, Lime-Rite, Roswell, GA) and received granular nutrients (3 $lb \cdot yd^{-3}$, Micromax Micronutrients, ICL, Tel Aviv, Israel) and mixed again.

A total of 32 nursery containers (C600, Nursery Supplies, Inc., Kissimmee, FL, USA) were separated into two groups of 16. One group of the pots were spray painted white (Rust-Oleum, Hawthorn Parkway, Vernon Hills, IL) and the other 16 were left black. Thus 8 replicates of HF in black pots, 8 replicates of HF in white pots, 8 replicates of HM in black pots, and 8 replicates of HM in white pots.

Substrate Physical Properties. Static physical properties including container capacity (CC), air space (AS), bulk density (D_b), and total porosity (TP) were determined on both substrate blends via NCSU porometer analysis as described by Fonteno and Bilderback (1993) on three replicates from each treatment. Particle size distribution was determined on both substrate blends by shaking 100 g of oven dried substrate through sieves consisting of 6.3, 2.0, 0.7, 0.5, 0.3, and 0.1 mm with a catch pan at the bottom using a Ro-Tap shaker (Rx-29; W.S. Tyler, Mentor, OH, U.S.A.) for five minutes. The contents of each tray were weighed and classified into four size classifications: extra-large (> 6.3 mm), large (2.0 – 6.3 mm), medium (2.0 – 0.7 mm), and fine (< 0.7 mm).

Growth Trial. All pots were filled to the top with one of the two substrates, tamped down three times, and leveled to ensure uniform compaction. A temperature sensor (HOBO data loggers, MX2201, Bourne, MA) was buried half-way down the pot's height into the substrate, and half-way between the wall of the pot and where the plug would be placed. Double Red Knockout Rose plugs (*Rosa* x 'Radtko') were planted in all 32 experimental units and fertilized by top-dressing with a 3 month 16-6-12 (16% N, 6% P_2O_5, 12% K_2O) controlled-release fertilizer (Harrell's, Lakeland, FL). Half the replicates receiving a low rate (L; 19 $g \cdot 2\ gal^{-1}$) and the other half receiving a high (H; 39 $g \cdot 2\ gal^{-1}$) rate. Thus, a multifactorial completely randomized design was used for this experiment, where the treatments consisted of black (B) or white (W) containers, L or H fertilizer rate, and disc-refined or hammermilled wood fiber (example of treatment label: B:L:HW representing black pot: low fertilizer rate: hammermilled wood).

Irrigation. Container units were hand-watered to CC using a water hose on the day of transplanting, and then set to irrigate for 12 min (~157 $mL \cdot min^{-1}$) at 0800 every day targeting a leaching fraction of 10%. On d 25, slight heat stress was observed in the crops (**Fig. 1**) and irrigation was set to run for 10 min (~130 $mL \cdot min^{-1}$) at 0800 and 5 min (~65 $mL \cdot min^{-1}$) at 1500 for the remainder of the study.

Data Collection. Growth index (GI; average of the height of the plant from substrate surface, width of the widest part of the plant, and width perpendicular to widest part of the plant), chlorophyll content estimated via SPAD meter (SPAD 502 Plus, Spectrum Technologies, Aurora, IL, USA), and substrate pH and electrical conductivity (EC) using a non-destructive pour-through method described by Wright (1986) utilizing a hand-held pH meter (GroLine HI9814, Hanna Instruments, Smithfield, RI, USA) were collected bi-weekly for a total of 3 measurement events throughout the study. At culmination of the study (45 d), all units were destructively harvested by cutting the shoots at the base of the substrate, collecting the substrate from the roots, and drying the shoots and roots in an oven set at 68°C for 7 d to be weighed for accumulated plant biomass.

Data Analysis. All data presented in tables and figures with corresponding statistical analysis was analyzed in JMP Pro (17.0; SAS Institute, Inc.; Cary, NC, U.S.) utilizing Analysis of variance (ANOVA) and Tukey's Honestly Significant Difference at the $\alpha = 0.05$ significance level.

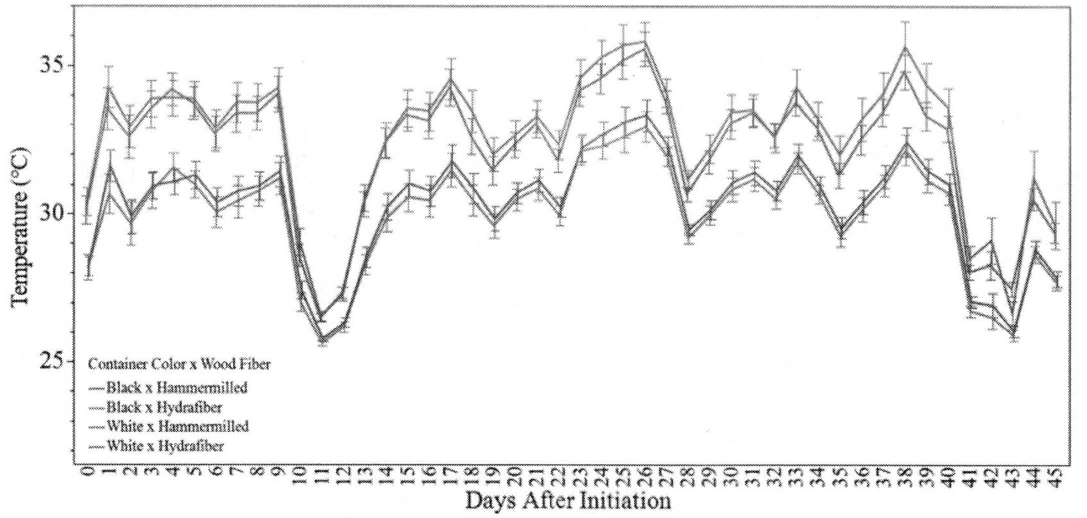

Figure 1. Temperature fluctuations of substrates amended with hammermilled wood (HW) or hydrafiber (HF) in white (W) or black (B) containers.

RESULTS AND DISCUSSION

Substrate Static Physical Properties. There were significant differences across all physical properties between the substrates amended with HW and HF (**Table 1**). This was hypothesized to be the case, as the physical properties of differently processed wood fibers have been shown to be variable (Poleatewich et al., 2022). The HF blend had a higher AS ($p = 0.0014$) and TP ($p = 0.0016$) compared to the HW blend, which had a higher CC ($p = 0.0255$) and D_b ($p = <.0001$). The HF fibers are fibrous and airy compared to the coarse HW; therefore, it was hypothesized that the HF substrate would have a higher AS and lower D_b. However, this substrate was not in the recommended range of 10-30% AS or 0.19-0.70 $g \cdot cm^{-3}$ D_b set by Yeager et al. (2007). The HW substrate had overall more suitable physical properties compared to the HF substrate. The treatments had different proportions of large, medium, and fine particles (**Table 1**). The substrate with HF amendments had a higher proportion of large particles ($p = <.0001$), while the HW-amended substrate had higher proportions of medium ($p = <.0001$) and fine particles ($p = 0.0093$). These results may be misleading, as usually substrates with smaller particles hold more water (Bilderback et al., 2005). Shown in the TP from the static physical properties, the HF blend held more water as opposed to the HW blend, which had more medium and fine particles than the HF blend. The HF particles are long and thin fibers that tend to clump together regardless of how well blended they are into a substrate. Whereas the HW fibers are shorter and wider particles and do not clump. Therefore, during the particle size measurement process, the HF fibers may have stayed on a larger size sieve due to the length of the fibers and clumps holding them there, even though they technically have a smaller particle size than the hammermilled wood.

Table 1. Static physical properties and particle size distribution of substrates consisting of bark, peat, and hammermilled wood fibers or hydrafiber.[z]

Static Physical Properties				
Treatment	Air Space[y] ($cm^3.cm^{-3}$)	Container Capacity[v] ($cm^3.cm^{-3}$)	Total Porosity[w] ($cm^3.cm^{-3}$)	Bulk Density[x] ($cm^3.cm^{-3}$)
Bark:Peat:HW[s] (60:10:30)	0.17 b[u]	0.53 a	0.70 b	0.18 a
Bark:Peat:HF[r] (60:10:30)	0.33 a	0.45 b	0.78 a	0.13 b
P-value[t]	0.0014	0.0255	0.0016	<.0001
Particle Size Distribution				
Treatment	Extra-large (>6.3 mm) ($g.g^{-1}$)	Extra-large (>6.3 mm) ($g.g^{-1}$)	Extra-large (>6.3 mm) ($g.g^{-1}$)	Extra-large (>6.3 mm) ($g.g^{-1}$)
Bark:Peat:HW[s] (60:10:30)	21.30 a	21.30 a	21.30 a	21.30 a
Bark:Peat:HF[r] (60:10:30)	24.27 a	24.27 a	24.27 a	24.27 a
P-value[t]	0.0655	0.0655	0.0655	0.0655

[z]Analysis performed using North Carolina State University porometer method (Fonteno et al., 1995).

[y]Air space = volume of water drained from sample ÷ volume of sample.

[x]Bulk density = oven dry weight of sample ÷ volume of sample

wTotal porosity = container capacity + air space

[v]Container capacity = (wet weight of sample – oven dry weight of sample) ÷ volume of sample.

[u]Means within columns separated using Tukey's HSD test (P = 0.05; n=3). Values followed by the same letter are not significant.

[t]Measures of substrate treatment effects using analysis of variance (P = 0.05).

[s]Bark:Peat:Hammermilled wood (v/v/v) substrate treatment.

[r]Bark:Peat:Hydrafiber (v/v/v) substrate treatment.

Rose Growth and Development. There were significant differences in crop GI across treatments (p = 0.0005; **Fig. 2**). Fertilizer treatment did not have a significant effect on crop growth (p = 0.6765). However, pot color had the most significant effect on growth (p = <.0001) followed by substrate blend (p = 0.0013). Plants had significantly higher growth when grown in white containers with the hammermilled wood substrate blend in high fertilizer (20.3 cm) and low fertilizer (20.2 cm) treatments. Crops grown in black containers with the HF substrate had the lowest growth in low fertilizer (12.8 cm) and high fertilizer (12.5 cm) treatments. All other treatments had no significant differences in growth.

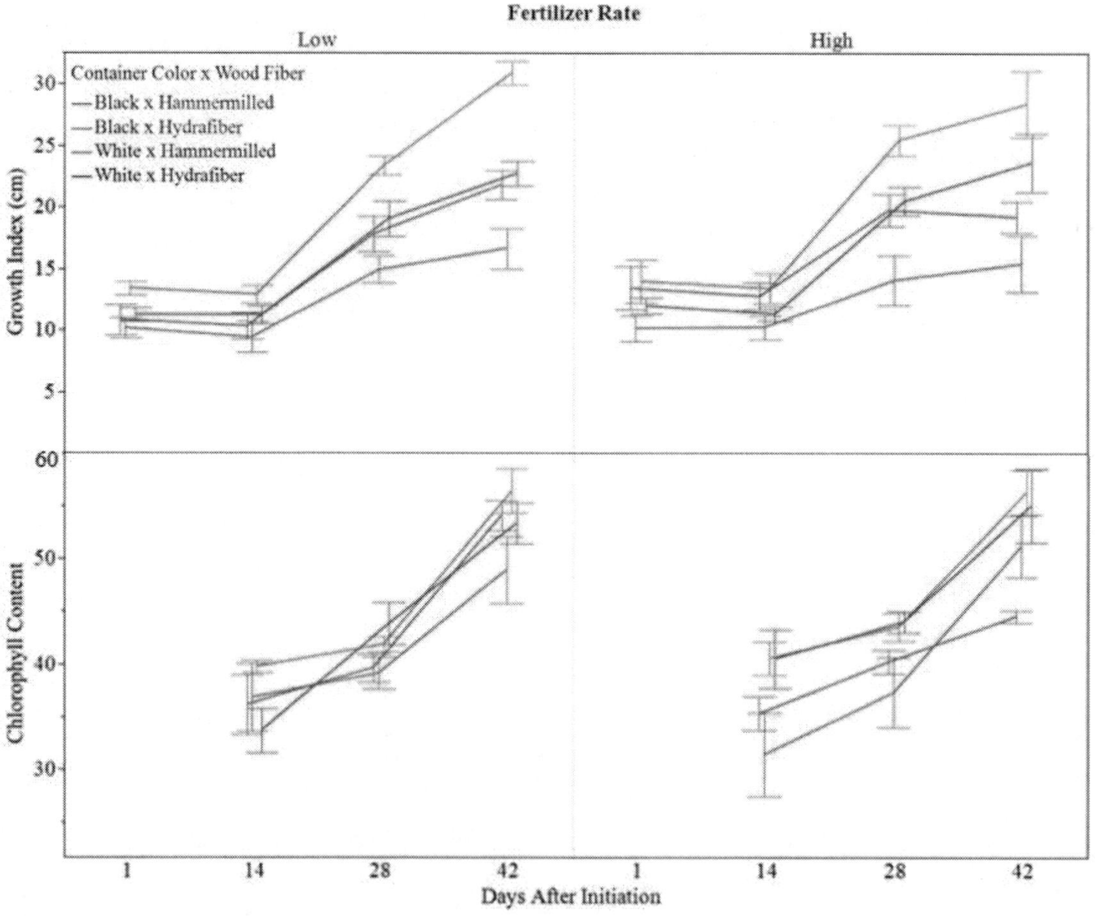

Figure 2. Growth index (GI) and chlorophyll content of roses grown in white (W) or black (B) containers, substrates amended with hammermilled wood (HW) or hydrafiber (HF), and at high (H; 39 g·2 gal^{-1}) and low (L; 19 g·2 gal^{-1}) fertilizer rates over time.

An important takeaway from these results is that the black pots remained at significantly higher temperatures than the white pots (p = <.0001), which could have caused extensive root damage. The temperatures in the black pots were, on average, around 32° C, with maximum temperatures reaching the upper 40°C, which has been shown to stop root growth completely (Mathers, 2003). Additionally, with the heavy rain events that occurred throughout the study, the HF substrate treatments held onto water much longer than the HW substrate blends, as would be assumed to be the case when considering the physical properties of the two substrates. The HF blend holding onto more water could have led to root rot in the roses, which tend to prefer a drier environment.

Furthermore, there were significant differences in the R:S ratios of accumulated biomass of the crops (**Table 2**, p = 0.0071), where the B:H:HF treatment had the highest R:S ratio and the W:L:HW treatment had the lowest. All other treatments had statistically similar R:S ratios.

Table 2. Ratios of Dried Root and Shoot Biomass of Crops grown in white (W) or black (B) containers, substrates amended with hammermilled wood (HW) or hydrafiber (HF), and at high (H; 39 g·2 gal^{-1}) and low (L; 19 g·2 gal^{-1}) fertilizer rates.

Root:Shoot Ratio of Biomass			
Container color	Fertilizer rate	Fiber type	Root:Shoot ratio
Black	Low	Hydrafiber	0.53 ab[z]
Black	High	Hydrafiber	0.57 a
Black	Low	Hammermilled wood	0.37 ab
Black	High	Hammermilled wood	0.44 ab
White	Low	Hydrafiber	0.25 ab
White	High	Hydrafiber	0.23 ab
White	Low	Hammermilled wood	0.17 b
White	High	Hammermilled wood	0.23 ab
P-value[y]			0.0071

[z]Means within columns separated using Tukey's HSD test (P = 0.05; n=4). Values followed by same letter are not significant.
[y]Measures of treatment effects using analysis of variance (P = 0.05).

Plants grown in W containers had overall higher chlorophyll content (45.7) compared to plants grown in B containers (41.0, p = 0.0071). It has been shown that high root zone temperatures can lead to chlorosis of plants and interfere with nutrient uptake (Ingram et al., 1989), which could have caused these results. There were no statistical differences in chlorophyll content of crops between substrate treatments (p = 0.4375) or fertilizer rates (p = 0.7478).

Substrate Fertility. There were no significant differences in substrate pH between the individual treatments (p = 0.3450). Fertility rate did effect substrate EC (p = 0.0066), with H fertilizer yielding increased EC (1.93 mS/cm) compared to low fertilizer treatments (1.58 mS/cm, p = 0.0001; **Fig. 3**).

CONCLUSION

The results from this study emphasize the importance of choosing materials that best suit the specific crop being grown and the region in which the crop is growing in, as substrate and container color can greatly influence crop growth and productivity. In terms of this experiment, which was conducted in southeastern Louisiana during the summer months where temperature spikes and heavy rain events are prominent, roses grown in a better-draining substrate in combination with a white pot out-performed crops grown in a substrate that holds more water and in a black pot. In a region where temperatures are cooler and rain events are less of an issue, this temperature difference may not be as influential. Another takeaway from this study was that crops did not perform any better with a higher fertilizer treatment compared to a low treatment. This study was relatively short, and likely did not allow enough time for the variation of fertilizer rate to make a difference. However, this may potentially allude to the little N-drawdown occurrence in ornamental crops grown with wood fibers which have been properly processed. This research indicates that perhaps nursery crops may not need a higher rate of fertilizer when being produced in a wood fiber-amended substrate. This, in combination with the cost reduction that comes with using wood fiber amendments, can substantially help growers save on production costs.

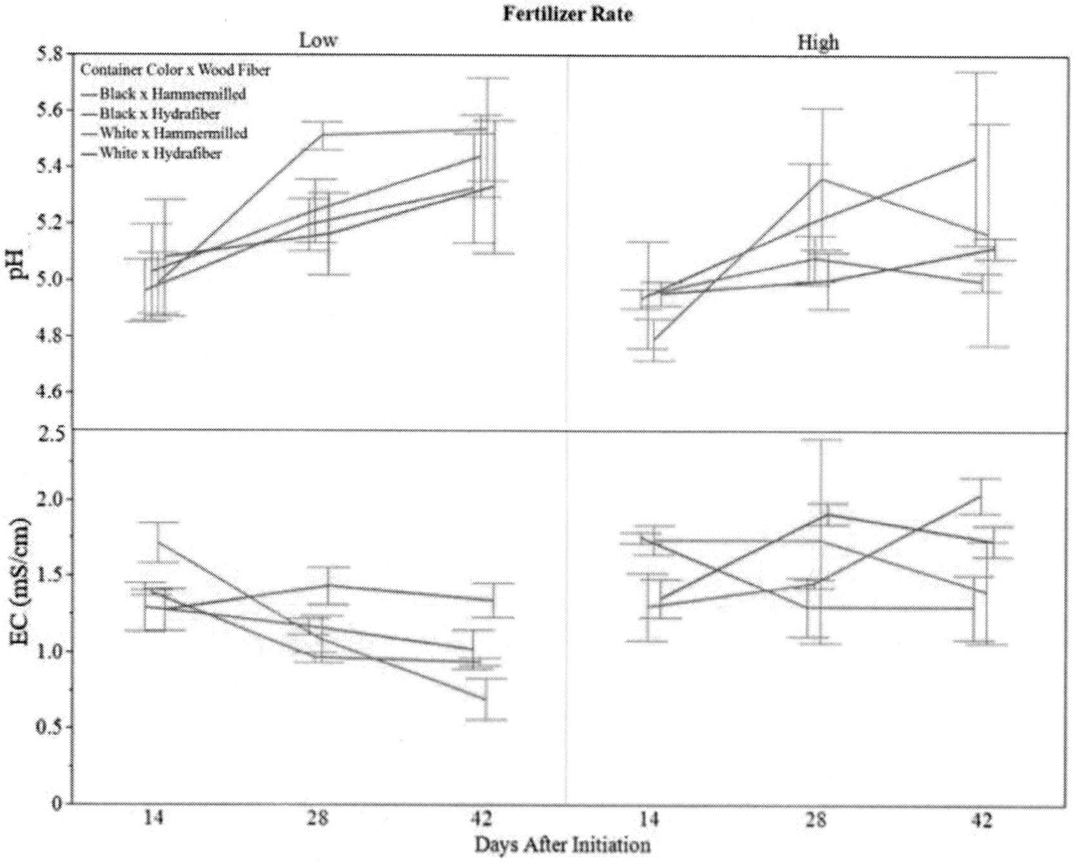

Figure 3. Electrical conductivity (EC) and pH of substrates consisting of (60:10:30; by vol.) bark, peat, and hammermilled wood fibers or hydrafiber over time

LITERATURE CITED

Bilderback, T.E., Warren, S.L., Owen, .S., and Albano, J.P. (2005). Healthy substrates need physicals too. HortTech. *15*: 747-751.

Bilderback, T.E., Riley, E.D., Jackson, B.E., Owen, J.S., Kraus, H.T.J., Fonteno, W.C., Altland, J., Fain, G.B. (2013). Strategies for developing sustainable substrates in nursery crop production. Acta Hortic. *1013*: 43–56.

Bobo, J.G. and Jackson, B.E. (2024). North American and European conifer species evaluated for use as wood components in growing media: a mini review. Acta Hortic. *1389*.
DOI: 10.17660/ActaHortic.2024.1389.16

Boyer, C.R. (2008). Evaluation of clean chip residual as an alternative substrate for container-grown plants. (Doctoral Dissertation). Auburn University.

Bunt, A.C. (1988). Media and mixes for container grown plants. 2nd Ed Unwin Hyman, London, UK.

Dickson, R.W., Helms, K.M., Jackson, B.E., Machesney, and Lee J.A. (2022). Comparison of fertilizer nitrogen availability, nitrogen immobilization, substrate carbon dioxide efflux, and nutrient leaching in peat-lite, pine bark, and pine tree substrates. HortScience 57(2):304–311.
https://doi.org/10.21273/HORTSCI16177-21

Dunn, C. and Freeman, C. (2011). Peatlands: our greatest source of carbon credits? Carbon Manag. *2*: 289-301.
https://doi.org/10.4155/cmt.11.23

Fain G.B., Gilliam, C.H., Sibley, J.L. and Boyer, C.R. (2008). Whole tree substrates derived from three species of pine in production of annual vinca. HortTech. *18*:13-17.

Fields, J., Owen, J., Lamm, A., Altland, J., Jackson, B., Oki, L., Samtani, J., Zheng, Y., and. Criscione, K. (2023). Surveying North American specialty crop growers' current use of soilless substrates and future research and education needs. Agricul. *13*: 1727. https://doi.org/10.3390/agriculture13091727.

Fonteno, W.C., Hardin, C.T., and Brewster, J.P. (1995). Procedures for determining physical properties of horticultural substrates using the NCSU Porometer. Horticultural Substrates Laboratory, North Carolina State University.

Gruda, N. and Schnitzler, W.H. (1999). Influence of wood fiber substrates and N application rates on the growth of tomato transplants. Advan. Hort. Sci. *13*: 20-24. https://www.jstor.org/stable/42881976

Ingram, D.L., Martin, C., and Ruter, J. (1989). Effect of heat stress on container-grown plants. Comb. Proc. Int. Plant Propagators Soc. *39*:348–353.

Jackson, B.E., Wright, R.D., and Barnes, M. (2008). Pine tree substrate, nitrogen rate, particle size, and peat amendment affect poinsettia growth and substrate physical properties. HortScience. *43*:2155–2161.

Jackson, B.E., Wright, R., and Alley, M. (2009). Comparison of fertilizer nitrogen availability, nitrogen immobilization, substrate carbon dioxide efflux, and nutrient leaching in peat-lite, pine bark, and pine tree substrates. HortScience *44*:781–790.

Mathers, H. (2003). Summary of temperature stress issues in nursery containers and current methods of protection. HortTech. *13*:617–624.

McClellan, M. (2022). Looking ahead. https://www.nurserymag.com/article/state-of-the-industry-2023-economic-outlook/

Poleatewich, A., Michaud, I., Jackson, B., Krause, M., and DeGenring, L. (2022). The Effect of peat moss amended with three engineered wood substrate components on suppression of damping-off caused by *Rhizoctonia solani*. Agricul. *12*: 2092. https://doi.org/10.3390/agriculture12122092.

Wright, R.D. (1986). The pour through nutrient extraction procedure. HortScience *21*:227–229.

Yeager, T.H., Gilliam, C.H., Bilderback, T.E., Fare, D.C., Niemiera, A.X., and Tilt, K.M. (1997). Best management practices guide for producing container grown plants. Southern Nursery Assn., Marietta, GA.

Combined Proceedings IPPS 74:351-359. 2024.

Rooted in Plastics: Cultivating Sustainability in Horticulture

Landon Erbrick[a], Mel Hill, and Paul Bartley

Department of Horticulture 101 Funchess Hall, Auburn University, Alabama 36849

USA

lbe0008@auburn.edu

[a]Second Place – Charles Student Research Paper Competition

Keywords: sustainability, plastic, floriculture, nursery, regulations, census, extended producer

Summary

This paper examines the critical role of plastics in the horticultural industry, particularly in plant production, packaging, and distribution. As global plastic consumption increases, the floriculture industry faces growing pressure from stricter regulations aimed at reducing environmental impact. To address the need for updated insights, this study analyzes recent census data to identify the types and quantities of plastics in use and assess their compliance with current regulatory standards. The findings aim to equip industry stakeholders with the information needed to ensure compliance, enhance sustainability, and maintain long-term profitability in an increasingly eco-conscious market.

IPPS Vol. 74 – 2024

INTRODUCTION

Since the invention of synthetic polymers in 1869, over 9 billion tons of virgin plastics have been produced, leading historians to term the current era the "Plastic Age" (Geyer et al., 2017; Pinto da Costa et al., 2020). In 1960, 390 thousand tons of plastic were produced in the US, increasing to more than 35 million tons by 2018 (EPA 2023). Of the estimated 35 million tons, containers and packaging were responsible for 14.5 million tons of plastic (EPA 2023). It is estimated that 79% of plastic produced has accumulated in landfills or in the natural environment, while 12% was incinerated, and 9% has been recycled (Geyer et al., 2017).

Plastics are deeply integrated into the horticultural industry, playing a ubiquitous role in plant production as glazing materials, irrigation systems, fertilizer coatings, and packaging. During the last half-century, plastic containers have become standard for producing plants for retail and landscape use due to their ease of processability, shipping, and marketing, in addition to their durability, low cost, and the variety of available sizes, shapes, and types of plastic containers. As a result, over 800,000 tons of petroleum-based, single-use plastics are estimated to be consumed each year by the greenhouse and nursery industries in the US (Schrader, 2013).

As plastic consumption increases, the floriculture industry faces a rapidly changing regulatory landscape, threatening its sustainability and profitability. Regulations, such as New Jersey's recycled content mandate, California's SB-343, and various Extended Producer Responsibility (EPR) laws in Oregon and Colorado, impose new standards for plastic use, recyclability, and reporting (**Table 1**).

Table 1. Current and emerging plastic regulations in the United States.

Regulation	Effective	Impact
New Jersey Recycled Content – Phase-in	Jan 2024	Require 10% post-consumer recycled content in 2024, ramping up to 50% by 2035
California SB-343 – Truth in advertising	Fall 2025	Removal of chasing arrows, restrictions on use of "recyclable" claim
Oregon SB-582 – EPR	July 2025	Pay fees for plastics sold into state – reporting requirements
Colorado HB22-1355 – EPR	Jan 2025	Pay fees for plastics sold into state – reporting requirements
Maine 2146 – EPR	2027	Pay fees for plastics sold into state – reporting requirements
California SB-54 – EPR	2028-2032	Pay fees for plastics sold into state – reporting requirements, required recovery rates for plastic materials
Maine LD-1504 PFAS - Phase-out / Ban	2030	Ban the sale of intentionally added PFAS except for "unavoidable use" by Jan 2030.

Oregon requires producers to pay 6¢/lb of paper/fiber, 24¢/lb of rigid plastic, and 34¢/lb of flexible plastic sold. Similarly, Colorado is expected to implement a dues schedule modeled after Canadian Producer Responsibility programs, charging 13¢/lb for PET bottles, 22¢/lb for plastic bags and films, and 36¢/lb for polystyrene containers. These regulations vary by plastic type and introduce complexities that the industry is not yet fully equipped to manage. While some current regulations state agricultural companies with less than $5 million in gross revenue from in-state sales are exempt from these fees, new legislation or amendments could be introduced at any time. Without a deeper understanding of the scale of our industry's plastic use and the specific plastics used across market segments, the industry may struggle to adapt, potentially leading to higher costs, stricter compliance demands, and potential supply chain disruptions.Moreover, estimates of horticultural plastic consumption are outdated, with the most recent estimates derived from the USDA Census of 2009 utilizing data from the year 2007 (Schrader, 2013). With pending regulations posing a risk to profitability, the objective of this paper is to provide the industry with the essential data and insights needed to effectively navigate the evolving regulatory landscape. By examining the most recent census data regarding horticultural container plant units sold, this study aims to identify the quantities of plastics consumed between the years 2007 and 2017 (the most recent USDA census data) and assess the industry's potential risk with current regulatory requirements. This information is crucial for assessing the industry's regulatory risk exposure and identifying necessary adjustments to ensure compliance and sustainability.

MATERIALS AND METHODS

To quantify the amount of plastic utilized in container plant production, the number of units sold within various horticultural market segments from the USDA Census of Horticultural Specialties for the years 2009, 2014, and 2019 were extracted and reproduced (**Table 2**; USDA, 2009; 2014; 2019). These censuses report data from the years 2007, 2012, and 2017. The data used from these censuses is the total number of units sold per horticultural specialty (e.g. Landscape palms, Annual bedding/garden plants sold as flats, Coniferous evergreens) as well as the total number of units sold per container size range per horticultural specialty (e.g. foliage plants for indoor/patio use – pots sold in containers >5-in (12.7cm), <12-in (30.5), and >12-in (30.5). The total number of units sold in each category were inflated by 20% to reflect loss and/or discarded plants.

The information that follows are assumptions made concerning container size and type, such as the fraction of containers molded by injection, thermoformed, and blow molded equipment. All units reported to be sold in containers >5-in (12.7 cm) were assumed to be 4-in (10.2 cm) in diameter. All units reported to be sold in containers 5-in (12.7 cm) or greater were assumed to be 6-in (15.2 cm) in diameter. Census data for "annual bedding/garden plants sold – pots" were assumed to be composed of half injection molded and half thermoformed containers. For "annual bedding/garden plants sold – flats," flats were assumed to all be thermoformed 1206 inserts and injection molded trays. "annual bedding/garden plants sold – hanging baskets" data were assumed to be produced in equal quantities of 10-in (25.4 cm) and 12-in (30.5 cm) injection molded baskets.

Table 2. Total units sold +20% to account for losses, reported in the USDA Census of Horticultural Specialties in 2009, 2014, and 2019 and trends between census years.

USDA horticultural market segment	Total units sold 2009	Total units sold 2014	Total units sold 2019	2009-2014 Trend	2014-2019 Trend	2009-2019 Trend
Floriculture sector						
Annual bedding/garden plants sold as flats	106,558,800	104,265,600	78,219,600	-2%	-25%	-27%
Annual bedding/garden plants sold as hanging baskets	52,570,800	59,082,000	47,257,200	+12%	-20%	-10%
Annual bedding/garden plants sold as pots	674,228,400	717,412,800	604,032,000	+6%	-16%	-10%
Potted herbaceous perennial plants	320,104,801	332,008,800	290,300,400	+4%	-13%	-9%
Potted flowering plants for indoor or patio use	236,610,295	265,603,900	291,138,124	+12%	+10%	+23%
Foliage plants for indoor or patio - pots	165,622,800	229,802,400	171,728,400	+39%	-25%	+4%
Foliage plants for indoor or patio - hanging baskets	13,229,023	13,818,014	13,806,709	+4%	0%	+4%
Plug seedlings	1,086,607,066	1,057,822,574	1,198,981,63	-3%	13%	10%
Nursery sector						
Broadleaf evergreens	136,576,396	107,441,898	106,227,192	-21%	-1%	-22%
Coniferous evergreens	291,190,597	72,336,062	53,615,185	-75%	-26%	-82%
Deciduous flowering trees	24,913,787	19,899,404	19,899,404	-20%	0%	-20%
Deciduous shade trees	41,365,841	25,351,537	21,310,021	-39%	-16%	-48%
Deciduous shrubs	138,874,715	118,503,336	101,483,441	-15%	-14%	-27%
Fruit and nut plants	112,947,600	166,769,270	154,473,068	48%	-7%	37%
Ornamental grasses	42,356,830	59,176,337	51,603,144	40%	-13%	22%
Landscaping palms	9,586,908	7,312,284	8,153,758	-24%	12%	-15%
Other woody ornamentals	100,003,130	53,991,382	54,604,208	-46%	1%	-45%
Liners sold	516,814,042	708,051,228	671,452,668	37%	-5%	30%

"Potted herbaceous perennials" were reported in "Chrysanthemums" and "Other plants" units sold. Chrysanthemums were reported in total units sold, units sold in pots <5-in (12.7 cm), and units sold in pots 5-in (12.7 cm) or greater. Other herbaceous perennial plants were reported in total units sold, units sold in containers less than 1-gal, units sold in containers from 1- to 2-gal, and units sold in containers 2-gal or greater. For chrysanthemums, it was assumed that half of the containers of each size are thermoformed and half of the containers of each size are injection molded. For all other herbaceous perennials, it was assumed that half of the containers were injection molded and half of the containers are blow molded for all container sizes. Data reported in "potted flowering plants for indoor or patio use" were assumed to be injection molded containers. "Foliage plants for indoor or patio use - pots" data is reported in total units sold, units sold in pots less than 6-in (15.2 cm), units sold in pots from 6-in (15.2 cm) to 13-in (33 cm), and units sold in containers 14 (35.6 cm) or greater; and all of these containers are assumed to be injection molded. Census data for "foliage plants for indoor or patio use – hanging baskets" were assumed to be produced in equal quantities of 10-in (25.4 cm) and 12-in (30.5 cm) injection molded baskets. "Broadleaf evergreens," "coniferous evergreens," "deciduous shade trees," and "deciduous shrubs" were assumed to be produced in 3-gal containers with equal quantities injection and blow molded. "Fruit and nut plants" and "other woody ornamentals" were assumed to be produced in 2-gal containers, half injection molded and half blow molded. "Ornamental grasses" were reported in total units sold in 1-gallon containers and it was assumed that half were injection molded and half were blow molded. "Landscaping palms" were assumed to be produced in a 10-gal blow molded container. Weights of representative containers of each container size and type were used to calculate the approximate total weight of plastic consumed by the horticulture industry and within each market segment (Schrader, 2013).

RESULTS AND DISCUSSION

Based on these estimations, the horticulture industry consumed 832,080 tons of plastic in 2009 (**Table 3**). In 2014, the industry's consumption of plastic increased 2.8% to 854,960 tons. However, between 2014 and 2019, plastic consumption decreased 2.3% to 835,856 tons of plastic. Over the decade spanning 2009 to 2019, the greenhouse and nursery industry has consumed an average of 840,965 tons of plastic per year from containers alone. However, some discrepancies were observed among the census data. Coniferous evergreen sales decreased 82% from 73,665 tons to 13,563 tons between 2009 and 2019 (Table 2). In 2009, the USDA reported 291,190,597 coniferous evergreens were sold. In 2014, the number of coniferous evergreens sold decreased to 72,336,062. In 2019, even fewer coniferous evergreens, 53,615,185, were sold. This significant decrease in sales could reflect incorrect assumptions concerning container size. It was assumed that all of the coniferous evergreens were grown in 3-gal containers, however a large portion of those reported in 2009 were likely seedlings grown in smaller containers and, thus, reported incorrectly. Since the number of coniferous evergreens reported in 2014 and 2019 were more similar, it could be assumed that those numbers are more accurate. The average coniferous evergreens sold in 2014 and 2019 was 62,975,623 plants. If that average replaced the amount reported in 2009, there would have only been a 15% decrease in units sold between 2009 and 2019. Furthermore, the total amount of plastic reported in 2009 would have decreased by 57,734 tons to 774,344 tons of plastic consumed.

Table 3. Tons of plastic containers consumed in the production of horticultural live goods derived from the USDA Census of Horticultural Specialties in 2009, 2014, and 2019.

USDA horticultural market segment	Tons of Plastic 2009	Tons of Plastic 2014	Tons of Plastic 2019
Floriculture sector			
Annual bedding/garden plants sold as flats	27,251	26,664	20,004
Annual bedding/garden plants sold as hanging baskets	12,140	13,644	10,913
Annual bedding/garden plants sold as pots	12,703	13,879	11,715
Potted herbaceous perennial plants	31,559	33,227	28,005
Potted flowering plants for indoor or patio use	31,794	35,490	37,216
Foliage plants for indoor or patio - pots	73,076	81,247	79,728
Foliage plants for indoor or patio - hanging baskets	3,055	3,191	3,188
Plug seedlings	201,227	195,896	222,037
Total floriculture consumption:	*392,805*	*403,238*	*412,806*
Nursery sector			
Broadleaf evergreens	34,551	27,181	26,873
Coniferous evergreens	73,666	18,300	13,564
Deciduous flowering trees	6,303	5,034	5,034
Deciduous shade trees	10,465	6,413	5,391
Deciduous shrubs	35,133	29,979	25,673
Fruit and nut plants	21,166	31,251	28,947
Ornamental grasses	2,965	4,142	3,612
Landscaping palms	7,271	5,546	6,184
Other woody ornamentals	18,740	10,118	10,232
Liners sold	229,015	313,758	297,540
Total nursery consumption	*439,275*	*451,722*	*423,050*
Total Plastic Consumption:	**832,080**	**854,960**	**835,856**

The USDA Horticultural Census market segments can be grouped into floriculture or nursery specific sectors. The market segments assumed to be floricultural are the following: annual bedding/garden plants sold as flats, annual bedding/garden plants sold as hanging baskets, annual bedding/garden plants sold as pots, potted herbaceous perennial plants, potted flowering plants for indoor or patio use, foliage plants for indoor or patio – pots, foliage plants for indoor or patio – hanging baskets, and plug seedlings. Within the floriculture sector, total plastic consumption in each of the three censuses were 392,801 tons, 403,236 tons, and 412,803 tons for 2009, 2014, and 2019, respectively (Table 3). Plug production was responsible for 54% of plastic consumed in the floriculture sector in 2019. Additional market segments contributing significantly to plastic consumption within this sector in 2019 were foliage plants for indoor or patio – pots (19%) and potted flowering plants for indoor or patio use (9%). Within each market segment, the floriculture crop consuming the most plastic per unit sold was foliage plants for indoor or patio – pots produced in 14-in (35.6 cm) containers, contributing 2.9 lbs. (1.3 kg) of plastic per unit. These pots are often more durable, ornamental containers, weighing more than a typical trade container. Among the plants in that segment, only 2% were assumed to be sold in 14-in (35.6 cm) containers. The number of units of annual bedding/garden plants sold as flats decreased by 27% between 2009 and 2019. Annual bedding/garden plants sold as hanging baskets and pots decreased by 10% between 2009 and 2019 (**Table 2**). Despite these downward trends, total plastic consumption in the sector increased due to expanding sales in indoor and patio plants and plug seedlings. Indoor and patio plant sales increased by 15%, and plug seedlings increased by 10% between 2009-2019.

The market segments assumed to be nursery crops are the following: broadleaf evergreens, coniferous evergreens, deciduous flowering trees, deciduous shade trees, deciduous shrubs, fruit and nut plants, ornamental grasses, landscaping palms, other woody ornamentals, and liners. Within the nursery sector, total plastic consumption in each of the three censuses were 439,275 tons, 451,722 tons, and 423,050 tons for 2009, 2014, and 2019, respectively (Table 3). Liners were the largest plastic consuming segment in the nursery sector, responsible for 70% of the plastic consumed in 2019. Other market segments with significant contribution to plastic consumption within this sector in 2019 were fruit and nut plants (7%) and broadleaf evergreens (6%). Within each market segment, the nursery crop consuming the most plastic per unit sold is landscaping palms, assumed to be produced in a 10-gal container (C4000; CREO Group, Kissimmee, FL), consuming 1.5 lbs. (0.7 kg) of plastic per unit. On average, 33% of the plastic consumed annually in the horticulture industry was due to liner production. The number of units of deciduous shade trees decreased by 48% between 2009 and 2019 (Table 2). Deciduous shrub sales between 2009 and 2019 also decreased by 27%. Despite this decline, the nursery sector saw a 37% increase in fruit and nut plant sales and a 30% increase in liners sold between the same time span.

The horticulture industry employs a wide variety of container types, shapes, and sizes, each with different levels of durability and reusability. This diversity in container use makes it challenging to accurately categorize and quantify the extent of plastic reuse across the industry. Furthermore, data presented in this study assumes that all containers are single-use, not accounting for the complexities associated with the lifecycle of these products. Specifically, the practice of retailers returning containers to producers for reuse is not consistently tracked or reported, leading to significant gaps in our understanding of plastic recovery and reuse within the industry. As a result, the exact percentage of plastics that are recovered and reused remains uncertain, highlighting the need for more detailed data collection and analysis to better understand and promote sustainable practices.

The extended producer responsibility (EPR) to the industry is based on the type of single-use plastic sold. Different containers are comprised of different polymers [high-density polyethylene (HDPE) and polypropylene (PP)] may be subject to different fees. While some legislation targets specific types of polymers (e.g. polystyrene), others are more general and apply fees to broad characterizations of plastics (e.g. rigid vs. flexible plastics). While EPR laws are only in effect in a few states, broader regulations may be enacted. If states adopt similar EPR laws to Oregon, charging 24¢/lb of rigid plastic, and 34¢/lb of flexible plastic, the industry could be subject to upwards of $571,856,200 annually.

Understanding what type and how much plastic consumed within these market segments is critical in identifying less expensive or more sustainable materials. Conducting field audits to directly observe and document the types and quantities of plastics in use would help verify the accuracy of self-reported data and provide a more detailed understanding of plastic consumption patterns.

CONCLUSION

The integration of plastics has been pivotal in the development of the horticulture industry. However, reliance on petroleum-based, single-use plastics has led to significant environmental challenges and an increasingly complex regulatory landscape. This study underscores the urgent need for the industry to adapt by understanding the specific plastics used across market segments and complying with new standards and EPR laws. To navigate these challenges, conducting a comprehensive audit of plastic usage is crucial. The industry must not only comply with regulations but also embrace sustainability as a competitive advantage by reducing plastic use, enhancing recyclability, and adopting innovative materials. Collaboration among growers, manufacturers, and policymakers is key to driving the industry toward a sustainable future.

LITERATURE CITED

Cal Recycle. (2022). Plastic pollution prevention and packaging producer responsibility act. https://calrecycle.ca.gov/packaging/packaging-epr/

CalRecycle. (2021). Accurate recycling labels. https://calrecycle.ca.gov/wcs/recyclinglabels/

Colorado General Assembly. (2022). Producer responsibility program for recycling. https://leg.colorado.gov/bills/hb22-1355

EPA. (2023). Plastics: Material-Specific Data. https://www.epa.gov/facts-and-figures-about-materials-waste-and-recycling/plastics-material-specific-data

Geyer, R., Jambeck, J.R., Law, K.L. (2017). Production, use, and fate of all plastics ever made. Science Advances *3*:25–29. https://doi.org/10.1126/sciadv.1700782

Maine Legislation. (2021). Stewardship program for packaging. https://legislature.maine.gov/statutes/38/title38sec2146.html

New Jersey Department of Environmental Protection. (2021). New Jersey's recycled content law. https://www.nj.gov/dep/dshw/recycled-content/recycled_content_faqs.pdf

Oregon Legislative Assembly. (2021). Senate bill 582. https://olis.oregonlegislature.gov/liz/2021R1/Downloads/MeasureDocument/SB582

Pinto da Costa, J., Rocha-Santos, T., Duarte, A. (2020). The environmental impacts of plastics and micro-plastics use, waste and pollution: EU and national measures. https://www.europarl.europa.eu/thinktank/en/document/IPOL_STU(2020)658279

Schrader, J.A. (2013). Report on the annual consumption of plastics for specialty-crop containers in the United States 1:567. https://doi.org/10.13140/RG.2.2.29992.39689/1

USDA National Agricultural Statistics Service. (2022). Census of horticultural specialties. https://nass.usda.gov/Surveys/Guide_to_NASS_Surveys/Census_of_Horticultural_Specialties/index.php (Accessed 30 August 2024)

Combined Proceedings IPPS 74:360-367. 2024.

Effects of Population and Temperature on Seed Germination of *Garberia*: A Florida Native with Ornamental and Ecological Value

Grace Carapezza[a] and Sandra B. Wilson

Department of Environmental Horticulture, University of Florida, PO Box 110690, Gainesville, Florida 32611 USA

gcarapezza@ufl.edu

[a]Third Place – Charles Student Research Paper Competition

Keywords: Asteraceae, pollinator plants, propagation

Summary

Seed propagation is the primary means of reproducing many native and endemic species, including garberia [*Garberia heterophylla* (W. Bartram) Merrill & F. Harper]. This sandhill species, belonging to a monotypic genus of Asteraceae, boasts an attractive display of purple inflorescences favored by an array of diverse pollinators. Yet it is scarcely found in nursery production and largely unknown to the gardening community. To better understand the seed biology of garberia, a series of experiments were conducted to evaluate the effects of population (North and Central Florida) on seed viability and storability as well as germination response to four seasonal temperatures that included 11/22 °C (winter), 15/27 °C (fall), 19/29 °C (spring), and 24/33 °C (summer). Initial seed viability was 49% and 60% for Central and North Florida populations, respectively. Seeds germinated readily across populations and temperatures (reaching 50% of final germination within 3-10 days), revealing a lack of physical dormancy. After 28 days meaningful germination responses were observed for temperature and population effects.

When placed in different seasonal temperatures, seeds collected from the northern population had higher maximum germination than seeds collected from the central population, except in the winter temperature where no difference was observed. For the central population, maximum seed germination was greatest in winter (53.4%), followed by spring (53.0%), fall (48.2%), and summer (35.8%). For the northern population, maximum seed germination was similar among summer and fall temperature treatments (55.4-55.6%) compared to spring and winter (53.4-54.0%). Further, it was observed that garberia seeds are intolerant of long-term, conventional dry storage, revealing a 9.8 and 36.3% reduction in germination after 3- and 6-months post storage, respectively. These findings contribute to the overall understanding of seed biology of underutilized species such as garberia, key to the development of efficient and reliable propagation systems for the nursery industry.

INTRODUCTION

Florida is the second largest producer of ornamental plants in the United States. Of total plant sales in 2019, approximately 15% were of species native to Florida (Khachatryan et al., 2021), meaning they occurred within the state boundaries prior to European contact, according to the best available scientific and historical documentation. In response to the many positive attributes native species can bring to urban gardens, growth in consumer interest and demand for residential and commercial native landscaping is expected to continue on an upward trajectory, with younger homeowners more likely to incorporate native plants in their landscapes (Gillis and Swim, 2020, Torres et al., 2024). To address this need, efforts are underway to support a widely available and diverse palette of native plants for commercial scale production (Rupp et al., 2018) and to develop efficient propagation systems for novel species (Wilson, 2020).

One such species with potential for widened use in the landscape is garberia (*Garberia heterophylla*). Garberia is a perennial shrub which blooms in Fall, boasting showy purple ray flowers as well as year-round foliar interest (**Fig. 1**). A pollinator plant, this species attracts a range of butterflies, bees and moths and naturally occurs in xeric plant communities. Vouchered as far north as Clay County and as far south as Highlands County in Florida (Wunderlin et al. 2024), garberia spans across cold hardiness zones 9A, 9B and 10A. Despite its many desirable traits, this species is carried by less than a handful of nurseries in the state (FANN, 2024) and its seed viability and germination requirements are largely unknown. Thus, the overall goal of this study was to develop optimal seed propagation practices for this underutilized species. Specific objectives were to: 1) determine the influence of population (seed origin) and temperature on seed viability and germination; and 2) assess seed storage longevity over time.

Figure 1. Images of garberia featuring: (A) upright habit and general form, (B) flower buds, (C) gray green foliage and two-ranked leaf arrangement, (D) purple inflorescence, (E) Achenes showing persistent pappi.

MATERIALS AND METHODS

Seed collection and initial viability. Seeds were collected from two natural populations situated on the periphery of Chiappini Farm Native Nursery (North Florida, USDA cold hardiness zone 9a) and The Natives Nursery (Central Florida, USDA cold hardiness zone 9b) within 2 weeks of each other in mid- to late- December 2023. A representative subsample of seeds from each population was sent to an independent seed testing facility to confirm initial pre-germination seed viability and embryo fill (US Forest Service National Seed Laboratory, Dry Branch, GA). First, seeds were non-destructively x-rayed using an Ultra Focus x-ray system with embryo fill calculated using Faxitron Vision software. Then, two replicates of 100 seeds per population were cut laterally and stained overnight at 37 C (98.6 F) in a 1.0% TZ (2,3,5-triphenyl-2H-tetrazolium chloride) solution in accordance with the Association of Official Seeds Analysts (AOSA) rules for TZ testing (AOSA, 2010). Seeds were considered viable when firm embryos stained evenly red under 10× magnification.

Seed germination. Seeds from each location were visually inspected and surface sterilized with 10% bleach solution (0.75% a.i. NaClO) for 5 min., then triple rinsed with sterile deionized (DI) water. Four replications of 100 seeds were placed in 10.9 x 10.9 cm transparent polystyrene germination boxes with lids (Hoffman Manufacturing, Albany, OR) containing two sheets of white blotter paper underneath one sheet of unbleached crepe germination paper (Hoffman Manufacturing, Albany, OR). The germination boxes were moistened with 20 mL of sterile DI water, then placed in light and temperature-controlled incubators (Percival

I30VL, Percival Scientific, Perry, IA) set at 11/22 °C (winter), 15/27 °C (fall), 19/29 °C (spring), and 24/33 °C (summer) to mimic seasonal temperatures of Florida. Light was provided by cool white, fluorescent bulbs providing at an average of 40 μmol·m-^{2}s^{-1} at each shelf level for 12 hr, followed by 12 hr of darkness. Germination boxes were opened only as needed to provide moisture and prevent seeds from desiccating. Germination progress was recorded at the first sign of radical emergence every three days. At the end of a 28-day period, final germination percentage (FGP) and $T50_{FG}$ (days to 50% of FGP) were determined for each treatment.

To determine how long seeds could be stored prior to losing viability, a subsample of the same seed lot collected from the central Florida population was placed in paper bags at room temperature (RT, 22-25°C) or in a refrigerator set to 4°C. Seeds were germinated after 0- 3- and 6-month storage times in an incubator set to 19/29°C (spring) using the same methods as previously described. Final germination was recorded after 28 days.

Experimental design and statistical analysis. Experiments utilized a modified randomized block design, with each of the four shelves of each incubator considered as a pseudo-block. Seed germination data was analyzed using generalized linear mixed model procedures through a 3-paremeter logistic model in SAS (version 14.1, SAS Institute, Cary, NC).

RESULTS

Initial pre-germination tetrazolium tests of garberia revealed seed viability was 49% and 60% for Central and North Florida populations, respectively (data not presented). Significant effects of population, temperature and their interaction were observed for seed germination that ranged from 37.8 to 62.8% after 28 days. Population responses to temperature differed from each other among all temperature treatments except for winter (**Fig. 2 A and B**). For seeds collected from Central Florida, the greatest final germination occurred in the winter temperature (53.4%), followed by spring (53.0%), fall (48.2%) and then summer (37.8%). For seeds collected from North Florida, the greatest final germination occurred in the summer and fall temperatures (62.8-63.8%) compared to the spring and winter temperatures (60.0-60.8%).

In addition to germination percentage, the rate of seed germination was also influenced by population and temperature (**Fig. 2 A and C**). For the inflection point ($T50_{FGP}$) of spring and summer treatments, a population effect was not observed. Population effects were however observed for fall and winter treatments. Significant responses in germination time were also observed between each temperature treatment. For both populations, seeds germinated earlier in the summer temperature (day 3) than the spring (day 4). In the fall, seeds collected from Central Florida germinated slightly earlier (5 d inflection point) than seeds collected from North Florida (5.5 d inflection point). Likewise, in the winter, Central Florida seeds had an earlier inflection point (8.9 d) than North Florida seeds (10.2 d).

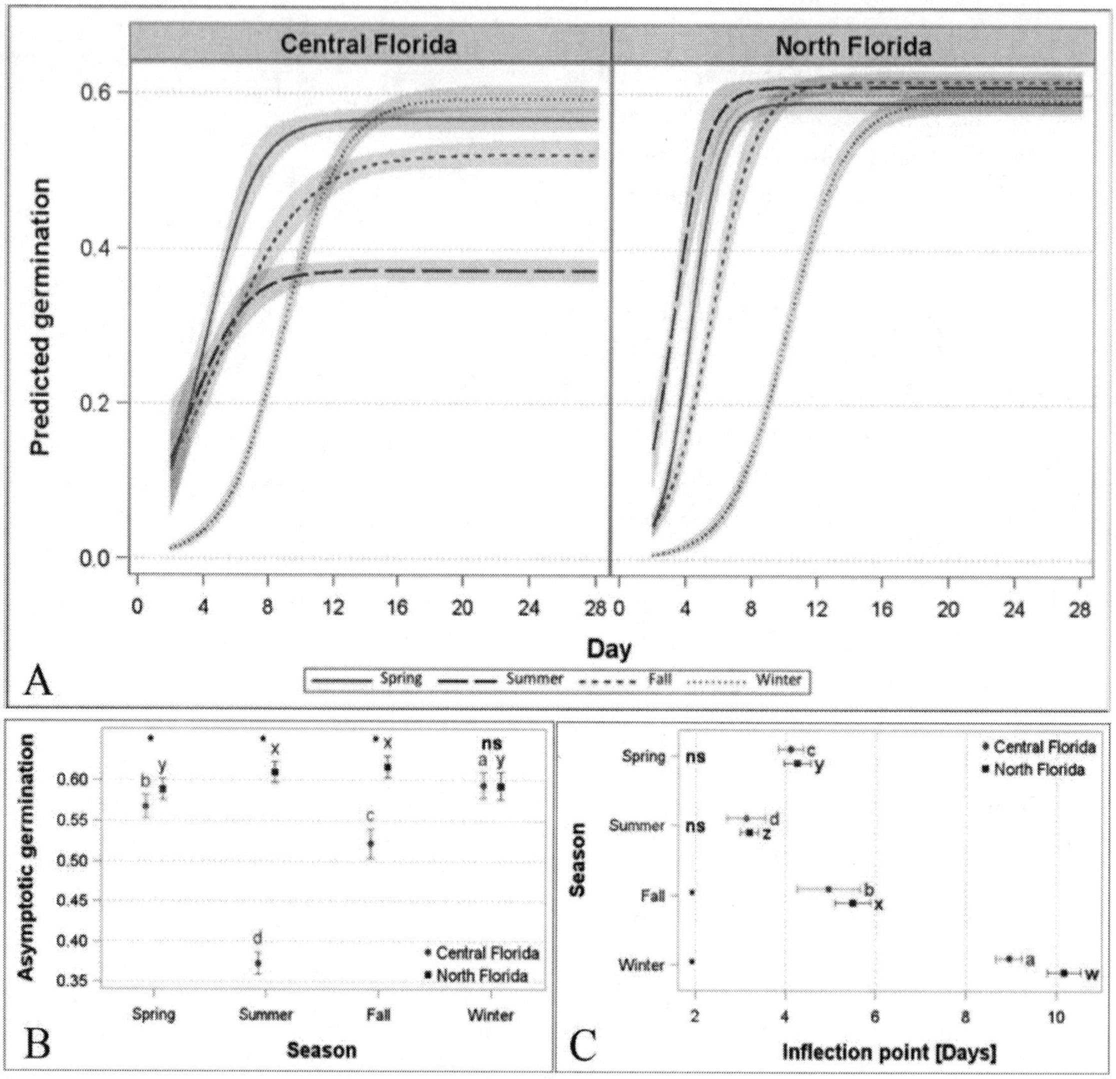

Figure 2. Seasonal germination response of garberia seed collected from North and Central Florida populations. **A.** Predicted germination proportion (symbols), fitted non-linear regression line (line) and 95% CI for the predicted response (colored areas). **B.** Asymptotic germination plus 95% CI. **C.** Inflection point (days to half-maximum germination) plus 95% CI. The colored letters indicate statistically meaningful differences among seasons within location ($\alpha = 0.05$). The * symbol indicates statistically meaningful differences among locations within season ($\alpha = 0.05$); ns=nonsignificant.

In seed longevity experiments, a significant effect of storage time and temperature was observed for seed germination of garberia. After three months of storage, a 9.8% decrease in seed germination was observed, regardless of whether seeds were stored at room temperature or refrigerated conditions (**Fig. 3**). After six months, seed germination decreased by 38.3% and 34.3% when stored at room temperature or refrigeration, respectively.

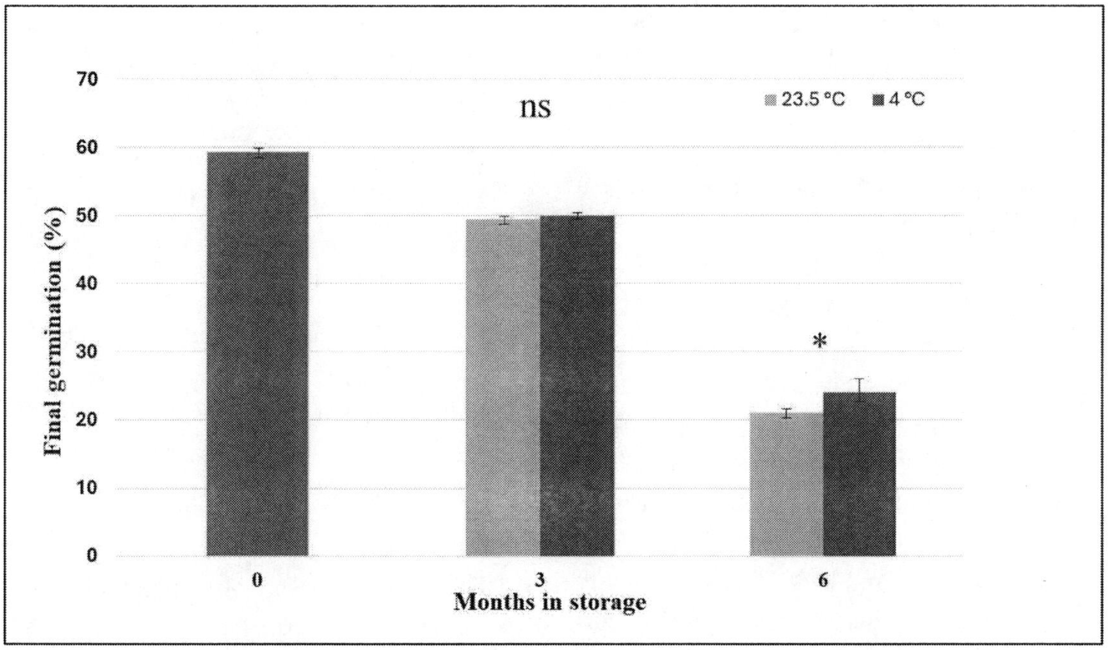

Figure 3. Final germination percentage of garberia seed that was freshly collected (Central Florida population) and dry stored at room temperature (23.5 °C) or refrigerated conditions (4.0 °C) for a period of 0-, 3-, and 6-months. After each storage interval, a subset of seeds was removed and placed in germination incubators set at 19/29 °C for 28 days. The vertical bars above the columns represent ± standard error of the mean. The * symbol indicates statistically meaningful differences among storage conditions within each storage interval ($\alpha = 0.05$); ns= nonsignificant.

DISCUSSION

Understanding the seed biology of native species allows nursery growers to anticipate the effect of seed viability for determining the sowing rate and to predict the varying degree of germination that may be influenced by seed provenance and season of planting. Likewise, knowledge of a species' seed storability index relative to deterioration can help inform decisions of collection times and production scheduling. Our results showed that garberia produces a considerable proportion of seeds that lack viability (≤ 60%) even when freshly collected from robust plant communities. This is not uncommon for wildflower species, particularly those of Asteraceae (Baskin and Baskin, 2014). Producers can account for this by using specialized seed separation equipment or simply by increasing the seed planting density per cell (Davies et al., 2018).

Results also revealed that germination responses of garberia varied not only by the geographic location (population) in which the seeds were collected but also by the temperature they were subjected to. It is of interest to note the lower germination of the Central Florida seed when germinated in the summer temperature (24/33 °C). Dell et al. (2021) also observed this effect when germinating a similar Asteraceae species, Eggert's sunflower (*Helianthus eggertii*), under warm temperatures (20/35°C), as did Wilson (2020) for other wildflower species. However, it remains unclear why the North

Florida seeds did not have the same response to the warmer temperature. This underscores the importance of considering the role of population effects when evaluating the seed propagation of native species for nursery production. Lastly, this study provided important insight into the storability of garberia. Unlike some species of Asteraceae that can survive at least a year of dry storage (Jiménez-Vázquez et al., 2021), garberia seed began to lose viability within 3 months of storage. Additional studies are warranted to determine if seeds are tolerant of longer-term storage options such as cryopreservation. Given the narrow seed collection window and lack of seed longevity for garberia, further studies exploring cutting propagation of this species may be worthwhile to ensure year-round nursery availability.

CONCLUSION

Research exploring germination responses of underutilized species is vital in developing propagation protocols and establishing best practices in the collection and storage of native seeds. Results presented herein suggest garberia is an excellent candidate for nursery production by seed. Using freshly collected seed from known populations, sowing at 3-4 seeds per cell, and planting in late winter or early spring will ensure best germination responses.

LITERATURE CITED

[AOSA] Association of Official Seed Analysts. (2010). Association of Official Seed Analysts/Society of Commercial Seed Technologists Tetrazolium Testing Handbook, 2010 edition.

Baskin, C. and Baskin, J. (2014). Seeds: Ecology, Biogeography and Evolution of Dormancy and Germination. Academic Press, Cambridge, MA.

Davies, F.T., Geneve, R.L., and Wilson, S.B. (2018). Hartmann and Kester's Plant Propagation. Principles and Practices. 9th ed. Pearson Education, NY.

Dell, N.D., Albrecht, M.A. and Long, Q.G. (2021). Effects of light and temperature on germination of Eggert's sunflower (*Helianthus eggertii*). Am. Midland Nat. *185*:49-56. doi:10.1637/0003-0031-185.1.49.

[FANN] Florida Association of Native Nurseries. (2024). Native plant and service directory. <https://fann.org/plants> (Accessed 29 Aug. 2024).

Gillis, A. and Swim, J.K. (2020). Adding native plants to home landscapes: the roles of attitudes, social norms, and situational strength. J. Environ. Psychol. *72*:1–11. doi: 10.1016/j.jenvp.2020.101519

Jiménez-Vázquez, A.M., Flores-Palacios, A., Flores-Morales, A., Perea-Arango, I., Gutiérrez, M.D., Arellano-García, J.D. and Valencia-Díaz, S. (2021). Seed longevity, viability and germination of four weed-ruderal Asteraceae species of ethnobotanic value. Bot. Sci. *99*:279-290. DOI:10.17129/botsci.2743

Khachatryan, H., Knuth, M., Hodges, A. and Hall, C. (2021). Florida nursery and landscape industry characteristics report: FE1108, Electronic Data Information Source (EDIS) of University of Florida. 2021(6). https://doi.org/10.32473/edis-fe1108-2021

Rupp, L.A., Anderson, R.M., Klett, J., Love, S.L., Goodspeed, J. and Gunnell, J. (2018). Native and adapted plant introduction for low-water landscaping. HortTech. *28*:431-435. DOI:10.21273/HORTTECH04044-18

Torres, A.P., Rihn, A.L., Barton, S.S. and Behe, B.K. (2024). Perceptions and socio-economic status influence purchases of native plants. HortTech. *34*:153-160. DOI:10.21273/horttech05345-23

Wilson, S.B. (2020). Expanding our plant palette: The role of natives and non-invasive cultivars. Proc. Florida State Hort. Soc. Keynote address, 133:xx-xxiv.

Wunderlin, R.P., Hansen, B.F., Franck, A.R. and Essig, F.B. (2024). Atlas of Florida Plants. Institute for Systematic Botany, University of South Florida, Tampa. https://florida.plantatlas.usf.edu Accessed 15 Aug. 2024.

Combined Proceedings IPPS 74:368-375. 2024.

Influence of Thiamethoxam Application Method, Timing, and Rate on Contamination of Floral Resources in Lantana

Mia Cabrera[1a], Sandra B. Wilson[1], Vanesa Rostán[2] and P. Christopher Wilson[2]

[1]Department of Environmental Horticulture, University of Florida, PO Box 110675, Gainesville, Florida 32611 USA; [2]Department of Soil, Water, and Ecosystem Sciences, University of Florida, PO Box 110290, Gainesville, Florida 32611 USA

mia.cabrera@ufl.edu

[a]Fourth Place – Charles Student Research Paper Competition

Keywords: neonicotinoids, bee pollinators, Bloomify™ rose, *Lantana strigocamara*

Summary

Pollinators are critical contributors to the natural world as well as to humans. However, their population numbers have been rapidly declining, in part due to pesticide exposure. Using the systemic insecticide thiamethoxam and the ornamental species Bloomify™ rose lantana (*Lantana strigocamara* 'UF-1011-2'), this study investigated the influence of application method (drench vs. spray), rate (low, medium, high), and timing (relative to flower bud maturity) on contamination of nectar in container grown plants. Results of nectar analysis showed meaningful differences between treatments. While spray applied thiamethoxam was not observed at quantifiable concentrations in nectar, drench applied thiamethoxam concentrations in nectar ranged from 87.7 to 1163.8 ng/mL, surpassing published LC50's for several bee species even at the lowest application rate. The timing and rate of drench application also affected thiamethoxam concentrations detected in nectar, with concentrations being the highest for applications at the highest rate and at the latest timing. These results provide insight into the development of nursery guidelines to help limit pesticide risk to pollinators before plants go to market.

IPPS Vol. 74 – 2024

INTRODUCTION

Pollinators play a critical part in global agricultural crop production and the health of natural ecosystems, their services being of great value ecologically and economically (Siviter et al., 2023). As such, there is marked interest in pollinator friendly gardening and the use of diverse flowering ornamental species to provide important nutritional resources like pollen and nectar for sustained pollinator health (Kalaman et al., 2022). Despite such efforts, pollinator populations are in peril with 40% of insect pollinators highly threatened worldwide, and nearly a quarter of native bee species at risk of extinction in North America (Kopec and Burd, 2017). Pesticide exposure is among a suite of contributors to declining bee populations, especially through necessary practices used by the horticultural industry to control pests and diseases (Halsch et al., 2022). Yet best management practices for the nursery industry in relation to pesticide method, timing, and rate are largely unknown. Prior work by Rostán et al. (2024) showed that drench pesticide applications to container grown indigo spires salvia (*Salvia* × 'Indigo Spires') were recovered in floral nectar samples at levels toxic to bees. To develop comprehensive guidelines for the industry, further studies are necessary to address other flowering species and additional timings and rates. Thus, this study's goal was to characterize the potential contamination of nectar in lantana due to insecticide treatment (thiamethoxam) during container nursery production. A repeat bloomer, Bloomify ™ rose lantana produces many umbel inflorescences throughout the year that attract diverse bee pollinators (Kalaman et al., 2022) and is useful as a model species. Thiamethoxam, the pesticide of interest, is highly toxic to pollinators because it binds to the nicotinic acetylcholine receptors (nAChRs) of insects, affecting the pollinators' ability to function (Simon-Delso et al., 2015). Our specific objectives were to determine the pesticide impacts of 1) spray vs. drench application, 2) timing of applications relative to anthesis (no flower buds, immature flower buds, mature flower buds), and 3) application rates (low, medium or high) on contamination of nectar.

MATERIALS AND METHODS

Plant material and pesticide treatments. Bloomify™ rose lantana plants were purchased as rooted liners (Ball Horticulture Company, West Chicago, IL) and up potted into 4.5-in (11.4 cm) containers filled with Pro-Mix HP Mycorrhizae media (Premier Tech Ltd., Quakertown, PA). Plants were maintained for 8 weeks in an environmentally controlled greenhouse with periodic pruning (3x) to promote branching and control the onset of flowering. Afterwards, plants were repotted into 2-gal. containers using the same media, top dressed with 1Tbs of 14N-14P-14K of slow-release fertilizer per plant and then moved to a Quonset style house covered with shade cloth for the duration of the experiment **(Fig. 1-A)**. Plants were drip irrigated twice a week for 15 minutes or as needed throughout the experiment.

Eight replicate plants were randomly assigned to pesticide treatments (including controls), utilizing a 2x3x3 factorial statistical design to explore relationships between pesticide application method (2 levels: spray and drench), timing (3 levels: no flower buds, immature flower buds, mature flower buds) and application rate (3 levels: low- 4.0 oz/100 gal, medium- 6.25

oz/100 gal, and high- 8.5 oz/100 gal). Commercially available Flagship 25WG (a water dispersible granule containing 25% thiamethoxam) was mixed and applied as a soil drench at half saturation (650 mL per pot) according to the labeled rates for ornamentals. Spray treatments were applied using a hand-operated spray bottle. In this case, plants were sprayed to the point where runoff just started to occur, taking care to wet as much foliage as possible. Due to the floral development of lantana, two weeks lapsed in between each pesticide application. The first treatment of plants with no flower buds occurred the day after transplanting. The second treatment occurred on plants where the buds were starting to form, and the last after the flower buds had fully developed on the plant but before they opened.

Figure 1. Images of (A) Bloomify™ rose lantana grown in a shade house during the study with (B) a closeup of the indeterminate inflorescences, and (C) nectar extraction from a single floret using a 20 μL glass microcapillary tube.

Nectar sampling and data analysis. Once all the flowers were blooming (**Fig. 1-B**), nectar samples were collected using 20 μL glass microcapillaries (**Fig. 1-C**). Nectar from each capillary tube was transferred into separate Eppendorf tubes, stored in a cooler on ice, and transported to a -80°C freezer until analysis. The samples were diluted with 180 μL of H_2O:ACN (9:1), then thoroughly vortexed and centrifuged (14,800 RCF, 8 min) before analysis. Thiamethoxam was analyzed using an Agilent 1290 Infinity II ultra high pressure liquid chromatography system (uHPLC) equipped with a C18 reversed-phase column (Zorbax Eclipse C18, Rapid resolution HD, 100 × 2.1 mm, 1.8 μm) and coupled to an Agilent 6495 tandem mass spectrometer for detection. The analysis method used gradient solvents as described in **Table 1** with transitions quantified as shown in **Table 2.** External calibration curves were used to determine the concentrations in the samples. Data were subjected to a three- and two-way analysis of variance (ANOVA) using the statistical software JMP (SAS Institute Inc., Cary NC) with significant effects separated using Tukey's honestly significant difference test at $P = 0.05$.

Table 1. Mobile phase gradients developed for analysis of thiamethoxam in the nectar of treated lantana.

Time (min.)	Solvent A (%)[z]	Solvent B (%)[y]	Flow (mL/min)
0.00	90	10	0.400
1.00	90	10	0.400
7.00	10	90	0.400
7.50	90	10	0.400

[z]Solvent A: 95% Optima LC-MS water, 5% Optima LC-MS ACN, with 0.1% Optima formic acid, 5 mM ammonium formate.
[y]Solvent B: 95% Optima LC-MS ACN, 5% Optima LC-MS water, with 0.1% Optima formic acid, 5mM ammonium formate.

Table 2. Multiple reaction monitoring (MRM) transitions (m/z) for identification and quantification of thiamethoxam in nectar as described by Rostán et al (2024).

Analyte	Precursor m/z	Quantifier m/z	Qualifier m/z
Thiamethoxam	292.03	211.1	181.1

RESULTS

Significant effects of thiamethoxam application method, timing and rate were observed in the nectar of lantana. The concentrations of thiamethoxam in nectar of spray-treated plants ranged from below detection limits (MDL= 0.1 ng/mL) to 14.42 ng/mL. Only 27 of the 72 spray treatment samples had detectable concentrations, with only three of those concentrations being above method quantification limits (MQL= 0.5 ng/mL). Given the lack of quantifiable concentrations in most samples, this treatment was removed from future analysis. However, drench-applications resulted in significant contamination of nectar with thiamethoxam relative to application rate (P= 0.0001) and timing (P= 0.0007). The interaction between rate and timing for the drench applications was also significant (P= 0.0015). Thiamethoxam concentrations in nectar increased as the flower bud development progressed and as application rate increased. When applications were made before buds had formed, concentrations ranged from 171.4 to 418.9 ng/mL with no difference between the low and medium application rates (**Fig. 2**). Likewise, applications made to plants with immature buds resulted in higher nectar contamination (ranging from 352.0 ng/mL to 723.4 ng/mL) as rates increased, with concentrations from the low and medium application rates being similar. Applications made to plants when mature buds had formed and just before the florets started to open resulted in the highest concentrations of thiamethoxam in nectar ranging from 276.5 to 951.9 ng/mL, with the concentration of thiamethoxam in nectar for each rate being statistically different from one another.

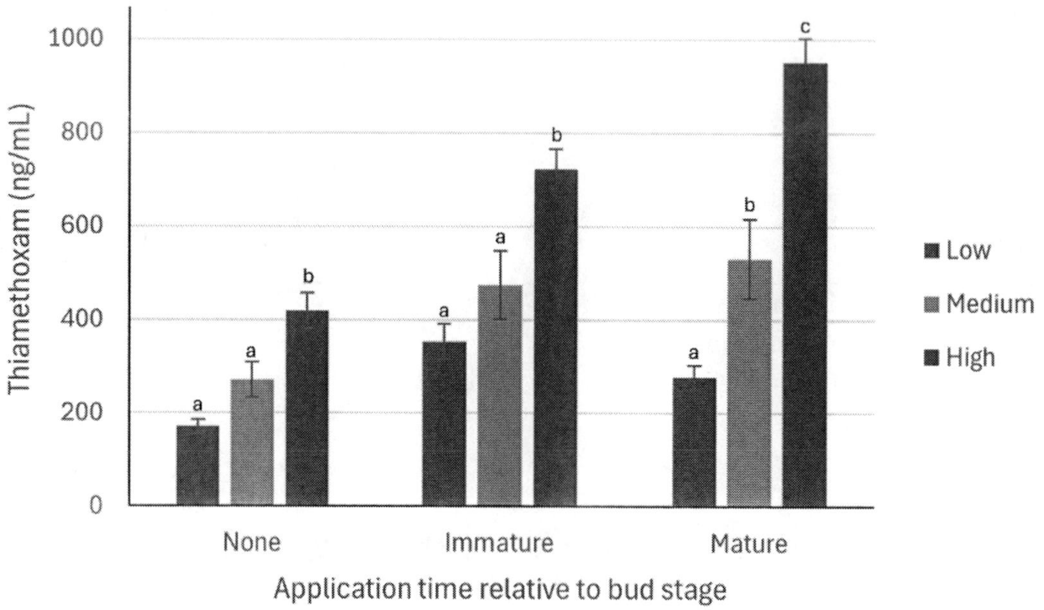

Figure 2. Mean thiamethoxam concentrations (±standard deviation) in nectar associated with three drench application times prior to anthesis (no flower buds, immature flower buds, mature flower buds) and three drench application rates (low, medium, and high). Different letters for each application timing indicate significant responses among application rates at $P \leq 0.05$ confidence.

Comparisons of thiamethoxam concentrations in nectar at each application rate relative to flower developmental stage were also of interest to note from this study. When the low application rate was applied, concentrations in nectar were lowest in applications made before flower buds were present (171.4 ng/mL) compared to applications made when flower buds were immature (352.0 ng/mL) or mature (276.5 ng/mL) (**Fig. 3**). Likewise, applications made to plants at the medium rate resulted in the lowest nectar contamination when made before flower buds were present (271.5 ng/mL) compared to applications made when flower buds were mature (531.2 ng/mL). Most dramatically, applications at the high rate made to plants with mature flower buds (latest timing) resulted in the highest concentrations of thiamethoxam in nectar, followed by the immature flower bud timing and then the earliest timing when plants did not have flower buds.

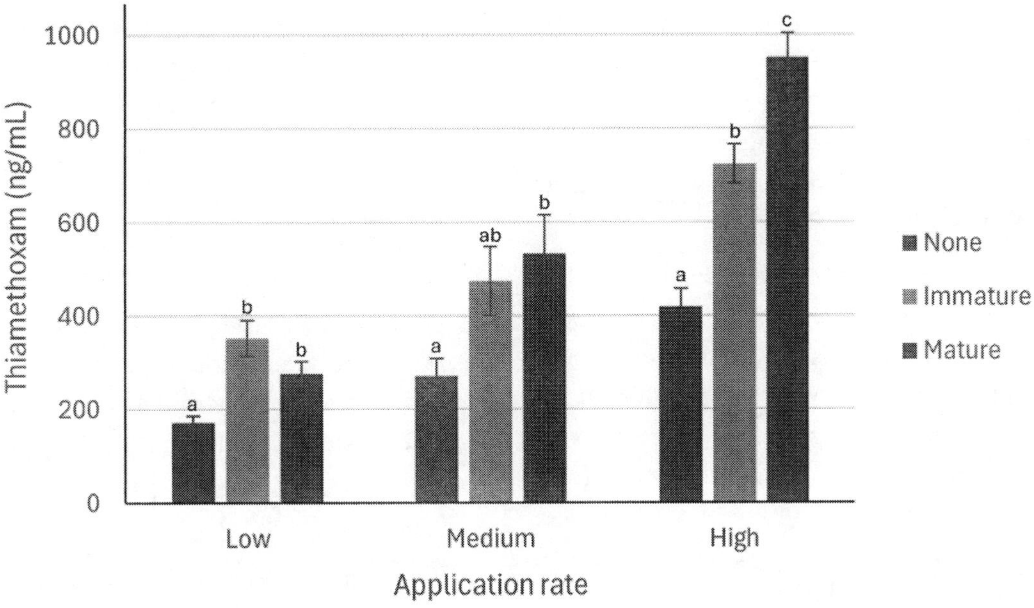

Figure 3. Mean thiamethoxam concentrations (±standard deviation) in nectar associated with three drench application rates (low, medium, and high) and three drench application times prior to anthesis (no flower buds, immature flower buds, mature flower buds). Different letters for each application rate indicate significant responses among application times at $P \leq 0.05$ confidence.

DISCUSSION

Results indicate that pesticide application method, timing, and rate all can influence contamination of nectar with the neonicotinoid insecticide thiamethoxam. As application rates increased, thiamethoxam concentrations in nectar increased from 1.5x to 3.4x depending on the flower bud stage. Likewise, as flower bud maturity increased, contamination of nectar increased by 1.3x to 2.3x depending on the application rate. Interestingly, thiamethoxam was rarely detected in the spray-applied treatments (low concentrations when detected) regardless of the timing or rate; whereas higher and more frequent detections occurred in the drench-applied treatments. These results are counter to those of Rostán et al. (2024) who reported that spray treatments of thiamethoxam applied to indigo spires salvia (*Salvia* × 'Indigo Spires') resulted in detectable concentrations in 100% of nectar samples collected, though concentrations were 1-2 orders of magnitude lower than concentrations from drench treatments. The lack of detections from the spray-applied thiamethoxam in lantana nectar may be attributed to its thick epidermal cuticle and high frequency of both glandular and non-glandular trichomes on the adaxial leaf surface potentially obstructing pesticide absorption (Sultana, 2016). With the high-rate drench treatment, as the application timings approached flowering, concentrations of thiamethoxam in nectar increased significantly. These differences in concentrations between timings likely resulted from less time for biodegradation of thiamethoxam to occur within the plants before sampling (Mach et al., 2018).

In addition, root density was most pronounced at the latest stage (mature flower buds), which would increase interception of the thiamethoxam molecules by the roots and result in higher concentrations within the plants (and presumably the nectar) (Namiki, 2022).

To screen for ecological risks of thiamethoxam, concentrations in nectar were compared to published median lethal concentrations (LC_{50}) in nectar for pollinators. For every drench treatment, thiamethoxam concentrations were found to exceed LC_{50} values of 54.3 ng/mL for the native bee, *Melipona scutellaris*. For every drench treatment except the low rate with no flower buds, thiamethoxam concentrations in nectar also exceeded LC_{50} values of 227 ng/mL for the European honeybee (*Apis mellifera*), indicating significant risks for acute toxicity (Miotelo et al., 2021).

CONCLUSION

The results presented herein indicate care should be taken for this species when drench-applying thiamethoxam as opposed to spray applications that present low risks to pollinators. From a pollinator-protection standpoint, applications should be restricted to the lower rates when possible and pre- to early-bud formation application window. Application of more pollinator-friendly (less toxic) insecticides should be considered if insect control is needed closer to the time when plants will go to market. Additional studies are needed to determine the amount of time it takes for lethal concentrations of pesticide in nectar to dissipate once plants are installed in the landscape.

Future research is being conducted to evaluate other ornamental species and pesticides to aid in the development of best management practices for the ornamental industry.

LITERATURE CITED

Halsch, C.A., Hoyle, S.M., Code, A., Fordyce, J.A., and Forister, M.L. (2022). Milkweed plants bought at nurseries may expose monarch caterpillars to harmful pesticide residues. Biol. Conserv. *273*, 109699. https://doi.org/10.1016/j.biocon.2022.109699

Kalaman, H., Wilson, S.B, Mallinger, R.E, Knox, G.W., and van Santen, E. (2022). Evaluating native and non-native ornamentals as pollinator plants in Florida: I. Floral abundance and insect visitation. HortScience *57*:126-136. https://doi.org/10.21273/HORTSCI16123-21

Kopec, K. and Burd, L.A. (2017). A systematic review of North American and Hawaiian native bees. Center for Biological Diversity, USA. https://biologicaldiversity.org/campaigns/native_pollinators/pdfs/Pollinators_in_Peril.pdf (Accessed 21 Aug. 2024).

Mach, B.M., Bondarenko, S., Potter, D.A. (2018). Uptake and dissipation of neonicotinoid residues in nectar and foliage of systemically treated woody landscape plants. Environ. Toxicol. Chem. *37*:860-870. doi.org/10.1002/etc.4021

Miotelo, L., Mendes dos Reis A.L., Malaquias J. B., Malaspina, O., and Roat, T.C. (2021). *Apis mellifera* and *Melipona scutellaris* exhibit differential sensitivity to thiamethoxam. Environ. Pollut. *268* (Pt A): 115770.
DOI: 10.1016/j.envpol.2020.115770

Namiki, S. (2022). Studies on the abilities of uptake and translocation from root to shoot of pesticides in soil. J. Pestic. Sci. *47*:131-138. Doi: 10.1584/jpestics.J22-01

Rostán, V., Wilson, P.C., Wilson, S.B., and van Santen, E. (2024). Influence of pesticide application method, timing, and rate on contamination of nectar with systemic and non-systemic pesticides. J. Environ. Toxicol. Chem. *43*:2616-2627. DOI: 10.1002/etc.5989

Simon-Delso, N., Amaral-Rogers, V., Belzunces, L. P., Bonmatin, J. M., Chagnon, M., Downs, C., Furlan, L., Gibbons, D.W., Giorio, C., Girolami, V., Goulson, D., Kreutzweiser, D.P., Krupke, C.H., Liess, M., Long, E., McField, M., Mineau, P., Mitchell, E.A., Morrissey, C.A., Noome, D.A., Pisa, L., Settele, J., Stark, J.D., Tapparo, A., Van Dyck, H., Van Praagh, J., Van der Sluijs, J.P., Whitehorn, P.R. and Wiemers, M. (2015). Systemic insecticides (neonicotinoids and fipronil): Trends, uses, mode of action and metabolites. Environ. Sci. Pollut. Res. Int. *22*:5-34. DOI:10.1007/s11356-014-3470-y

Siviter, H., Fisher, A., Baer, B., Brown, M. J. F., Camargo, I. F., Cole, J., . . . Harrison, J. F. (2023). Protecting pollinators and our food supply: Understanding and managing threats to pollinator health. Insect. Soc., *70*:5-16. https://doi.org/10.1007/s00040-022-00897-x

Sultana, R. (2016). Stem and leaf anatomy of *Lantana camara* L. - A plant of the verbenaceae family. Int. J. Curr. Res. Biosci Plant Biol. 3:27-31.

http://dx.doi.org/10.20546/ijcrbp.2016.301.002

Combined Proceedings IPPS 74:376-383. 2024.

Investigating the Effect of Hydrafiber and Biochar as a Substitute for Peat-Based Substrate for Zinnia and Snap-dragon Production

Lilin Chen[1], Julia Rycyna [1,2] and Ping Yu [1]

[1]Department of Horticulture, University of Georgia, Griffin, Georgia 30223,USA

[2]Environmental Horticulture Department, University of Florida, Gainesville, Florida 32607 USA

Lilin.chen@uga.edu

Keywords: container, growth index, growing media, floriculture

Summary

Increasing environmental and economic concerns necessitate the research for peat moss alternatives, aiming to balance ecological sustainability with cost-effectiveness. This study assessed whether biochar (BC) and hydrafiber (HF) could be a partial replacement for peat moss as substrate components. Twelve substrates were formulated by either mixing BC (20%, 40%, and 60%, by vol.) with HF (20%, 40%, and 60%, by vol.), with the remaining being peat moss or mixing BC (0%, 20%, 40%, and 60%, by vol.) with the commercial substrates (CS) to grow zinnia (*Zinnia elegans*) and snapdragon *(Antirrhinum majus)* plants in containers. Plant growth parameters included growth index (GI) and leaf greenness (indicated with SPAD), biomass, and number of flowers - measured biweekly. The results showed all the substrate mixes had similar SPAD. Treatment with 20% BC and 80% CS yielded the highest GI, biomass, and numbers of flowers in both zinnia and snapdragon. In conclusion, BC could be used to partially (20%) replace commercial substrate mix for container-grown zinnia and snapdragon.

INTRODUCTION

The escalating environmental concerns and cost of peat moss has underscored the urgency of identifying viable alternatives for container substrate. Peat moss has been a controversial substrate in greenhouse/nursery production (Yu et al., 2023). It has excellent chemical and physical properties for plant growth and development (Barrett et al., 2016). However, the extensive exploitation of peatlands for agricultural and horticultural purposes, particularly as a primary component of soilless substrates, has raised significant environmental concerns (Savvas and Gruda, 2018). These factors have led researchers to explore alternative substrates that can fulfill the role of peat moss without its associated environmental downsides (Sradnick et al., 2023).

Biochar (BC) is rich in carbon and made of a variety of renewable feedstocks that undergoes pyrolysis at 400°C to 1200°C with absence or limited oxygen (Barrett et al., 2016). Substituting BC from peat moss could reduce peat moss harvesting, and protect the peatland ecosystem (Page et al., 2002). Biochar enhances plant nutrient uptake by bolstering cation exchange capacity (CEC) and water use efficiency while also mitigating nutrient leaching, all at a lower cost than peat moss (Ding et al., 2016).

Hydrafiber (HF) is an innovative wood- and bark-based fiber product, as a viable alternative to peat in substrate composition ("HydraFiber Hub," n.d.). Hydrafiber is made through a specialized process that refines wood pulp into long, thin fibers using mechanical and thermal techniques, resulting in a porous, durable material ideal for horticulture ("HydraFiber Hub," n.d.). Those unique properties are highlighted by the manufacturer as reducing the risk of overwatering, an advantage that underscores its potential as a partial peat substitute (Tomczyk et al., 2020).

Researchers have tested the effect of BC or HF on plant growth separately. There are limited studies on the co-effects of BC and HF. Thus, the objective of this research was to: 1) compare BC and HF as a container substrate component; and 2) investigate the co-effects of BC and HF mixture as container substrate component for zinnia and snapdragon.

MATERIALS AND METHODS

Snapdragon (*Antirrhinum majus*, Madame Butterfly Cherry Bronze F1, hybrid Snapdragon) and Zinnia (*Zinnia elegans*, Giant Dahlia Flowered Orange) seeds (Johnny's Selected Seeds, Winslow, ME, USA) were sown in 128-cell propagation trays (cell depth: 5.7 cm; cell top length and width: 54.0 cm and 28.6 cm; volume: 25.1 cm^3) on 21 February 2023, with propagation media (Pro-Mix FPX Bio-fungicide media, Quakertown, PA, USA). Uniform zinnia and snapdragon seedlings were transplanted into 6-in. (15.2 cm) azalea pots (depth: 10.8 cm; top diameter: 15.5 cm; bottom diameter: 11.3 cm; volume: 1330 mL) on 9 March 2023, after two true leaves emerged. Plants were fertilized weekly with 400 mL of 240 mg L^{-1} water-soluble fertilizer [20 mg L^{-1} N, 8.6 mg L^{-1} P, and 16.6 mg L^{-1} K; Plantex Master Plant (Prod Inc., Leipsic, OH, USA)]. Each substrate was irrigated at the same scheduled time and with the same amount of greenhouse tap water (pH at 6.6, and EC at 20.0 mS m^{-1}) with 10–20% leaching rate and maintained at a greenhouse located at the University of Georgia, Griffin, Georgia. The average humidity and temperature during the experiment were 61% and 26.1 °C, respectively.

Twelve substrates were formulated by either mixing BC (20%, 40%, or 60%, by vol.) with HF (20%, 40%, or 60%, by vol.), with the remaining being peat moss (P) or mixing BC (0%, 20%, 40%, and 60%, by vol.) with the commercial substrate (CS). The twelve treatments were:

T1 - 20BC:20HF:60P	T7 -20BC:60HF:20P
T2 - 40BC:20HF:40P	T8 - 40BC:60HF
T3 - 60BC:20HF:20P	T9 - 20BC:80CS
T4 - 20BC:40HF:40P	T10 - 40BC:60CS
T5 - 40BC:40HF:20P	T11 - 60BC:40CS
T6 - 60BC:40HF	T12 - 100CS, control

Substrate mix components used in the study included the mixed hardwood BC (Proton Power, Inc., Lenoir City, TN, USA), HF (HF Ultra 160WB from HF Advanced Substrate, Buffalo Grove, IL, USA), P (Peat: THE GOLD Canadian Sphagnum Peat Moss by Fertilome, Worth, TX, USA), and CS (Jolly Gardener Pro-line C/25 Growing Mix, Oldcastle Lawn & Garden Inc. Atlanta, Georgia, USA). The BC was made from fast pyrolysis with a pH of 10.6 and EC of 1010 mS m^{-1}. The pH of HF, P, and CS was 4.9, 5.0, and 5.7, respectively, and EC was 112 mS m^{-1}, 179 mS m^{-1}, and 2383 mS m^{-1}, respectively, measured with pour-through methods. The CS was used as the control and consisted of 55% aged pine bark and the remaining 45% was composed of Canadian sphagnum peat moss, perlite, and vermiculite.

Plant height and two perpendicular widths were measured biweekly starting at 0 WAT. Growth index (GI) was calculated using the formula: GI= [height/2+ (width1 + width 2)/4]. Leaves greenness (SPAD) was recorded, then average from three mature leaves of each plant with a chlorophyll meter (SPAD-502 Minolta Camera Co., Osaka, Japan) measured biweekly starting at 2 WAT for zinnia and 4 WAT for snapdragon, respectively. When the plants started flowering, the numbers of flowers were recorded biweekly. Plants dry weights were determined at the end of 8 WAT for zinnia plants and 10 WAT for snapdragon plants by placing the shoots into the air-forced dry oven for 48 hours.

RESULTS

For zinnia plants, all the SPAD values were similar to or significantly lower than those of the control at 2, 4, 6, and 8 WAT (**Fig. 1a**). AT 8 WAT, all the treatments had similar SPAD values, and there was no significant difference among treatments. Similarly, for snapdragon plants, all the SPAD values were similar to or significantly lower than those of the control at 4, 6, 8, and 10 WAT **(Fig. 1b).**

For both zinnia and snapdragon plants (**Fig. 2 and 3**), all the treatments had a similar GI to the control except for T1 (20BC:20HF:60P) and T6 (60BC:40HF) for snap-dragon plants, which had a significantly lower GI than the control at 10 WAT. Snapdragon plants grown in T1 (20BC:20HF:60P) mixes had the lowest GI (14.2), while in T3 (60BC:20HF:20P), T5 (40BC:40HF:20P), T9 (20BC:80CS), T10 (40BC:60CS), T11 (60BC:40CS), and T12 (100CS), they had significantly higher GIs (62.0, 64.4, 59.2, 60.6, 62.2 and 63.8, respectively).

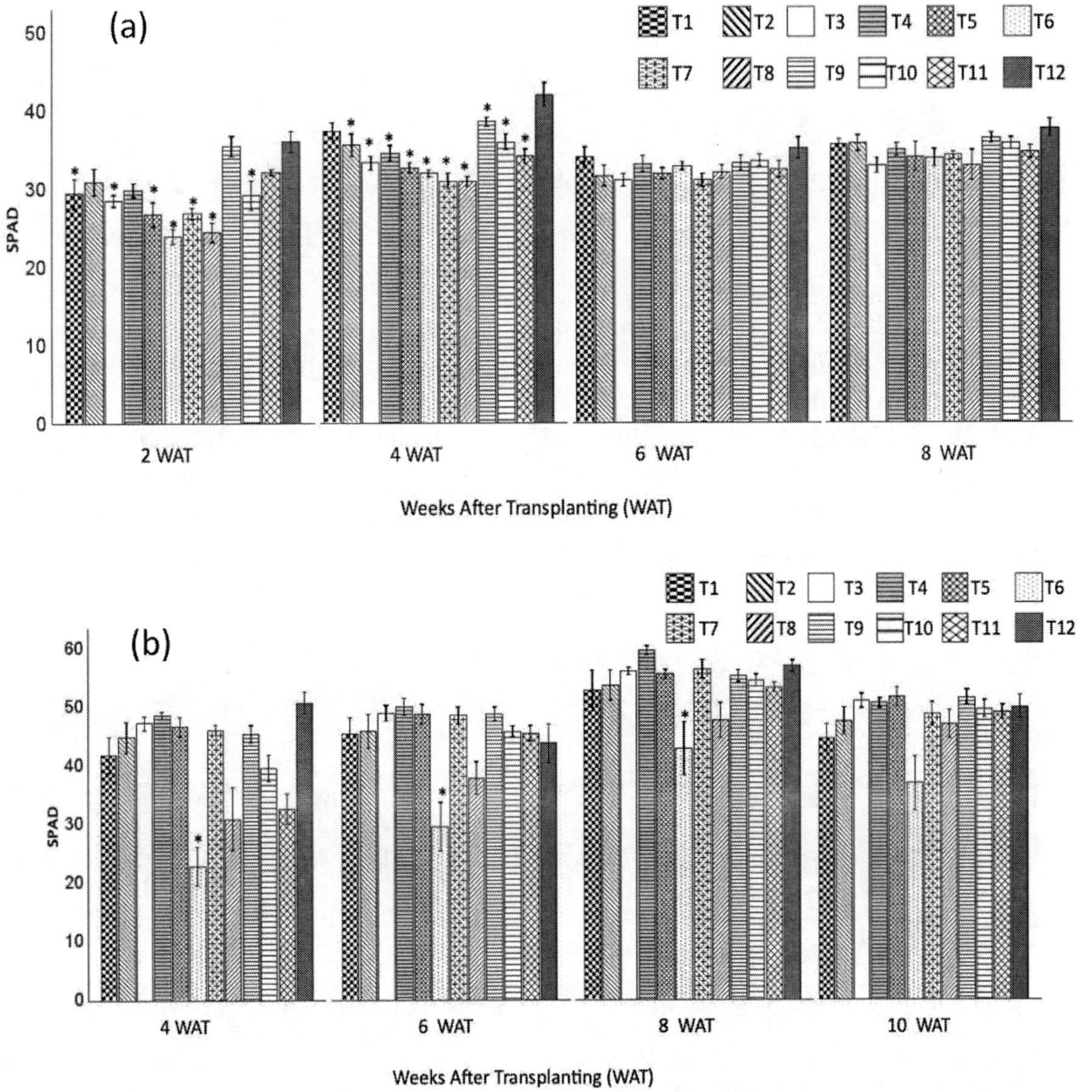

Figure 1. The SPAD values (mean ± standard error) of zinnia (**a**) and snapdragon (**b**) plants grown in twelve substrates at 2, 4, 6, 8, and 10 (snapdragon only) weeks after transplanting (WAT). Treatment 1 (20BC:20HF:60P), T2 (40BC:20HF:40P), T3 (60BC:20HF:20P), T4 (20BC:40HF:40P), T5(40BC:40HF:20P), T6 (60BC:40HF), T7 (20BC:60HF:20P), T8 (40BC:60HF), T9 (20BC:80CS), T10 (40BC:60CS), T11 (60BC:40CS), and T12 (100CS, control). * Indicates that means are significantly different from the control using Dunnett's test at $p \leq 0.05$.

Zinnia plants grown in T9 (20BC:80CS) mixes had the highest shoot dry weights (26.09 g), while in T7 (20BC:60HF:20P) mixes had the lowest dry weights (11.1 g, **Fig. 4a**). Snapdragon plants grown in T9 (20BC:80CS) and the control T12 (100CS) had the highest dry weights (24.0 g) while those grown in T1 (20BC:20HF:60P) had the lowest dry weights (1.5 g, **Fig. 4b)**.

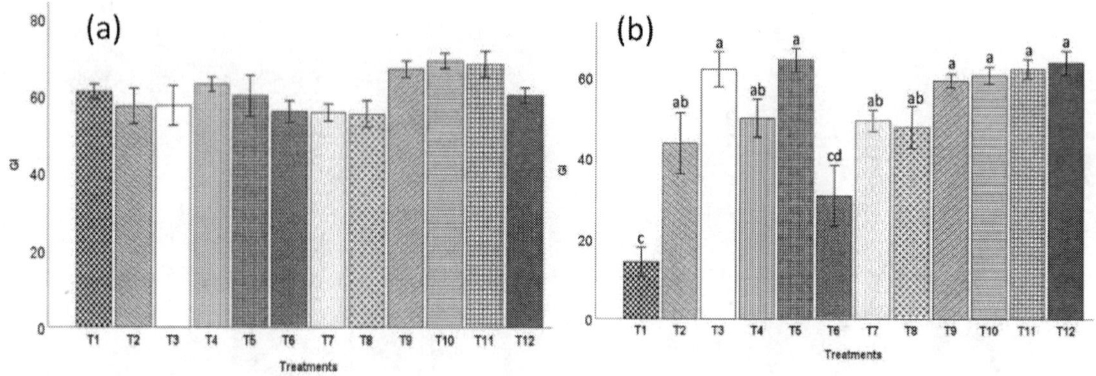

Figure 2. Growth indexes (mean ± standard error) of zinnia (**a**) and snapdragon (**b**) plants grown in twelve substrates at 8 and 10 weeks after transplanting (WAT), respectively. There were no significant differences among treatment for zinnia. Treatment 1 (20BC:20HF:60P), T2 (40BC:20HF:40P), T3 (60BC:20HF:20P), T4 (20BC:40HF:40P), T5(40BC:40HF:20P), T6 (60BC:40HF), T7 (20BC:60HF:20P), T8 (40BC:60HF), T9 (20BC:80CS), T10 (40BC:60CS), T11 (60BC:40CS), and T12 (100CS, control). Means indicated by the same alphabet letters are not significantly different according to Tukey–Kramer's HSD test at $p \leq 0.05$.

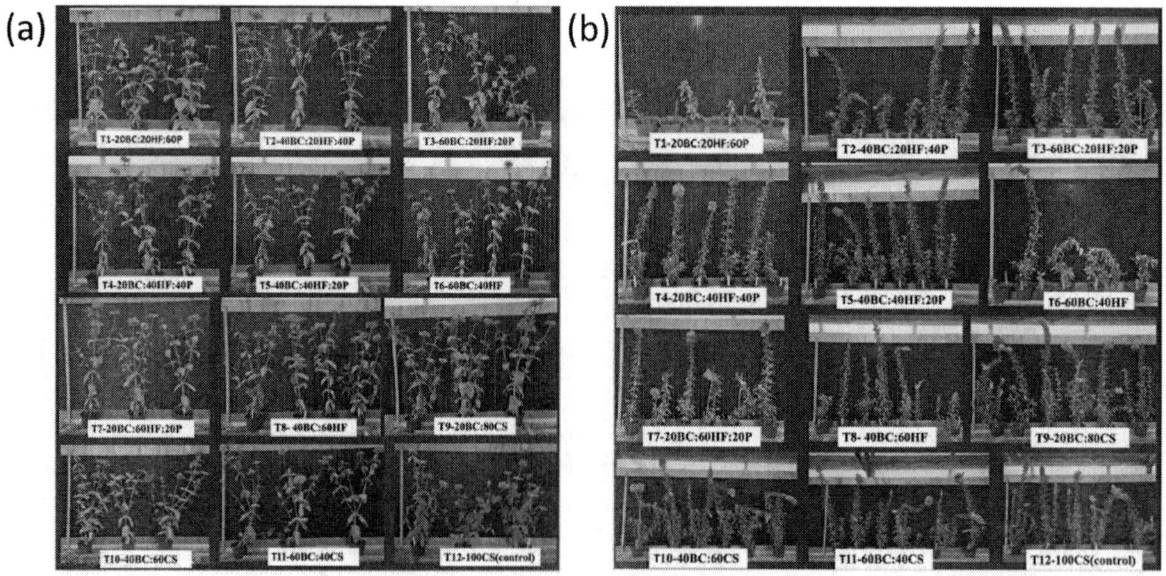

Figure 3. Plant growth of zinnia (**a**) and snapdragon (**b**) plants at 8 and 10 WAT, respectively. Treatment 1 (20BC:20HF:60P), T2 (40BC:20HF:40P), T3 (60BC:20HF:20P), T4 (20BC:40HF:40P), T5 (40BC:40HF:20P), T6 (60BC:40HF), T7 (20BC:60HF:20P), T8 (40BC:60HF), T9 (20BC:80CS), T10 (40BC:60CS), T11 (60BC:40CS), and T12 (100CS, control).

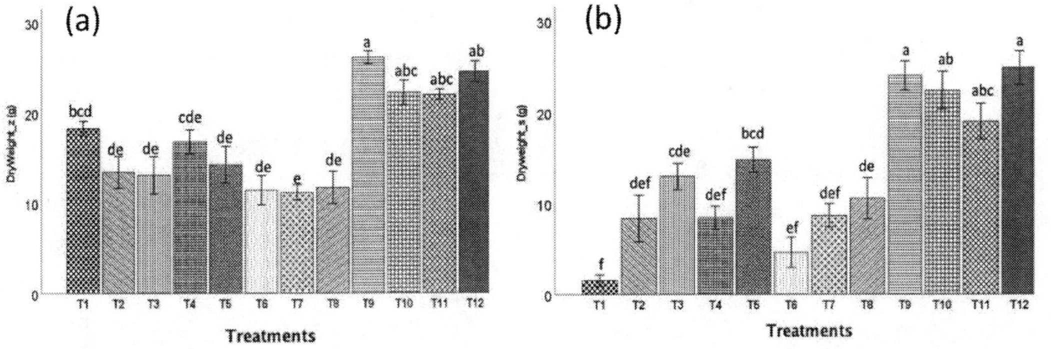

Figure 4. Shoot dry weight (mean ± standard error) of zinnia (**a**) and snapdragon (**b**) plants harvested at 8 and 10 weeks after transplanting, respectively. Treatment 1 (20BC:20HF:60P), T2 (40BC:20HF:40P), T3 (60BC:20HF:20P), T4 (20BC:40HF:40P), T5 (40BC:40HF:20P), T6 (60BC:40HF), T7 (20BC:60HF:20P), T8 (40BC:60HF), T9 (20BC:80CS), T10 (40BC:60CS), T11 (60BC:40CS), and T12 (100CS, control). Means indicated by the same alphabet letters are not significantly different according to Tukey–Kramer's HSD test at $p \leq 0.05$.

For zinnia plants, the control (100CS) had the highest numbers of flowers (11), whereas T6 (60BC:40HF) had the least numbers of flowers on average (3.5, **Fig. 5a**).

For the snapdragon plants, T9 (20BC:80CS) had more flowers (3.8) than the control (3.3), while T1 (20BC:20HF:60P) and T6 (60BC:40HF) had the least numbers of flowers (0.17 and 0.33, **Fig. 5b**).

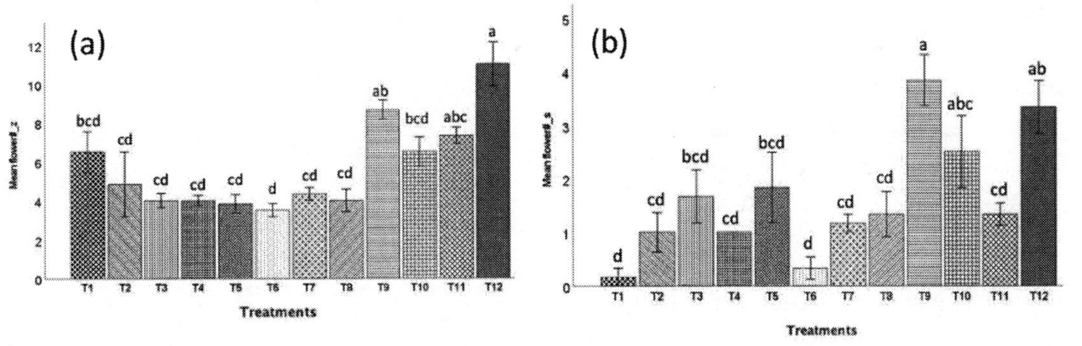

Figure 5. Numbers of flowers (unopened and opened flowers) (mean ± standard error) of zinnia (**a**) and snapdragon (**b**) plants harvested at 8 or 10 weeks after transplanting, respectively. Treatment 1 (20BC:20HF:60P), T2 (40BC:20HF:40P), T3 (60BC:20HF:20P), T4 (20BC:40HF:40P), T5 (40BC:40HF:20P), T6 (60BC:40HF), T7 (20BC:60HF:20P), T8 (40BC:60HF), T9 (20BC:80CS), T10 (40BC:60CS), T11 (60BC:40CS), and T12 (100CS, control). Means indicated by the same alphabet letters are not significantly different according to Tukey–Kramer's HSD test at $p \leq 0.05$.

DISCUSSION

This study found no significant difference in SPAD value associated with increasing HF percentage, aligning with previous research findings across various plant species and substrates. For example, HF percentage had no effect on SPAD for 'Supertunia Vista Bubblegum' petunia (*Petunia hybrida*) compared with other treatments including hammer-milled pine wood or coconut (*Cocos nucifera*) coir (Harris et al., 2020). In addition, the SPAD value was not significantly different from the control (100% peat) when treating plants with 10%, 20%, and 30% wood fiber with 90%, 80%, and 70% peat respectively in geranium (*Interspecific geraniums*)(Zawadzińska et al., 2021). However, 40% wood fiber in peat-based substrate led to smaller and fewer flowers, and lower SPAD value (Zawadzińska et al., 2021). This may be explained by two reasons. First, wood material may reduce nutrient availability and uptake, and produce phytotoxic compounds. Second, the use of wood material may need additional N fertilizer for geranium which requires substantial N to achieve optimal growth in plants (Zawadzińska et al., 2021). In our study, there were no negative effects associated with increasing HF percentage in leaf greenness because zinnia and snapdragon plants need low to medium levels of N during plant growth (Whipker et al., 2018).

Our study's findings of plant growth, plant weight, and number of flowers align with those of previous research. The study found a reduction in the fresh weight of geranium (*Interspecific geraniums*) with an increased proportion of pine wood fiber (Zawadzińska et al., 2021). Furthermore, a study discovered that a high BC concentration (70%) diminished flowering and plant growth, whereas a lower BC content (30%) did not negatively impact the flowering or growth of pelargonium plants (Conversa et al., 2015).

CONCLUSION

In conclusion, this study recommends 20% BC with CS or 100% CS for growing zinnia and snapdragon. While HF did not prove to be the most effective substrate component for cultivating zinnia and snapdragon, it still holds potential for partial peat substitution given its impact on biomass, plant growth, and floral yield.

LITERATURE CITED

Barrett, G.E., Alexander, P.D., Robinson, J.S., and Bragg, N.C. (2016). Achieving environmentallysustainable growing media for soilless plant cultivation systems – A review. Scien. Hort. *212*: 220–234. https://doi.org/10.1016/J.SCIENTA.2016.09.030

Conversa, G., Bonasia, A., Lazzizera, C., and Elia, A. (2015). Influence of biochar, mycorrhizal inoculation, and fertilizer rate on growth and flowering of Pelargonium (*Pelargonium zonale* L.) plants. Front. Plant Sci. *6*: 127299. https://doi.org/https://doi.org/10.3389/fpls.2015.00429

Dickson, R.W., Helms, K.M., Jackson, B.E., Machesney, L.M., and Lee, J.A., (2022). Evaluation of peat bended with pine wood components for effects on substrate physical properties, nitrogen immobilization, and growth of petunia (*Petunia* ×*hybrida* Vilm.-Andr.). HortScience *57*: 304–311. https://doi.org/10.21273/HORTSCI16177-21

Ding, Y., Liu, Y., Liu, S., Li, Z., Tan, X., Huang, X., Zeng, G., Zhou, L., and Zheng, B. (2016). Biochar to improve soil fertility. A review. Agron. Sustain. Develop. *36*: 36, 1–18. https://doi.org/10.1007/S13593-016-0372-Z

Guo, Y., Niu, G., Starman, T., Volder, A., and Gu, M. (2018). Poinsettia growth and development response to container root substrate with biochar. Horticul. *4*:1. https://doi.org/10.3390/HORTICULTURAE4010001

Harris, C.N., Dickson, R.W., Fisher, P.R., Jackson, B.E., and Poleatewich, A.M. (2020). Evaluating peat substrates amended with pine wood fiber for nitrogen immobilization and effects on plant performance with container-grown petunia. HortTech. *30*:107–116. https://doi.org/10.21273/HORTTECH04526-19.

HydraFiber Hub [WWW Document], n.d. https://www.hydrafiber.com/ (accessed 2.26.24).

Jahromi, N.B., Walker, F., Fulcher, A., Altland, J., and Wright, W.C. (2018). Growth response, mineral nutrition, and water utilization of container-grown woody ornamentals grown in biochar-amended pine bark. HortScience *53*: 347–353. DOI:10.21273/HORTSCI12643-17

National Minerals Information Center (2022). Peat - Historical Statistics (Data Series 140) [WWW Document]. U.S. Geological Survey. https://www.usgs.gov/media/files/peat-historical-statistics-data-series-140(accessed 10.4.23).

Page, S.E., Siegert, F., Rieley, J.O., Boehm, H.D. V., Jaya, A., Limin, S. (2002). The amount of carbon released from peat and forest fires in Indonesia during 1997. Nature *420*(6911): 61–65. https://doi.org/10.1038/nature01131

Savvas, D., and Gruda, N.S. (2018). Application of soilless culture technologies in the modern greenhouse industry-A review. Eur. J. Hortic. Sci. *83*: 280–293. https://doi.org/10.17660/eJHS.2018/83.5.2

Sradnick, A., Werner, M., and Körner, O. (2023). Make a choice: A rapid strategy for minimizing peat in horticultural press pots substrates using a constrained mixture design and surface response approach. PLOS ONE 18. https://doi.org/10.1371/JOURNAL.PONE.0289320

Tomczyk, A., Sokołowska, Z., and Boguta, P. (2020). Biochar physicochemical properties: pyrolysis temperature and feedstock kind effects. Rev. Environ. Sci. and Biotech. *19*: 191–215. https://doi.org/10.1007/S11157-020-09523-3/TABLES/3

Whipker, B.E., Henry, J., and Owen, W.G. (2018). Nutritional monitoring series zinnia (*Zinnia elegans*). Nutrit. Monitor. Series 2018 1.

Yu, P., Qin, K., Niu, G., and Gu, M. (2023). Alleviate environmental concerns with biochar as a container substrate: a review. Front. Plant Sci. *14*:1176646. doi.org/10.3389/fpls.2023.1176646

Zawadzińska, A., Salachna, P., Nowak, J.S., and Kowalczyk, W. (2021). Response of *Interspecific geraniums* to waste wood fiber substrates and additional fertilization. Agricul. *11*:119. https://doi.org/10.3390/agriculture11020119

Combined Proceedings IPPS 74:384-390. 2024.

Assessing the Impact of Plant Hormones on *Osmanthus* spp. Cutting Propagation

Yulong Chen[1], Julia Rycyna [1,2] and Ping Yu [1]

[1]Department of Horticulture, University of Georgia, Griffin, Georgia 30223 USA

[2]Environmental Horticulture Department, University of Florida, Gainesville, Florida 32607 USA

Yulong.Chen@uga.edu

Keywords: Indole-3-butyric acid; potassium indole-3-butyric acid, seaweed, IBA, K-IBA

Summary

Osmanthus is a genus of ornamental plants with valuable qualities such as pest resistance, evergreen foliage, and aromatic flowers. However, different species respond differently to plant hormones in regard to propagation success. This study evaluated five different plant hormones/rates (3,000 mg/L of indole-3-butyric acid (IBA) powder, 500 mg/L and 2,000 mg/L of potassium indole-3-butyric acid (K-IBA) solution, 10% seaweed extract solution effects on six different *Osmanthus* species (*Osmanthus heterophyllus* 'Kaori Hime', *Osmanthus armatus* 'Jim Porter', *Osmanthus* × *fortunei* 'Patty's Secret', *Osmanthus heterophyllus* 'Rotundifolius', *Osmanthus delavayi, and Osmanthus* × *fortunei* 'Fruitlandii')cuttings. The water dip was used as the control. Cuttings' rooting rates, root length, and survival rate were measured. The results showed the *Osmanthus heterophyllus* 'Kaori Hime' exhibited the highest rooting rate, root length, and survival rate, while the *Osmanthus heterophyllus* 'Rotundifolius' had the lowest survival rate. The species significantly influence the rooting percentage, root length and survival.

INTRODUCTION

Osmanthus, a genus consisting of about 30 species, mostly native to Eastern and southeastern Asia (Qian et al., 2023), is widespread across several Asian countries, including Japan, Korea, southwestern China - the Himalayas, and North America (Wang et al., 2022). *Osmanthus* has a wide range of growth habits, is planted in landscapes, and is adaptable to different environmental conditions.

Stem cuttings are the most commonly used asexual propagation method for woody plants (Nair et al., 2008). For a successful stem cutting propagation, different hormones were used to promote rooting rate, root length, and uniformity (Lewis et al., 2020). For instance, IBA (Indole-3-butyric acid) at 2,500 mg/L was used to improve rooting of *O. heterophyllus* 'Ilicifolius' hardwood cuttings from 80.6% (the control) to 86.1% (Blazich and Acedo, 1989). Additionally, hardwood cuttings of *Asclepias tuberosa* (Butterfly weed) treated with K-IBA (Potassium Indole-3-butyric acid) at 3000mg/L had greater root quality (11.9 ± 1.4) compared to the control (9.2 ± 1.3) (Lewis et al., 2020).

Seaweed extract Kelpak®, made from the kelp *Ecklonia maxima*, contains several phytohormones, including auxins, cytokinins, gibberellins, and abscisic acid (Stirk et al., 2014). It has been widely used by agriculture to enhance plant health and rooting rate (Stirk et al., 2014). For instance, using 10% Kelpak® has increased the rooting rate of hybrid tea rose (*Rosa* × *hybrida*) stem cuttings by 11% compared to the control (distilled water) (Traversari et al., 2022).

Additionally, different species (even the types of stem cutting) of *Osmanthus* respond to hormones differently. For instance, semi-hardwood cuttings of *O. heterophyllus* 'Ilicifolius' (>80%) had a significantly higher rooting rate compared to *O. heterophyllus* 'Rotundifolius' (0%). In the same study, 2,500mg/L IBA significantly increased the rooting of softwood cuttings for *O.* × *fortunei* from 63.9% (control) to 91.7% (Blazich and Acedo, 1989). However, research on different species of *Osmanthus* response to hormones is still insufficient, and there are few relevant studies on *O. delavayi* and *O. armatus*. Thus, this study aims to provide insights into the different *Osmanthus* species propagation responses to hormones.

MATERIALS AND METHODS

Six *Osmanthus* species including *O. heterophyllus* 'Kaori Hime', *O. armatus* 'Jim Porter', *O.* × *fortunei* 'Patty's Secret', *O. heterophyllus* 'Rotundifolius', *O. delavayi*, and *O.* × *fortunei* 'Fruitlandii' were used in this study. All the cuttings were semi-hardwood cuttings obtained on 16 March 2023, from Atlanta Botanical Garden (Gainesville, GA, USA). Stems were cut into 2-in (5.1 cm) cuttings with the top 1 or 2 leaves remaining and treated with three hormones: indole-3-butyric acid (IBA) powder (Hormodin® 2, at 3,000 mg/L), potassium indole-3-butyric acid (K-IBA) at 500 mg/L and 2,000 mg/L, and Kelpak® at a 10% solution; the water dip treatments were used as the control. For the IBA powder treatment, the basal ends of the cuttings were first moistened to facilitate adherence to the powder and then dipped into the IBA powder to ensure comprehensive coverage. The excess powder was removed by lightly tapping the cuttings. For liquid formulations treatments (K-IBA, Kelpak®, and water), the cuttings basal (2 cm) were immersed for

5 seconds and air dried for at least 10 minutes before being inserted in the media.

Cuttings were inserted into 72-cell plug trays (21.25×11×2.25 inches) filled with 7:3 (by vol.) peat to perlite media. The media components used in this study included peat (The Gold Canadian Sphagnum Peat Moss; Voluntary Purchasing Groups, Inc., Fort Worth, TX, USA) to perlite (Dicaperl Hydraged Aluminum by Dicalite Management Group; Dicalite Management Group, Bala Cynwyd, PA, USA). Rooting percentage and root length (cm) were measured on April 26 (week 6). The rooted cuttings were then transplanted into square pots (1½ inches square by 2¼ inches deep) filled with Jolly Gardener Pro-line C/25 Growing Mix (Jolly Gardener; Oldcastle Lawn & Garden Inc., Atlanta, GA, USA). Survival rates were recorded on April 26 (week 6), May 16 (week 9), and June 6 (week 12) after treatments.

The cuttings were placed in a glass-covered greenhouse on the University of Georgia, Griffin campus. During the experiment, day and night temperatures were recorded at 29.3±2.8°C and 20.8±1.0°C, respectively. Relative humidity high and low levels were approximately 90.54±0.97 and 52.70±2.06%. Intermittent mist operated 4 seconds every 6 minutes daily.

The experimental design was a complete randomized design, each experimental unit consists of a single treatment applied to an individual species, with 12 replicates. Data were analyzed with One-way and Two-way Analysis of Variance using R program software (version 4.3.1; RStudio, Boston, MA, USA) to test the effect of different levels of hormones and different species on the rooting rate, average root length, and survival rate. Duncan's multiple range tests were used to compare means among treatments at $P< 0.05$.

RESULTS

There were no two-factor interactions for root length or survival rates, while the interaction between species and treatment significantly impacted the rooting rate (**Table 1**). The hormone did not significantly affect the rooting rate, root length, or survival rate. Conversely, the *Osmanthus* species significantly influenced the rooting percentage, root length, and survival rate.

Table 1. ANOVA of cuttings rooting rate, root length, and survival rate as influenced by hormone treatments and *Osmanthus* species.

Source	Rooting rate (%)	Root length (cm)	Survival rate (%)
Hormone	NS	NS	NS
Species	*	***	***
Hormone × Species	*	NS	NS

[z] *, **, *** show significant difference at $P \leq 0.05$, 0.01 and 0.001 respectively; NS, not significant at $P > 0.05$.

Table 2. The influence of *Osmanthus* species on the rooting rate, root length, and survival rate of cuttings.

Species	Rooting rate (%) [z, y]	Root length (cm) [z, y]	Survival rate (%) [z]		
			Week 6	Week 9	Week 12
O. heterophyllus 'Kaori Hime'	21.7±5.4a	3.950±1.242a	86.7±4.2a	78.3±5.7a	68.3±10.3a
O. armatus 'Jim Porter'	0.0±0.0b	0.0±0.0b	80.0±4.2a	56.7±7.2b	35.0±3.1bc
O. × *fortunei* 'Patty's Secret'	3.3±2.3b	0.333±0.235b	58.3±7.9b	36.7±8.6bc	20.0±5.7c
O. heterophyllus 'Rotundifolius'	——	——	0.0±0.0d	0.0±0.0d	0.0±0.0d
O. delavayi	1.7±1.7b	0.017±0.017b	41.7±7.9c	28.3±3.3c	21.7±4.2bc
O. × *fortunei* 'Fruitlandii'	6.7±3.2b	1.600±1.294b	70.0±5.0ab	53.3±10.1b	38.3±5.7b

[z] Each value is based on 60 cuttings.
[y] Each value is based on the number of cuttings rooted for a particular treatment.
[x] Measurements of survival rate were taken on April 26, 2023 (week 6), May 16, 2023 (week 9), and June 6, 2023 (week 12). The rooting rate and root length were measured on April 26, 2023 (week 6).
[w] Data are shown as Mean ± SE. Different letters indicate significant differences among the species at $p < 0.05$.

Different species significantly impacted the rooting rate, root length, and survival rates (**Table 2**). The *O. heterophyllus* 'Kaori Hime' had a significantly higher rooting rate (21.7 ± 5.7%) and longer root length (4.0±1.2cm) than the other species. The rate and length of *O.* × *fortunei* 'Patty's Secret' (3.3±2.3% and 0.3±0.2cm), *O.* delavayi (1.7±1.7% and 0.02±0.02 cm), and *O.* × *fortunei* 'Fruitlandii' (6.7±3.2% and 1.6±1.3 cm) were significantly lower than that of *O. heterophyllus* 'Kaori Hime'. *O. heterophyllus* 'Kaori Hime' consistently exhibited significantly higher survival rates than other species on weeks 6, 9, and 12, which was 86.7±4.2%, 78.3±5.7%, and 68.3±10.3%, respectively. *O.* delavayi showed the second-lowest survival rate in week 6 (41.7±7.9%) and week 9 (28.3±3.3%). While in week 12, *O.* × *fortunei* 'Patty's Secret' had the second-lowest survival rate, at 20.0±5.7%. In week 6, all the *O. heterophyllus* 'Rotundifolius' were decreased, and *O. armatus* 'Jim Porter' did not produce any roots.

Although different hormones did not significantly impact the rooting rate, root length, or survival rate (**Table 3**), the cuttings treated by the 500 mg/L K-IBA had the highest rooting rate (11.7±4.2%) and root length (2.4±1.4 cm) among all the hormones used in this study. The cuttings treated with 3,000 mg/L IBA powder, 10% Kelpak, and the control exhibited the lowest rooting (5.0±2.8%), with the 3,000 mg/L IBA powder treatment having the shortest root length (0.3±0.2 cm).

Table 3. The influence of the types and rates of plant growth hormones on the rooting rate, root length, and survival rate of cuttings.

Treatment	Rooting rate (%) [z, y]	Root length (cm) [z, y]	Survival rate (%) [z]		
			Week 6	Week 9	Week 12
3,000 mg/L IBA powder	5.0±2.8a	0.300±0.231a	59.7±13.5a	52.8±13.6a	29.2±7.7a
500 mg/L K-IBA	11.7±4.2a	2.433±1.382a	47.2±13.7a	34.7±11.9a	29.2±11.5a
2,000 mg/L K-IBA	6.7±3.2a	1.783±0.957a	61.1±13.9a	40.2±13.7a	30.5±14.7a
10% Kelpak	5.0±2.8a	1.067±0.697a	55.6±14.7a	37.5±11.5a	31.9±10.0a
Control (water dip)	5.0±2.8a	0.317±0.204a	56.9±12.8a	45.8±10.0a	31.9±7.9a

[z] Each value is based on 72 cuttings.
[y] Each value is based on the number of cuttings which rooted for a particular treatment.
[x] Measurements of survival rate were taken on April 26, 2023 (week 6), May 16, 2023 (week 9), and June 6, 2023 (week 12). The rooting rate and root length were measured on April 26, 2023 (week 6).
[w] Data are shown as Mean ± SE. Different letters indicate significant differences among the species at $P < 0.05$.

DISCUSSION

At week 6 there was complete mortality of *O. heterophyllus* 'Rotundifolius' semi-hardwood cuttings – which failed to root. However, Blazich and Acedo (1989) reported that hardwood cuttings of 'Rotundifolius' rooted in high percentages without auxin. Seasonal timing could be a key factor in the rooting of this cultivar. The lack of success of semi-hardwood cuttings in our study align with Faust et al. (2016) who attributed cutting failure in different species to difficulties in root initiation and susceptibility to environmental stress during propagation. (Faust et al., 2016).

Moreover, this study found no significant difference between chemical treatments and the control on rooting rate, root length, and survival rate. The results can be attributed to the seasonal effects on the efficacy of hormone treatments (the experiment lasted from March 16 to June 6, 2023) – and differences among cutting types; hardwood and softwood cuttings were not tested. According to Southworth and Dirr (1996), March to June is less conducive for K-IBA treatment in promoting the rooting rate of *Cephalotaxus harringtonia* (Japanese plum-yew) cuttings. This suggests that environmental factors like temperature, humidity, and light conditions play a crucial role in plant propagation success and may even overshadow the influence of hormonal treatments.

The genotype and growth regulators are the main factors influencing rooting rate and root length in cuttings (Metaxas et al., 2004). In this study, we found that the different *Osmanthus* species had a significantly higher impact on the rooting rate, root length, and survival rate of cuttings than hormones. The results align with the findings of Gomes et al. (2010), who reported that when considering the number of roots formed per explant, factors such as the genotype and the periods of root induction and development have a significant influence.

Similarly, Oboho and Iyadi's study showed that different species treated with water dip had different survival rates with *Garcinia kola* (bitter kola) having the highest, at 85%, followed by *Gambeya albida* (white star apple) at 75%, *Irvingia gabonensis* (wild mango) at 25%, *Annona muricata* (soursop) at 15%, and *Triplochiton scleroxylon* (African whitewood) at 0% (Oboho and Iyadi, 2013).

CONCLUSION

In conclusion, the chemical treatments of auxins and Kelpak® -seaweed extract - did not influence rooting. However, different species responded significantly differently in rooting percentage, root length, and survival rate. Future research should cover a broader range of species, different environmental conditions, seasonal variation with different cutting types, and different types and rates of rooting hormones.

LITERATURE CITED

Blazich, F. A. and Acedo, J. R. (1989). Propagation of *Osmanthus heterophyllus* 'llicifolius' and 'Rotundifolius' by stem cuttings, J. Environ. Hort. *7*:133-135. DOI:10.24266/0738-2898-41.4.133

Faust, J. E., Dole, J. M. and Lopez, R. G. (2016). The floriculture vegetative cutting industry. Horticul. Rev. *44*: 121-172.

Gomes, F., Simões, M., Lopes, M. L. and Canhoto, J. M. (2010). Effect of plant growth regulators and genotype on the micropropagation of adult trees of *Arbutus unedo* L. (strawberry tree). New Biotech. *27*:882-892. https://doi.org/10.1016/j.nbt.2010.02.009

Lewis, M., Chappell, M., Thomas, P. A., Zhang, D. and Greyvenstein, O. (2020). Development of a vegetative propagation protocol for *Asclepias tuberosa*. Native Plants J. *21*:27-34.

Metaxas, D. J., Syros, T. D., Yupsanis, T. and Economou, A. S. (2004). Peroxidases during adventitious rooting in cuttings of *Arbutus unedo* and *Taxus baccata* as affected by plant genotype and growth regulator treatment. Plant Growth Regul. *44*:257-266. DOI: 10.1007/s10725-004-5931-7

Nair, A., Zhang, D., Smagula, J. and Hu, D. (2008). Rooting and overwintering stem cuttings of *Stewartia pseudocamellia* Maxim. relevant to hormone, media, and temperature. HortScience *43*:2124-28. DOI: 10.21273/HORTSCI.43.7.2124

Oboho, E. and Iyadi, J. (2013). Rooting potential of mature stem cuttings of some forest tree species for vegetative propagation. J. Appl. Natur. Sci. *5*:442-446. https://doi.org/10.31018/jans.v5i2.350

Qian, Y., Shan, L., Zhao, R., Tang, J., Zhang, C., Chen, M., Duan, Y. and Zhu, F. (2023). Recent advances in flower color and fragrance of *Osmanthus fragrans*. Forests *14*:1403.
DOI: 10.3390/f14071403

Southworth, A. L. and Dirr, M. A. (1996). Timing and K-IBA treatments affect rooting of stem cuttings of *Cephalotaxus harringtonia*. HortScience *31*:222-223.

Stirk, W. A., Tarkowská, D., Turečová, V., Strnad, M. and Van Staden, J. (2014). Abscisic acid, gibberellins and brassinosteroids in Kelpak®, a commercial seaweed extract made from *Ecklonia maxima*. J. Appl. Phycol. *26*: 61-567.
https://doi.org/10.1007/s10811-013-0062-z

Traversari, S., Cacini, S. and Nesi, B. (2022). Seaweed extracts as substitutes of synthetic hormones for rooting promotion in rose cuttings. Horticult. *8*:561. https://doi.org/10.3390/horticulturae8070561

Wang, B., Luan, F., Bao, Y., Peng, X., Rao, Z., Tang, Q. and Zeng, N. (2022). Traditional uses, phytochemical constituents and pharmacological properties of *Osmanthus fragrans*: A review, J. Ethnopharmacol. *293*:115273. DOI: 10.1016/j.jep.2022.115273

Combined Proceedings IPPS 74:391-398. 2024.

Leaf Sap Analysis for Plant Resilience

David Knaus

Apical Crop Science LLC, 1382 SE 3rd Ave, Suite 4, Canby, Oregon 97013 USA

david@apical-ag.com

Keywords: plant sap analysis, nutrient management, abiotic and biotic stress, disease and insect resistance, diagnostics, data science, precision agriculture

Summary

Leaf sap analysis, also known as plant sap analysis, is a diagnostic tool that assesses plant health by measuring mineral levels and other analytes in plant sap. This method provides an immediate snapshot of micro- and macro-elements (e.g., nitrogen, phosphorus, potassium) being transported within the plant, as well as additional compounds such as total proteins, phenolic compounds, ethanol, and carbohydrates (measured via Brix analysis). These metrics help evaluate plant stress and susceptibility to pests and diseases before visible symptoms emerge. By integrating diagnostics, data science, and crop biofeedback, growers gain real-time insights into nutritional imbalances, enabling informed adjustments to fertilization and management practices. This approach enhances plant resilience, optimizes resource use (e.g., fertilizers, water), and reduces reliance on pesticides and fungicides, supporting sustainable agriculture.

INTRODUCTION

Leaf sap analysis, a subset of plant sap analysis, has emerged as a cornerstone of precision nutrient management, offering applications in foliar spray design, fertilizer efficiency evaluation, biostimulant testing, nutrient stress diagnostics, and systemic issue identification (**Fig. 1**).

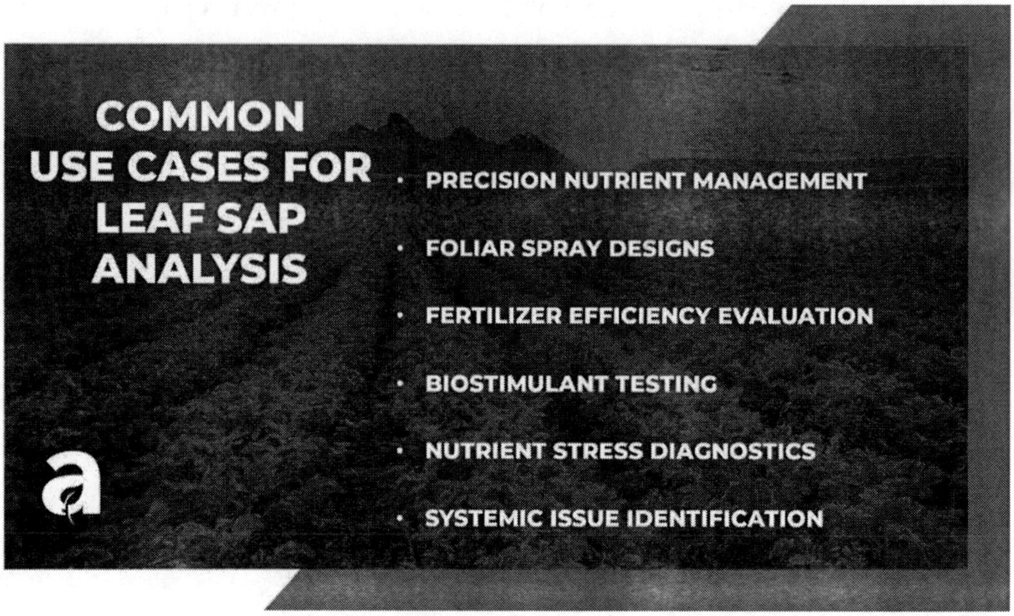

Figure 1. Common Uses for Leaf Sap Analysis. This figure illustrates key applications of leaf sap analysis, including precision nutrient management, foliar spray optimization, fertilizer efficiency assessment, biostimulant evaluation, nutrient stress diagnostics, and systemic issue detection. Visual elements include icons representing plants, nutrient charts, and analytical tools (e.g., spectrometers), emphasizing its multifaceted role in crop care.

Unlike traditional soil or tissue analysis, which provide static snapshots, leaf sap analysis captures the dynamic translocation of nutrients and metabolites within the plant, offering a real-time assessment of physiological status. This enables growers to detect deficiencies or excesses (e.g., nitrogen, phosphorus, zinc) before physical symptoms manifest, facilitating proactive management decisions that bolster crop health and resilience against abiotic (e.g., drought, salinity) and biotic (e.g., pathogens, insects) stresses.

The scientific foundation for leaf sap analysis is well-established, with studies demonstrating its efficacy in monitoring macro- and micro-element levels in plant tissue (Esteves et al., 2021). Recent research by Fan et al. (2021) highlights the biochemical and physiological cross-talk between macro- (e.g., N, P, K) and micro-nutrients (e.g., Zn, Fe), emphasizing molecular mechanisms such as nutrient stress signaling and phytohormone interactions. This understanding is critical for sustainable intensification, a strategy that optimizes fertilizer and input efficiency while minimizing environmental impact (Tilman et al., 2011). Additionally, advancements in analytical techniques—such as inductively coupled plasma mass spectrometry (ICP-MS) and high-performance liquid chromatography (HPLC)—have enhanced the precision of sap analysis, making it a valuable tool for modern agriculture (Reuter & Robinson, 1997).

SOIL, LEAF TISSUE AND SAP ANALYSIS

The practice of analyzing soil, leaf tissue, and plant sap for mineral determination dates back to the 1920s, pioneered by agricultural chemists like Treub (1923), who correlated sap nutrient levels with plant growth **(Fig. 2)**. However, it was not until the 2010s that leaf sap analysis gained prominence with the integration of advanced technologies, such as RGB imaging, SPAD chlorophyll meters, and near-infrared spectroscopy (NIRS). These tools enable detailed assessments of plant biomass, chlorophyll content, nitrogen status, and pest/disease indicators (Gitelson et al., 2003). Today, specialized laboratories equipped with ICP-MS, gas chromatography-mass spectrometry (GC-MS), and automated sap extraction systems provide comprehensive insights into plant-sap correlations with soil health and crop quality (Jones, 2012).

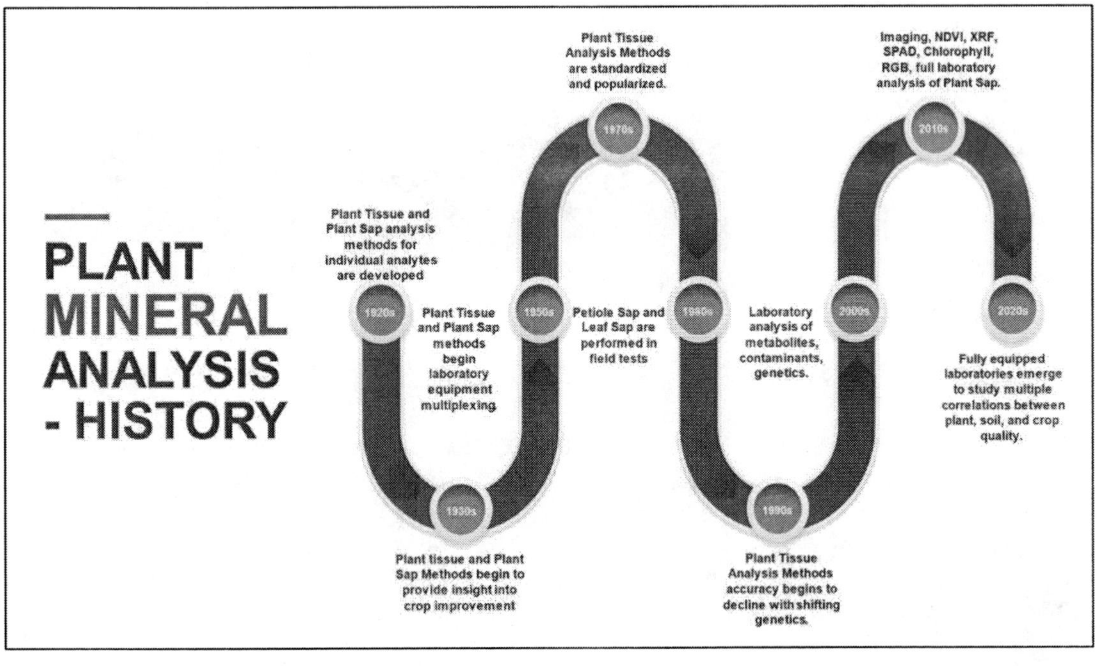

Figure 2. A History of Plant Mineral Analysis Using Plant Tissue and Plant Sap Methods. This timeline chart traces the evolution of plant mineral analysis from the 1920s (initial soil and tissue studies) to the 2010s (integration of RGB imaging, SPAD meters, and NIRS). Key milestones include Treub's early sap analysis (1923), the advent of tissue testing in the 1940s, and modern precision tools, highlighting technological advancements.

A COMPARISON OF PLANT ANALYSIS METHODS

Plant analysis methods—utilizing satellites (RGB imaging), drones (Normalized Difference Vegetation Index, NDVI), tissue analysis, handheld spectral instruments, and leaf sap analysis—offer distinct benefits and limitations (**Fig. 3**). Leaf sap analysis stands out for its immediate snapshot of nutrient translocation, cellular precision, and precise sampling, making it ideal for real-time decision-making. Satellite and drone methods provide broad spatial coverage but lack cellular resolution, while tissue analysis, though detailed, is retrospective and labor-intensive (Marschner, 2012). Handheld spectral tools offer portability but are less specific than sap analysis. The integration of these methods with data science enhances diagnostic accuracy, as demonstrated by machine learning models predicting nutrient deficiencies (Singh et al., 2020).

SOME COMMON TOOLS FOR ASSESSING MINERAL IMBALANCE

Some common tools for assessing mineral imbalance (deficiency and excess) in plants include identifying plant macro- and micro-elements based on leaf maturity (older vs. newer leaves) and visual observations (**Figs. 4 and 5**). Mulder's Chart is based on the interaction of plant macro- and micro-elements in plant nutrition and fertility programs (Mulder, 1953).

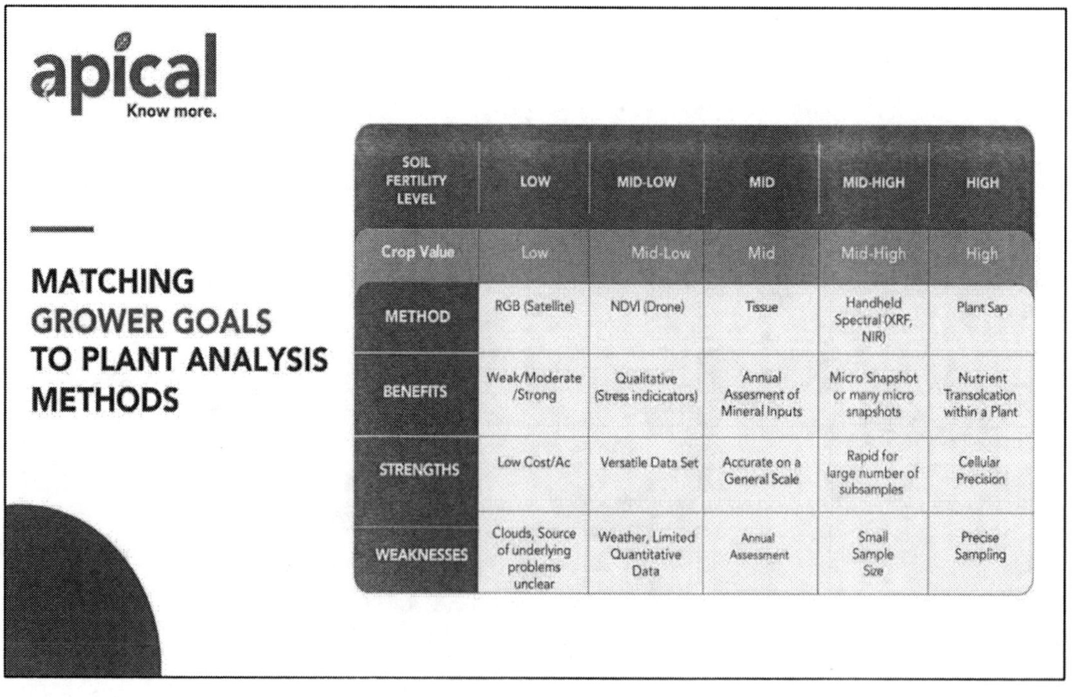

SOIL FERTILITY LEVEL	LOW	MID-LOW	MID	MID-HIGH	HIGH
Crop Value	Low	Mid-Low	Mid	Mid-High	High
METHOD	RGB (Satellite)	NDVI (Drone)	Tissue	Handheld Spectral (XRF, NIR)	Plant Sap
BENEFITS	Weak/Moderate /Strong	Qualitative (Stress indicicators)	Annual Assesment of Mineral Inputs	Micro Snapshot or many micro snapshots	Nutrient Transolcation within a Plant
STRENGTHS	Low Cost/Ac	Versatile Data Set	Accurate on a General Scale	Rapid for large number of subsamples	Cellular Precision
WEAKNESSES	Clouds, Source of underlying problems unclear	Weather, Limited Quantitative Data	Annual Assessment	Small Sample Size	Precise Sampling

Figure 3. A Comparison of plant analysis methods utilizing satellites (RGB), drones (NDVI), tissue analysis, Handheld Spectral Instruments, and Plant Sap Analysis. The table compares methods based on coverage (e.g., field-wide vs. cellular), timeliness (real-time vs. retrospective), precision (high vs. moderate), and cost (low vs. high). Leaf sap analysis is highlighted for its real-time nutrient translocation data, while drones excel in spatial mapping, and tissue analysis provides historical insights.

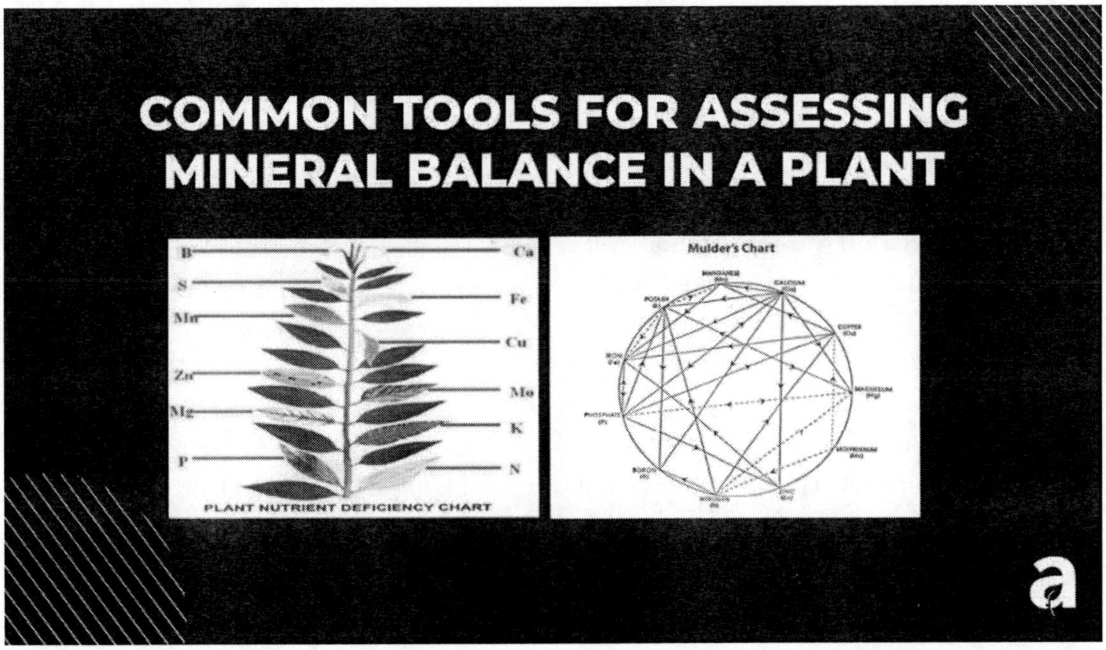

Figure. 4. Common tools for assessing mineral imbalance (deficiency and excess) in plants. This figure shows (left) a diagram of leaf maturity analysis (older vs. newer leaves) with color-coded deficiency symptoms (e.g., yellowing for N deficiency), and (right) Mulder's Chart illustrating nutrient interactions (e.g., P-Zn antagonism).

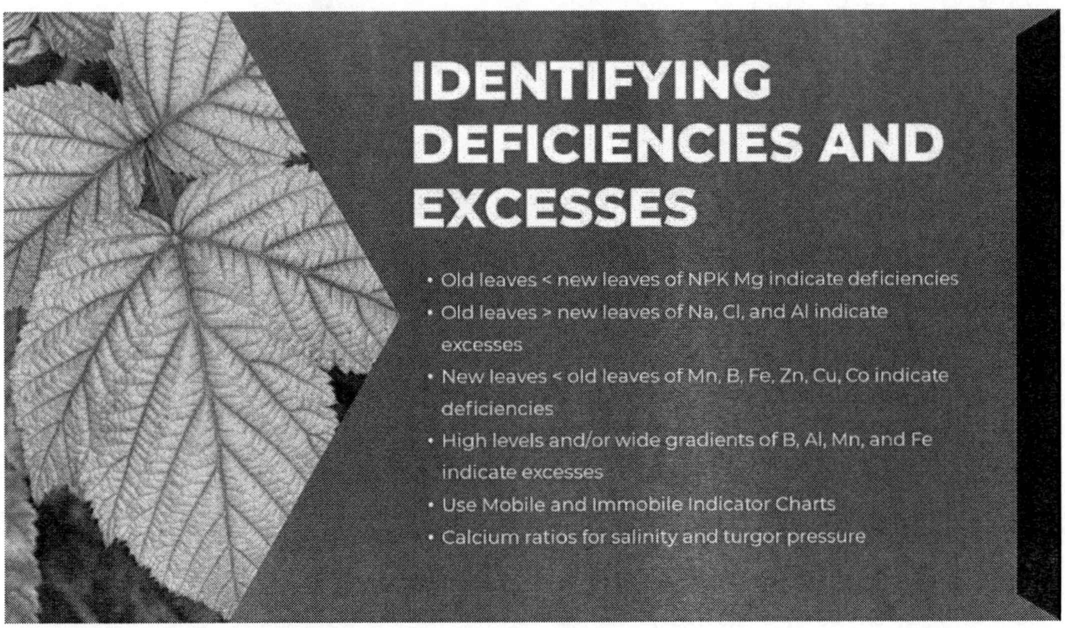

Figure 5. Identifying plant mineral deficiencies and excesses in older and new leaves. This image depicts leaf samples with labeled deficiencies (e.g., N in older leaves, Zn in new leaves) and excesses (e.g., K-induced Mg deficiency), providing a visual guide for field diagnosis.

UTILIZING LEAF SAP ANALYSIS TO MANAGE FERTILIZATION AND PLANT EFFICACY

Leaf sap analysis eliminates guesswork in fertilization by detecting imbalances that cause stress, such as macro- and micro-element deficiencies or over-fertilization leading to defoliation or stunted growth (**Fig. 6**). Mineral imbalances weaken plant immunity, increasing susceptibility to biotic (e.g., aphids, fungi) and abiotic (e.g., heat, drought) stresses, often linked to carbon deficiency from disrupted photosynthesis (**Fig. 7**). By monitoring sap nutrient levels (e.g., P at 0.2–0.5% dry weight, Zn at 20–50 ppm), growers can adjust fertilization to optimize photosynthetic efficiency and immune response (Reuter & Robinson, 1997).

This approach, part of comprehensive nutrient management, integrates diagnostics, data science, and crop biofeedback to recommend tailored interventions (**Fig. 8).** For instance, machine learning algorithms analyzing sap data with soil and weather inputs can predict stress thresholds, reducing input waste by 10–15% (Kamilaris and Prenafeta-Boldú, 2018). Leaf sap analysis thus supports sustainable intensification, aligning with global goals to enhance food security while minimizing environmental footprints (Godfray et al., 2010).

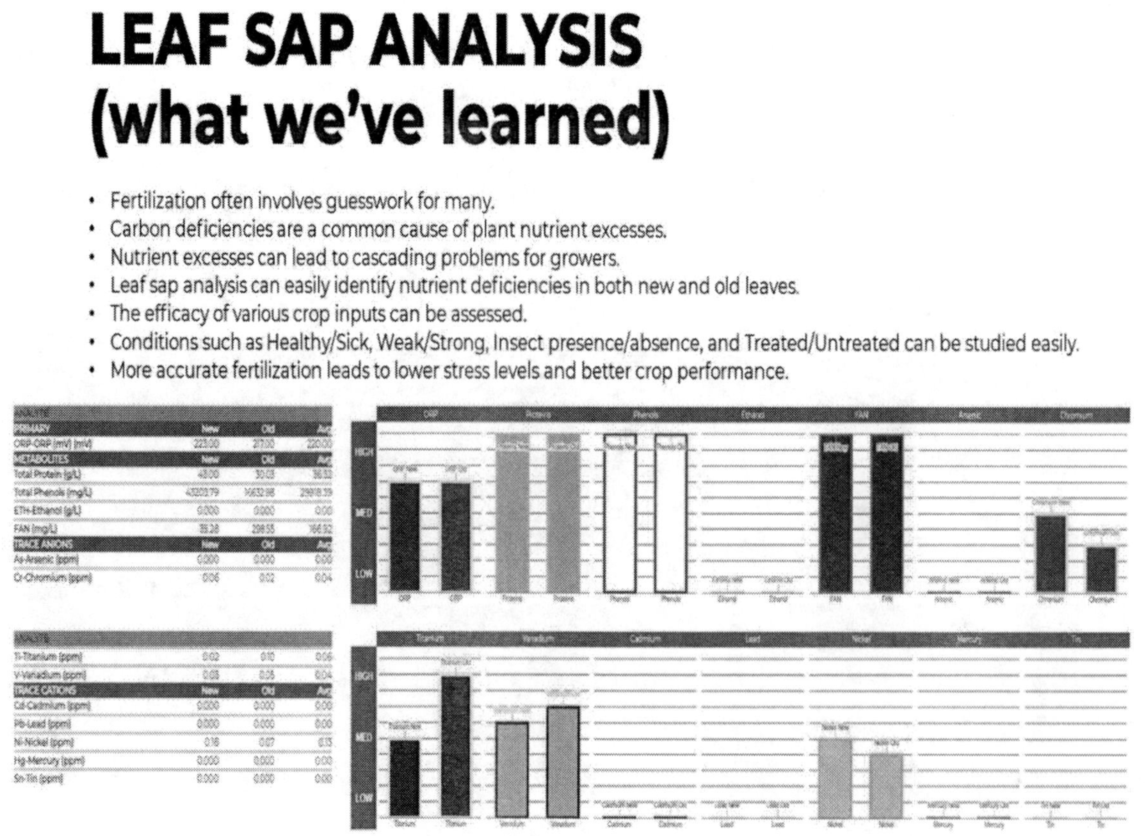

Figure 6. Utilizing leaf sap analysis to manage fertilization and plant efficacy. This flowchart shows sap analysis detecting imbalances (e.g., low P), triggering fertilizer adjustments, and improving resistance to pests, with icons for sap extraction tools and healthy plants.

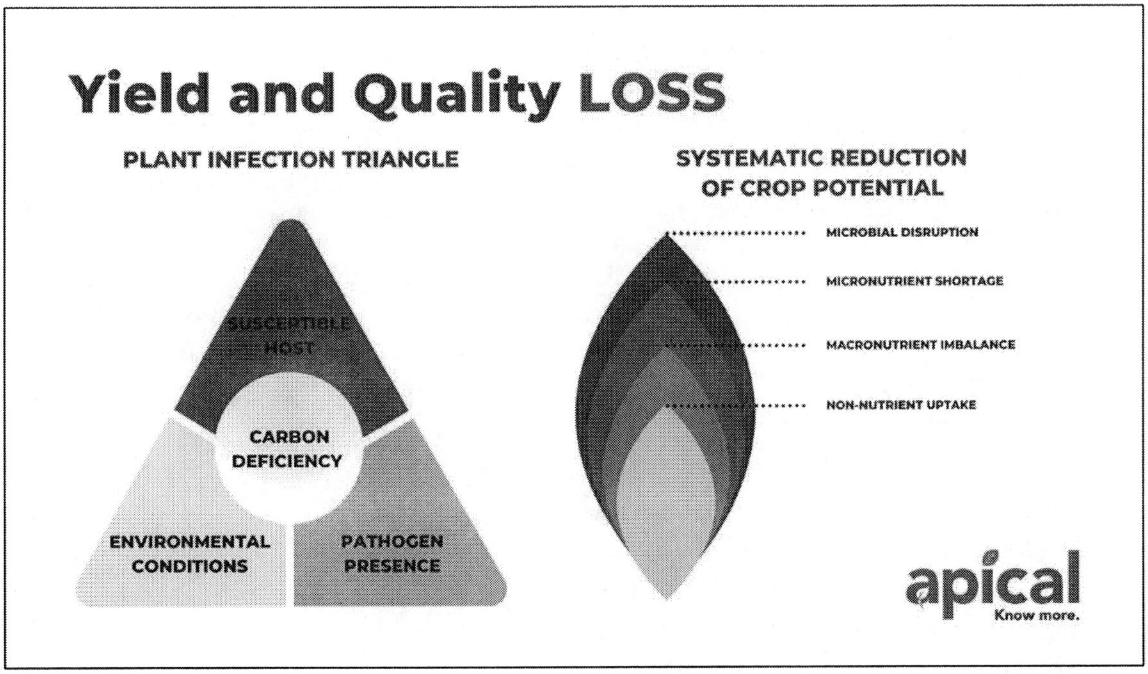

Figure 7. Carbon deficiency (reduced photosynthesis and subsequent carbon production) due to mineral imbalance. This diagram illustrates how mineral shortages (e.g., Mg for chlorophyll) reduce photosynthesis, weakening pest resistance, with arrows linking nutrient uptake to carbon metabolism.

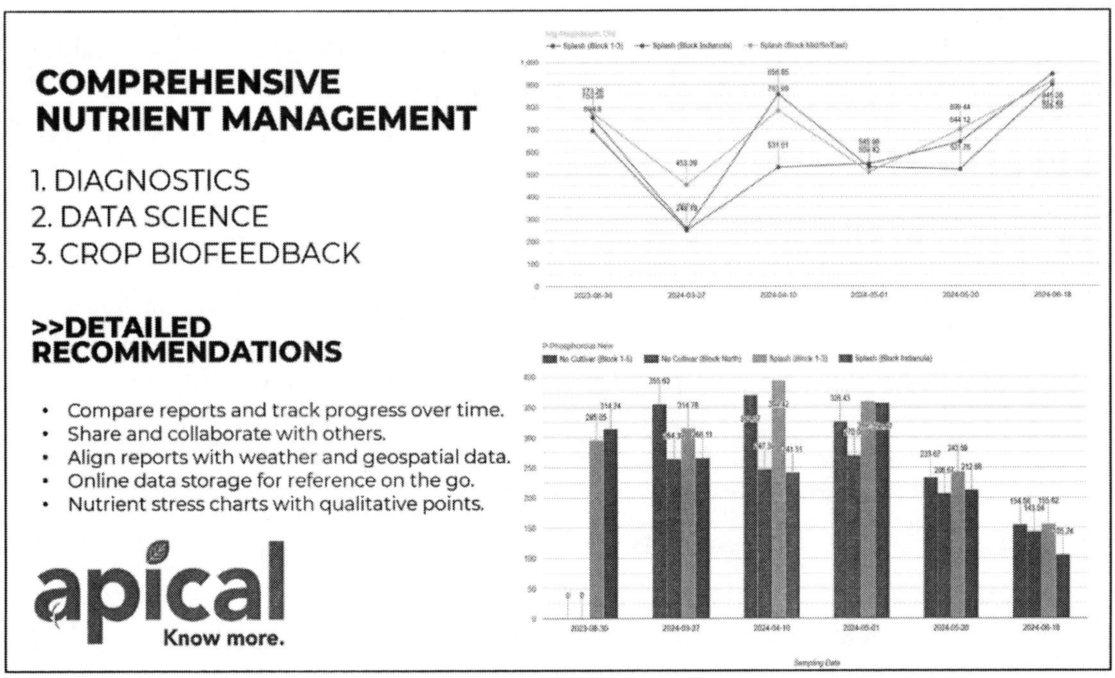

Figure 8. Comprehensive nutrient management includes diagnostics, data science, crop biofeedback, and subsequent crop recommendations. This cycle diagram depicts sap analysis feeding into data models, generating biofeedback, and providing fertilizer recommendations, with feedback loops to refine management practices.

LITERATURE CITED

Esteves, E., Locatelli, G., Alcon Bou, N., and Soranz, R. (2021). Sap analysis: A powerful tool for monitoring plant nutrition. Horticulture 2021, *7*(11): *26* https://doi.org/10.3390/horticulturae7110426

Fan, X., Zhou, X., Chen, H., Tang, M., and Xie. X. (2021). Cross-talk between macro- and micronutrient uptake and signaling in plants. Front. Plant Sci. *12*:663477. doi: 10.3389/fpls.2021.663477

Gitelson, A. A., Gritz, Y., and Merzlyak, M. N. (2003). Relationships between leaf chlorophyll content and spectral reflectance across a wide range of species. Remote Sensing of Environment 87: 337–352. https://doi.org/10.1016/j.rse.2003.06.003

Godfray, H. C. J., Beddington, J. R., Crute, I. R., et al. (2010). Food security: The challenge of feeding 9 billion people. Science *327*: 812–818. https://doi.org/10.1126/science.1185383

Jones, J. B. Jr. (2012). Plant Nutrition and Soil Fertility Manual. CRC Press.

Kamilaris, A., and Prenafeta-Boldú, F. X. (2018). Deep learning in agriculture: A survey. Computers and Electronics in Agriculture *147*:70–90. https://doi.org/10.1016/j.compag.2018.02.016

Marschner, P. (2012). Marschner's Mineral Nutrition of Higher Plants. Academic Press.

Mulder, D. (1953). Les elements mineurs en culture fruitiere. Convegno Nazionale Frutticoltura *118*:98.

Pohl, P., Dzimitrowicz, A., and Lesniewicz, A. (2019). Advances in elemental analysis of plant sap using portable XRF spectrometry. *Talanta*, *195*:–317. https://doi.org/10.1016/j.talanta.2018.11.051

Reuter, D. J., and Robinson, J. B. (1997). Plant Analysis: An Interpretation Manual. CSIRO Publishing.

Singh, A., Ganapathysubramanian, B., Singh, A. K., and Sarkar, S. (2020). Machine learning for high-throughput plant phenotyping. Plant Methods *16*:87. https://doi.org/10.1186/s13007-020-00620-2

Tilman, D., Balzer, C., Hill, J., and Befort, B. L. (2011). Global food demand and the sustainable intensification of agriculture. Proceedings of the National Academy of Sciences *108*: 20260–20264. https://doi.org/10.1073/pnas.1116437108

Treub, M. (1923). On the mineral composition of plant sap and its relation to growth. Annals of Botany *37*: 123–135.

Combined Proceedings IPPS 74:399-404. 2024.

Improving Your Success with Granular Pre-emergence Herbicides

Steve Larson

Envu Environmental Science US, 5000 CentreGreen Way, Suite 400, Cary, North Carolina 27513 USA

steve.larson@envu.com

Keywords: sequential split herbicide applications, weed control program, seasonal weed pressure

Summary

More time and money are spent in nursery weed control than other pest group. The biggest nursery cost is not being proactive – and implementing a well-designed, effective weed control program. All pre-emergence herbicide active ingredients slowly degrade over time – affecting a product's ability to offer optimal weed control. Applying split applications of granular herbicide can improve herbicide performance and significantly extend its longevity. Split applications are applied at lower rates than just one singular application. The practice of sequential split applications has proven to be as safe on treated crops as a single, full-rate treatment. Split application techniques of granular herbicide can increase effectiveness and longevity on multiple weed species: bittercress, large crabgrass, doveweed, eclipta, longstalk phyllanthus, common purslane, common groundsel, oxalis and spotted spurge.

IPPS Vol. 74 – 2024

INTRODUCTION

Nursery weed control is time-consuming, labor intensive, and costly. Nursery growers spend more resources – time and money – on controlling weeds than any other pest group (insects/mites or diseases) - when you factor in chemical costs, labor, hand-weeding, and periodic spot spraying for escaped weeds. When it comes to nursery weed control, I would argue that the biggest cost is the cost of neglect. Growers that fail to create and execute an effective weed control strategy are likely to spend the most money on nursery weed management.

Weed control is hard **(Fig 1)**. It requires a year-round commitment to timely chemical applications, periodic hand weeding, and constant scouting to keep weeds under control. Pre-emergence herbicides are an important tool growers have at their disposal to minimize expensive hand weeding. When deciding on which re-emergence herbicide to use, growers must consider both the 'regionality' and 'seasonality' of the weed activity at their nursery. Selecting the proper preventative herbicide for the predominant weeds in your geographic region is critical. Regional weed species differ significantly – South Alabama and Northern Ohio nurseries have different weed species at different times of the year. It is essential that a grower knows which weeds they will encounter in their geographic region and select effective pre-emergence herbicide options accordingly.

Each season of the year (spring, summer and fall) offers unique and different types of weed pressure. In the southern US, the Spring and Summer seasons typically bring stronger spurge and eclipta weed pressure, while the Fall season brings higher bittercress, chickweed and Oxalis weed activity. Each nursery requires a unique weed management program that considers it's regionality and seasonality of local weed activity.

Figure 1. Weeding is hard, tedious work – and an expensive labor cost.

Growers using granular pre-emergence herbicides in their weed management program are always looking for ways to improve product performance, especially improve herbicide longevity. Once applied, all pre-emergence herbicide active ingredients begin to slowly lose activity over time (**Fig. 2**). Products vary in their performance longevity due to a combination of factors such as: microbial degradation, natural half-life, photo degradation, volatilization, soil adsorption, application rate, and environmental factors (temperature and rainfall). All or some of these factors play a role in a products' ability to offer optimal weed control over time.

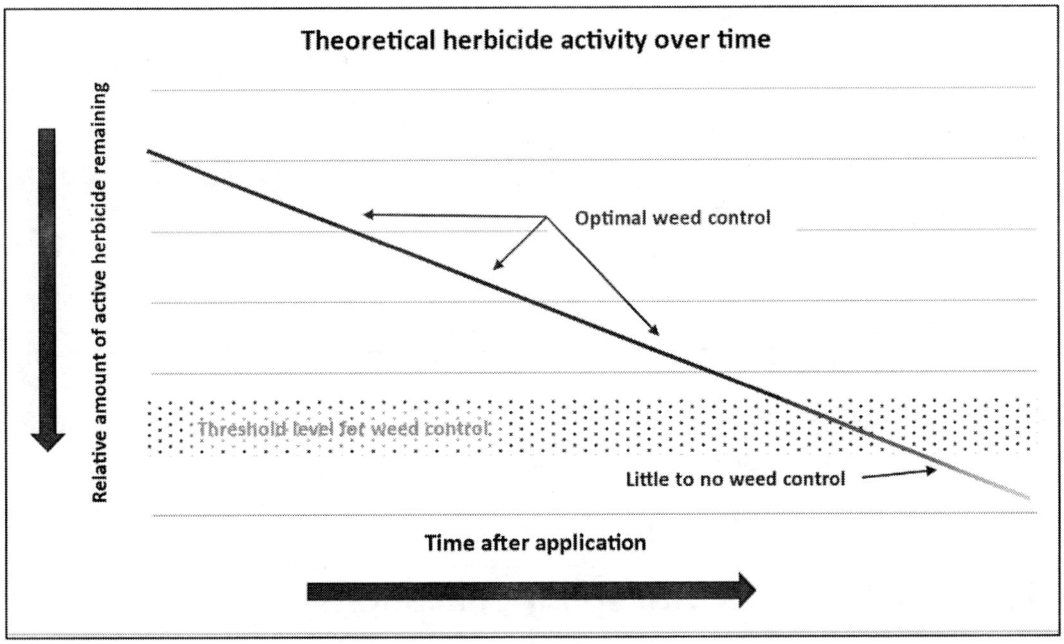

Figure 2. Theoretical herbicide activity over time.

Through recent research, there is a proven technique to improve herbicide performance and significantly extend herbicide longevity: the use of split applications. Turfgrass managers (golf course superintendents and commercial lawn care managers), have used split applications of their pre-emergence herbicides for decades. Rather than apply one single, high rate, they split the applications to two, reduced-rate treatments 4-6 weeks apart.

For a nursery using granular pre-emergence herbicides, a split application might look like this: rather than a single 200 lb. per acre application, consider splitting the application into two treatments of one application of 150 lbs. per acre, followed by a sequential treatment 4-6 weeks later with a second, 150 lb. per ace treatment of the same herbicide (**Figs. 3, 4 and 5**). This technique is legal (within label guidelines) for most pre-emergence herbicides, reduces potential injury to sensitive crops, and the two sequential applications combine to significantly extend herbicide longevity. The practice of sequential split applications has proven to be as safe on treated crops as a single, full rate treatment.

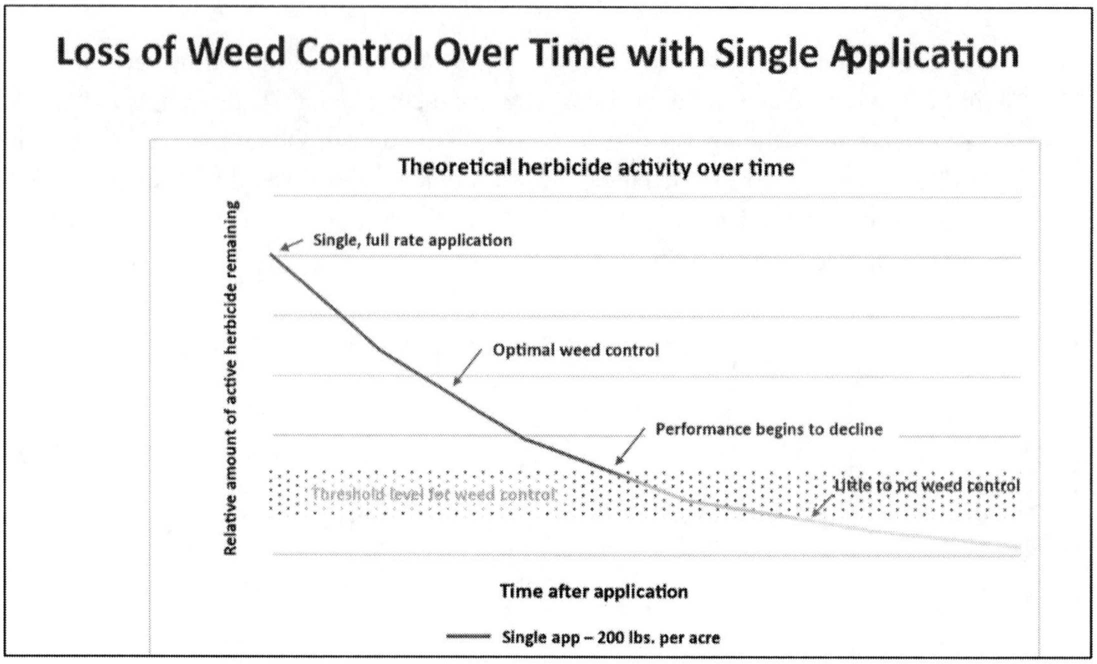

Figure 3. A single application of recommended herbicide treatment at higher concentration (200 lbs/acre) – showing gradual loss of weed control.

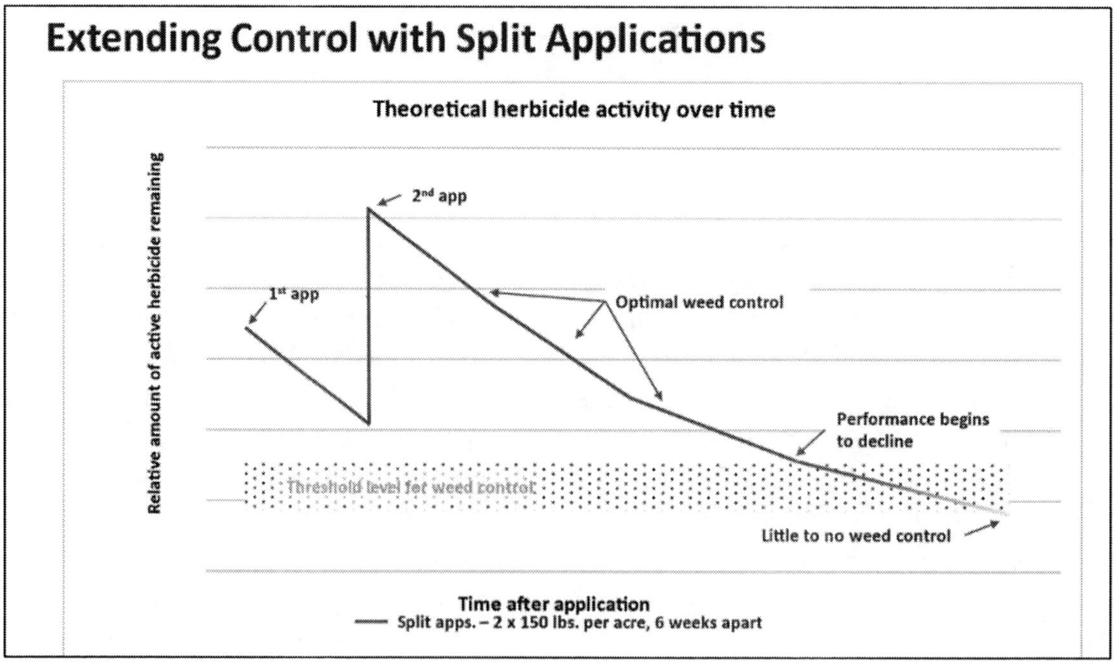

Figure 4. Split applications of two appications of lower concentration [2 x (150 lbs./acre)] applied 6-weeks apart – which extends the period of weed control.

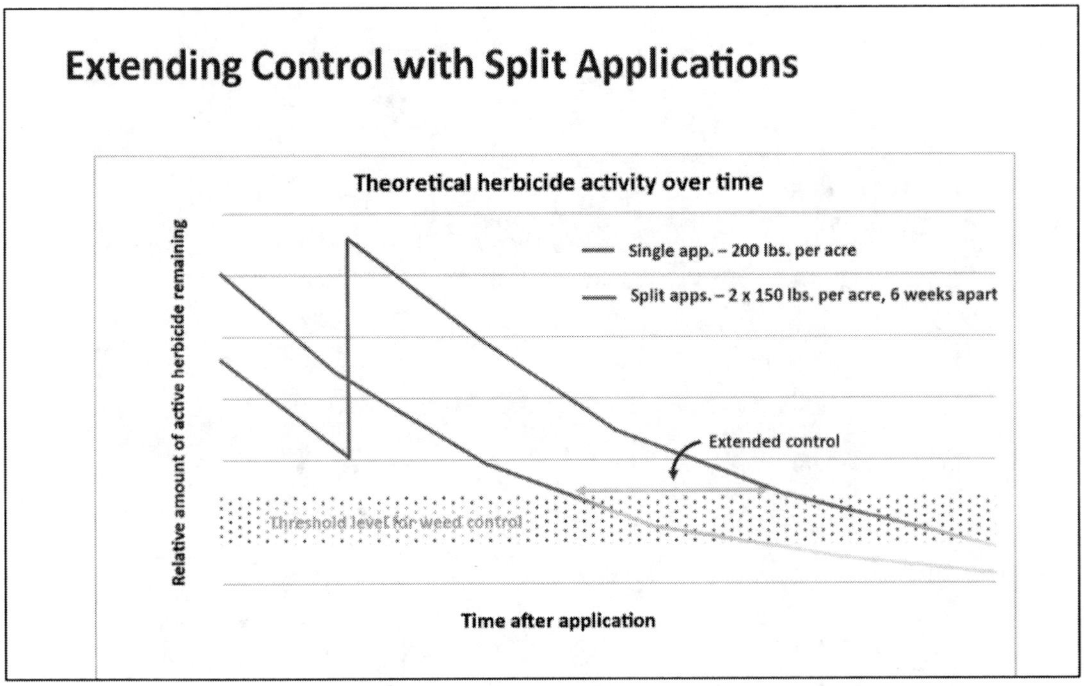

Figure 5. Comparison of a single application (200 lbs./acre) – compared to a split applications at lower concentration [2 x (150 lbs/acre)] applied 6-weeks apart; this leads to an extended period of weed conrol.

To test the theory of split application treatments on container-grown ornamentals in a nursery setting, research was conducted by Dr. Chris Marble (University of Florida) to test the effectiveness of split applications of Marengo G and a competitive granular herbicide. The results showed that the split application technique increased herbicide effectiveness and longevity on multiple tested weed species: bittercress, large crabgrass, doveweed, eclipta, longstalk phyllanthus, common purslane, common groundsel, oxalis and spotted spurge (**Figs. 6 and 7**). Subsequent use of the split-app technique by several prominent container growers confirms Dr. Marble's research findings.

Figure 6. Extending weed contorl with split herbicide applications on eclipta, 16 weeks after treatment (WAT).

Figure 7. Extending weed control with split herbicide applications on Oxalis/woodsorrel, 16 weeks after treatment.

Combined Proceedings IPPS 74:405-412. 2024.

Improving Irrigation Water Quality for Crop Health

Justin Leavitt

Moleaer Inc., 3232 W El Segundo Blvd, Hawthorne, California 90250 USA

Justin@moleaer.com

Keywords: water quality, irrigation, oxygenation, nanobubbles, dissolved oxygen (DO), oxidation-reduction potential (ORP), pathogen management, water testing

Summary

Water quality is a crucial factor in achieving optimal farm and crop health, yet rigorous and frequent testing is often overlooked as a priority. High-quality water can enhance plant health and productivity, while poor water quality can harbor pathogens and create conditions conducive to crop disease, uneven irrigation and loss of crops. This paper explores the importance of monitoring key water quality metrics such as dissolved oxygen (DO), oxidation-reduction potential (ORP), and pathogen levels - discussing the challenges of maintaining these parameters and the importance of prioritizing regular, comprehensive testing. By leveraging advanced oxygenation technology, growers can address water quality challenges with precision. Through targeted water quality improvements and proactive testing, growers can enhance crop health, reduce risks, and optimize resource use. Moleaer's role in providing efficient oxygenation technology is also discussed, emphasizing the benefits of oxygen-rich water in reducing pathogens and promoting healthier crop growth.

IPPS Vol. 74 – 2024

INTRODUCTION

Water is the most essential resource used in farming. Yet the quality and testing of irrigation water is often not prioritized. It is not just a question of water scarcity, but also water's ability to support plant health and growth. Poor-quality water can introduce harmful pathogens, such as *Pythium* and other water-borne pathogens, creating a domino effect of negative impacts on crop yield and farm efficiency. This paper delves into the importance of water quality, understanding the metrics that matter, and methods to use that improve water quality for the benefit of crops. Technological advancements, such as the use of nanobubbles to increase oxygen levels in water, provide additional tools to improve water conditions. By prioritizing water quality, growers can gain insights into what is happening within their irrigation systems and make informed decisions to safeguard their crops.

WHY WATER QUALITY MATTERS

Water quality plays a pivotal role in determining overall crop health. While many growers focus on water availability, the importance of water quality in supporting plant growth and minimizing disease is often overlooked. High-quality water can create an environment that promotes healthy root systems and nutrient absorption while poor-quality water can introduce pathogens, inhibit plant growth, and contribute to system inefficiencies like clogged irrigation emitters. Additionally, oxygen-rich water promotes aerobic conditions that suppress harmful pathogens, ensuring healthier root development and plant growth. Algae and biofilm buildup can reduce irrigation uniformity, leading to uneven water distribution and poor crop performance. Addressing water quality is not just about preventing these issues but also about creating a foundation for long-term success that results in healthier crops and higher yields.

KEY WATER QUALITY METRICS

While algae is often favored as an indicator of water body health, it does not sufficiently capture what's really going on – some healthy water bodies have algae. There are three primary water quality metrics that every farm should regularly test and monitor: Dissolved oxygen (DO), oxidation-reduction potential (ORP), and pathogen levels. Together, these provide comprehensive insights into the health of the water source and its ability to support optimal crop growth.

Dissolved Oxygen (DO): DO is critical for promoting aerobic conditions that suppress harmful anaerobic pathogens such as Pythium. Low DO levels can stunt plant growth and increase susceptibility to diseases. Water with DO levels below 5 parts per million (ppm) is considered unacceptable, while levels between 7-9 ppm are acceptable, and 9-18 ppm is excellent. High DO levels promote healthy root development and reduce the risk of overwatering.

Oxidation-Reduction Potential (ORP): ORP measures a water source's ability to cleanse itself through oxidation, with higher ORP levels indicating a reduced risk of pathogen growth. In irrigation systems, ORP values above 300 millivolts (mV) are ideal, as they indicate a water environment that is hostile to pathogens and supports more efficient nutrient absorption.

Pathogen Testing: Testing for bacterial and fungal pathogens is essential for understanding the health risks present in irrigation water. Total Colony Forming Units (CFU) for bacteria and fungi provide a snapshot of microbial communities in the water, helping growers identify potential threats before they reach their crops. Regular pathogen testing at both the water source and at the end of irrigation systems ensures that water treatment strategies are effective.

IMPROVING WATER QUALITY

Maintaining high water quality requires proactive management and regular testing. Growers can take several steps to improve their irrigation water, starting with addressing biofilm and algae buildup. Filtration systems can help reduce clogging in emitters, while oxygenation treatments can improve overall water quality by shifting microbial communities from anaerobic to aerobic.

Oxygenation, specifically, can be a game-changer for irrigation systems. By increasing DO levels, growers can reduce pathogen loads, promote healthier plant growth, and improve irrigation efficiency. Moleaer's nanobubble technology is a solution for boosting DO levels and transforming water quality across the system. A study conducted by Biosabor in collaboration with Agrocolor FL https://www.agrocolor.de/ examined the impact of nanobubble-enriched irrigation water on greenhouse tomatoes cultivated in native soil. The research involved comparing tomato crops irrigated with nanobubble-enriched water against those using untreated water. The findings demonstrated that the use of nanobubbles not only elevated dissolved oxygen levels but also enhanced the overall quality and yield of crops. Specifically, the nanobubble-enriched irrigation water exhibited a higher and more stable DO concentration averaging 20.8 ppm, compared to 6.3 ppm in the control group **(Figs. 1 and 2)**.

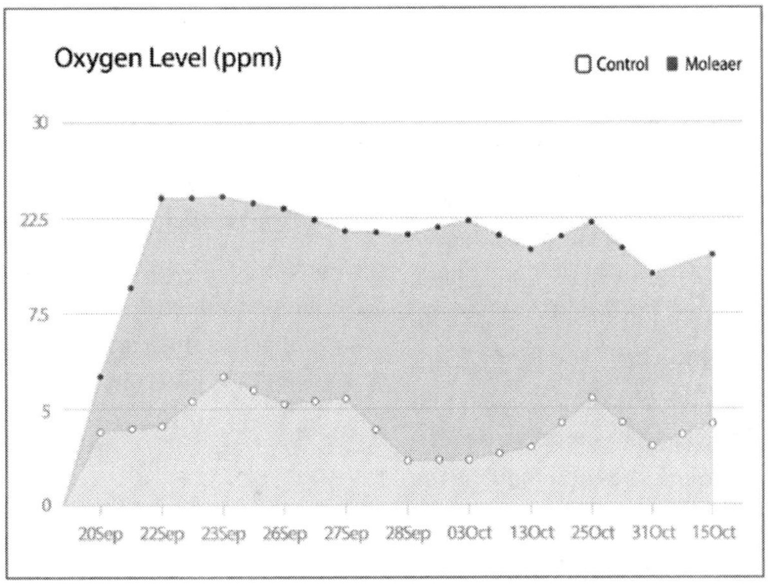

Figure 1: Tomatoes increased yield by 9.7% with nanobubble-Enriched irrigation and saw over 200% increase in dissolved oxygen (Moleaer, 2022a).

Figure 2. Biosabor, a distributor of Moleaer's nanobubble technology in Spain, together with Agrocolor SL, studied the effects of nanobubble-enriched irrigation water on greenhouse tomatoes grown in native soil. Oxygen nanobubbles improved crop yield by creating better root zone conditions for more robust root development. https://www.moleaer.com/resources/case-study-biosabor-tomatoes

THE ROLE OF OXYGEN IN WATER QUALITY

Oxygen plays a critical role in water management, particularly in determining whether a water environment is conducive to pathogen growth. Lower oxygen levels promote anaerobic conditions, leading to an increase in harmful bacteria and fungi. Conversely, increasing oxygen levels creates an aerobic environment that supports beneficial microbial communities and reduces the risk of disease.

Aardbeienkwekerij Penninx, a strawberry nursery in the Netherlands, https://www.chielkesaardbeien.nl/ implemented a trial with Moleaer's nanobubble technology to address organic contamination in their irrigation system. Before the treatment, the nursery faced challenges, particularly during early spring and summer, when stagnant water led to increased organic contamination. After introducing nanobubble-enriched water, measurements showed significantly higher dissolved oxygen (DO) levels in both the irrigation water tank and drip lines, enhancing plant resilience against diseases and improving nutrient uptake. Notably, levels of harmful pathogens like *Pythium* and *Phytophthora* were reduced, and biofilm contamination in the water system diminished, resulting in visibly clearer water and healthier substrate conditions **(Fig. 3)**.

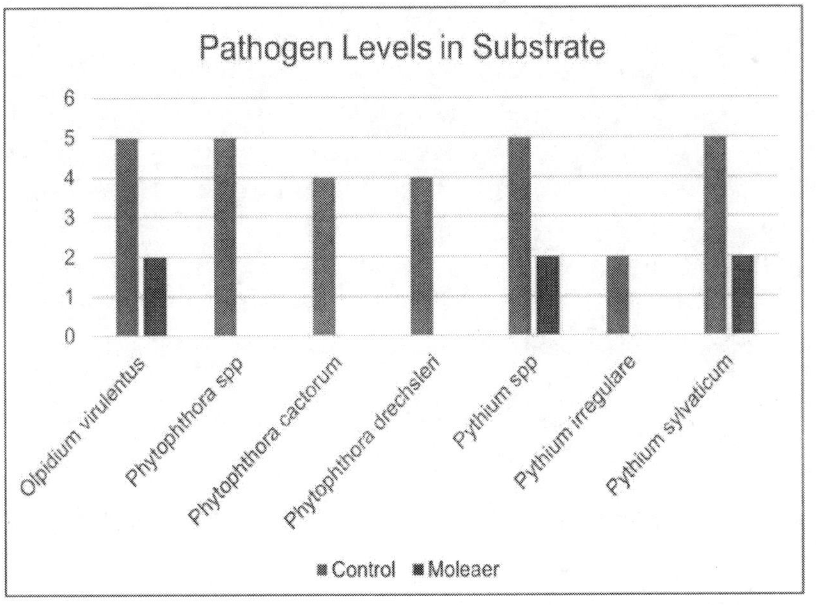

Figure 3: Pathogen levels were rated on a scale of 1-6, 1 being the lowest concentration and 6 the highest. There was a significant reduction in *Pythium* and *Phytophthora* (Moleaer, 2022b).

PATHOGEN REDUCTION AND IRRIGATION WATER

Waterborne pathogens are a significant challenge in agriculture, especially as water temperatures rise and pathogen counts increase in irrigation ponds. By improving water quality through oxygenation and ORP management, growers can reduce the pathogen load in their irrigation systems, making fungicides more effective and protecting crops from diseases.

Regular testing of both pathogen levels and water quality metrics are key to identifying and addressing potential issues before they impact crop health. By testing at both the source and at the end of irrigation lines, growers can ensure that their water is free from harmful pathogens throughout the entire irrigation process.

UNPACKING THE SCIENCE OF NANOBUBBLE TECHNOLOGY

Nanobubbles are incredibly tiny gas bubbles, measuring less than 2,500 times smaller than the size of a grain of salt **(Fig. 4)**. Due to their minuscule size and significantly lower buoyancy compared to ordinary bubbles, they remain suspended in water for extended periods **(Fig. 5)** This unique property allows nanobubbles to alter the physical characteristics of water, creating a reservoir of entrained oxygen. In fact, nanobubbles can increase dissolved oxygen (DO) levels by up to 20% beyond the normal gas saturation point, resulting in more stable oxygen levels over time.

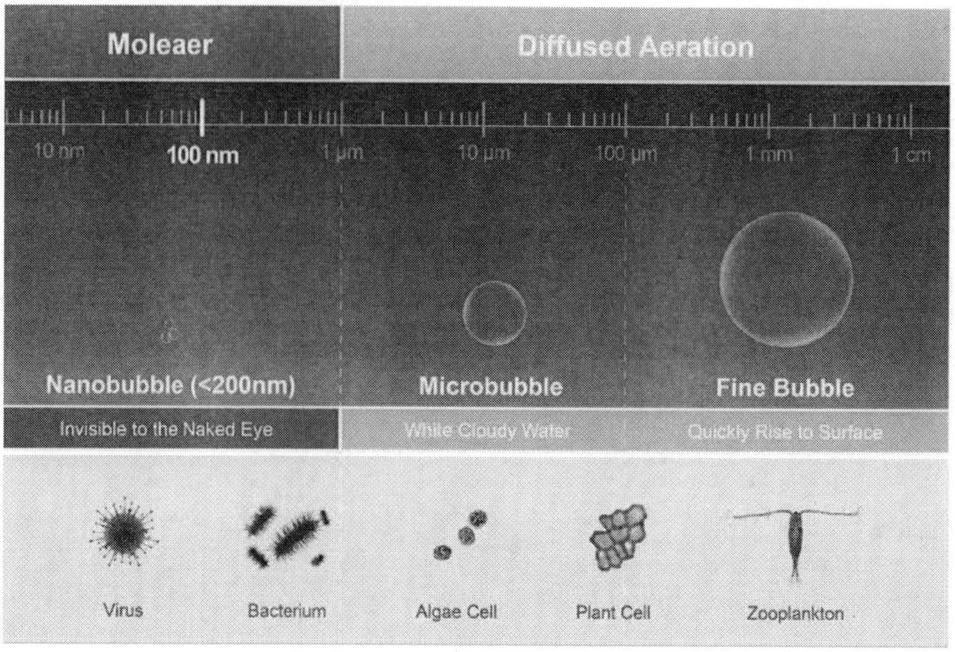

Figure 4: Nanobubbles are not visible to the naked eye and are about the size of a virus or small bacteria. They are 2,500x smaller than a grain of salt.

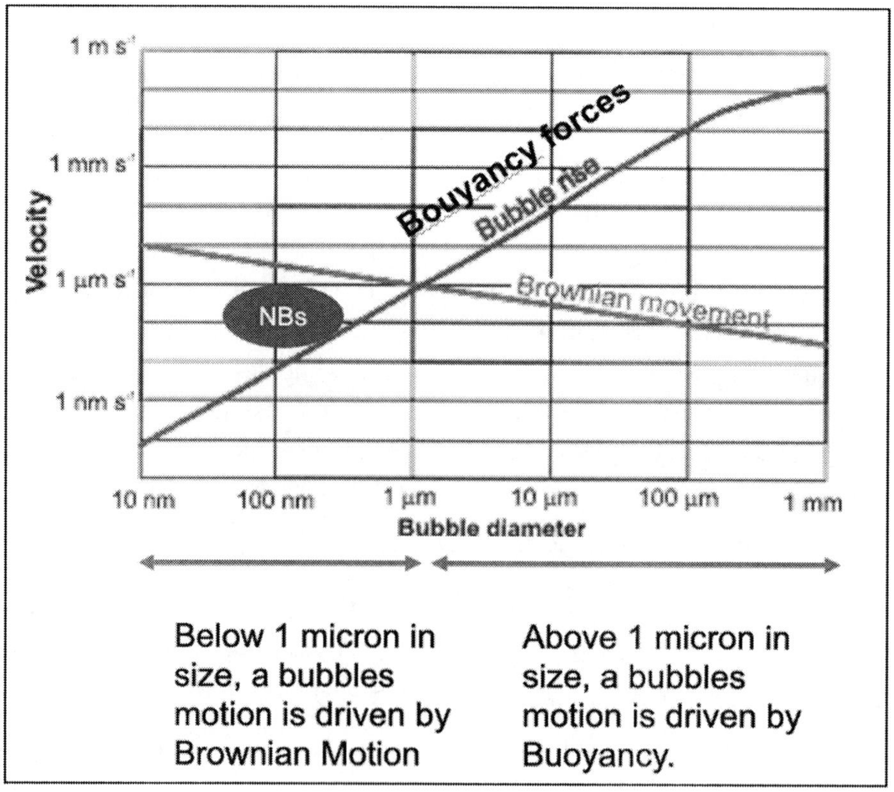

Figure 5. The Forces of Brownian Motion are stronger than buoyancy - so nanobubbles do not rise to the liquid surface.

Additionally, nanobubbles create conditions that reduce the growth of pathogens and algae. That is, nanobubbles significantly increase DO and the ORP of irrigation water, which suppresses pathogen growth while promoting the growth of beneficial microbes such as mycorrhizae. In independent testing, Moleaer technology was found to provide the highest proven oxygen transfer rate in the aeration/gas infusion industry (Dr. Michael Stenstrom, University of California-Los Angeles).

Moleaer's nanobubble technology harnesses this efficiency to deliver superior gas-to-liquid transfer, injecting trillions of nanobubbles into water. This process not only replaces energy-intensive gas transfer methods but also eliminates the need for harmful chemical oxidants and surfactants. The chemical-free and energy-efficient nature of nanobubble technology makes it an ideal solution for improving water quality, reducing biofilm buildup, and preventing clogging in irrigation systems. By incorporating nanobubbles, growers can optimize water conditions to promote healthier crops.

With a hard, stable surface, nanobubbles naturally scour biofilm and scale from irrigation systems to keep surfaces clean and reduce pathogenic growth. The natural oxidation capacity of nanobubbles also degrades biofilm and prevents its buildup. As mentioned, biofilm is pervasive on most surfaces in frequent contact with water and acts as a potential refuge for a wide range of plant root pathogens. For example, scientists at the University of California and China Agricultural University recently found that nanobubbles effectively control biofouling in irrigation pipelines.

They are "detrimental to the mutualistic interactions among microbial species, destabilizing the [molecular ecological] network complexity and size...decreasing extracellular polymers and biofilm biomass."

CONCLUSION

Improving irrigation water quality is essential for optimizing crop health, enhancing productivity, and mitigating risks associated with pathogens. The ability to measure and manage critical water quality metrics, such as dissolved oxygen (DO), oxidation-reduction potential (ORP), and pathogen levels, is key to maintaining a healthy growing environment. By embracing advanced oxygenation technologies, like Moleaer's nanobubble technology, growers can create an oxygen-rich water environment that not only promotes healthier root development but also inhibits harmful pathogens and reduces biofilm buildup in irrigation systems.

Incorporating regular and comprehensive water testing allows producers to identify potential issues early and take corrective action before they impact crop performance. The benefits of improving water quality are twofold: growers can reduce reliance on chemical treatments while simultaneously boosting the overall efficiency of their irrigation systems. Nanobubbles, with their ability to deliver higher levels of dissolved oxygen and improve ORP, represent a significant advancement in water treatment, offering a chemical-free, energy-efficient solution to long-standing agricultural challenges.

Ultimately, the health of the crops is intricately linked to the quality of the water they receive. By prioritizing water quality improvements, growers can increase yields, reduce the risks of disease, and ensure long-term sustainability in their farming operations. Through the adoption of cutting-edge technologies, the agricultural industry can continue to advance toward a future of more resilient and productive farming practices.

LITERATURE CITED

Moleaer. (2022a). Tomatoes: increased yield with nanobubble-enriched irrigation water. https://www.moleaer.com/resources/case-study-biosabor-tomatoes

Moleaer. (2022b). Nanobubbles reduce pathogens and improve root development in strawberries. https://www.moleaer.com/resources/ case-study-penninx-strawberries

Combined Proceedings IPPS 74:413-427. 2024.

Plastic Horticulture Containers: Environmental Impacts and Regulatory Trends

Tom Marting

The HC Companies, 2450 Edison Blvd, Suite 3, Twinsburg, Ohio 44087 USA

tmarting@hc-companies.com

Keywords: plastic pollution regulation, human health, microplastics, environmental impacts, recovery/recycling, post-consumer recycled (PCR) materials, circular economy, extended producer responsibility (EPR), producer responsibility organization (PRO), consumer packaged goods (CPG)

Summary

A growing body of evidence and public concern over the environmental impacts of plastic pollution and the potential human health impacts of microplastics is driving regulatory actions at international, federal, state, and local levels. These regulations are intended to control the composition and application of plastic products, drastically increase recovery and recycling rates, and prevent misleading or false environmental claims about plastics. These current and coming regulations can apply to plastic manufacturers, propagators, growers, retailers, and marketers in various geographical regions in many different ways. It will become increasingly important for all companies in the horticulture value chain to understand the potential environmental impacts of plastics and how their operations must adjust to how plastics will be managed in the future.

IPPS Vol. 74 – 2024

INTRODUCTION

Plastics in Horticulture. Plastics are an essential part of our modern economy. They are lightweight, durable, versatile, strong, abundant, and inexpensive materials often with no viable technical alternatives for many applications. Plastics are a foundational technology for our modern society and a key material for realizing a sustainable future; a necessity for technologies like electric vehicles, solar panels, wind turbines, energy efficient buildings, and so on (Hepburn, 2023). This is no different for the horticulture industry. Plastics are essential materials for horticulture propagation - from containers and trays to films and wraps to tags and labels to irrigation systems, and even as the materials of construction of the greenhouses themselves (Orzolek, 2017).

Yet many of the same properties that make plastic products beneficial during their use contribute to environmental hazards and challenges after their useful life is finished (**Fig. 1**). Plastic's strength, durability, and resistance to the elements means they persist in nature without biodegrading and so accumulate in the environment when not disposed of properly (Greenpeace, 2022). Plastics are mass produced from chemicals derived from petroleum production and so are inexpensive and manufactured in high volumes, about 400 million tons are produced annually (as of 2022) to satisfy growing demand for plastics (Statistica, 2024), (Hofmann, 2023).

Figure 1. There are increasing environmental and regulatory challenges in the disposal of horticultural plastic pots, containers and production materials.

The growing concerns over plastic pollution are driving consumer product companies and retailers to set sustainability goals to reduce or eliminate the use of single use plastics, require the use of increasing amounts of post-consumer recycled (PCR) materials, and for manufacturers to take responsibility for how products sold can be recycled or disposed of after their use (Association, 2020). These same concerns are driving plastic pollution regulations at city, state, federal, and international levels, many of which are now in effect or will be within the next 2 to 7 years (Fowler, 2023). While many of these regulations initially targeted industries like food services or consumer packaging, now plastics used in the horticulture markets are being included and growers are going to have to comply with these new regulations or face financial penalties and/or exclusions from certain retailers or markets.

Plastics are becoming a regulated material. Companies across the horticulture value chain will need to understand their compliance obligations with this new regulatory environment. Understanding how to use and market plastics in a responsible manner will be vital to the success of everyone now and in the next few years.

Decarbonization / renewable energy transition

There is an accelerating global trend to transition the world economies away from fossil fuels and petroleum over the next 10 to 15 years. The Paris Agreement signed in 2015 by 194 nations is a non-binding international agreement to limit the rise in the average global temperature by the year 2100 to less than 2°C by reducing global CO_2 emissions from the burning of fossil fuels by 45% by 2040 based on a 1990 baseline (UN, 2015). The latest UN IPCC update report shows collective global progress is falling short of achieving this goal. At the COP 27 conference delegates called for an international non-proliferation treaty for fossil fuel production and development (IPCC, 2022). Nations are committing to accelerating decarbonization and reduction in fossil fuel consumptions to try to meet these commitments (Krishnan, 2022).

Many countries, states, cities, and automakers are pledging to stop the production of gasoline and diesel-powered vehicles. Some 30 countries including Canada, UK, and the European Union have banned the sale of gasoline powered passenger cars by 2035 (Plumer, 2021). California, Maryland, Massachusetts, New Jersey, New York, Oregon, and Washington have also committed to the ban by 2035 (Grieve, 2023). Automakers Ford, General Motors, Mercedes-Benz, Volvo, Jaguar Land Rover, and BYD have all pledged to stop the sale of gasoline and diesel-powered cars by 2040 (Miller, 2021). These shifts away from fossil fuels are leading to accelerating growth in renewable energy investments (Ellerbeck, 2023), which is resulting in increased pricing and volatility for electrical and natural gas energy (Frangoul, 2022).

Environmental impacts of plastics

Plastic products impact the environment in many ways, but the primary concern is that they do not degrade but are present in the environment for centuries after they were manufactured and used. Many factors can affect the rate of decomposition of materials in the environment, from the thickness of the finished parts to the aerobic, anaerobic, or aquatic conditions it is exposed to, to how much mechanical shearing and stress it experiences. The American Chemistry Council (ACS) published a study detailing how long it takes different polymers to break down in the environment under different conditions (Chamas, 2020). According to that study a polypropylene product with an 800μm thickness would take 780 years to degrade in landfill and 87 years to degrade in the ocean. An HDPE product with a thickness of 500μm will take 190 years to break down in a landfill and 58 years in the ocean.

The majority of plastic materials are not collected and recycled; most plastic waste goes to a landfill, incineration, or ends up loose in the environment making its way into surface waters and the ocean. Based on a 2016 McKinsey report annually 330 million metric tons of polymer products were produced globally and about 12% of that volume is recovered, mechanically recycled, and returned to the production process (Hundertmark, 2018). Some 40% is landfilled, 25% is incinerated, and 19% is unmanaged and released into the environment **(Fig. 2)**.

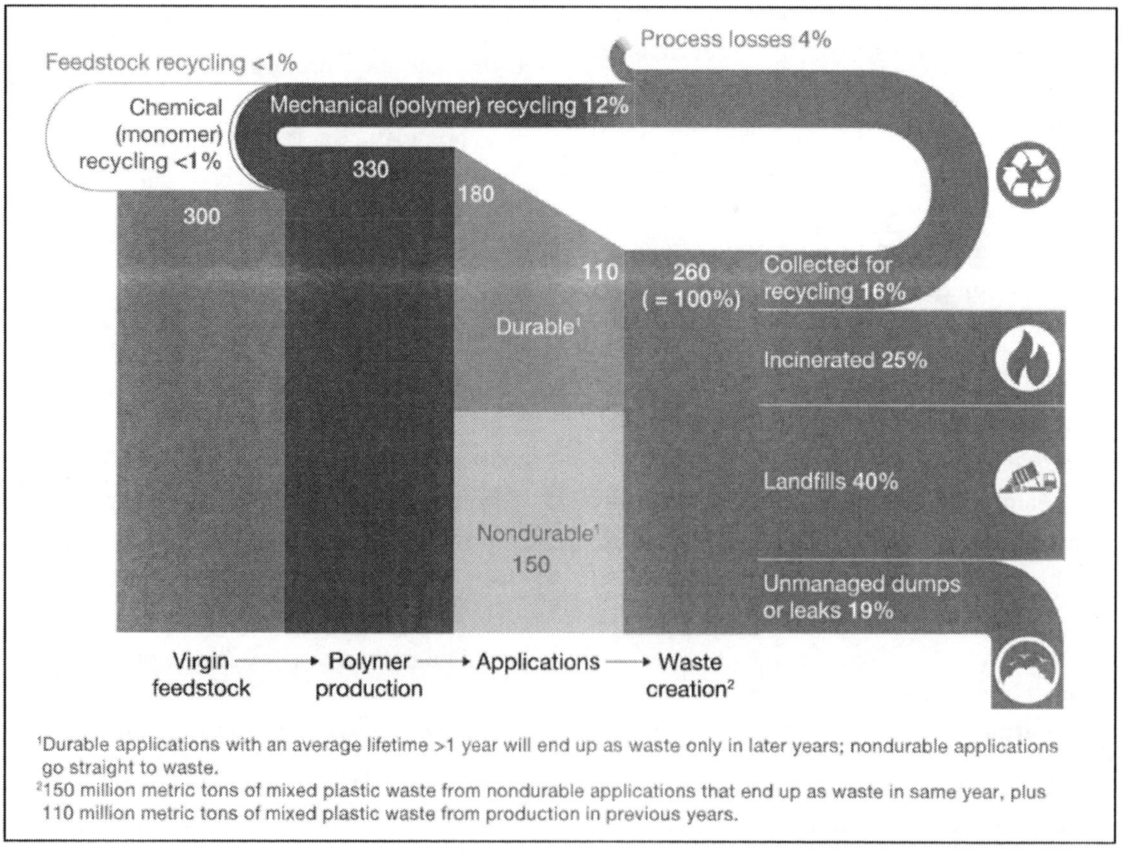

Figure 2: Global polymer flows – millions of metric tons, 2016 (Hundertmark, 2018).

The plastic released into unregulated dumps or released into the environment is accumulating in the world's ocean gyres, which are large areas of little water movement that are the result of the ocean's circulating currents. These collection areas produce what are called "garbage patches," the largest of which is in the Pacific Ocean (NOAA, 2023). A 2015 research paper published in Science found that 8 million metric tons a year are released from land into the oceans, primarily from high-risk geographic areas with unregulated dumps and unenforced environmental standards (Jambeck, 2015). These numbers are increasing annually with both increased plastic production volumes and population growth in these at-risk areas. According to the World Economic Forum's Global Plastic Action Partnership if the rate of release of plastics into the world's oceans continues at the current rate by 2050 the total amount of plastic in the ocean by weight will be greater than the total weight of fish in the ocean (GPAP, 2023).

As a result of these environmental releases microplastic particles are being detected in the environment including in drinking water and food (Koelmans, 2019). Another 2021 study published in Environmental Science and Technology estimated that humans inhale or ingest between 0.1mg and 10mg of microplastics per day, which is the equivalent to eating at minimum a credit card amount of plastic every month (Nor, 2021).

The long-term human health impacts from microplastics are an area of emerging study and concern, with some studies finding microplastics accumulating in human tissues. Recent studies have shown the presence of microplastics in 60% of fatty plaque deposits in human heart arteries (Liu, 2024), the presence of microplastics in arteries correlated to a 400% increase in the risk of stroke or heart attack in those who had microplastics in their bodies (Marfella, 2024), and 100% of human placentas tested showed the presence of microplastics (Braun, 2021).

These emerging concerns over the environmental and human health impacts of plastic pollution has made plastics very unpopular among the public and has produced widespread public support for regulations that hold manufacturers and producers responsible for taking actions to reduce the potential plastic pollution impacts of their products. A worldwide public opinion poll of 20,000 respondents across 28 countries conducted by IPSOS in 2022 found that of those surveyed 85% want manufacturers and retailers to be held responsible for reducing, reusing and recycling plastic packaging, 82% favor product with less plastic packaging, 90% favored an international treaty that would address plastic pollution (IPSOS, 2022). Another poll conducted by IPSOS on behalf of Oceana in 2023 of 1,000 US voters across all 50 states found that 80% of voters are concerned about single-use plastic products and are in favor of requiring companies to reduce plastic packaging (Oceana, 2023). These growing concerns over the environmental and human health impacts of plastic pollution are driving the public aversion to single use plastics and accelerating plastic pollution regulatory actions.

Plastic pollution prevention regulations

There is a lot of activity at the international, federal, state, and local levels to address plastic pollution concerns and transition plastics to a circular economy where recovery and recycling is drastically increased compared to the present. Some of the relevant regulations for horticulture plastics are listed (**Table 1**). These regulations can be grouped into four categories:

Material Requirements These are regulations that control or restrict the material properties or compositions of plastic products and their applications. Examples of these kinds of regulations are bans for the use of plastics in certain markets like shopping bags or food service containers, eliminations of components or chemicals of concern like PFAS or expanded polystyrene, and requirements for minimum recycled material content percentages. Examples of these include New Jersey S-2515 and Maine LD-1504.

Extended Producer Responsibility (EPR) These are regulations that intend to pass costs for the collection and disposal of plastic products up the value chain to producers and manufacturers. This requires companies to disclose the amount of single use or packaging plastics sold into those states to regulators and then pay fees based on those amounts. These fees are structured to encourage transition to more sustainable materials or lower weights of materials. Examples of these include Colorado HB22-1355 and Oregon SB-582

Truth in Advertising These are regulations that restrict and control the environmental benefit claims that producers, marketers, and manufacturers can use for their products. The goal is to eliminate misleading or unsupported claims for terms such as "recyclable" or "compostable" and associated symbols like the use of the "chasing arrows" recycle symbol. Examples of this include California SB-343 and the FTC Green Guides

Environmental Risk or Impact Disclosure Requirements These regulations require larger corporations to assess and disclose environmental impacts and risks such as their climate change related risks to operations and their greenhouse gas emissions. These risks can include purchased goods or value chain risks in addition to the risks of the operations of the company itself. Examples of these regulations include California SB-253 and SEC's ESG final disclosure rules.

One of these growing trends is for states to regulate single use plastic products and phase out their use in specific applications or markets to address impacts from plastics accumulating in the environment. The United Nations passed a resolution in 2022 committing to create a binding treaty to eliminate plastic pollution and begin the process of ratifying it by 2024 (Nations, 2022). Canada has banned the use of plastic in specific food service items and identified "carbon black" as a plastic colorant specifically as unacceptable due to recyclability challenges (Canada, 2023). Connecticut, California, Delaware, Hawaii, Maine, New York, Oregon, and Vermont and many cities including Miami, FL and Charleston, SC have all banned different kinds of single use plastic products like drinking straws or plastic shopping bags (Richard, 2023).

Canada and states like New Jersey are enacting laws requiring minimum percentages of *post-consumer recycled material (PCR)* in plastic products (Quinn, 2022). These laws include ramp up schedules with requirements for manufacturers to demonstrate to state regulators that their plastic products contain increasing proportions of PCR in future years. The intent of these laws is to generate demand and revenue for curbside recycling facilities, thereby increasing recovery rates and reducing the amount of plastic landfilled or entering the environmental uncontrolled.

Table 1: Pending plastics legislation relevant to the horticulture industry.

Regulation	Effective Date	Implication to Producers
New Jersey S-2515	Jan 2024	Require 10% post-consumer recycled content in 2024 ramping up to 50% by 2035. Registration completed by Jan 1, 2024.
California SB-343	Fall 2025	Removal of chasing arrows, restrictions on use of "recyclable" claim.
Colorado HB22-1355	Jan 2025	Pay fees for plastics sold into state – reporting requirements
Oregon SB-582	July 2025	Pay fees for plastics sold into state – reporting requirements
Canada Plastics Registry	2025	Register and report quantity of plastics sold into the country. Removal chasing arrows, qualifications for recyclability claims. Prohibit claim of degradable or compostable without 3rd party cert.
California SB-54	2026 thru 2032	Rules to be finalized Jan 1, 2025, pay fees for plastics sold into state – reporting requirements, required recovery rates for plastic materials
Canada Plastics Registry	2026	Requires 20% PCR in rigid containers by 2026, ramping up to 60% PCR.
Maine 2146	2027	Pay fees for plastics sold into state – reporting requirements
Maine LD-1504	Jan 2030	Ban sale of intentionally added PFAS of any kind from all products sold in the state, except "unavoidable use" by Jan 2030.
UN Plastic Pollution Treaty	2030+	Negotiations are continuing. Ratification and timelines not yet available

Another plastic pollution trend is *Extended Producer Responsibility (EPR)* legislation that requires producers to report to state agencies on the quantity of plastics sold in the states and pay fees to compensate the states for managing the plastic waste. The framework for these laws includes the selection of a non-governmental organization called a *Producer Responsibility Organization (PRO)* to represent industry to the state regulators and work to set reduction and recovery targets, collect fees from manufacturers, and report on progress against industry goals to the states (Packaging, 2024). Maine, Colorado, Oregon, and California have already passed EPR laws and are in the process of enacting rules while 15 other states have EPR laws submitted to state legislatures for consideration (Felton, 2022).

The Federal Trade Commission, Canada, and several states like California are taking regulatory action against false or misleading product environmental claims called greenwashing (Brooks, 2019). These actions require producers to provide documented support for environmental claims like a claim that a plastic product is "recyclable" and avoid the use of misleading symbols or language. One target of this is the "chasing arrows" symbol around the Resin Identification Code (RIC), often embossed by manufacturers on plastic products (California, 2023). The intention of the RIC symbol is to indicate to consumers and recyclers what type of plastic the product is made of. The problem is that most consumers identify the "chasing arrows" symbol as meaning the product is recyclable regardless of the resin type or how commonly the products are accepted for recycling, and so has become misleading (Davis, 2024). Another target is the claim that products are "recyclable" without evidence that the products are actually being recovered and recycled (Keller Heckman LLC, 2023). So in decades past the term "recyclable" might have meant that the plastic material was technically capable of being recycled regardless of the economics of recycling, and now these regulations are re-defining "recyclable" as meaning that the products are actually being recycled in significant percentages and with large proportions of the population having access to recycling resources for those products (Mallen, 2021).

The Retailer Industry Leaders Association (RILA) has a Retail Compliance Center that is monitoring and communicating plastics regulations that are material to retailers. Many of the coming regulations are listed in their December 2022 Fact Sheet (RILA, Dec 2022). The Association of Plastic Recyclers (APR) are also tracking developments in these regulations and regularly communicating these on their website in blogs and guidance documents (APR, 2023).

The result of these coming changes is that plastics are essentially becoming regulated materials like toxic chemicals or hazardous wastes (Polman, 2023). This will create a need for regulatory compliance resources and infrastructure for plastic producers and their suppliers. This could also mean increasing competition for recycled materials that follows the mandated PCR content requirements, which would result in price increases and volatility in demand increasing before supply can catch up.

Growers are already expecting the prices of plastic container to increase significantly and impact their businesses. According to 2023 the industry survey done by Greenhouse Growers 93% of growers expect an increase in plastic container costs with 36% of respondents anticipating an increase greater than 11% (Greenhouse Grower, 2023). The 2023 price index report done by AmericanHort found costs for inputs increased 59.8% between 2007 and 2022 and project a 3.5% increase in overall costs for growers, stating that "the leading input cost increases are for containers and other plastics, freight and trucking, propagative materials, fertilizers, fuel and energy, and, of course, labor" (Hall, 2023).

Retailers have been making significant commitments in the past 3 to 5 years with respect to single use plastics. Many retailers have set goals to remove undesirable plastics like expanded polystyrene (EPS) or polyvinyl chloride (PVC) from their packaging, reducing weight and amounts of plastics, and increasing PCR content for plastic packaging. The Consumer Brands Association (CBA) in 2020 compiled the sustainable packaging commitments of the top 25 consumer brands (Association, 2020). The goals of some of the more relevant consumer brands are summarized **(Table 2).**

The Home Depot published their Responsible Product Standards in late 2021 (Depot, Oct 2021). In this guidance document they make say that plastic packaging should be made with rPET, PET, HDPE, or PP and should avoid EPS or PVC. They state that they will require a minimum of 5% PCR content in plastics and a minimum of 30% PCR content in corrugated / paper products. They also state in this guideline that suppliers should utilize bio-based plastics and materials where packaging efficacy will not be compromised. The Home Depot highlights several 3rd party ecolabel certifications that they recognize including ones for recycled content and biobased content.

Lowe's stated in their 2021 Corporate Responsible Report (Lowe's, 2021) that they are forming internal task forces to look at removing EPS and PVC from their packaging, that will include sustainability claims and features in their buying guidance documents, and "are exploring ways to communicate [their] sustainability values to [their] customers and showcasing how customers can enhance their home's sustainability profile, including ways to reduce their carbon footprint." They also state "[They] are also increasing [their] communication with suppliers at the onset of product development to increase recycled content going into products and packaging."

Table 2: Consumer Brand Association - consumer packaged goods (CPG) sustainability commitments.

CPG Companies	Packaging Sustainability Commitments
Newell Brands	Use at least 20% non-virgin (recycled content) in plastic packaging for Newell manufactured goods by 2025.
Procter & Gamble	P&G committed to achieve 100% recyclable or reusable packaging by 2030. They will also reduce global use of virgin petroleum plastic in packaging by 50% by 2030.
Nestlé	Nestlé has committed to 100% recyclable or reusable packaging by 2025 and will reduce the use of virgin plastics by one third by 2025.
Unilever	Unilever committed to making all its plastic packaging fully reusable, recyclable or compostable and increasing recycled plastic material content to 25% by 2025.
Henkel AG	By 2025, 100% of Henkel's packaging will be recyclable or reusable and the use of virgin plastics will be reduced by 50%.
Keurig Dr Pepper	By 2025, 100% of Keurig Dr Pepper's packaging will be recyclable or compostable and use 30% post-consumer recycled content.
RB	By 2025, all RB packaging will be 100% recyclable or reusable and include at least 25% recycled plastic content.
PepsiCo, Inc.	By 2025, PepsiCo strives to design 100% of packaging to be recyclable, compostable or biodegradable and reduce virgin plastic use across its beverage portfolio by 35%.
SC Johnson	Make 100% of plastic packaging recyclable or reusable by 2025, increase the percentage of PCR plastic used in our North American and European bottles from 20% today to 40%.

Horticulture challenges with plastics

The future trends for growers and producers using plastic products is for those companies to be required to be increasingly held accountable for the environmental impacts of those products (Giordano, 2022). Producers will need to take active roles in ensuring that the plastic product they purchase and market are compliant to various regulations, include recycled content including required amounts of PCR, that the products are easily recycled in most geographic areas, and that their environmental related marketing claims are credible and transparent.

Many challenges to the recyclability of plastic products exist for horticultural products include:

- The use of mixed materials or laminates such as the integration of reinforcing fibers that do not separate during mechanical recycling.
- The inclusion of labels with adhesives that do not easily detach and separate during recycling.
- Vacuum or thin film products that cannot easily be mechanically chipped or flaked (Greenpeace, 2022).
- Inclusion of chemicals of concern like PFAS in inks or coatings (Whitehead, 2023).
- Colorants like carbon black that are commonly used in horticulture products make the plastic invisible to optical sorting equipment used by recyclers to sort materials. (Cheippo, 2020).
- Ensuring symbols and marketing messages regarding recyclability and environmental impacts are compliant to regulations, credible, and supported with data (Brooks, 2019), (Americover, n.d.)

As the horticulture value chain moves to keep up with the need to be more sustainable, more transparent, and more coordinated as an industry, it will be increasingly important for everyone to take an active role in managing what they purchase, produce or manufacture, and sell is complying with these ever changing and sometimes confusing regulations and customer / stakeholder expectations (Hudson, 2023).

CONCLUSION

In the coming years, the horticulture industries in the United States and Canada, which utilize plastics, will encounter numerous challenges and opportunities. Current trends indicate over the long term that plastics will encounter supply disruptions, price volatility, consumer desire for more sustainable products, and increased regulatory pressure. Plastic pollution has become a major concern for government regulators and consumers in the last few years and continues to grow in significance. Many new laws and regulations are being enacted that will both increase the costs of plastic products and require recyclability and recycled content. Consumers, particularly Millennials and Gen Z, are increasingly concerned over the environmental impacts of plastics and are viewing sustainability as a "have to have" feature or aspect of purchased products.

To be ready for the coming years all companies in the horticulture value chain will need to develop internal capabilities, systems, and business practices to manage the challenges and risk associated with plastic products. They will need to be able to quantify and report specific environmental data to customers, stakeholders, and regulators, and lastly, they will need to take steps to maximize the recyclability and minimize the environmental impacts of their products to meet consumer and evolving customer requirements. In the end these necessary changes will help to reduce the environmental impacts plastic products can have and help move us to a more sustainable future.

LITERATURE CITED

Americover. (n.d.). Goodbye recycling code arrows and hello triangles. Americover: https://www.americover.com/blog/goodbye-recycling-code-arrows-and-hello-triangles/

APR. (2023, Mar 16). Association of Plastic Recyclers. plasticsrecycling.org: https://plasticsrecycling.org/news-and-media/recycled-plastic-content-requirements-are-here-and-more-are-coming-soon-here-s-what-you-need-to-know

Association, C. B. (2020). Top 25 CPG companies packaging sustainability commitments. CBA. https://consumerbrandsassociation.org/wp-content/uploads/2020/06/Top-25_Sustainability-Commitments.pdf

Braun, T. (2021). Detection of microplastic in human placenta and meconium in a clinical setting. Pharmaceutics *13*(7):921. doi.org/10.3390/pharmaceutics13070921

Brooks, M. (2019, October). Corporate compliance, professional perspective - navigating FTC guidance and green marketing litigation. https://www.bloomberglaw.com/external/document/XBN3G6PK000000/corporate-compliance-professional-perspective-navigating-ftc-gui

California. (2023). Accurate recycling labels. https://calrecycle.ca.gov/wcs/recyclinglabels/

Canada. (2023, April 18). Single-use plastics prohibition regulations - overview. https://www.canada.ca/en/environment-climate-change/services/managing-reducing-waste/reduce-plastic-waste/single-use-plastic-overview.html

Chamas, A. (2020, February 3). Degradation rates of plastics in the environment. ACS Sustainable Chemical Engineering, 3494-3511. https://pubs.acs.org/doi/10.1021/acssuschemeng.9b06635

Cheippo, M. (2020). Plastic Pots and the Green Industry. Association of Professional Landscape Designers. https://ecoplantplans.com/wp-content/uploads/2021/10/APLD-Plastic-Pots-White-Paper.pdf

Davis, A. (2024). The fraud of plastic recycling. Center for Climate Justice. https://climateintegrity.org/plastics-fraud

Depot, H. (Oct 2021). Responsible Product Standards. Home Depot. https://ecoactions.homedepot.com/wp-content/uploads/Responsible-Product-Standard-October-2021.pdf

Ellerbeck, S. (2023, March 16). IEA: More than a third of the world's electricity will come from renewables in 2025. https://www.weforum.org/agenda/2023/03/electricity-generation-renewables-power-iea/

Felton, D. (2022, Sept 21). Four states enact extended producer responsibility laws for packaging. https://www.packworld.com/news/sustainability/article/22419036/four-states-enact-packaging-epr-laws

Fowler, S. (2023, Jan 17). Efforts to regulate plastic pollution likely to increase in 2023. Pillsbury. https://www.pillsburylaw.com/en/news-and-insights/regulate-plastic-pollution-increase-2023.html

Frangoul, A. (2022, Nov 9). Energy markets are facing 'one or two years of extreme volatility,' Enel CEO says. https://www.cnbc.com/2022/11/29/energy-markets-facing-one-or-two-years-of-extreme-volatility-ceo.html

Giordano, G. (2022, Dec 12). The plastics police had a banner year. https://www.plasticstoday.com/packaging/the-plastics-police-had-a-banner-year

GPAP. (2023). More plastic than fish. https://www.plasticsoupfoundation.org/en/plastic-problem/plastic-soup/more-plastic-than-fish/

Greenpeace. (2022). Circular claims fall flat again. https://www.greenpeace.org/usa/reports/circular-claims-fall-flat-again/

Grieve, P. (2023, Mar 16). 7 states now pan to ban gas-powered car sales. *Money*. https://money.com/states-banning-gas-powered-cars/

Hepburn, P. C. (2023, Sept 18). The plastics dilemma: towards a sustainable future. https://www.smithschool.ox.ac.uk/news/plastics-dilemma-towards-sustainable-future

Hofmann, T. (2023, Sept). Plastics can be used more sustainably in agriculture. Communications Earth and Environment, 4. www.nature.com/articles/s43247-023-00982-4

Hudson, G. (2023, July/August). Kicking the plastic pot habit. The American Gardner, pp. 30-33. https://ahsgardening.org/about-us/news-press/category/the-american-gardener/

Hundertmark, T. (2018, December). How plastics-waste recycling could transform the chemical industry. https://www.mckinsey.com/industries/chemicals/our-insights/how-plastics-waste-recycling-could-transform-the-chemical-industry#/

Insight, F. (2021). Consumer expectations for sustainable retail. Philidelphia: University Pennsylvania - Wharton School of Business.

IPCC. (2022). UN IPCC AR6 Report Policymaker Guidance. https://www.ipcc.ch/sr15/chapter/spm/

IPSOS. (2022). Attitudes towards single use plastics. IPSOS. https://www.ipsos.com/sites/default/files/ct/news/documents/2022-02/Attitudes-towards-single-use-plastics-Feb-2022.pdf

Jambeck, J. (2015). Plastic waste inputs from land into the ocean. Science, 768-771. https://www.science.org/doi/10.1126/science.1260352

Keller Heckman LLC. (2023, January 31). California begins formal rulemaking on changes to its recycling and disposal system. https://www.natlawreview.com/article/california-begins-formal-rulemaking-changes-to-its-recycling-and-disposal-system

Koelmans, A. (2019). Microplastics in freshwaters and drinking water: Critical review and assessment of data quality. Water Resources, 410-422. https://www.ncbi.nlm.nih.gov/pmc/articles/PMC6449537/

Krishnan, M. (2022, Jan 25). The net-zero challenge: Accelerating decarbonization worldwide. McKinsey Sustainability.

Liu, S. (2024, May 5). Microplastics in three types of human arteries detected by pyrolysis-gas chromatography/ mass spectrometry (Py-GC/MS). Journal of Hazardous Materials. *469*. https://www.sciencedirect.com/science/article/abs/pii/S0304389424004345

Lowe's. (2021). Corporate Responsibility Report. Lowe's. https://corporate.lowes.com/sites/lowes-corp/files/CSR-reports/Lowes_2021_CSR_06.03.22.pdf

Mallen, D. (2021, Oct). New california law limits sse of recycling 'chasing arrows' symbol. https://www.loeb.com/en/insights/publications/2021/10/new-california-law-limits-use-of-recycling-chasing-arrows-symbol

Marfella, R. M. (2024, Mar 6). Microplastics and nanoplastics in atheromas and cardiovascular events. New England Journal Medicine, 390 (10). https://www.nejm.org/doi/full/10.1056/NEJMoa2309822?logout=true

Miller, C. (2021, November 10). Six major automakers agree to end gas car sales globally by 2040. https://www.caranddriver.com/news/a38213848/automakers-pledge-end-gas-sales-2040/

Nations, U. (2022, Mar 2). Historic day in the campaign to beat plastic pollution: Nations commit to develop a legally binding agreement. https://www.unep.org/news-and-stories/press-release/historic-day-campaign-beat-plastic-pollution-nations-commit-develop

NOAA. (2023, 1 23). What is the great pacific garbage patch? https://oceanservice.noaa.gov/facts/garbagepatch.html

Nor, N. H. (2021). Lifetime accumulation of microplastic in children and adults. Environmental Science and Technology, 5084-5096. Retrieved from https://pubs.acs.org/doi/pdf/10.1021/acs.est.0c07384?src=getftr

Oceana, I. (2023). *Ocrana National Plastic Pollution Survey.* IPSOS. Retrieved from https://www.ipsos.com/en-us/three-four-americans-support-national-policies-reduce-single-use-plastic

Orzolek, M. (2017). A Guide to the Manufacture, Performance, and Potential of Plastics in Agriculture . In M. Orzolek, A Guide to the Manufacture, Performance, and Potential of Plastics in Agriculture. ISBN: 9780081021705.

Packaging, S. (2024, Jan 10). Circular Action Alliance Appointed as California's Producer Responsibility Organization. Retrieved from Packaging World: https://www.packworld.com/sustainable-packaging/article/22883569/circular-action-alliance-appointed-as-californias-producer-responsibility-organization

Plumer, B. (2021, November 9). 6 Automakers and 30 countries say they'll phase out gasoline car sales. New York Times. https://www.nytimes.com/2021/11/09/climate/cars-zero-emissions-cop26.html

Polman, P. (2023, May 24). Plastics are in our air, food, and water. A reckoning is coming–and smart businesses can see it. https://fortune.com/2023/05/24/plastics-air-food-water-reckoning-business-environment-paul-polman/

Quinn, M. (2022, Jan 11). New Jersey governor signs ambitious recycled content bill into law. https://www.wastedive.com/news/new-jersey-recycled-content-plastic-murphy/616975/

Richard, J. (2023, March 21). The U.S. progress with single-use plastic bans. Seaside Sustainability: https://www.seasidesustainability.org/post/the-u-s-progress-with-single-use-plastic-bans

RILA. (Dec 2022). Retail Compliance Center Mandatory Recycled Content Regulations. RILA. https://www.rila.org/getmedia/37c42c3f-44f6-4a67-ab21-7e2a13e6ad2e/Mandatory-Recycled-Content-Laws-for-Packaging-1-23.pdf?ext=.pdf

Statistica. (2024, Jan). Annual production of plastics worldwide from 1950 to 2022. https://www.statista.com/statistics/282732/global-production-of-plastics-since-1950/

UN. (2015). Paris Agreement. https://www.un.org/en/climatechange/paris-agreement

Whitehead, H. (2023, March 7). Directly fluorinated containers as a source of perfluoroalkyl carboxylic acids. Environ. Sci. Tech. Letters. https://pubs.acs.org/doi/10.1021/acs.estlett.3c00083

Combined Proceedings IPPS 74:428-433. 2024.

Air-Root Pruning: A Great Step Forward in Propagation and Liner Production

Chris Murphey

Rediroot, PO Box 191, Boring, Oregon 97009 USA

chris@rediroot.com

Keywords: root architecture, transplant shock, fibrous & coarse roots, auxin/cytokinin balance, root apex

Summary

Air-root pruning systems have considerable propagation and production advantages in producing superior plants of greater commercial value. In air-root pruning the liner pots and propagation flats redirect roots to leave the growing media and desiccate which stimulates greater branching of new roots. Advantages of air-root pruning include improved root architecture with denser, more fibrous roots, reduced transplant shock during upcanning/transplanting, and greater stress resistance in the finished crop with a more extensive root system. While air-pruning was first developed to improve growth and transplanting of coarse-rooted taxa – it can also be beneficial for greenhouse-grown vegetables, herbs, and flowering plants. It all begins in propagation with air-pruning of primary roots of the liner crop.

INTRODUCTION

Air-root pruning can significantly enhance root system development of woody plant species (trees, shrubs) during propagation and liner production. When propagated and grown in standard plastic flats and liner pots - coarse-rooted taxa propagated from seed - can form crooks and root circling of primary roots (Gilman, 1990; Miller and Bassuk, 2018). This can lead to root girdling and a poor-quality crop (**Fig. 1**).

Specialized containers/pots that air root prune can redirect root architecture – leading to a more fibrous, easily transplanted, desirable plant. It all begins in propagation with air-pruning of primary roots of the liner crop. Air root pruning has also been beneficial for greenhouse-grown vegetables, herbs, and flowering plants.

Figure 1. A problem with coarse rooted woody taxa can be root girdling (arrows) that leads to an undesirable, poor-performing plant in the landscape.

WHAT IS AIR-ROOT PRUNING?

In air-root pruning the liner pots and propagation flats redirect roots to leave the growing media and desiccate which stimulates greater branching of new roots (**Fig. 2**). Roots, like shoots, have apical dominance – controlled by auxin. When primary roots are redirected to leave the propagation, liner or container media - the root apices are "pruned" after they protrude and come in contact with air. This leads to a decrease in auxin production and a relative increase in cytokinin levels, thereby triggering the plant to produce more lateral roots near the pruned area. This altered auxin/cytokinin hormonal balance favors lateral root initiation (Aloni et al., 2006). Subsequently, apical root dominance is broken, encouraging secondary roots to regenerate from primary roots – creating a more fibrous root system.

Figure 2. In air-root pruning the liner pots redirect roots to leave the growing media and desiccate which stimulates greater branching of new roots.

ADVANTAGES OF AIR-ROOT PRUNING

Some of the advantages of air-root pruning include: reduction or elimination of circling roots, reduced incidence of root disease, increased oxygen exchange with roots, more rapid growth of the root system, increased rooting success with recalcitrant plants, faster turn-over of propagation space, increased root surface area for greater access to water (Jacobs et al., 2009) and nutrients, a denser, healthier root system for reduced transplant shock, and a quality root system – resulting in fewer plant loses during transplanting (**Fig. 3**). Air-root pruning can also be done with larger containerized plants (**Fig. 4**).

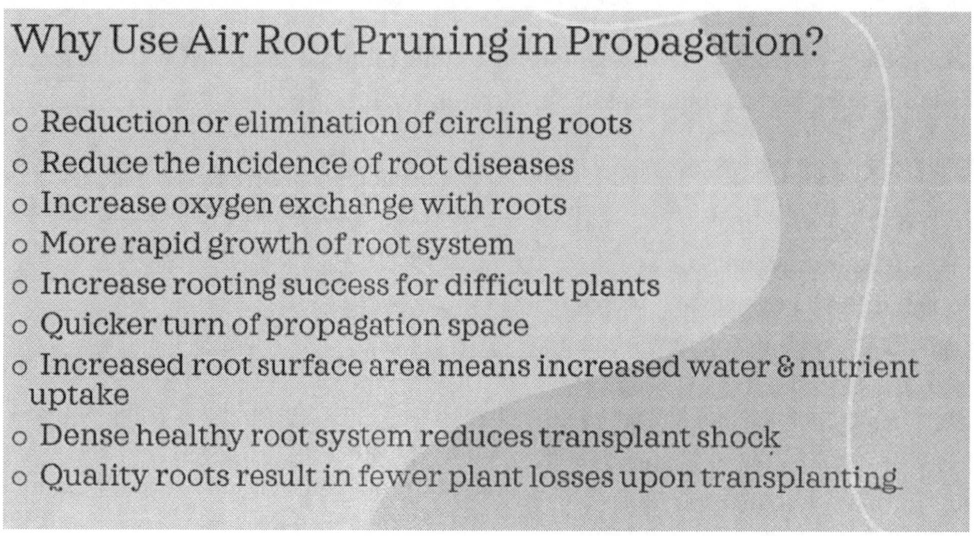

Figure 3. Some advantages of air-root pruning.

Figure 4. Air-root pruning can also be done with larger containerized plants. Notice the fibrous white roots.

DOWNSIDES AND ADJUSTMENTS NEEDED WITH AIR-ROOT PRUNING SYSTEMS

There are few downsides to utilizing air-root pruning systems. However, water-scheduling will need to be adjusted since there is greater evaporative loses from the containers – and greater water use. It is best not to mix different systems with different irrigation demands in the same growing area **(Fig. 5)**.

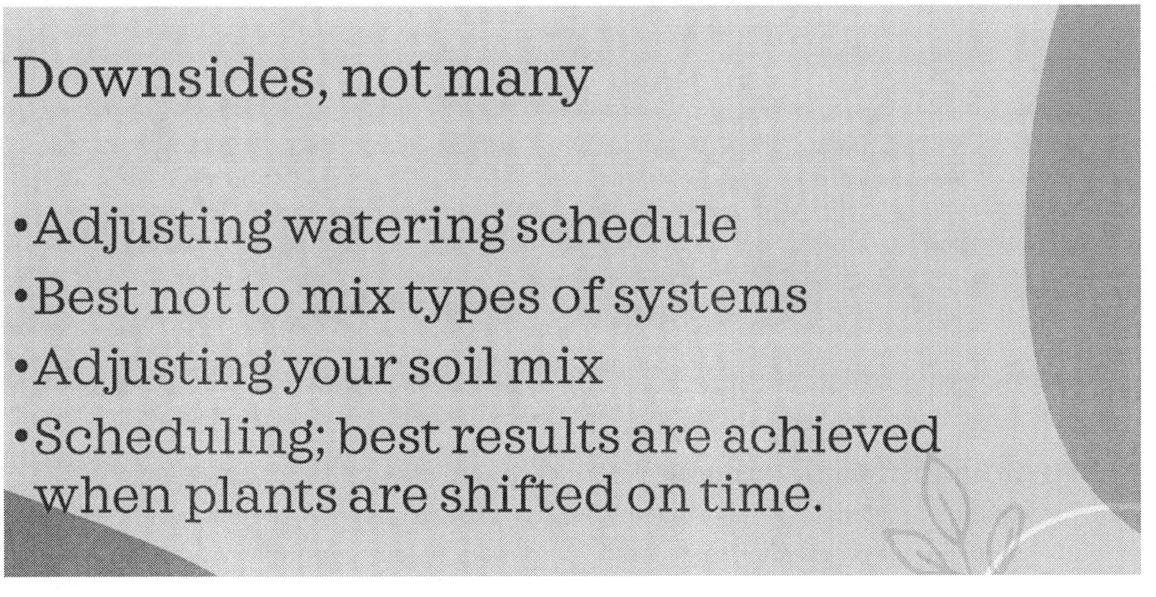

Figure 5. Potential downsides and adjustments needed with air-root pruning systems.

Soil media will need to be adjusted. Adhering to production schedules during propagation and liner production – shifting or upcanning when needed is important. Propagation flats, liner pots and containers may be more costly – but the enhanced growth and development – and production of a superior plant can justify the expense.

CONCLUSION

Air-root pruning systems have considerable propagation and production advantages in producing superior plants of greater commercial value. Air-root pruning has been utilized for more than 50-years (Carlson, 1974). However, Whitcomb (2006) with his Rootmaker® propagation flats/containers/liner pots was instrumental in getting the nursery/green industry to adapt commercial air-root pruning systems https://rootmaker.com/rootmaker-system.

A number of other commercial companies have since introduced their air-root pruning products (**Figs. 6 and 7**) including:

Rediroot® https://rediroot.com/ ,

Proptek® https://www.proptek.com/,

Anderson Band® www.andersonpots.com, Ellepot® https://www.ellepot.com/,

Fibercell® https://bccfibercell.com/,

Accelerator® https://plantproducts.com.

Air-pruning is an important component in modern propagation and production systems.

Figure 6. (left) Growcoon is a unique cutting and seed plug holder, composed of a biodegradable material that holds the media, and (right) Fertilpot propagation cups are for use in our RediRoot® propagation trays for the purpose of reducing transplant shock without compromising air-pruning technology.

Figure 7. (left) Rootmaker® and (right) Rediroot® are other commercially available air-pruning, liner container systems.

LITERATURE CITED

Aloni, R., Aloni, E., Langhans, M., and Ullrich, C.I. (2006). Role of cytokinin and auxin in shaping root architecture: regulating vascular differentiation, lateral root initiation, root apical dominance and root gravitropism. Ann. Bot. *97*:883–893. doi: 10.1093/aob/mcl027

Carlson, W.C. (1974). Root initiation induced by root pruning in northern red oak. Forest Res. Rev. Ohio Agr. Res Develop. Center, Wooster, Ohio, USA

Gilman, E.F. (1990). Tree root growth and development. I. Form, spread, depth and periodicity. J. Environ. Hort. *8*:215-220.

Jacobs, D.F., Salifu, K.F., and Davis, A.S. (2009). Drought susceptibility and recovery of transplanted *Quercus rubra* seedling in relation to root system morphology. Ann. For. Sci. *66*:1-12.

Miller, B. and Bassuk, N. (2018). The effects of air-root pruning on seedlings of species with taproots. Comb. Proc. Intl. Plant Prop. Soc. *68*:145-151.

Whitcomb, C.E., (2006). Plant Production in Containers II. 3rd Ed. Lacebark Publications Inc. Stillwater, OK ISBN-13978-0961310912

Combined Proceedings IPPS 74:434-438. 2024.

Chemical Free Water Treatment: Unleashing the True Potential and Power of Water

Jeff Nunes

AQUA4D US LLC, 7740 W. Sunnyview Ave., Visalia, California 93921 USA

nunes@aqua4d.com

Keywords: water quality, water scarcity, water & soil salinity, surface tension, water absorption, hydration

Summary

AQUA4D is a global leader in chemical-free water treatment solutions. The system electrically treats water - affecting water structure and hydration. Treated water has improved hydration with smaller clusters of water molecules – increasing salts (minerals, fertilizers) in solution. There is reduced surface tension leading to better water absorption in soils.

INTRODUCTION

AQUA4D https://www.aqua4d.com/ has spent over 20 years researching and developing resonant frequencies and the effects of electrically treated water. The company was founded in 2004 by a team of PhDs and engineers who originally sought to solve issues related to calcified pipes in commercial buildings. This breakthrough technology spilled over into agriculture when a tomato farmer, plagued by calcified irrigation lines, used AQUA4D to unclog his system. Not only were his irrigation lines restored, but his plants also showed remarkable physiological changes—requiring less water and fertilizer, which opened the door to broader agricultural applications.

Today, AQUA4D has over 5,000 installations globally, spanning 40-plus countries across five continents. Our ISO 9001-certified technology has been recognized with awards for its innovative approach to water treatment. Since entering the U.S. market in 2017, we have helped growers in nut, stone fruit, table grape, and ornamental horticulture achieve greater efficiency in water and fertilizer use (**Fig 1**).

Figure 1. AQUA4D water systems in California agriculture production.

The core focus of AQUA4D is tackling two of agriculture's most pressing challenges: **water scarcity** and **soil salinity**. These issues are especially critical in the world's most agriculturally productive regions. AQUA4D offers a solution by improving water use efficiency, reducing fertilizer inputs, and restoring soil health.

Our system is easy to install, modular, and energy-efficient—perfectly adaptable to any irrigation setup (**Fig. 2**). By altering the structure of water at a molecular level, AQUA4D enables water to carry essential minerals more effectively and helps soils overcome hydrophobic properties caused by salt saturation and other factors (**Fig. 3**). This leads to better water retention, improved nutrient uptake, and healthier crops.

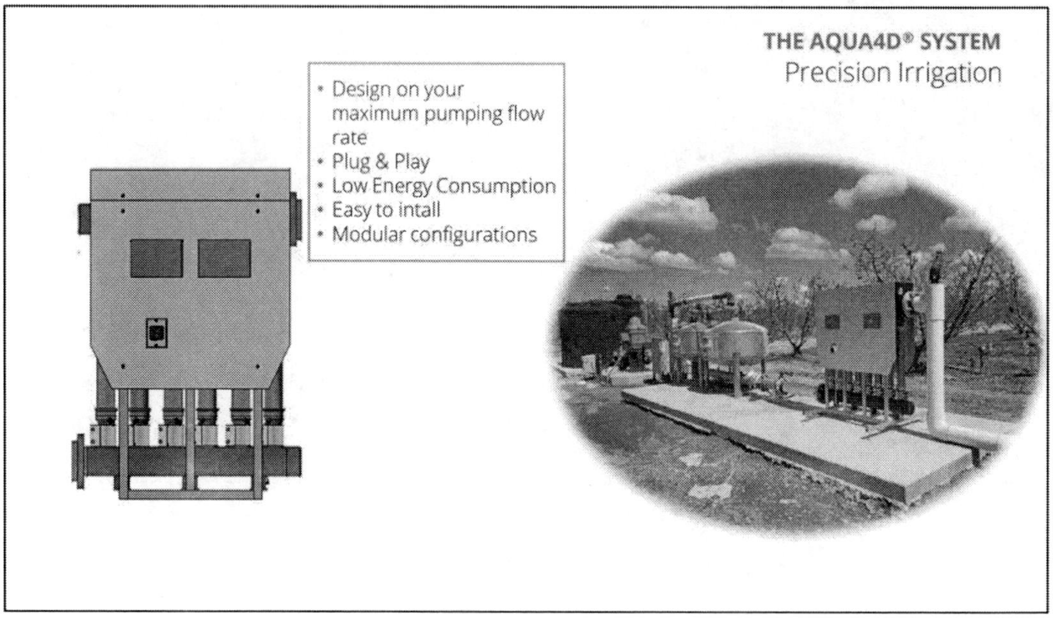

Figure 2. An AQUA4D precission irrigation system with modular configurations.

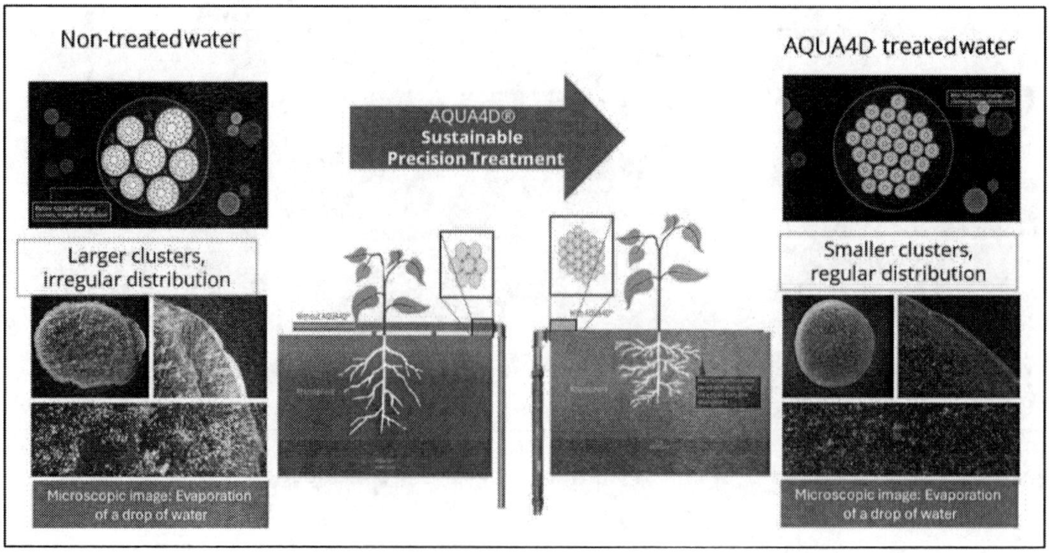

Figure 3. Non-treated water with larger clusters, compared to electrically-treated water with smaller clusters of water molecules – potentially leading to better root systems.

AQUA4D's impact extends beyond soil and crop health. In hydroponic, aeroponic, and aquaponic systems, the technology drastically reduces biofilm and calcification, leading to healthier irrigation systems and further reductions in input costs—up to 30% on average.

In a world where water scarcity is increasingly critical, AQUA4D allows farmers to save 20-30% on water usage while improving soil health through salinity management. Our technology mobilizes unwanted salts in the soil, pushing them away from root zones and promoting healthier plant growth (**Fig. 4**).

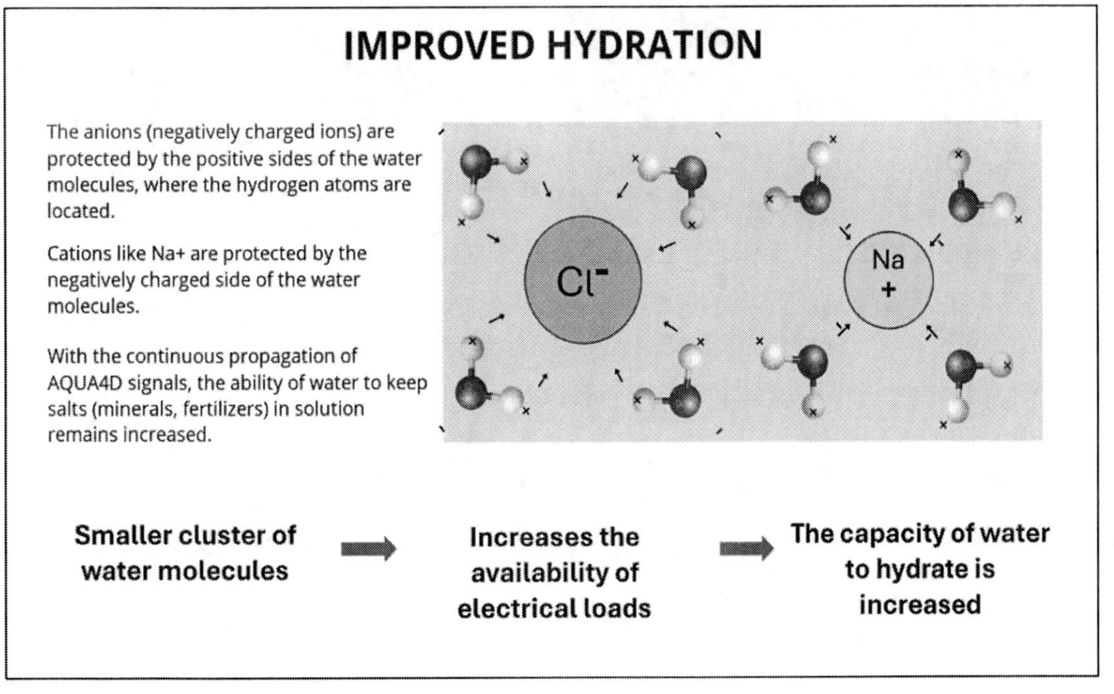

Figure 4. The electrically treated water has improved hydration with smaller clusters of water molecules - increasing salts (minerals, fertilizers) in solution.

Additionally, AQUA4D enhances irrigation system performance, reduces clogging, and increases the efficiency of fertilizers by ensuring they are fully absorbed by plant roots (**Fig. 5**). This makes it an indispensable tool for any crop.

By enhancing soil properties, reducing water use, and improving plant vigor, AQUA4D is at the forefront of sustainable agriculture. Enhancing the potential of water, helps farmers reduce input costs, conserve water, and minimize environmental impact.

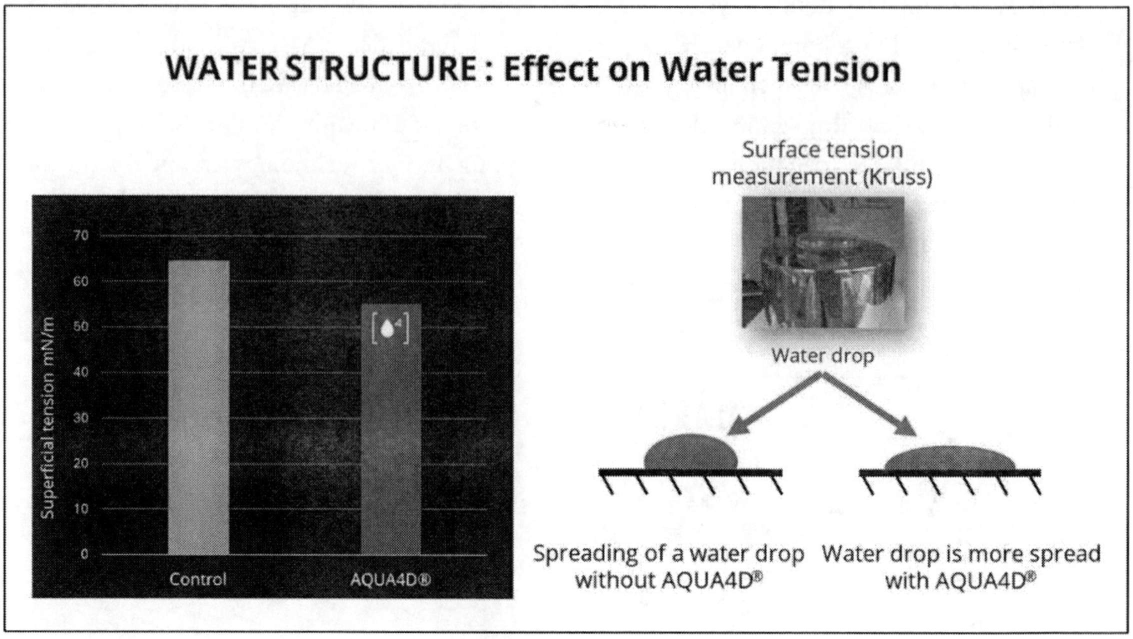

Figure 5. Treated water has reduced surface tenion leading to better water absorption in soils.

Combined Proceedings IPPS 74:439-445. 2024.

Rose Rosette Disease – Where From? What Now?

Kevin Ong

Department of Plant Pathology & Microbiology, Texas A&M University,

College Station, Texas 77843 USA

kevo@tamu.edu

Keywords: rose rosette disease (RRD), rose rosette virus (RRV), eriophyid mite, *Phyllocoptes fructiphilus*, RNA virus, *Emaravirus,* genomic based tests, Raman spectroscopy, virus-free propagation, virus indexing

Summary

Rose rosette disease (RRD) is a destructive viral disease of roses that was first observed in North America in 1940s. The disease has since spread across the United States. RRD is caused by rose rosette virus (RRV), that is transmitted by the microscopic eriophyid mite, *Phyllocoptes fructiphilus*. RRD can results in one or more of the following symptoms: deformed, excessive growth known as "witches' brooms," reddening of leaves, excessive thorniness, malformed leaves, and death of plant. There are current research efforts to seek knowledge and better understanding to help in the development of resistant rose varieties, better RRV detection methods and improved management strategies and practices against the disease. Despite it devastating effects, RRD has remained relatively obscure until the last 20-25 years when it became more widespread and significantly impacted cultivated rose varieties. Since there are no known rose disease resistance, nor effective miticidal control – diseased plants must be rogued and destroyed to limit disease spread.

IPPS Vol. 74 – 2024

INTRODUCTION

Historical origins. Initial observations describing rose rosette disease (RRD) can be traced back to 1940 in a Canadian plant disease survey. More details were later provided on additional findings in Wyoming and California of this disease occurring on native rose species (Thomas and Scott, 1953). There were not much reported about this plant disease issue for the next decades, until the 1970s and 80s it began to spread to various parts of the United States (Soto et al., 2020) (**Fig. 1**). This was when it became more of a problem on cultivated rose varieties.

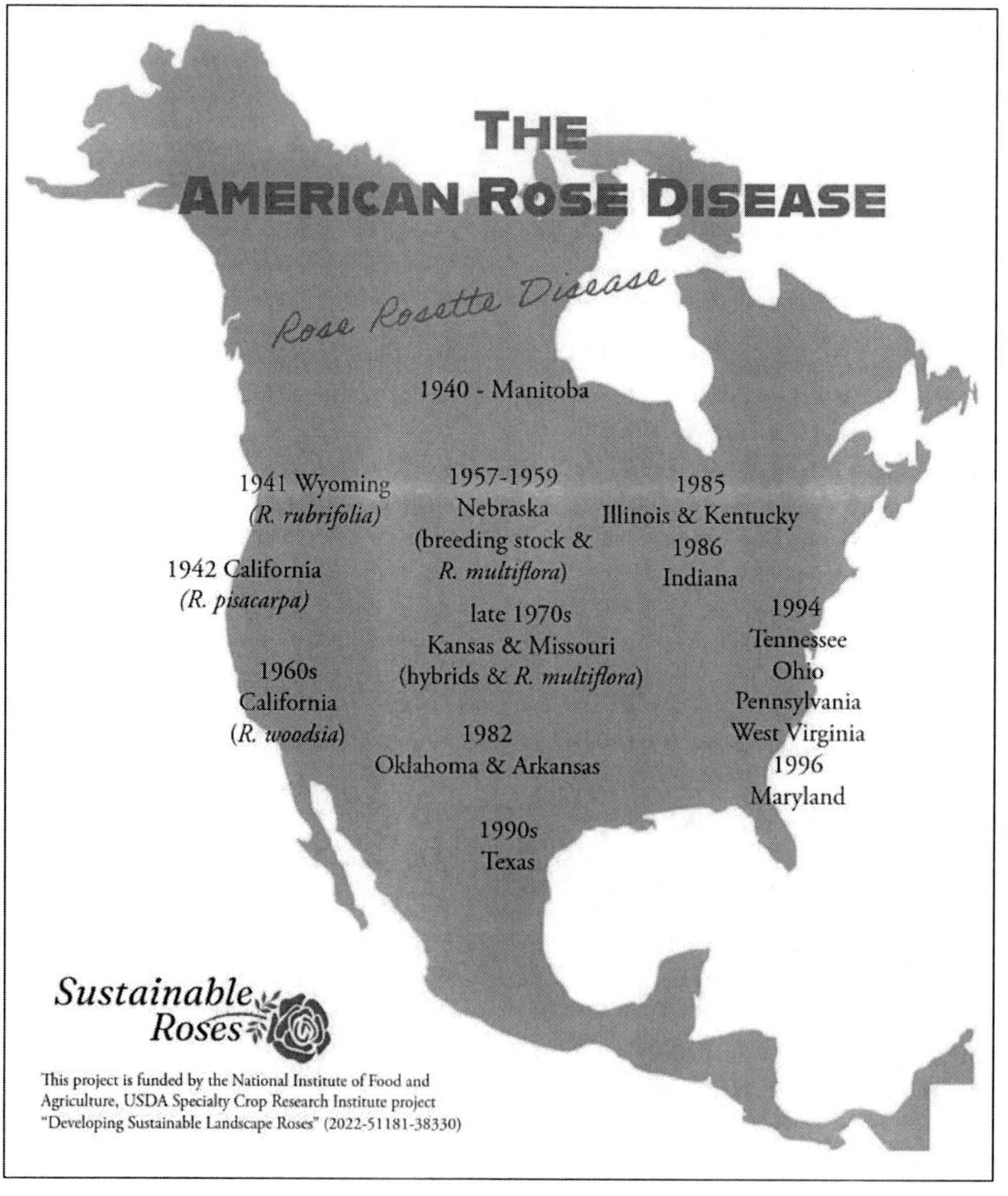

Figure 1. History of rose rosette disease (RRD) in the USA.

BIOLOGY AND PATHOLOGY OF ROSE ROSETTE DISEASE (RRD)

As early at 1953, RRD was suspected to be caused by a virus. However, there exact causal agent was not clearly known. But through some studies, an eriophyid mite (*Phyllocoptes fructiphilus*) was implicated as the vector of the RRD pathogen (Allington et al., 1968) (**Fig. 2**). These mites are microscopic, wingless, and travel between plants either by walking, wind dispersal or by attaching themselves to other insects and animals. These mites are believed to acquire the virus from feeding on infected roses, and then transmit the virus to healthy plants at subsequent feeding (Amrine et al., 1988).

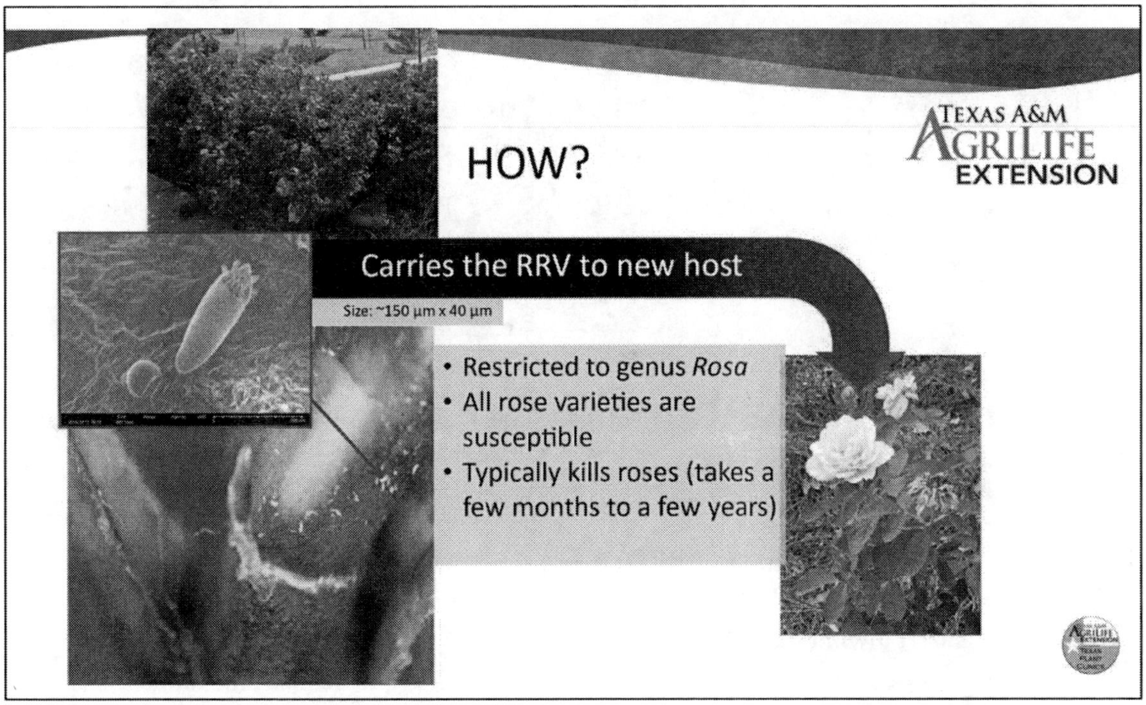

Figure 2. The microscopic, eriophyid mite, *Phyllocoptes fructiphilus*, that carries the rose rosette virus (RRV) can be spread by wind and contaminated clothing and equipment.

The first conclusive report demonstrating the causal agent to be a virus, rose rosette virus (RRV) was in 2011 by Laney et al. (2022). RRV is a negative-sense RNA virus belonging to the genus *Emaravirus* (**Fig 3**). This virus appears to only infect roses (*Rosa spp.*). Current understanding suggests that RRV can systemically spread throughout the infected plant. Infected plants exhibit characteristic symptoms such as abnormal red pigmentation, distorted leaves, elongated stems, excessive thorn production, and "witches' brooms" (dense, bushy clusters of growth) (Doudrick et al., 1983) (**Fig. 4**). Over time, infected plants suffer reduced vigor and eventually die prematurely: naturally or from other factors such as cold/freeze incidences) (Epstein et al., 1995).

RRD poses a significant threat to rose cultivation, especially in landscapes and commercial rose production (Solo et al.,2020). RRV can spread rapidly and may have a long incubation period before visible symptoms appear, making early detection challenging. Once infected, plants cannot be cured, and the only control method is to remove and destroy the affected plants to prevent the virus from spreading. Integrated management strategies, such as vigilant monitoring, utilizing resistant rose varieties, managing the mite population would be essential in mitigating the impact of this disease. Unfortunately, we know very little about the resistance to roses to this disease or effective miticide treatments to manage the eriophyid populations.

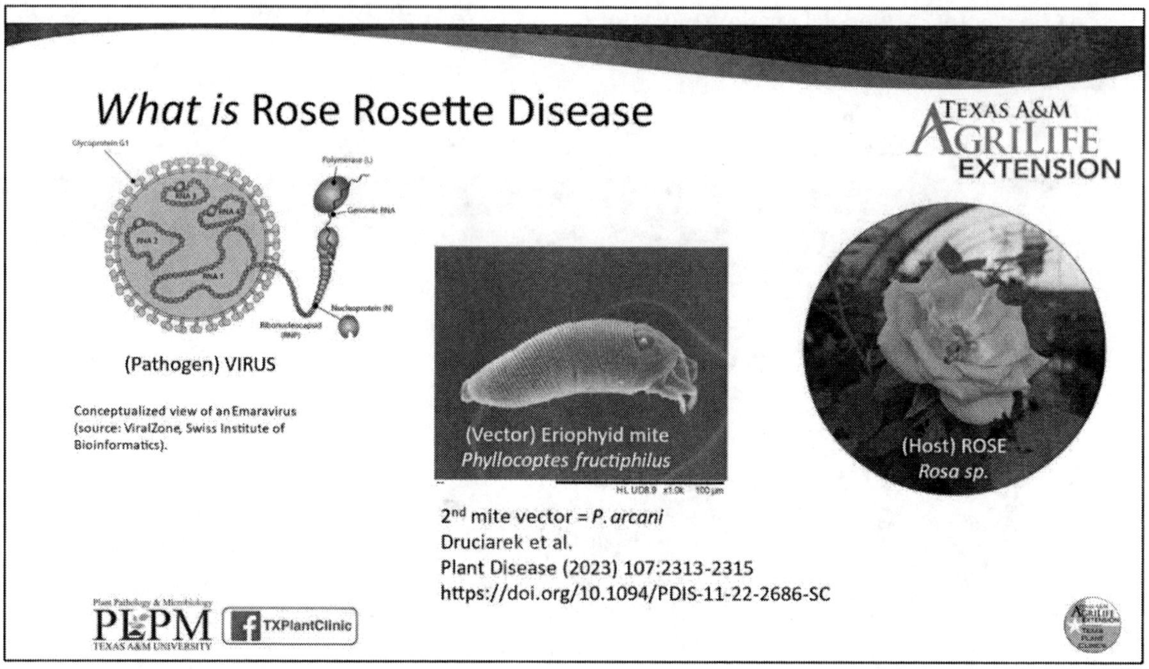

Figure 3. In rose rosette disease (RRD), there are three players: 1) Susceptible rose cultivars, 2) the rose rosette virus (RRV) [negative-sense RNA virus (genus *Emaravirus*)] that causes the disease; and 3) a wind-dispersed eriophyid mite (*P. fructiphilus)* that serves as the vector of the disease – infecting rose plants. The wild rose is believed to be instrumental in the epidemic of this disease and spread of the pathogen.

DIAGNOSTICS

Diagnostics of RRD in the early years relied on symptomology. Later, the presence of the eriophyid mite along with visible disease symptoms were indication of RRD. Only in 2011, a genomic based tool was available to use to detect RRV (Laney et al., 2011). Various research groups have furthered our knowledge and ability to detect RRV using genomic techniques.

Today, there are a number of different genomic based tests that are available to detect RRV in suspect plants. Progress is continually made in this area to find cheaper, accurate and easy to execute test procedures (Claros et al., 2022). Other technologies are explored as RRV detection tools. For example, Raman spectroscopy has been shown (proof of concept) to be able to detect infection in rose plants prior to RRD symptoms appearing (Faber et al., 2019).

Figure 4. There may be rose rosette disease (RRD) with only a single symptom.

CLEAN PLANT PRODUCTION

Efforts to produce virus-free roses in commercial operations against Rose rosette virus (RRV) have intensified due to the devastating impact of rose rosette disease (RRD) on the ornamental rose industry. Since there is no cure for infected plants, prevention and virus-free propagation are critical strategies. Even then, there is always the challenge to maintain the plants in a manner to keep them free of RRV.

Virus indexing, a process of testing plant material for the presence of RRV, is routinely employed during the propagation phase. This involves molecular techniques (such as PCR) to detect the virus even when symptoms are not visible. More recent, high throughput sequencing (HTS), a molecular technique which allows for generating large genomic data sets that can provide insight into looking for RRV or other viruses within the rose plant.

Mitigating the spread of *Phyllocoptes fructiphilus*, the eriophyid mite vector responsible for transmitting RRV, is another essential component of producing virus-free roses. Mite management includes chemical control measures, like miticides, as well as cultural practices such as spacing plants properly to reduce contact and using physical barriers to prevent mites from spreading between plants.

Currently, there is ongoing efforts by rose breeders and researchers to develop resistant varieties through traditional breeding and genetic research. Identifying genetic traits associated with resistance and incorporating them into new cultivars offers a potentially promising long-term solution to the problem.

INFORMATION GAP AND CONTINUED CHALLENGES

Despite advances in understanding RRD, several critical information gaps remain. One major gap is the complexity of the relationship between the RRV and its vector, the eriophyid mite *Phyllocoptes fructiphilus*. While we know the mites transmit the virus, the precise mechanisms of how they acquire, harbor, allow for virus multiplication, and spread the virus within rose populations are not fully understood. This lack of detail hinders the development of effective mite control strategies.

Another gap lies in the variability of symptoms among rose species and cultivars (Epstein and Hill, 1999). Symptoms of RRD can vary significantly, ranging from subtle changes to severe deformities, and may take months to appear. Understanding why certain roses show delayed or less severe symptoms, or why some are more resistant than others, is key to breeding and selecting more resilient varieties.

Additionally, research into the genetic basis of resistance to RRV is still in its early stages. While some rose species show partial resistance, the genes and mechanisms behind this resistance remain poorly understood. Furthermore, the nature of RRV in the rose plant, such as movement and distribution to infection points, is also poorly understood. These contribute to limiting our ability to develop resistant cultivars.

The irregular incubation period of the virus complicates early detection strategies, and current diagnostic tools need further refinement to ensure more accurate and rapid identification of the virus in asymptomatic plants (early detection).

There remains much to be learned of this virus and the disease it causes. Research is still being done and we can look forward to new information in the near future.

LITERATURE CITED

Allington, W. B., Staples, R., and Viehmeye, G. (1968). Transmission of rose rosette virus by eriophyid mite *Phyllocoptes fructiphilus*. J. Econ. Entomol. *61*:1137.

Amrine Jr., J. W., Hindal, D. F., Stasny, T. A., Williams, R. L., and Coffman, C. C. (1988). Transmission of the rose rosette disease agent to *Rosa multiflora* by *Phyllocoptes fructiphilus* (Acari: Eriophyidae). Entomol. News (USA).

Boonham, N. (2020). Facing rose rosette virus: a risk to European rose cultivation. Plant Path. *69*:1603-17.

Claros, N. A., Shires, M., Mollov, D., Hammond, J., Jordan, R., Ochoa-Corona, F., Olson, J., Ong, K., and Salamanca, L. R. (2022). Rose rosette disease: a diagnostic guide. Plant Health Prog. *23*:482-491.

Doudrick, R. L., and Millikan, D. F. (1983). Some etiological and symptomological aspects of rose rosette. Phytopath. *73*:840-840.

Epstein, A. H., and Hill, J. H. (1995). The Biology of rose rosette disease - a mite-associated disease of uncertain etiology. J. Phytopath. *143*:353-360.

Epstein, A. H., and Hill, J. H. (1999). Status of rose rosette disease as a biological control for multiflora rose. Plant Dis. *83*:92-101.

Farber, C., Shires, M., Ong, K., Byrne, D., and Kurouski, D. (2019). Raman spectroscopy as an early detection tool for rose rosette infection. Planta *250*:1247-1254.

Laney, A. G., Keller, K. E., Martin, R. R., and Tzanetakis, I. E. (2011). A discovery 70 years in the making: characterization of the rose rosette virus. J. Gen. Virol. *92*:1727-1732.

Solo, K. M., Collins, S. B., Shires, M. K., Ochoa, et al., (2020). A survey of rose rosette virus and eriophyid mites associated with roses in the Southeastern United States. HortScience *55*:1288-1294.

Thomas, H. E., and Scott, C. E. (1953). Rosette of rose. Phytopath. *43*:218-219.

Combined Proceedings IPPS 74:446-454. 2024.

Drone Applications in Nursery Production

James Robbins[1] and Joe Mari Maja[2]

[1]Ornamental Hort Solutions, 300 N Pine St, Little Rock, Arkansas 72205 USA

[2]Center for Applied Artificial Intelligence for Sustainable Agriculture, South Carolina State University, 810 Goff Ave., Orangeburg, South Carolina 29117 USA

jrobbins4930@gmail.com

Keywords: small unmanned aircraft systems (sUAS), unmanned aerial vehicle (UAV), object based image analysis software (OBIA), marketing/sales, asset tracking & management, plant inventory management, chemical/nutrient applications, crop monitoring, radio-frequency identification tags (RFID), remotely piloted aerial application systems (RPAAS)

Summary

Small Unmanned Aircraft Systems (sUAS) or 'drones' are a technology that can be used to automate or augment certain operations in open field nursery production. They may be used for a variety of activities including: 1) marketing and sales, 2) asset tracking and management, 3) plant inventory, 4) application of chemicals and nutrients, and 5) crop monitoring. Today users have the option to perform these tasks in-house or use an outside provider.

INTRODUCTION

Drones are currently accepted as a safer and more economical solution for various situations. Non-agricultural applications such as law enforcement, utility inspection, bridge inspection, wildfire monitoring, construction and news reporting are commonplace. Agriculture is also benefiting from these affordable and technologically advanced aerial systems (Robbins, and Maja, 2021a,b).

We have identified at least five applications for drone use in open-field nursery production. Each application can be evaluated for its likelihood of being adopted by nurseries from immediate to long-term. Businesses also need to evaluate whether the process they want to automate with drones can be accomplished in-house, by an outside provider, or by a hybrid of those options. Today it is common for outside companies to provide images, image processing, and aerial application services.

The first application for drones in nursery production, considered low hanging fruit, is for **marketing and sales**. Nurseries of any size can immediately use a small drone outfitted with a camera to capture needed aerial photographs of the nursery or crops (**Fig. 1**). Still images or videos taken by the nursery can immediately be used for their website, social media, catalogs or trade show marketing. Alternatively, a local 3rd party provider could provide the same material for a fee.

1. Marketing/sales

Neil Marek, Magnolia Gardens Nursery

Figure 1. Drones can be used to capture images for marketing and sales

The second application, **asset tracking and management**, can also be immediately adopted by most nurseries. Examples of this application include safe inspection of gutter connected and retractable roof structures, estimation of bark pile volume, and inspection of other infrastructure, i.e. fences, buildings, etc. (**Fig. 2**). Routine inspection of general infrastructure can be accomplished in-house using an inexpensive small drone with a camera (**Fig. 3**). Estimation of bark pile volume is a bit more complicated. In this case, still images taken by either the nursery or an outside provider would require software processing by an outside provider, which includes fees.

Figure 2. Using drones for asset tracking and management.

Figure 3. An inexpensive, entry-level sUAS/ drone with remote control.

The next drone application is **plant inventory (Fig. 4)**. Within this category, we also lump the use of simple still photographs that can be used for crop insurance purposes. Plant inventory is the application our team has worked on the longest (2010). To date, we have identified three approaches to collecting plant inventory. The first and least complex method is to take aerial images of blocks of plants and then print out your images and simply count plants captured in the image.

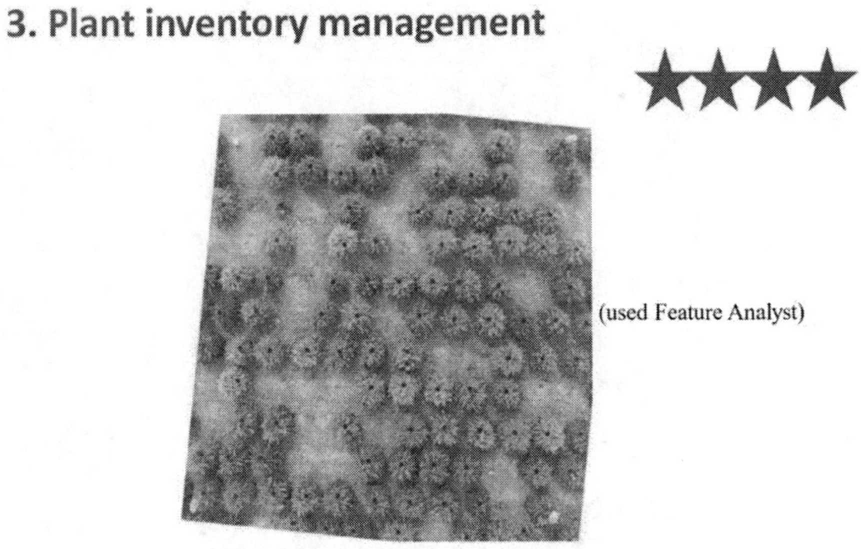

Figure 4. Plant inventory management utilizing drones.

An easy and inexpensive way to count is to use an inexpensive counter-pen marker which marks and counts each plant as you analyze the image (**Fig. 5**).

Method #1: manual counting from aerial photographs

Even if in 2024 we are not ready to count plants using software & RFID, a 'low hanging fruit' is to print-out your aerial RGB photograph and count plants in an AIR CONDITIONED office using this very inexpensive 'counter pen'!

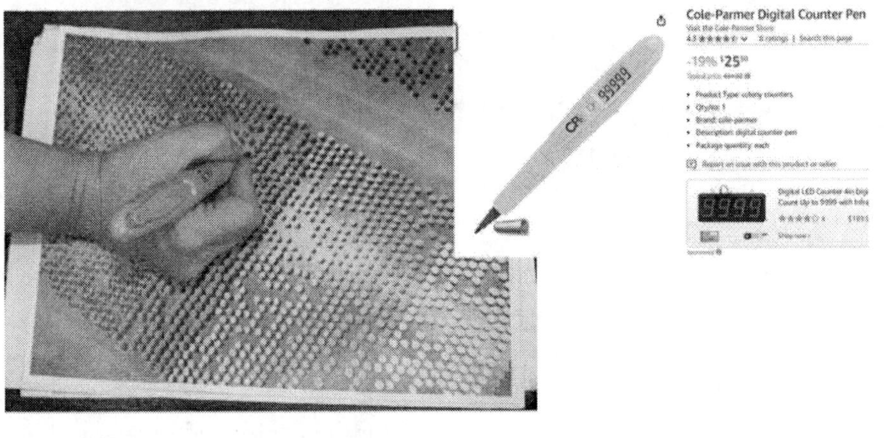

Fig. 5. Manual counting from aerial photographs.

Method #2 clearly requires advanced skill if done in-house or requires using an outside fee-based service. This method uses 'Object Based Image Analysis' (OBIA) software (i.e., Feature Analyst™, eCognition) to analyze a digital aerial image (**Fig. 6**).

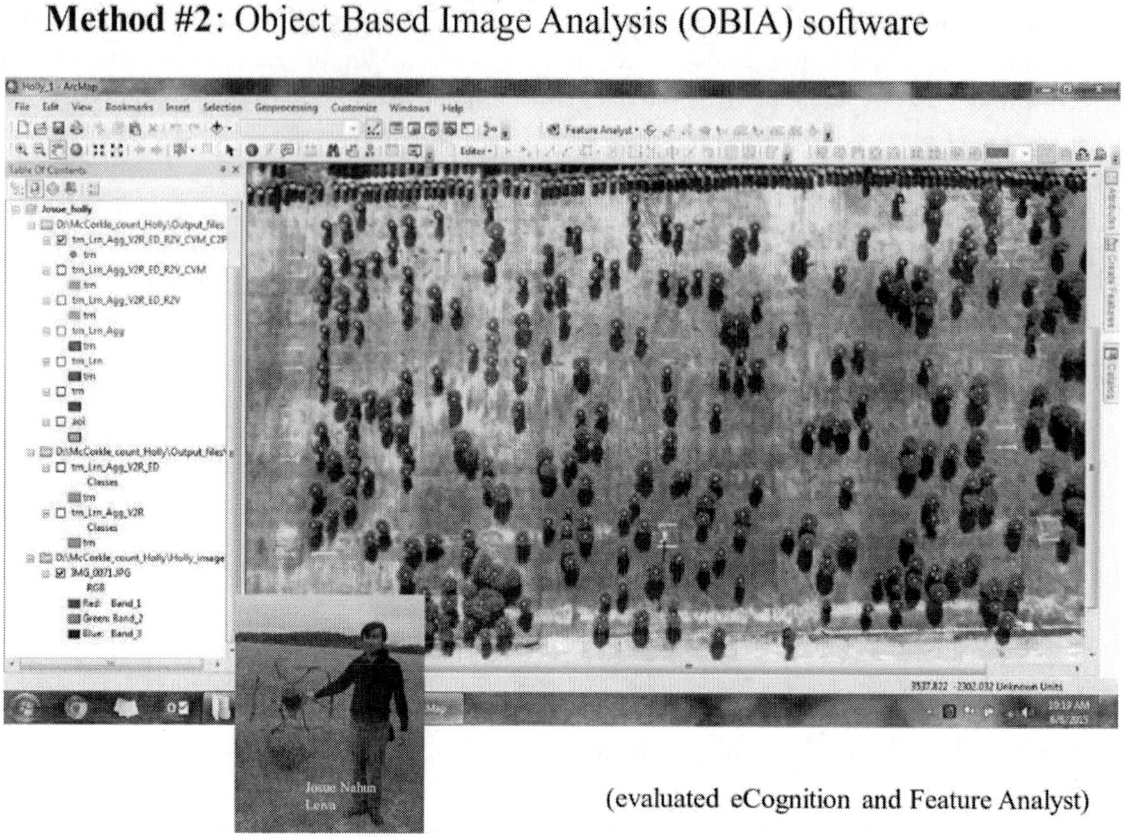

Figure 6. Object based image analysis (OBIA) software.

The software is 'trained' to extract geospatial features (in our case, plants). Most of these programs are tightly integrated with the EsriR ArcGISR platform, which would require additional cost and expertise for a majority of nurseries. In most cases, once you have created a training set for a specific plant, you may be able to apply it later in the production cycle. This method is a clear example of how a subscription to an outside service (e.g. Agremo, Solvi, DroneDeploy) to analyze the images may be a better business decision **(Fig. 7)**. Our publication in the Proceedings of the Southern Nursery Association Research Conference (Maja et al., 2015) - would be helpful if you are interested in this approach. Method #3 to automate plant inventory involves using radio-frequency identification (RFID) tags and a hand-held or drone-based reader (**Fig. 8**). This effort, in collaboration with Dr. Tom Fernandez at Michigan State University, is simple and can be adopted by nurseries of any size almost immediately (Maja et al., 2024). Our team feels strongly that the future of fast and cost-effective plant inventory rests in merging RFID tags with a small drone. The same RFID tag used during production is being designed to pass forward to the retailer for reprogramming for their sales and inventory purposes.

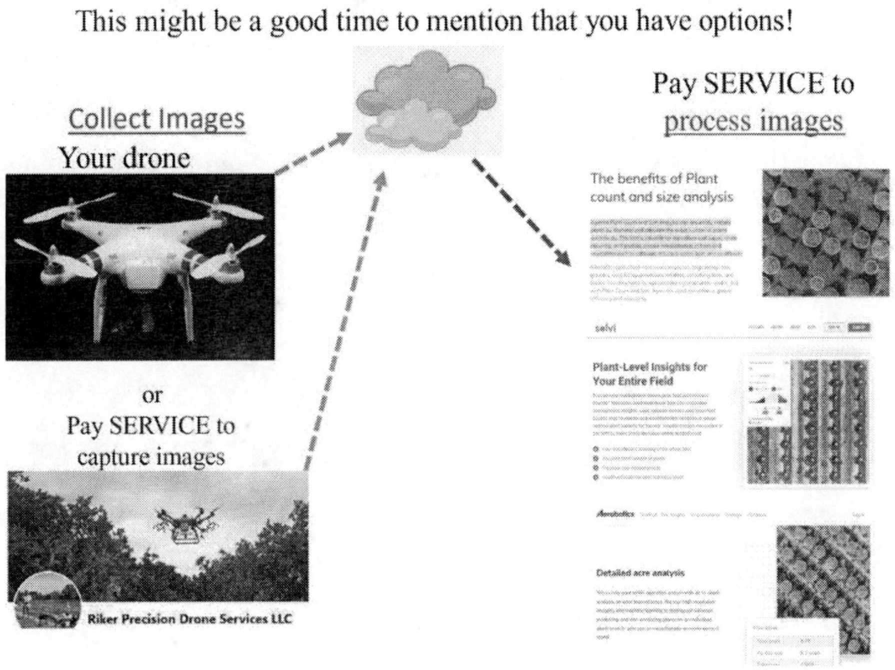

Figure 7. Options to use collect images via drones, pay outside to service to capture images – and subscription to an outside service (e.g. Agremo, Solvi, DroneDeploy) to analyze the images.

Method #3: Merging RFID tags with sUAS

Figure 8. Merging RFID tags with drones to track and manage assets and products. A microchip and antenna make up the "tag," which can be read by an RFID scanning device. The scanner turns the radio waves into digital data. RFID offers a major advantage over bar coding in that the scanning device does not need direct line of sight to read the tag.

The next nursery application is the **aerial application** of chemicals or nutrients (**Fig. 9**). Materials that can be applied using a drone can be liquid or granular (e.g., fire ant bait). Long term, this application may have the greatest impact on the nursery industry, but we are clearly in the early stages of ironing out all the details (i.e., flight regulations, pesticide label rules, application rates) (Robbins, et al., 2021). These small 'spray drones' offer tremendous potential to nurseries that grow hundreds of different crops in small blocks. A cottage industry is emerging to offer spray services to nurseries which would simplify the startup impact for a nursery trying to do this in-house. Currently, the Remotely Piloted Aerial Application Systems (RPAAS) working group is the best source for information on this specific topic.

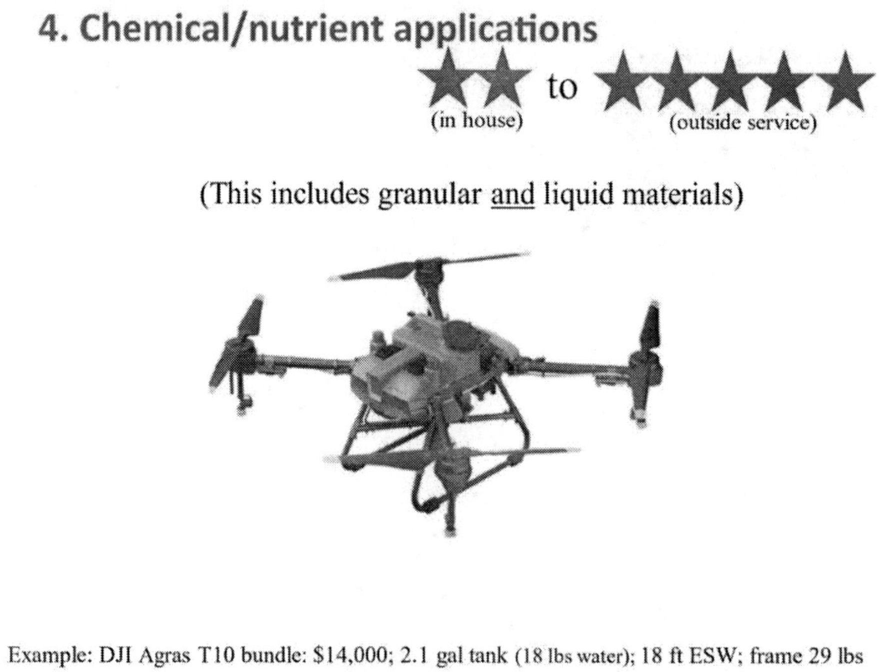

Figure 9. Chemical and nutrient applications using drones.

The final drone application is the broad category of **crop monitoring**, which includes monitoring for nutrients, water, insects, diseases, and general plant health (**Figs. 10 and 11**). Our team has worked in this area and it is our opinion that beyond a simple survey of general plant health, this application is years off from being widely applied to the nursery industry. The sheer diversity of plant types grown at a typical nursery makes this challenging. In the short term, crop monitoring using a drone is more likely in a monoculture like turfgrass or row crops (e.g., soybean, corn, rice, wheat). Only the largest nurseries will likely try this in-house due to the technical challenges with sensors and software to collect and process images. Special training is also required to correlate changes in spectral aspects of images with a specific plant response (e.g., water stress, mite infestation) that you are interested in. While drone companies currently offer crop monitoring services to nurseries, make sure that the provider has specific experience with the types of plants you grow.

5. Crop monitoring (nutrient, water, pest, health) ★

John Nowatzki, North Dakota State University

Figure 10. Crop monitoring for nutrition, water, pest, plant health.

J. Frank Schmidt Family Charitable Foundation
P. O. Box 189 | Boring OR 97009 | 503-663-4128

2017

Experimental approach to detect water stress in ornamental plants using sUAS-imagery

Ana.I. de Castro[1], Joe Mari Maja[2], Jim Owen[3], James Robbins[4], Jose M. Peña[5]

Figure 11 An experimental approach to detect water stress in nursery crops using drones.

CONCLUSION

In summary, there is no doubt that sUAS/ drones will be a useful technology in outdoor plant production. Still, users need to carefully evaluate their specific situation to determine what the best approach is. Some users may conclude that purchasing the sUAS/ drones and software is the best option, while others may find that hiring an outside company to acquire and process the images is their best option.

The authors are certified remote pilots (Robbins: #3952601; Maja: #3952164) – and have been using sUAS/ drones for 14 years.

Acknowledgment: We wish to acknowledge the collaborative help from many people and organizations, including the J. Frank Schmidt Family Charitable Foundation, Horticultural Research Institute (HRI), J. Frank Schmidt & Son Co., McCorkle Nurseries, Greenleaf Nursery Company, Bailey Nurseries, Cherrylake, Yule Tree Farm, Woodburn Nursery & Azalea, Willoway Nurseries, Hale and Hines Nursery and R.A. Dudley Nurseries.

LITERATURE CITED

Maja, J. M., Fernández, T., Robbins, J. (2024). A potent combo. Research explores the viability of drones and RFID tags for nursery and garden center inventory. Nurs. Manag. *40*: 37-42.

Maja, J.M., Leivab, J.N, Robbins, J. and Chappell, M. (2015) Comparison of manual inventory counts to an automated approach using an unmanned aerial system. South. Nurs. Assoc. Res. Conf. *60*:73-77. https://irp.cdn-website.com/5d757b5b/files/uploaded/15resprosec03.pdf

Robbins, J. and Maja, J.M. (2021a). Eyes in the skies. Nurs. Manag. *37*(1): 16-17.

Robbins, J. and Maja, J.M. (2021b). Make it count with drones. Nurs. Manag. *37*:24-26.

Robbins, J., Martin, D., Reynolds, B., and Maja, J.M. (2021). Aer. Chem. Applic. *37*: 22, 24-25.

Combined Proceedings IPPS 74:455-468. 2024.

Innovative Trends in Irrigation Technology: Enhancing Efficiency in Container Nurseries

Caleb Saunders

Canopy Company, 487 Thoreau St, Smyrna, Georgia 30082 USA

caleb.p.saunders@gmail.com

Keywords: precision irrigation, evapotranspiration (ET), leaching fraction measurement, demand-based irrigation, automation, nursery water conservation, operational efficiency, artificial intelligence (AI), machine learning (ML), internet of things (IoT)

Summary

This paper examines the shift from traditional timer-based irrigation to data-driven, precision irrigation systems that use real-time data such as evapotranspiration (ET) models and leaching fraction measurements to optimize water and resource use. Canopy www.CanopyGrow.Tech is a company focused on revolutionizing irrigation with advanced software and hardware solutions. Examples include Saunders Brothers Nursery's 43% reduction in water use and 33% decrease in fertilizer within their first year, as well as Holden Nursery's 50% water savings with improved plant health. These nursery irrigation practices - underscore how a focus on crop quality, paired with automation and real-time data integration, can enhance operational efficiency, reduce waste, and elevate crop outcomes—leading to more sustainable and profitable practices for nurseries.

IPPS Vol. 74 – 2024

INTRODUCTION

This paper includes innovative trends in irrigation technology and processes related to irrigation - and how one can leverage them to maintain perfectly watered plants, while making employees' jobs easier.

I grew up in the family nursery business at Saunders Brothers. After earning a mechanical engineering degree from Virginia Tech and returning to the farm, I assisted with implementing their current irrigation system. This experience ultimately led to my current project at Canopy, and I am excited to share these insights with you. Saunders Brothers' irrigation approach includes utilizing leaching fraction measurements (Owen et al., 2019), evapotranspiration-based irrigation, automation, and infrastructure investments

Why irrigation matters

To set the stage, consider the quote from Hunter et al. (2010) that "Without question, irrigation provides the life-blood of pot (container) plant production. Compared to in-ground culture the relatively small amount of available water in a pot quickly becomes a major limitiing factor to growth unless the water status is regularly reinstated. With the exception of sub-irrigation, all overhead application systems either apply too little or too much simply because no instrument can provide accurate needs-based information on total pot water deficit." (M. Hunter et al. 2010).

This captures the essence of why irrigation is critical, especially in container nurseries where precise water management can drastically impact plant health and resource efficiency. Additionally, it highlights the challenge that exists when it comes to precisely watering with overhead irrigation.

Moving toward smarter irrigation practices

This paper covers what to consider, how insights are improving efficiencies and adding value to nurseries today, and how it starts with a simple idea: smarter, not harder. Traditionally, nurseries have relied on fixed timers to control irrigation, but the "set and forget" method is outdated. Today, real-time data allows systems to make decisions based on weather and plant needs, making nurseries far more efficient.

My goal is to provide insights you can start implementing in your nursery. Let's discuss the kind of data needed and how to make informed irrigation decisions because, imagine trying to bake a cake without knowing the oven temperature. It is complete guesswork. Without the right information, you cannot make the right decisions on how much to water. From a show of hands at the 2024 IPPS Southern Region conference in Tulsa, Oklahoma - about 40% of responders use guesswork – based on the weather or seasoned intuition to determine the run-times for their daily irrigation.

Framework for choosing a modern irrigation system

In deciding what to consider to improve your irrigation system and approach, this quote lays out the criteria quite nicely for a container nursery:

"An irrigation model well-suited for the nursery industry would: (1) use a physiological basis to accurately estimate water use to prevent over- and under-irrigation, thus conserving water and minimizing leaching, (2) be demand-based, (3) be non-invasive, (4) easily configured to a large number of crops, (5) not increase production time over current irrigation scheduling, and (6) be automated" (Fulcher et al., 2012).

Key components of an optimized irrigation model include:

- *Demand-Based* - accurately estimates water needs using a physiological and demand-based approach.
- *Non-Intrusive and Configurable* - be non-intrusive to the existing operation and easily configurable for a variety of crops.
- *Maintains Production Timelines and is Automated* - maintains production timelines, while leveraging automation for efficiency.

Each component addresses key considerations in creating an irrigation system tailored to nursery needs.

Demand-based irrigation

Firstly, you have to get a handle on your water use and in turn, what each crop needs. No more guessing games. Key considerations include: evapotranspiration, leachate fraction, and real time adjustments.

Evapotranspiration (ET). Accurate Water Use. To manage water effectively, nurseries need data on evapotranspiration (ET) rates, which consider factors like temperature, humidity, and other weather conditions. Measured in millimeters, ET tells you how much water is lost from the soil surface through evaporation and through the plants via transpiration, helping you better understand your irrigation needs.

Leaching Fraction - Conserves Water and Minimizes Leaching. Leachate is like filling a glass with water until it overflows—the water spilling over is not being used. In a container nursery, this overflow represents the extra water draining out of pots after irrigation. This excess water, along with valuable nutrients, is essentially wasted if not managed properly. Measuring leachate allows growers to see how much water is not being absorbed by plants, helping them adjust irrigation to reduce waste and improve efficiency.

The leaching fraction, on the other hand, quantifies this process. It is calculated as the amount of water that drains out of a container after irrigation, divided by the total water applied. This metric is key to assessing the efficiency of an irrigation system (Saunders, 2024a).

Volume-Based Measure of Leaching Fraction. To measure the volume of water applied, an open container is placed near the plants being irrigated to capture the applied water. Place a similarly sized container (lined with plastic) or a bucket with tight seal under the test plants to make sure that no water enters either the empty or planted container from the sides. One hour after the entire irrigation cycle is completed (e.g., if using cyclic irrigation, wait until the last cycle is complete), measure the volume of the water leached from the planted container and the empty container. From these two measurements, the leaching fraction can be determined. For example, if 1,000 mL of water is applied via irrigation (empty container) and 250mL leaches from the bottom of the planted container – this is a leaching fraction of 0.25 (i.e., 25% of the total volume of water applied was leached from the container) (Stanley et al., 2019).

It is recommended that you maintain a leaching fraction of ~10-15%, but it depends on the crop and is ultimately up to the grower. Additionally, according to AgriFi.AI (2024), reducing water usage also lowers energy costs, while minimizing runoff helps protect aquatic ecosystems. With less water leaching out, the runoff is better managed.

Demand-Based Adjustments. Adjusts in Real-Time. A demand-based system operates dynamically, adjusting in real-time to the plants' evolving needs throughout the growing season and based on weather on a daily basis, rather than adhering to a fixed schedule. This approach minimizes both over- and under-watering, optimizing water usage and promoting plant health. In essence, a demand-based system responds directly to what the plant requires, rather than following a pre-set timer schedule.

Non-intrusive and configurable

Irrigation needs to be configurable. You need to be able to tweak it; set it up to handle a bunch of different crops all with their own quirks and water needs. This might mean setting up different irrigation zones, for example, to give each plant group exactly what it needs. Customization is king when you have so many different plant types.

Maintains production timelines and is automated

So, I am sure you are wondering about the bottom line here: does switching to a more modern system slow down finishing your crops on time? The last thing you want is for a new system to change your entire shipping schedule. The answer is no, not if it's done right. You have to find the balance between innovation and running a smooth operation. Switching to advanced irrigation methods should not delay existing production timelines. If implemented correctly, these systems will keep crop schedules intact and enhance overall productivity for your employees.

Automation - as a labor and cost saver Labor availability is a constant challenge in nurseries, often leaving little time for getting everything done in time. As one nursery mentioned, they have 20 workers applying plastic to the greenhouses to prepare for winter over a period of weeks. Automation presents a valuable opportunity to streamline these repetitive tasks, improving efficiency and free-up staff for more critical work **(Fig. 1)**.

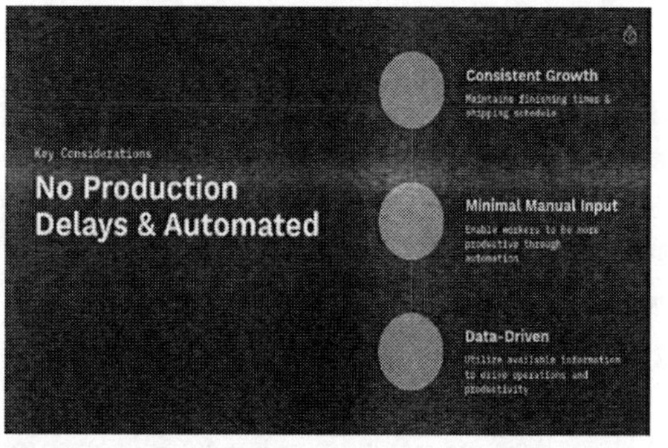

Figure 1. Last set of considerations include maintaining production timelines and being automated.

As noted by Fulcher et al. (2023), "Nursery production is inherently reliant on a scarce and unreliable manual labor workforce. Therefore, the nursery industry must take opportunities to continue developing automation strategies and innovations that address this critical issue." Tilt (2000) adds that "computerized systems can save miles of running each week and reduce the strain on nursery managers," further illustrating how automation benefits labor efficiency and irrigation effectiveness.

By reducing manual tasks, automation saves you time, reduces costs, and allows your team to focus on what truly matters: actual plant care, pest management, and propagation. This can make everyone happier and more productive in the long run. It's a win, win.

CURRENT TRENDS: From guesswork to data-driven irrigation

As mentioned earlier, 40% of nurseries utilize guesswork or intuition to adjust irrigation schedules. Additionally, 30% of the attendees at the IPPS conference mentioned they hand-water or manually turn valves on and off. These traditional methods, though familiar, are often inefficient, leading to higher water usage, labor demands, and inconsistent plant health.

Manually irrigating and guessing how long to irrigate is not optimal, because it ***fails to***:

- *Irrigate efficiently*, based on real-time changes to plant needs and weather conditions
- *Mitigate knowledge gaps*, when folks leave your nursery
- *Reduce costs*, associated with water, fertilizer, and labor
- *Limit water consumption* and deal with regulatory legislation

Ultimately, it's not optimal to efficiently produce and ship quality crops **(Fig. 2).**

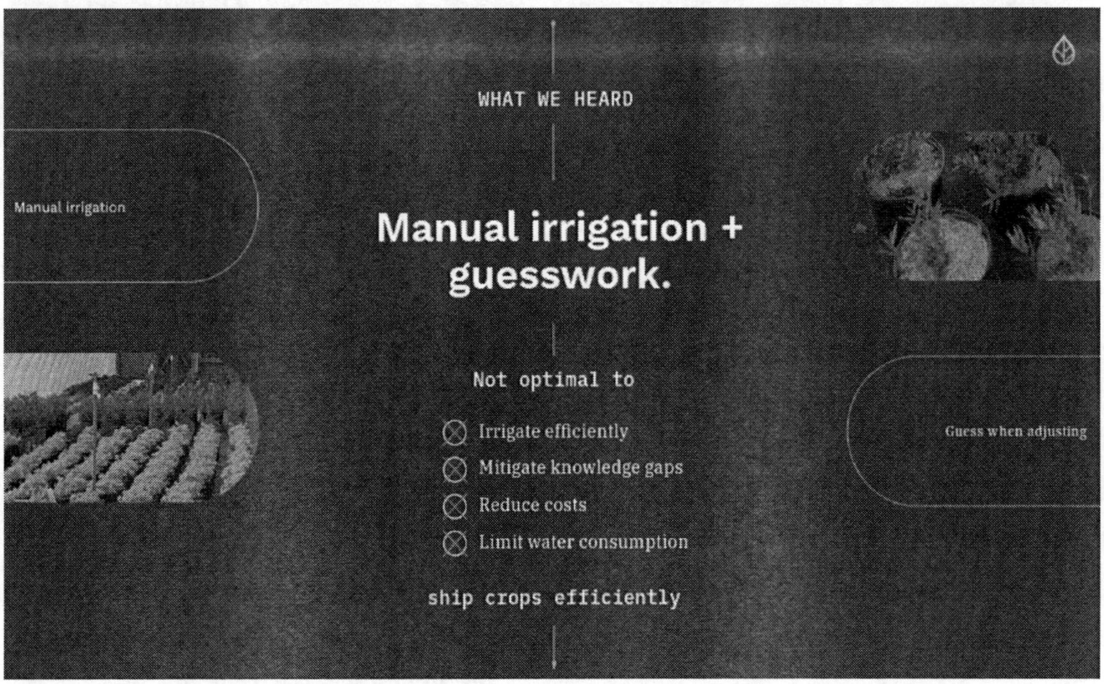

Figure 2. Manual irrigation and guesswork are not optimal to ship crops efficiently.

Adaptive change requires proper attention. As Million et al. (2010) mentions, "The container nursery industry is continuously seeking new irrigation and fertilization strategies to improve efficiencies and reduce negative environmental impacts… [while striking] a balance between the rewards of reduced … inputs and the risks of reduced plant growth and quality."

To make the case, much of the research points towards trends in a few key areas: from removing the guesswork, reducing the inputs by better managing the soil moisture, to using automation (**Fig. 3**).

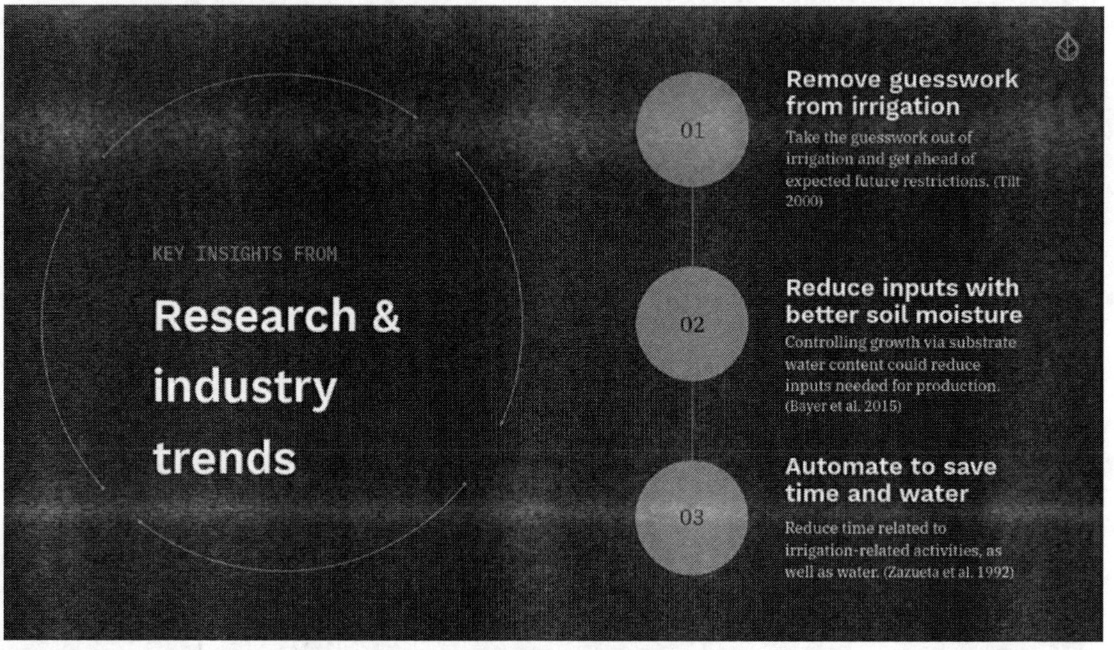

Figure 3. Research and industry trends point towards removing guesswork, reducing inputs, and automating.

SHAPING THE FUTURE: INNOVATIVE TRENDS

1. *Irrigation Delivery System Improvements*: Nurseries have adopted drip irrigation and manage overhead systems better using technology.
2. *Irrigation Schedule Improvements*: ET models and leaching fraction systems refine watering schedules.
3. *Technology and Infrastructure*: Making infrastructure investments enables fine-tuned control over water delivery per zone, along with the use of Artificial Intelligence (AI) / Machine Learning (ML) and Internet of Things (IoT) (see definitions below).

Focus on the irrigation delivery system improvements, such as drip or using technology to better manage your overhead irrigation. Secondly, improve your irrigation schedule itself, whether that be from using the ET based models mentioned and/or leaching fraction measurements.

You can also incorporate soil sensors, as Bayer et al. (2015) suggests, "Automated irrigation using soil moisture sensor-controlled systems ensures efficient water use, reducing waste while maintaining plant quality. This system can optimize irrigation by applying only the necessary water based on real-time soil moisture levels, leading to improved growth and reduced environmental impact." Soil moisture sensors are best used in places where there is minimal movement to keep in mind the consideration around being "non-intrusive", e.g. field grown crops, larger crops like trees, or other crops which will be fairly static for a longer period of time. Soil sensors are not optimal in 1–5-gal container nursery settings where the sensors can get lost, stepped on, or shipped when the plant is sold.

Lastly, adjust your technology and infrastructure. By better controlling each individual valve associated with individual irrigation zones or plant groups, you yield better quality and gain a return on your investment over time and reduce time related to irrigation related activities, as well as water, per Zazueta et al. (1992).

THE POWER OF TECHNOLOGY: AI, ML, AND IOT IN IRRIGATION

Technologies of great potential include artificial intelligence (AI), machine learning (ML), and the internet of things (IoT). Think of the thermostats in your home you can control from your phone as an example of an IoT device, which may use artificial intelligence and machine learning to determine when you normally get home to turn on the air conditioning in advance and how to save you electricity, while keeping you comfortable. The cool thing about AI is it is always learning, always getting smarter. The more data, the better it gets at figuring out what your plants need - and making those little changes that make all the difference in your production. This opens up the possibility of proactively sending irrigation schedules out in advance, based on historical weather data, plant information (such as container size, age, etc.) and being able to predict the run-time needed with accuracy.

Additionally, if you have your water-usage dialed in - you have the ability to experiment with using irrigation as a growth regulator, so you do not need as much production time producing a finished crops before shipping them. As one nursery mentioned, using a leaching fraction of 0% for hydrangeas.

WHY TRANSITION TO MODERN IRRIGATION SYSTEMS NOW?

The Old World of Irrigation. In this old world, irrigation decisions are made based on single tasks, and gut instincts rather than data (**Fig. 4**). This task-centric mindset leads to fragmented operations and missed opportunities for efficiency. With traditional watering methods, you tend to overwater and accidentally create the perfect breeding ground for fungus or disease. I am sure you can recall the image of someone dragging hoses around or moving those big clunky sprinklers… all of that is automated with modern irrigation. With traditional overhead irrigation, there is wasted water, fertilizer, and unhappy plants. And instead of spending hours on manual watering, those workers can focus on tasks that actually require a human touch. No more picking up pots to see if they are dry, adjusting irrigation based on a guess of how much the weather or pruning and spacing affected the water needs, or manually turning on and off valves in the Quonset house or greenhouse.

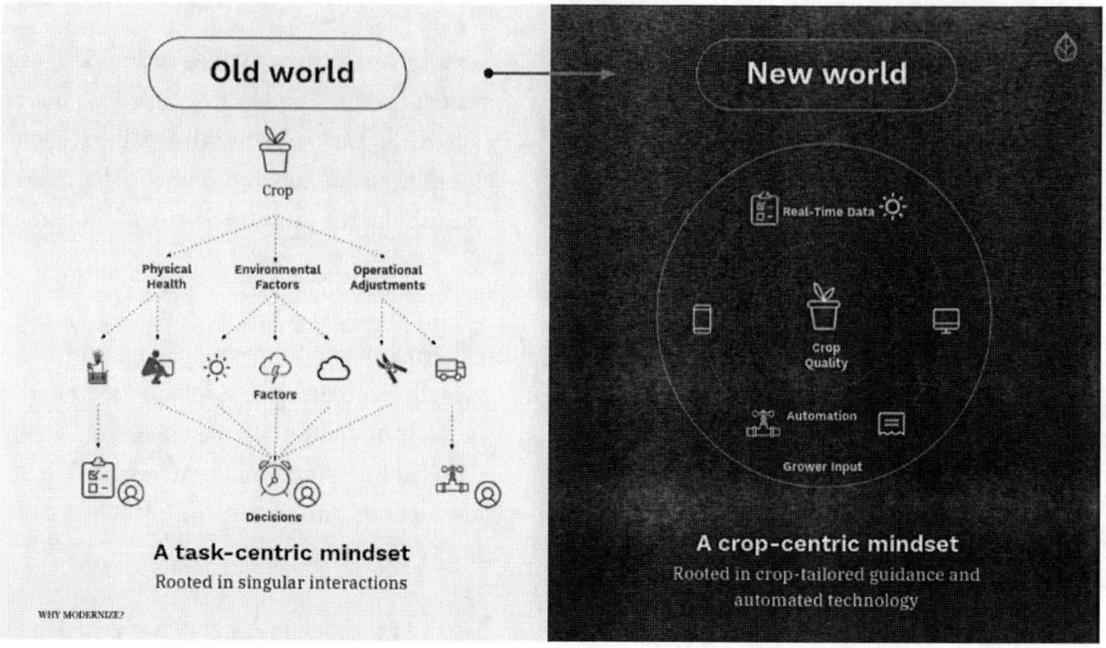

Figure 4. Old vs new world in irrigation, with the focus moving from tasks to crop quality.

IMAGINE THE NEW WORLD OF IRRIGATION – The future is already here

Imagine walking into your nursery in the morning. Instead of checking timers or manually adjusting irrigation, you know that real-time data from a weather station placed around your nursery has already been gathered overnight—measuring rainfall, humidity, wind speed, and even sunlight intensity. This data is automatically sent to a central system that analyzes it and makes real-time adjustments to your irrigation schedule. It's like having a 24/7 irrigation manager that never takes a break.

Now, think about what this means for your day. Instead of managing watering schedules manually or responding to plant issues after they have already occurred, you have time to focus on what really matters—propagation, quality checks, and optimizing crop yield to reduce shrinkage. Automation is not just a time-saver; it puts consistent crop quality at the center, using technology and real-time data to support your decisions. With your grower input holding it all together!

Picture this: no more guesswork, no more reacting to plant stress - too late! Instead, your crops receive the exact amount of water they need, when they need it. This system prevents both over- and under-watering, saving water, avoiding nutrient loss, and preventing root rot. The result? Healthier plants, better growth, and no wasted resources. So, what does a day in the life look like with modern irrigation? It is a day where you spend less time troubleshooting irrigation and more time nurturing healthier, more profitable plants, all while your system works in the background to optimize water-use and ensure your crops thrive.

So, why not invest in modern irrigation solutions? One can boost your efficiency with faster adjustments, improve your crop quality and nutrient retention, while lowering your operational costs, and enhancing sustainability (**Fig. 5**). Technology is already giving some nurseries a cutting edge, while allowing them to stay ahead of labor risks and regulatory decisions. Water, an increasingly scarce resource, is being consumed far faster than it can regenerate.

"The availability of water for nursery irrigation is expected to decline, as urban expansion and drought conditions increase competition for water resources, forcing nurseries to adopt more efficient irrigation technologies (Beeson et al., 2004)."

Figure 5. Making the case for investing in modern solutions.

Regulations around water use are tightening in many regions, with some areas mandating significant reductions. For example, a Canadian nursery we spoke with must reduce their water usage by 50% in 2024 compared to 2023. Why not take proactive steps to meet these challenges before they restrict your production? "The integration of advanced technologies such as AI, ML, IoT, and renewable energy solutions offers significant opportunities to improve irrigation efficiency. These innovations enable precise water management, reduce resource wastage, and enhance agricultural productivity (Ejaz et al., 2024)."

REAL-WORLD SUCCESS STORIES

Saunders Brothers. Diving into a few of the real-world success stories, let's start with Saunders Brothers, a nursery close to my heart! This is related to the project I mentioned at the start. They were really looking to tackle this issue of over-watering and fertilizer waste. This entailed using ET-based irrigation, the leaching fraction system, automation, and an infrastructure investment.

This resulted in a 43% water and chlorine savings in the first year alone. Fertilizer use dropped by 33% on some crops, showing the return on investment achievable through modern irrigation (Stanley et al., 2019).

The system Saunders Brothers uses was developed through a partnership with the University of Florida and an electrical engineer, custom building the components. Canopy (Canopygrow.tech) is focusing on taking knowledge from this implementation to provide similar solutions to other nurseries.

Holden Nursery. Holden Nursery also utilized the leaching fraction system, cutting their water usage by 50%. Less water has to have some trade off in quality, right? No! Not only did they use less water, but their plants are greener and healthier than before (Yeary et al., 2014).

Hibernia Nursery. The last example, Hibernia Nursery, also with leaching fraction testing and automation estimated $35,000 to $40,000 annually on labor costs because they need fewer people to manage the irrigation now. That savings does not include the electricity savings on top of that (Million and Yeager, 2019). These small changes can make a big difference, less waste, lower costs and it's better for the environment!

PRACTICAL STEPS FOR YOUR NURSERY

In candor, not every nursery can go from zero to fully automated overnight. It is definitely a process. But even making a few small changes, like adding leaching fraction measurements or tweaking your watering schedule based on the weather forecast, can make a difference.

Choosing a plug-and-play technology, one can tailor the irrigation to each crop - allowing a nursery to reach scale. By separating plant groups into irrigation zones, one can customize the amount of watering per each group throughout the growing season. And as those smaller steps are taken, look into automation to remotely control your irrigation system without needing to send a person around to turn the valves on and off. These changes might look different for each nursery, so depending on your current setup:

- *Valves > Solenoids + Timers* - changing from valves to solenoids with timers.
- *Solenoids + Timers > Automated* - upgrading from solenoids and timers to an automated system that turns the solenoid on and off based on weather and grower inputs.

In discussions with nurseries, it became evident that selecting the right solenoid is crucial to achieving optimal performance. Solenoids vary in functionality, with some electrically activated, such as Rainbird models, and others pressure-activated. Choosing a solenoid that aligns with your system's pressure availability ensures the desired irrigation outcomes and prevents operational inefficiencies. For instance, using a pressure-activated solenoid without sufficient pressure in the pipe can prevent the valve from opening, resulting in no water delivery to that zone. Ensuring the solenoid matches the system's pressure requirements is essential for reliable irrigation performance.

Summarizing the actions you should invest towards (Fig. 6):

1. *Start Small:* Implement leaching fraction testing or adjust schedules based on weather.
2. *Aim for Scalable Solutions*: Zone-based irrigation allows for growth, while meeting specific crop needs.
3. *Invest in Automation*: Remote controls for irrigation reduce manual intervention, ensuring consistency and efficiency.

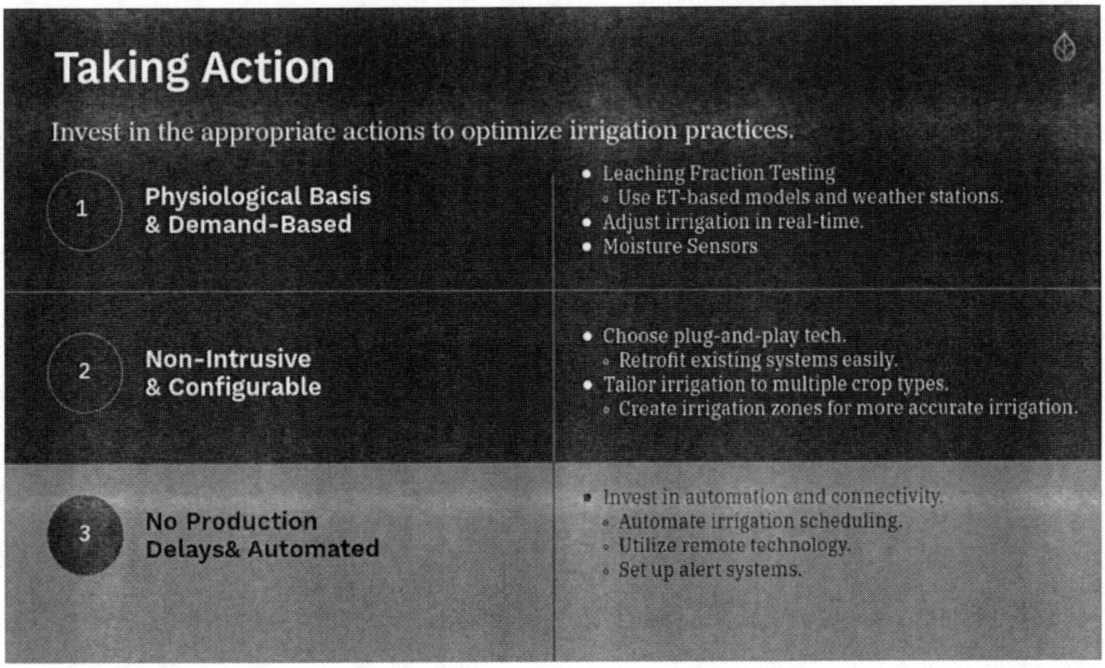

Figure 6. Taking the right actions can start small and build from there.

MEASURING OUTCOMES (Fig. 7)

Never be afraid to try new things and stay curious. That is what we are all about at IPPS. Learning from each other and sharing new ideas. As the latest *IPPS International Newsletter* advocated: becoming the beacon for fostering sustainable, resilient, and productive plant ecosystems.

I have covered water savings, healthier plants with reduced waste, versatility across your entire nursery, and ensuring no delay in finishing your crops. But what also needs to be emphasized are personnel – people who are out there doing the work day in and day out. You have to consider the human-side of things. Especially, when it comes to improving the growers' knowledge! By having a system inform them how the weather affects the irrigation, how the adjusted watering affects the salt levels when taking EC (electro-conductivity) measurements, how growth of the plant increases or decreases the amount of irrigation needed due to the canopy of the crop (after taking a leaching fraction measurement), - and what all of that means in the amount of time they should recommend to water that plant each day. That is a lot for a grower or irrigation manager to keep in their head!

It is also about making your jobs a little easier. When you can automate these tasks that are tedious and physically demanding - it frees up your team to use their brains and skills on tasks that are more interesting and rewarding. You know what happy employees mean?

They stick around, less turnover - which means less time and money spent on training new people. I am sure you have heard the saying that: The employee that opens and closes your valves controls the crop quality. Who is opening and closing your valves?

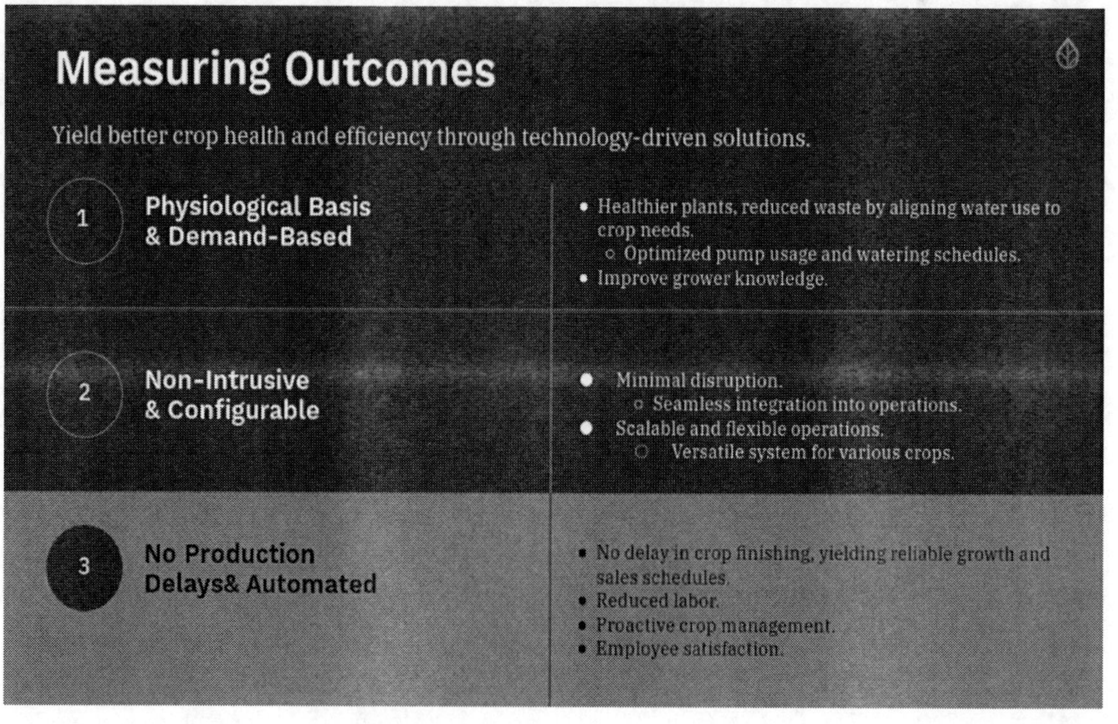

Figure 7. Measuring the outcomes related to your investments can lead to better crop health

CONCLUSION

There is a quality, smarter future for nurseries. "The future of irrigation lies in the continued integration of advanced technologies and effective collaboration across stakeholders to ensure sustainable water use (Ejaz et al., 2024)." Progress will depend on partnerships, universities driving research, nursery professionals sharing practical insights, and the commercial sector developing solutions tailored to industry needs. Hopefully this gives good insight into ways you can grow healthier plants, meaning you can get your products to market more effectively. Because in a seasonal business like a nursery, timing is everything!

And with these investments towards your future irrigation system of ET and leaching fraction-based systems and/or automation and technology, you can save time, water, fertilizer, and labor. Which in turn, you can use to grow your business by reinvesting those resources to get more of your crops in customers hands each season

LITERATURE CITED

AgriFi.AI. (2024). Smart irrigation systems: revolutionizing agriculture with data-driven solutions. AgriFi.AI Report. *15*: 45-59.

Bayer, D., Ruter, J. M., and van Iersel, M. W. (2015). Automated irrigation control for improved growth and quality of *Gardenia jasminoides*. HortTech. *25*:349-356.

Beeson, R.C., Arnold, M.A., Bilderback, T.E., Bolusky, B., Chandler, S., Gramling, H.M., Lea-Cox, J.D., Harris, J.R., Klinger, P.J., Mathers, H.M., Ruter, J.M., and Yeager, T.H. (2004). Strategic vision of container nursery irrigation in the next ten years. J. Environ. Hort. *22*:113-115. DOI: 10.24266/0738-2898-22.2.113

Ejaz, M., Godwin, H., and Cena, R. (2024). Innovative technologies and stakeholder collaboration boosting irrigation efficiency. Agricul. J. *78*: 123-135.

Fulcher, A., Buxton, J. W., and Geneve, R. L. (2012). Developing a physiological-based, on-demand irrigation system for container production. Scient. Hort. *138*: 221-226. https://doi.org/10.1016/j.scienta.2012.02.030.

Fulcher, A., Rihn, A.L., LeBude, A.V., Schexnayder, S, Altland, J.E., Bumbarner, N., Marble, S.C., Nackley, L., Palma, M., Velandia, M., Zhu, H., Gan, H. and Owen, J.S. (2023). Overcoming the nursery industry labor shortage: a survey of strategies to adapt to a reduced workforce and automation and mechanization technology adoption levels. HortScience. *58*:1513–1525. https://doi.org/10.21273/HORTSCI17230-23

Hunter, P., Scattini, C., Davis, M., Hickey, K., and Levander, J. (2010). The twin-pot water management system: demand (water) driven, automated container irrigation that increases crop turnover. J. Irrig. Drain Eng. *136*:105-112.

Million, J., and Yeager, T. (2019). Lose the leachate: an automated weather-based irrigation system helps Florida's Hibernia Nursery save water and labor. Nurs. Mgmt. https://www.nurserymag.com/article/automatic-irrigation-system-containers-hibernia-nursery/

Million, J., Yeager, T., and Albano, J. (2010). Evapotranspiration-based irrigation scheduling for container-grown *Viburnum odoratissimum* (L.) Ker Gawl. HortScience. *45*: 1741-1745. https://journals.ashs.org/hortsci/view/journals/hortsci/45/11/article-p1741.xml

Owen Jr., R. A., LeBude, A., Fulcher, A., Stanley, J., and Oki, L. (2019). Leaching fraction: a tool to schedule irrigation for container-grown nursery crops. Virginia Tech Cooperative Extension Publication SPES-128P. DOI:10.13140/RG.2.2.30534.22089

Saunders, C. (2024a). Understanding leaching fraction testing: A beginner's guide. Canopy. https://www.canopygrow.tech/blog/understanding-leaching-fraction-testing (Accessed: 18 November 2024).

Saunders, C. (2024b). Enhancing irrigation efficiency with weather-based systems. Canopy. www.canopygrow.tech/blog/weather-based-irrigation-efficiency (Accessed: 18 November 2024).

Saunders, C. (2024c). Reducing labor with automated irrigation systems. Canopy. https://www.canopygrow.tech/blog/reduce-labor-in-irrigatio n (Accessed: 18 November 2024).

Saunders, C. (2024d) Cut fertilizer and fungicide with smart irrigation schedules. Canopy. Available at: https://www.canopygrow.tech/blog/smart-irrigation-reduce-fertilizer (Accessed: 18 November 2024).

Saunders, C. (2024e) Increase nursery profits with smart irrigation strategies. Canopy. Available at: https://www.canopygrow.tech/blog/improving-profit-margins-within-container-nurseries (Accessed: 18 November 2024).

Saunders, C. (2024f) Irrigation and crop health - connecting the dots. Canopy. Available at: https://www.canopygrow.tech/blog/understanding-leaching-fraction-testing (Accessed: 18 November 2024).

Tilt, S. (2000). Small steps to improving your irrigation system, reducing labor and increasing your bottom line. Comb. Proc. Intl. Plant Prop. Soc. *50*:501-505.

Yeary, W., Fulcher, A., and Lieb, B. (2014). Nursery Irrigation: A Guide for Reducing Risk and Improving Production. University of Tennessee Extension publication PB 1836. https://ag.tennessee.edu/plantsciences/Documents/Fulcher-Sustainable%20Nursery/fulcher_irrigation_manual.pdf

Zazueta, F. S., and Smajstrla, A. G. (1992). Computer control of irrigation systems. J. Irrig. Drain. Eng. *118*:434-439.

Combined Proceedings IPPS 74:469-473. 2024.

Building an Ecosystem in Your Container

Mike Serant

San Jacinto/MicroLife, 2221 West 34th Street, Houston, Texas 77018 USA

mserant@sanjacsupply.com

Keywords: microorganisms, microbes, microbiome, rhizosphere, plant health

Summary

Evolution created a symbiotic relationship between plants and beneficial microbes that when properly implemented will produce healthier plants more economically and sustainably. By working with a more sustainable plant ecosystem – via incorporation of microbes to the plant rhizosphere – plants are healthier, more resistant to pests, require fewer chemical inputs, mitigate potential pollution problems, create a safer workplace – and ultimately, produce crops more economically.

INTRODUCTION

Microorganisms (microbes) are crucial for all ecosystems from the human body to farmlands, to all areas of the world and to earth itself. We humans would not exist without beneficial microbes. In fact, over half the cells in the human body are non-human – composed of bacteria and other organism in our microbiome. In humans, beneficial microbes are part of our immune system and have great influence on the way we think.

The same holds true for plants. Higher land plants could not have come into existence 460 million years ago without the aid of microbes (McNear, 2013). This is natural law at work. Evolution created a symbolic bond between plants and beneficial microbes that when appreciated, will grow healthier plants with less expense and less work (Harman, et al., 2021). Meaning, ensure that your soil ecosystem is thriving with beneficial microbes. That is easy to do. Just introduce them to your soil/media system and support them with good microbial foods, judiciously use chemical fungicides and make sure there is sufficient oxygen availability – for a healthy, aerobic root system environment.

All the science and data are there. While relatively new to horticulture, the study and use of microbes for plants and humans has become increasingly important (Davies, 2008). Simply put, without a good microbial system in place - all humans, animals and plants would not exist.

Unfortunately, most American do not have a healthy gut microbiome and therefore America leads the world in all categories of chronic illnesses. Poor health causes premature death, lingering sickness, huge medical bills and impairs the quality of life. The same holds true for most plants grown and maintained in America. The gut microbiome for the plants is the soil (plant rhizosphere); and if soil is not healthy - then plants are not healthy. Soil gut health, like human gut health, is determined by inputs - that determines whether good microbes or bad microbes flourish. Unsuitable inputs for humans include super-processed foods, pesticide sprayed fruits and vegetables, and foods with low nutrient density.

Bad inputs for the soil rhizosphere are unsustainable use of chemical pesticides, particularly chemical fungicides, and high salt chemical fertilizers. These will kill the good microbes, and the vacuum created will be filled with nothingness or dominated by bad microbes.

A soil system dominated by bad microbes will weaken a plant's ability to grow and defend itself - which means more inputs are required to keep them alive to the sellable stage.

We analyzed typical growers' soil/media (aged pine bark) for microbial activity. The untreated soil was very low in microbial activity. Then we added a bio-inoculant to the same soil and the numbers had a huge increase (**Table 1**). We then followed some of the plantings inoculated, measuring the microbial count at 30 days, 60 days and 90 days and found significant microbial counts. The numbers can be quite staggering (**Table 2**). A large number of microbes can be introduced to a 4-in (10 cm) square pot (**Table 3**).

Table 1. Soil microorganisms in inoculated and non-inoculated, aged pine bark.

Aged Pine Bark l Laboratory Count CFU Per Gram

ORGANISM	NO INOCULATION	WITH BIO INOCULANT
Bacillus - Paenibacillus	1.4×10^6	$32{,}000 \times 10^6$
Pseudomonas	0	85×10^7
Streptomycetes	0	3.0×10^6
Trichoderma	0	1.5×10^6
Glomus	0	930

Table 2. Two 90-day bioinoculant trials.

BIO INOCULANT TRIALS - 90 DAYS			
SAMPLE ID	TOTAL COUNT	NON-MYCORRHIZAL COUNT (BACILLUS, PSEUDOMONAS, STREPTOMYCES, TRICHODERMA)	MYCORRHIZAL FUNGI COUNT
Bio 1 - March	1.200×10^7 cfu/gm	1.160×10^7	4.000×10^5 cfu/gm
Bio 1 - April	3.000×10^8 cfu/gm	2.998×10^8	2.000×10^5 cfu/gm
Bio 1 - May	4.000×10^9 cfu/gm	3.99965×10^9	3.500×10^5 cfu/gm
Bio 2 - March	3.600×10^7 cfu/gm	3.550×10^7	5.000×10^5 cfu/gm
Bio 2 - April	6.000×10^8 cfu/gm	5.970×10^8	3.000×10^6 cfu/gm
Bio 2 - May	9.000×10^9 cfu/gm	8.995×10^9	5.000×10^6 cfu/gm

So, who are the main beneficial microbes we are talking about? They primarily will fall into 5 groups: bacteria, fungi, actinomycetes, protozoa and nematodes (**Table 4**).

Table 3. Organisms in the plant rhizosphere.

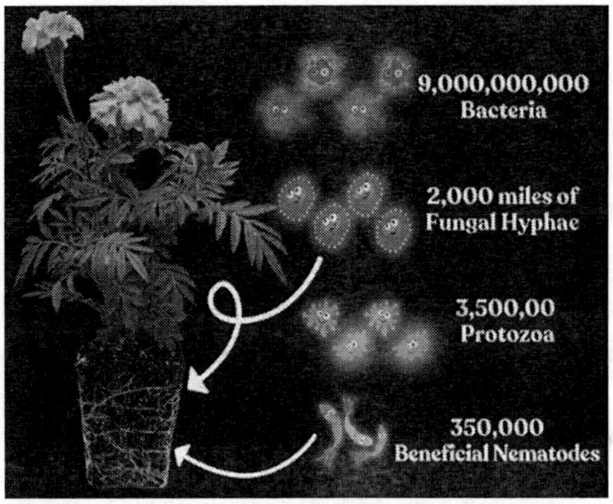

Table 4. Major plant rhizosphere microbial components.

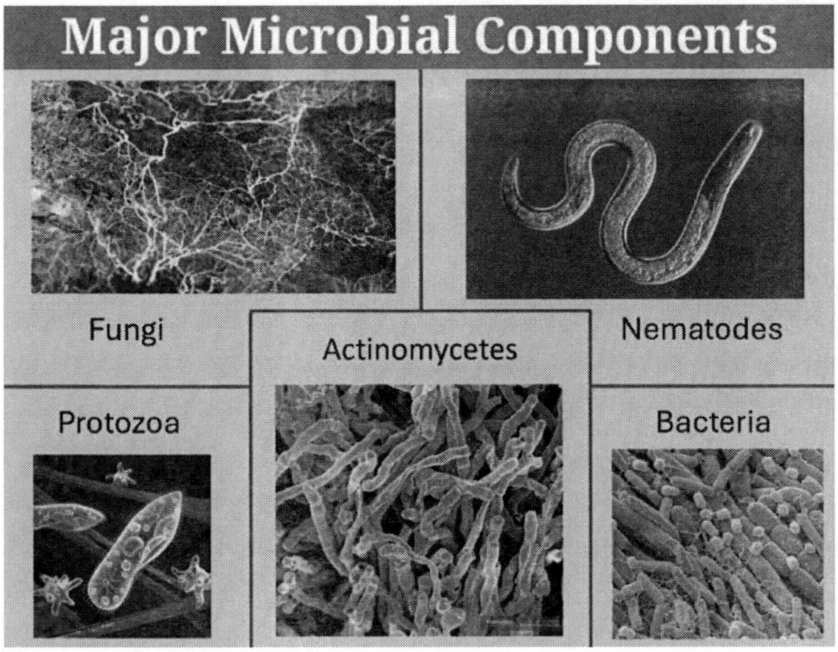

There are opportunities with using mycorrhizal fungi to benefit nursery propagation and production systems (Davies, 2008). These beneficial microbes are your 24/7 work force that have a symbiotic relationship with plants. The plant directly supports the microbes through root exudates (carbohydrates, amino acids, organic acids, enzymes) - in exchange for benefits such as solubilization of immobile elements, increased nutrient uptake (extraradical hyphae of mycorrhiza), drought resistance, pest resistance – and an overall - more healthy/resilient plant ecosystem.

Some of the essential benefits that plants will receive by having a healthy soil microbial population include:

- plant protection from soil disease and pest insects
- making more minerals available
- more extensive root system
- faster, healthier growth
- reduced production time & enhanced post-production longevity
- resistance to abiotic stress

A management tool to measure plant health is a Brix refractometer (**Fig. 1**.). Brix measures the amount of sugar in a plant which equates to plant energy/health (Roe, 2021). The more energy a plant has, the healthier it is. If you can get to a Brix reading of 15, the plant is basically bullet proof to most insect pests and diseases. Brix measurement is simple and fast. Plant leaves are pressed to drop liquid plant sap on the Brix refractometer or hydrometer - and the reading is immediate.

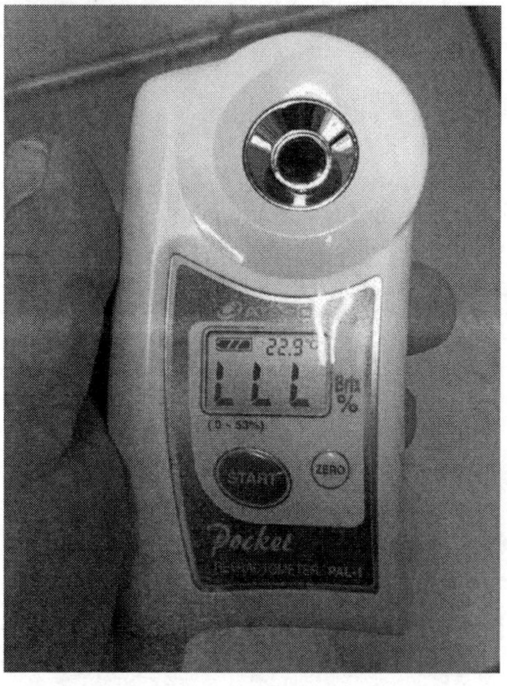

Figure 1. A refractometer is used to measure the leaf Brix which is the carbohydrate/sugar concentration as a percentage. Testing the Brix leaf sap with a refractometer is a quick way to determine plant health.

To achieve high Brix levels, these conditions must be met:

- only use high grade fertilizers, quality organics are preferred
- have a healthy soil ecosystem; use bioinoculants as needed.
- judicious use of chemical fungicides
- reduce stress, do not over-water, etc
- foliar spray with combinations of fish, seaweed, molasses, humic acid

Foliar spray with combinations of fish, seaweed, molasses, humic acid can raise the photosynthetic efficiency which equates to healthier plants and soils.

CONCLUSION

The world of organics, soil microbiology and nutrition is fascinating and provides a lifetime of exciting learning. By working with a more sustainable plant ecosystem – via incorporation of plant microbes to the rhizosphere – plants will be healthier, more resistant to pests, require less chemical usage, mitigate potential pollution problems, create a safer workplace – and ultimately, produce crops more economically.

LITERATURE CITED

Davies, F.T. (2008). Opportunities from down under: how mycorrhizal fungi can benefit nursery propagation and production systems. Comb. Proc. Intl. Plant Prop. Soc. *58*:539-548.

Harman, G., Khadka, R., Doni, Febri, and Uphoff, N. (2021). Benefits to plant health and productivity from enhancing plant microbial symbionts. Frontiers Plant Sci. 10.3389/fpls.2020.610065

McNear Jr., D. H. (2013) The rhizosphere - roots, soil and everything in between. Nat. Ed. Know. 4(3):*1*.

Roe, M. (2021). Is your fertility program stuck in the 1950's? Nutritional approach to passive plant pest immunity. Comb. Proc. Intl. Plant Prop. Soc. *71*:222-234.

Combined Proceedings IPPS 74:474-479. 2024.

Managing Plant Nutrition for Resistance to Pests and Diseases

Jason Stoll

Advancing Eco Agriculture, 4551 Parks West Rd, Middlefield, Ohio 44062 USA

jstoll@advancingecoag.com

Keywords: plant health, plant immune systems, disease and insect resistance, plant health pyramid, regenerative agriculture

Summary

Advancing Eco Agriculture (AEA) is a global leader in regenerative agriculture. The mission is to provide farmers and growers with innovative, science-backed solutions to regenerate and revitalize agriculture and food systems. We need to rethink what causes plant disease and pest outbreaks. It is the health of a plant! A nutritionally unfit plant lacks resistance to disease and pests. This paper covers plant immunity – and integrating the plant health pyramid into more sustainable farm management. The plant health pyramid model demonstrates how plant health/immunity progresses through different levels with greater resistance to pests (microbes to animals – disease & insects) via the plant's ability to produce more complex proteins, carbohydrates and phytonutrients. Healthier plants reach the top of the pyramid and have great immunity/resistance to pests.

IPPS Vol. 74 – 2024

INTRODUCTION

The Hidden Resilience of Healthy Plants. The presence of pests and diseases is often a symptom of deeper problems in plant health. While agricultural practices have traditionally responded to these issues with pesticides, fungicides, and other interventions, a growing body of evidence supports a different approach: building plant health to prevent susceptibility. At Advancing Eco Agriculture (AEA), we have seen that healthy plants - with the right nutrition - have natural defenses against many pests and diseases. This concept is more than theoretical—it has practical applications on the farm, allowing growers to foster robust, resilient crops.

UNDERSTANDING PLANT IMMUNITY

Plant immunity can be understood as a progressive journey through four distinct levels, each tied to specific physiological processes and nutrient requirements. This progression was effectively captured by John Kempf's "Plant Health Pyramid," which provides a framework for understanding how plants can incrementally achieve pest and disease resistance https://advancingecoag.com/plant-health-pyramid/ . As plants ascend each level, they gain new immunity, beginning with soil-borne fungal pathogens and advancing toward resilience against more complex threats, including insects and even viruses.

LEVEL 1: Complete photosynthesis

The foundation of plant health begins with complete photosynthesis (**Fig. 1**). This stage is critical, as it enhances the plant's capacity to produce sugars—vital energy sources that power growth and immune functions. A plant photosynthesizing at full capacity can boost its carbohydrate production by as much as 3-4 times, a shift that can be easily monitored in the field with a refractometer to measure Brix levels. When plants photosynthesize efficiently, their Brix readings can leap from an average of 3-5 up to 12-15 or higher, indicating healthier, more robust energy production.

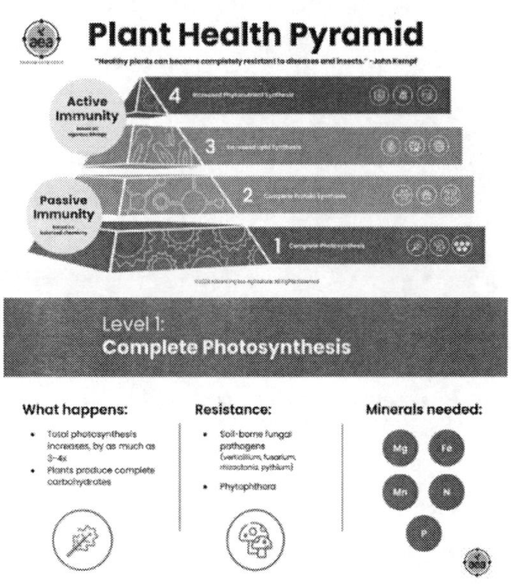

Figure 1. Level 1 on the plant health pyramid with increased photosynthesis and passive immunity to soil-borne fungal pathogens.

This increase in sugar quantity goes hand-in-hand with improved sugar quality. Plants at this stage produce complex carbohydrates instead of simple, non-reducing sugars, which changes the profile of carbohydrates in the plant's root exudates. This shift has significant implications for soil health, as complex carbohydrates support a microbial community in the rhizosphere that actively suppresses soil-borne fungal pathogens. At this level, plants become naturally resistant to soil-borne fungal diseases, such as verticillium, fusarium, and rhizoctonia.

To achieve Level 1, plants need five critical minerals involved in photosynthesis: magnesium, nitrogen, iron, manganese, and phosphorus. While magnesium and nitrogen are essential to chlorophyll synthesis, iron is crucial for chlorophyll assembly, and manganese is needed for water hydrolysis in photosynthesis. Phosphorus supports ATP production, which is essential for metabolizing the sugars generated. AEA products like MacroPak and MicroPak can supply these minerals - with MacroPak providing nitrogen, phosphorus, and magnesium and MicroPak offering iron and manganese.

LEVEL 2: Complete protein synthesis

The second level of plant health centers on the synthesis of complete proteins (**Fig. 2**). Plants at this stage efficiently convert nitrogen into complete proteins. At this point, plants become more resistant to pests with simple digestive systems, such as larval insects and aphids, as these pests do not have the digestive capacity to handle complete proteins. Growers can monitor a plant's protein synthesis through plant sap analysis, observing nitrate and ammonium levels, which should be close to zero if the plant is actively synthesizing proteins.

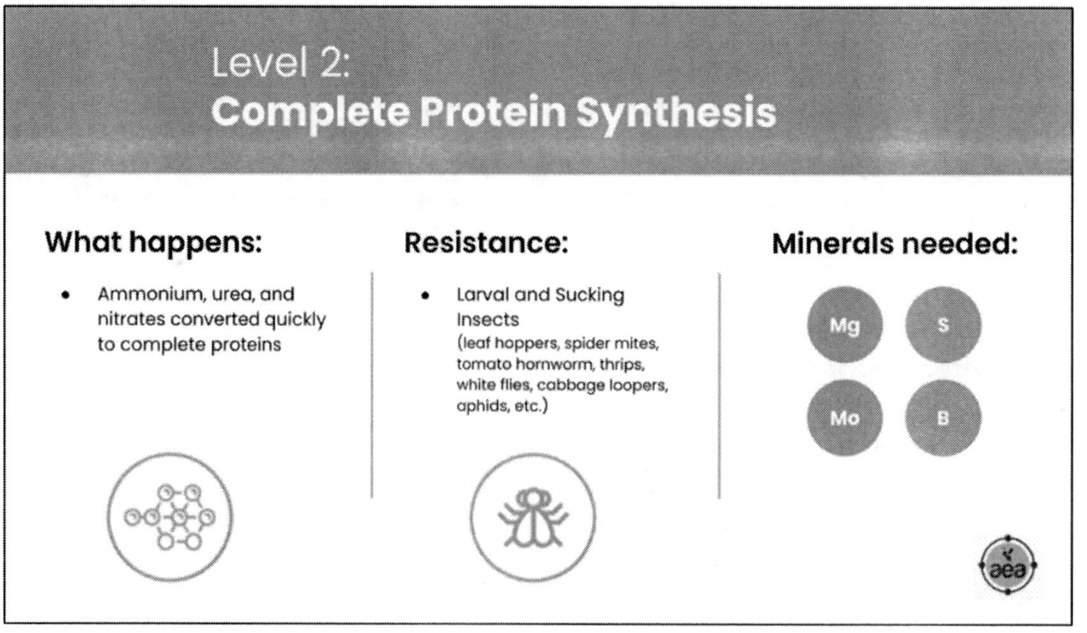

Figure 2. Level 2 on the plant health pyramid with with ammonium, urea and nitrates converting to complete proteins.

To advance to this level, plants need four essential minerals: magnesium, sulfur, molybdenum, and boron. Magnesium, sulfur, and molybdenum directly support nitrogen synthesis, while boron strengthens pest resistance by stabilizing cell walls. AEA's PhotoMag™ delivers this combination, or growers can use MacroPak and MicroPak together to cover these nutrient needs.

LEVELS 1 AND 2: Passive immunity through balanced chemistry

Levels 1 and 2 provide passive immunity to pests and diseases. In these stages, plants create metabolic conditions so that they are no longer a food source, deterring infestations before they begin. This type of immunity relies on balanced plant chemistry and can often be achieved quickly with targeted nutrient applications. Growers aiming to reach these levels can apply the required minerals through well-formulated foliar applications to address common nutrient imbalances.

LEVEL 3: Increased lipid synthesis

At Level 3, plants reach a state of energy surplus, allowing them to produce lipids, which they store as a waxy layer on their leaves **(Fig. 3)**. This layer becomes a physical shield, preventing airborne pathogens from penetrating plant tissues. Plants at this stage develop resilience against fungal and bacterial pathogens, including powdery mildew, downy mildew, and rust.

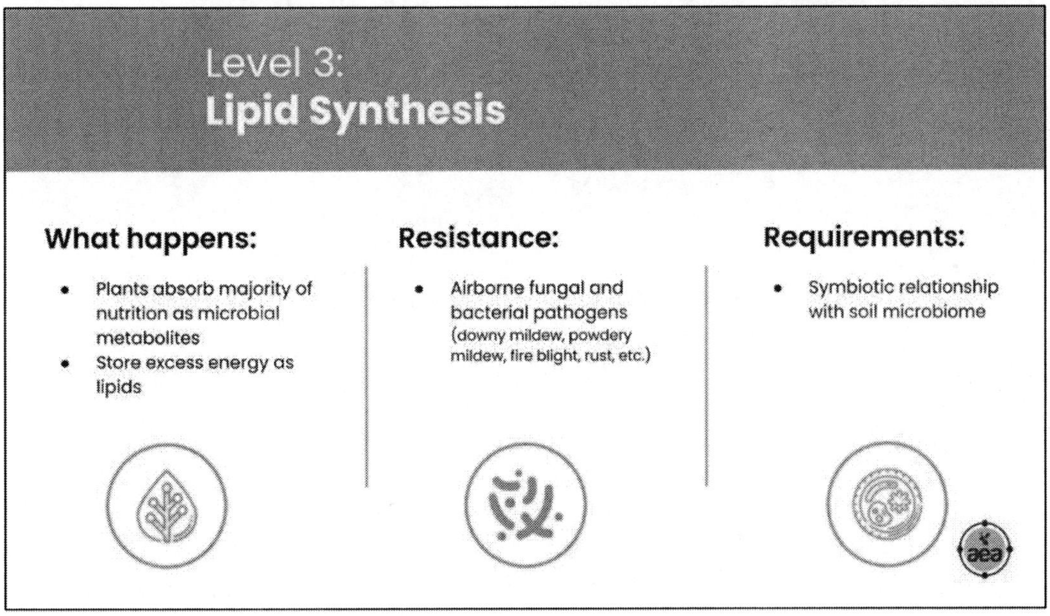

Figure 3. Level 3 on the plant health pyramid with plants storing excess energy as lipids.

Unlike Levels 1 and 2, which are largely chemistry-driven, Level 3 is influenced by the microbiome. As plants interact with soil microbes, they absorb complex nutrients, conserving energy that would otherwise go toward nutrient processing. This "prefabricated" nutrition enables sufficient energy reserves for lipid synthesis, creating the physical barriers essential for Level 3 immunity.

LEVEL 4: Elevated phytonutrient synthesis

The pinnacle of plant health, Level 4, is marked by the production of plant secondary metabolites—compounds such as flavonoids, terpenoids, and alkaloids **(Fig. 4)**. These phytonutrients act as potent defenses against a broad spectrum of threats, including chewing insects like beetles and, potentially, viruses. At this level, plants shift to active immunity, using biochemical defenses to repel pests and pathogens. To sustain this level of resistance, plants depend on a diverse and abundant microbiome, which stimulates systemic acquired resistance (SAR) and induced systemic resistance (ISR) within the plant.

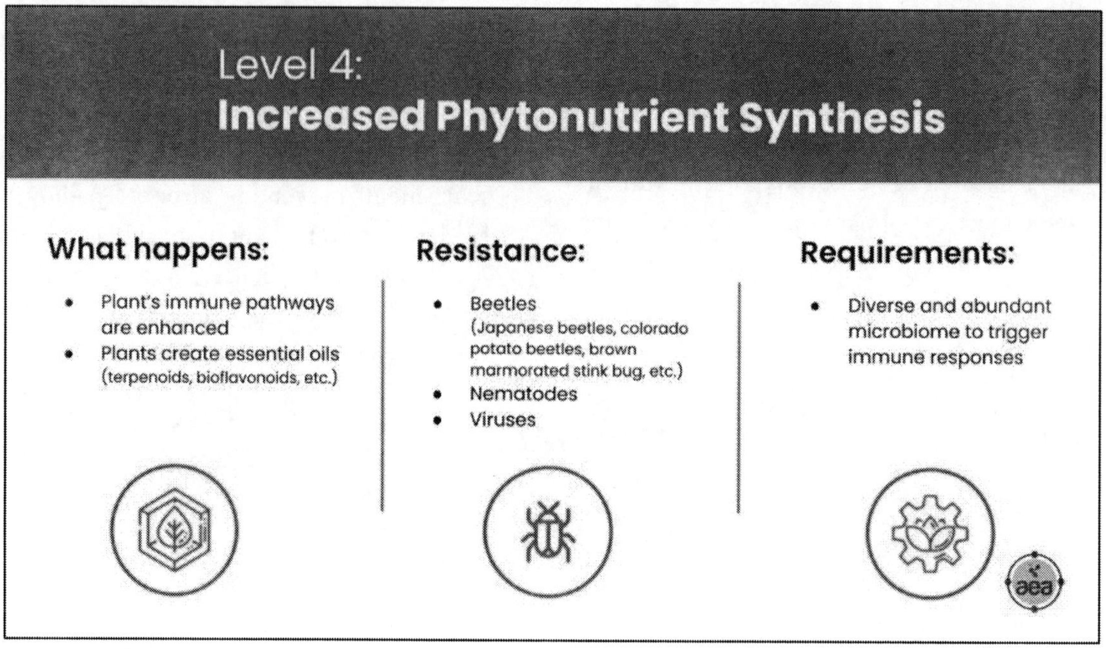

Figure 4. Level 4 on the plant health pyramid with increased phytonutrient synthesis.

INTEGRATING THE PLANT HEALTH PYRAMID INTO FARM MANAGEMENT

The Plant Health Pyramid provides a roadmap for growers to enhance crop resilience through balanced nutrition and microbiome management. By systematically progressing through each level, growers can support their plants in achieving new levels of health and immunity, reducing reliance on chemical interventions while producing healthier crops and contributing to a more sustainable agricultural ecosystem.

As growers strive to manage plant nutrition and microbiome health, they move beyond symptom treatment and embrace a holistic approach that supports plant resilience. This approach not only strengthens crop productivity but also promotes regenerative agriculture, where healthy plants, ecosystems, and communities thrive together **(Fig. 5)**.

HOW TO HACK THE SYSTEM

Degenerative Agriculture Cycle

Without an excellent nutritional program in place to feed soil biology and microbes, plant and soil vigor degrades, and fruit quality suffers.

1. Low plant nutrition leads to lower photosynthetic efficiency
2. Decreased photosynthetic efficiency means less sugars to the roots
3. This decreases the available food source for microbes in the soil
4. Decreased microbial activity lessens mineral availability
5. Poor mineral absorption by plants equals pest susceptibility and poor fruit quality

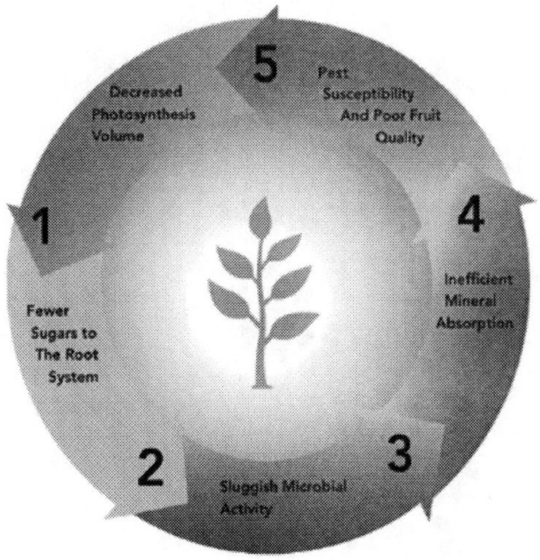

Regenerative Agriculture Cycle

Work with Advancing Eco Agriculture to turn soil activity, fruit quality, and profitability in a positive direction.

1. Nutritional foliar sprays increase photosynthetic efficiency
2. Increased photosynthetic efficiency increases volume of sugars to root system
3. Increased volume of sugars moved out through the root system increases microbial activity in the soil
4. Increased microbial activity in the soil increases mineral availability
5. Better mineral absorption by plants equals greater disease and pest resistance and higher quality and yield

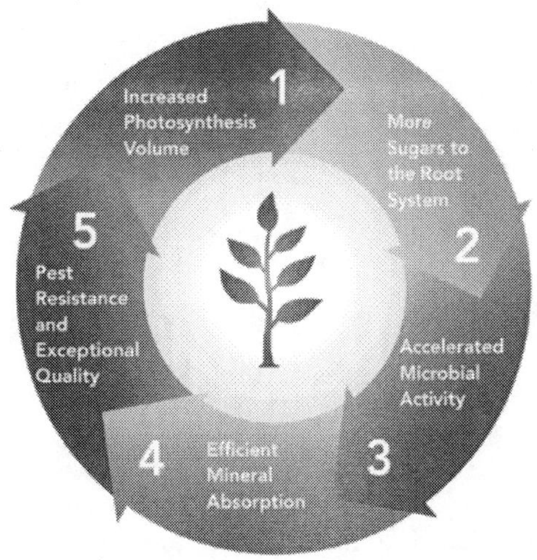

Figure 5. How to hack the system: (top) a degenerative agriculture cycle compared to (bottom) a more sustainable, regenerative agriculture cycle.

Combined Proceedings IPPS 74:480-495. 2024.

IPPS European Exchange 2023

Teagan H. Young

Department of Environmental Horticulture, University of Florida, Gainesville, Florida 32611 USA

Teagan.young@ufl.edu

Keywords: IPPS Southern Region of North America (IPPS-SRNA), Early-Career Professional International Exchange Program, IPPS European Region Conference, Danish and Swedish nurseries

Summary

This paper discusses the IPPS-SRNA Early-Career Professional International Exchange Program with the IPPS-European Region took place in 2023. The exchange proved to be an unparalleled, transformative journey—one I consider among the most profound and impactful in my life. The European Region's annual meeting took place in Køge, Denmark, encompassing other cities such as Copenhagen, Denmark and Lund, Sweden. I was able to tour nurseries in Germany, Denmark and Sweden.

IPPS Vol. 74 – 2024

INTRODUCTION

In 2021, I attended my first IPPS Southern Region Annual Meeting in Mobile, Alabama, and was honored to receive the Vivian Munday Young Horticulture Professional Scholarship. Engaging in various roles, from operating the registration desk and assisting during the auction to participating in bus tours and nursery visits - allowed me to connect with numerous members and explore the diverse offerings of IPPS. Conversations with Donna Foster and other committee members during the conference introduced me to the Charlie Parkerson Student Research Competition and the Early Career Professional International Exchange Program. By the meeting's end, I decided to become a lifelong IPPS-SRNA member.

In the following year, 2022, I was again awarded the Vivian Munday Young Horticulture Professional Scholarship. Additionally, I presented a poster and delivered an oral presentation in the student research competition, earning second place among the finalists. This experience heightened my confidence, motivating me to apply for the exchange program. In 2023, I was privileged to be chosen as the Southern Region representative for the Early-Career Professional International Exchange Program with the European Region. The exchange proved to be an unparalleled, transformative journey—one I consider among the most profound and impactful in my life. I extend deep gratitude to the organization and mentors who facilitated this opportunity.

In 2023, the European Region's annual meeting took place in Scandinavia, encompassing cities such as Copenhagen, Køge, and Lund. Embarking on my journey from Tampa International Airport on the evening of October 6, 2023, I had layovers in Chicago O'Hare and Frankfurt International before reaching my final destination, Hamburg Airport, in the late afternoon of October 7. I was scheduled to meet my host, Louise Heissel, whom I had only communicated with through WhatsApp—or so I thought. Upon arrival, we both realized we had met in person the previous year at the 2022 IPPS Southern Region conference, where she served as the European exchange professional. Sharing a laugh, we loaded her vehicle and departed for her home in Barmstedt, Germany.

While in route to Barmstedt, I could not help but observe the striking contrast in the landscape. Gone were the familiar pineland forests of North Florida that I had grown accustomed to over the past four years. Instead, I found myself immersed in a vibrant, green environment reminiscent of my childhood in Michigan. Regrettably, as we approached Barmstedt, the sun descended toward the horizon, leaving no time for stops on our journey to my new home. Consequently, I missed the chance to explore along the way but was compensated with the opportunity to savor an authentic homemade German meal. During this time, engaging discussions unfolded on topics ranging from education and politics to healthcare. Due to Louise's Danish background, I acquired a distinctive outlook from her experiences residing in both Denmark and Germany.

GERMANY

On Sunday morning, an unusual scene unfolded as families, individuals, and pets leisurely strolled, jogged, and cycled around town, a reflection of the widespread closure of stores in Germany on Sundays. Discussing this with Louise, I discovered that this practice is not rooted in religion, contrasting with my local town, where Sunday alcohol sales are restricted due to religious practices. With everything closed, we seized the opportunity to explore a place of personal interest—the Future Climate Tree Exhibit sponsored by the German nursery Lorenz Von Ehren (**Fig. 1**). This exhibition delved into the vital role of urban trees in cities, regulating microclimates, providing shade, and filtering air and soil emissions. Urban trees also double as habitats for wildlife. Despite their merits, urban trees grapple with challenges such as soil compaction, climate-related issues, pollutants, road salt, and artificial light. Climate change further compounds these difficulties, impacting established species and inviting harmful organisms.

Figure 1. A snapshot at the beginning of the Future Climate Tree Exhibit.

The exhibit highlighted notable urban tree species, including *Acer campestre* (field maple), *Carpinus betulus* (common hornbeam), *Cornus mas* (Cornelian cherry), *Liquidambar styraciflua* (sweet gum), *Quercus palustris* (swamp oak), and *Tilia europaea* (common linden). Recognizing the escalating importance of urban trees in addressing climate change, the exhibition advocates for a thoughtful approach to species and site selection, ensuring their adaptability to changing urban environments. This strategic focus promises long-term benefits and reduced municipal maintenance costs. The significance of this subject was deeply personal to me, given that my PhD research centers around these very issues.

Our second destination of the day was Deutsches Baumschul Museum, the German Tree Nursery Museum, in Penneberg, Germany, where we attended a presentation by Dr. Andreas Wrede on the impact of climate change on the future ranges of woody plants (**Fig. 2**). Dr. Andreas Wrede serves as the Head of Experimental Engineering at the Schleswig-Holstein Chamber of Agriculture (**Fig. 3**). I faced a language challenge since the presentation was conducted solely in German. However, in the field of horticulture, where scientific names hold significant importance, I was able to decode content by relying on these names and identifying words that resembled English. To my surprise, I was able to grasp the main points of the presentation and looked forward to discussing it further with Dr. Wrede after the session. In his presentation, he discussed potential tree selections suitable for urban environments, taking into account the impact of climate change and the promotion of biodiversity (**Table 1**). The emphasis was on trees that, while not necessarily native, had to exhibit resilience to withstand the challenges posed by climate change. Furthermore, the ranking of trees was influenced by the number of species they could support, with a higher ranking assigned to those fostering greater biodiversity.

Figure 2. The entrance to Deutsches Baumschul Museum (German Tree Nursery Museum).

Figure 3. Having a conversation with Dr. Andreas Wrede regarding his research.

Table 1. Trees being considered by Dr. Andreas Wrede regarding the influence of climate change on the future distribution of woody plants.

Urban Green North	
Trees Tested	
Acer buergerianum	Trident Maple
Acer monspessulanum	Montpellier Maple
Alnus x spaethii	Spaeth Alder
Carpinus betulus 'Lucas'	Upright Hornbeam
Celtis australis	European Nettle Tree
Fraxinus ornus 'Obelisk'	Manna Ash
Fraxinus pennsylvanica 'Summit'	Summit Green Ash
Ginkgo biloba 'Fastigiata'	Fastigiata Maidenair Tree
Gleditsia triacanthos 'Skyline'	Honey Locust
Liquidambar styraciflua	Sweet Gum
Magnolia Kobus	Kobushi Magnolia
Ostrya carpinifolia	European Hop Hornbeam
Parrotia persica	Persian Ironwood
Platanus orientalis	Old World Sycamore
Quercus cerris	Turkey Oak
Quercus frainetto	Hungarian Oak
Sophora japonica 'Regent'	Japanese Pagoda Tree
Tilia tomentosa 'Brabant'	Silver Linden
Ulmus 'Rebona'	Rebona Elm
Zelkova serrata 'Green Vase'	Green Vase Japanese Elm

Our final destination for the day was Baumschule Mohr, the initial German nursery where Louise lived when she first moved to Germany. Thomas, the owner, exclusively communicated in German and claims full credit for teaching Louise the language—a fact they found amusing every time he brought it up. This nursery specializes in tree and shrub production and operates a potting machine for efficiency. Their two-person potting machine handles tasks such as filling pots, drilling holes, incorporating bark as a weed suppressant, and loading carts. One individual placed the liner into the drilled-filled pot, while the second person compacted the soil around the liner. For the remainder of our visit, facilitated by Louise as a translator, we delved into discussions about the distinctions between the nursery industries in Germany and the United States. Topics ranged from employee wages, product pricing, and quality standards to production timing, horticulture education, and the aspects of selling or inheriting nursery companies. Mohr Nursery's ownership is currently transitioning from Thomas to his son, Torben. As Torben prepares to take the reins of Mohr Nursery, the transition marks both a generational shift and a continuation of the nursery's legacy.

On Monday morning, we traveled to Bunk Jungpflanzen Aus Samen, a multi-generational nursery specializing in cultivating young plants from seeds located in Elmshorn (**Fig. 4**).

Figure 4. Nursery and delivery trunk for Bunk Jungpflanzen Aus Samen.

Bunk stands out by concentrating on rare and unique species exclusively propagated by seed. Their mission goes beyond ornamental value, emphasizing the identification of species with additional attributes, including medicinal and edible qualities. The nursery is committed to producing high-quality, robust, and cold-hardy tree species. Operating on an extensive scale, the nursery religiously cultivates 56 species sourced from seeds worldwide. The germination process begins with seeds sown in a complete sand mixture on the ground within a hoop house. Certain species will sprout within weeks, while others require up to two years of winter dormancy before germinating. Following germination, the seedlings are moved to trays, allowing them to grow sufficiently before being transplanted into finished 4-in containers. Some selected seedlings may be designated for in-ground production. These finished 4-in containers are sold to growers, who up-pot them to 1 or 3 gal containers before distributing them to wholesale purchasers.

Our subsequent destination was Kordes Jungpflanzen, a multi-generational liner production nursery situated in Bislon. Their guiding principle was "quality from the beginning." As I strolled through the expansive 37 hectares of production, the meticulous time and care invested in their products became evident, ensuring a high standard of quality. Specializing in sustainable and environmentally friendly cultivation - they offer a diverse range of over 1000 varieties of ornamental trees and shrubs. A notable recent accomplishment includes the introduction of a new hydrangea named "Dolly Buster," named after the size of its flowers (**Fig. 5**). In conversation with one of the owners, Christian Kordes, I learned that the decision to name this hydrangea "Dolly Buster" was a significant and somewhat controversial move within the conservative German horticulture community, given that Dolly Buster is best known for her work in adult films. Nevertheless, despite the gamble, the benefits have been substantial.

Figure 5. Kordes Jungpflanzen newly released hydrangea “Dolly Buster”.

As we bid farewell to Germany and made our way to Denmark, I took a moment to contemplate my experiences. Our exploration of German nurseries provided a captivating journey into the intersection of horticulture, environmental consciousness, and cultural nuances. The Sunday morning stroll revealed not only distinctive closure practices but also underscored the profound impact of urban trees on the environment, a theme echoed throughout the enlightening Future Climate Tree Exhibit. Dr. Andreas Wrede's presentation added a layer of international collaboration, emphasizing the global importance of addressing climate change in the horticultural domain. Our visit to Baumschule Mohr marked a symbolic transition in ownership and horticultural approaches. Moving into Monday, Bunk Jungpflanzen Aus Samen and Kordes Jungpflanzen showcased the meticulous care and innovative spirit inherent in German nursery operations. The introduction of "Dolly Buster" stood as a testament to the industry's adaptability. This immersive experience not only enriched my understanding of global horticultural practices but also emphasized the interconnectedness of environmental stewardship on a global scale.

DEMARK: PART ONE

Continuing our journey to Denmark, we concluded our travels with a final visit to a local garden center (**Fig. 6**). Stepping inside, I was greeted by a scene entirely unlike any I had ever experienced. While garden centers in the United States typically feature a range of plants, from houseplants and annuals to perennials, fruit-bearing trees, and ornamental shrubs - the Danish Garden Center took me by surprise with its unique offerings. Alongside the expected variety of tulip bulbs, the establishment boasted an array of merchandise - including clothing, home goods, interior design items, and exotic species like palm trees. Still in a state of astonishment, we loaded our purchases into the car and continued on our way through Denmark.

Figure 6. Visiting a local garden center on our way to Denmark.

We spent the night at Louise's childhood home, a working farm dating back to the 1880s. I had the privilege of touring their property, discovering that they cultivate blackberries, raspberries, row crops, and, to my delight, Christmas trees. The evening unfolded with a traditional Danish dinner, during which I experienced bone marrow for the first time. Engaging in delightful conversation mirroring topics of my initial night in Germany, we enjoyed each other's company. As the night drew to a close, I joined Louise in crafting description sheets for the IPPS European plant auction. The next morning, we set out early with the aim of making a specific stop before heading to our hotel.

To my surprise, a visit to a seed company had been arranged. We explored Levinsen Treeseeds, a renowned establishment owned by Ulrik Nyvold **(Fig. 7)**.

Figure 7. The initial drawing shelf for pine species at Levinsen Treeseeds.

Levinsen A/S specializes in marketing seeds for the cultivation of high-quality plants for forestry, ornamental purposes, Christmas trees, and greenery. While its primary market is in northern Europe, the company operates globally, with activity spanning most parts of the world. They personally undertake the entire seed process—from collecting and cleaning to drying, stratifying, and conducting germination tests on all species. While touring their impressive facility, I gained a profound appreciation for the extensive efforts required to produce top-notch seeds. The most intriguing aspect was the cold stratification room, where the company uses custom-made containers for the stratification of specific seeds. Unfortunately, I cannot provide details about the equipment used in this process, and regrettably, photography was not permitted inside the facility.

Upon arriving at the hotel, we joined a small group of early arrivals for a quick tour of Køge. Our exploration led us to Køge Church Cemetery, an experience that might be perceived as somewhat unconventional in the United States (**Fig. 8**). However, in Denmark, cemeteries are not just considered resting places for the deceased but are seen as public spaces where people can appreciate gardens.

Figure 8. A traditional Danish cemetery offering views of plant varieties and gardens.

During this visit, I engaged in discussions with fellow society members about the distinctions between Danish and American cemeteries. One significant difference was the payment structure for plots. In the US, you purchase a plot and can retain it indefinitely. In Denmark, plots are rented for 25 years, and if payments cease after that period, individuals are removed from the plot. Another noteworthy difference was the incorporation of personalized gardens on each plot. On their rented plots, family members have the freedom to plant any species of their choice. An intriguing aspect of these cemeteries is that some individuals utilize them for species identification. Following our visit to the cemetery, we returned to the hotel to ensure a restful night's sleep before the IPPS 2023 European tour commenced.

IPPS EUROPEAN REGION CONFERENCE

The 2023 IPPS European Region Conference took place from October 11-13, featuring two days of bus tours and a third day dedicated to presentations. The conference's main theme was "It's All About the Roots," with both the tours and presentations focusing on the soils and nutrients essential for cultivating robust and healthy roots. One of the significant challenges confronting Scandinavia and other European nations is the widespread use of peat moss as a soil medium. Numerous nurseries continue to rely on 100% peat moss soils for plant production. In contrast, in the United States, we have transitioned to diverse soil mixtures that incorporate elements such as bark, sand, perlite, compost, and more. The exploration and experimentation with alternative mixtures in Scandinavia became apparent during our nursery tours.

The conference kicked off with our initial bus tour visit to Stångby Plantskola, a Swedish nursery that specializes in park trees, avenue trees, solitary bushes, and larger fruit trees. What stood out to me about this nursery was its distinctive approach to selecting plants for sale. They conduct extensive testing of tree species in south, central, and northern Sweden, ensuring survivability for all clients. Additionally, they employ two unique pots to enhance root growth: a root control container (**Fig. 9a**) and an open-bottom custom pot (**Fig. 9b**). Both containers were designed to provide roots with better access to air - resulting in a more robust root system at a faster pace. The key distinction is that the root control container has a fixed size, whereas the open-bottom custom pot comes in sheets and can be cut to fit specialty root balls.

Figure 9. A large root control container (A) and an open bottom custom container (B).

Our second stop in Sweden was a nursery with two distinctive specialties. Spender Plant is renowned for its sales of E-Plants (**Fig. 10a**) and the customized creation of compost tea (**Fig. 10b**). Additionally, they concentrate on technical solutions, leaf analysis, leaf fertilizers, organic fertilizers, and regenerative agriculture. You might be curious about E-Plants—are they electric plants? In fact, E-Plants are plants with seeds sourced, produced, and sold exclusively in Sweden. Their E-Plants encompass a wide variety, including trees, shrubs, perennials, and groundcovers. Regarding the compost tea, it takes three days to brew one batch. They incorporate kelp, calcium, and various acids to adjust the pH. Once completed, the compost tea is applied all at once, with applications occurring monthly.

Figure 10. A display of a few E-plants sold by Spendor Plant (A) and the compost tea container (B).

Our final destination in Sweden was Klinta Trädgård, owned by designer and nurseryman Peter Korn (**Fig. 11**). If you seek a garden and nursery that will leave you astounded, this is the place to visit. Peter's mission revolves around the mindful use of plants in harmony with the nature of the soil. How does he achieve this? He cultivates all his perennials and grasses in a soil mixture of sand and mycorrhiza, with no additional nutrients added. Initially, this approach may seem contrary to conventional teachings, as many of our members mentioned. However, this nursery serves as proof that it can be successfully accomplished. Peter utilizes these plants as bare-root transplants for his landscape designs, personally overseeing the installation. Moreover, he conducts research on dependable plants for green roofs and green walls (**Fig. 12**), primarily focusing on perennials and succulents throughout his experiments.

Figure 11. Plants at Peter Korn's Klinta Trädgård growing in a soil mix of complete sand and mycorrhiza.

Figure 12. Examples of Peter's researching involving green walls and green roofs.

Our bus tour in Denmark commenced with a visit to Gunnar Christensens Planteskole, where we explored the nursery and engaged in a series of workshops. The workshops covered topics such as alternative growing media, fertilizer, pot covers, willow compost, peat & green transition, peatless growing, and CO_2 reduction in the production of pots. At Gunnar Christensens, they cultivate over 1.5 million plants annually across a 25-ha area and employ around 70 individuals during peak season. Their plant range includes container-grown varieties such as berry bushes, ornamental shrubs, perennials, herbs, and strawberries. The nursery places a strong emphasis on water conservation, collecting rainwater from greenhouse roofs and extensive potting areas for reuse. With significant irrigation capacity, they can water the entire nursery during the early morning hours, minimizing the impact of wind and evaporation. The workshops provided insights into container trials experimenting with various soil mixtures, the use of eco-friendly plant tags and recyclable pots, and the application of biostimulants to enhance soil and substrate biology. During the garden tour, we explored the greenhouse, walked the property layout, and observed the current plant varieties under cultivation. The highlight of the tour was strolling through their trial garden (**Fig. 13**), where I encountered plants typically associated with warmer climates like Florida, including agave, southern magnolias, and coreopsis.

Our next stop was Nordic Harvest (**Fig. 14**), a vertical hydroponic farm with a mission to establish a genuinely sustainable food production system and return agricultural land to nature. They achieve this by daily care for the plants, guiding them from seed to harvest to ensure optimal crispness and flavor. Sowing and reaping take place every single day. The team consists of experts in nutrition, water management, biology, as well as cooks, technicians, and computer scientists. The plants are grown on 14 floors in nutrient-enriched water under LED lights with controlled temperature, humidity, and CO_2 levels. The closed environment is free from pests, spores, bacteria, and pollution, eliminating the need for harmful chemicals. This enables the plants to focus exclusively on energy-efficient, large-scale growth, resulting in flavorful and nutritious produce. Their product range includes crisp salad, mixed green salad, baby iceberg, baby romaine, thyme, baby kale, and baby arugula. Currently, Nordic Harvest is gearing up to expand the farm, aiming to quadruple its production capacity.

Figure 13. The species included in Gunnar Christensens.

Figure 14. The growing room of Nordic Harvest.

The day's final destination was Pometet, a garden and research center affiliated with the University of Copenhagen's Department of Plant and Environmental Science, dedicated to preserving varieties for NordGen – the Nordic Genebank. During my visit, I discovered that "pomet" refers to a diverse assortment of fruit tree and shrub varieties, taking its name from the Latin term "pomum," signifying "fruit on trees." Pometet hosts an extensive array of varieties, with a notable focus on collecting apple varieties locally and internationally (**Fig. 15**). However, their collections span a wide range, covering all of Denmark's commonly grown fruit species. Beyond research and cultivation, their mission extends to fostering awareness and knowledge about the varieties and fruit species present on their property.

Figure 15. A display of all the apple varieties grown on Pometet property.

On the final day of the conference, October 13th, presentations took center stage. Poul Petersen of Overdam Planteskole offered insights into Naturalistic planting methods, with a particular emphasis on utilizing grasses. Mette Buw Lorensen from Byblomst delved into collaborative landscape design approaches between growers and designers. Katrine Turner from Vilskab explored tactics for enriching biodiversity in new urban plantings. The event drew to a close with a preview of next year's conference, slated to take place in England.

DEMARK: PART TWO

Although the conference had concluded, my journey continued as I transitioned to my final destination in Sorø, Denmark, where I stayed with Bent and his wife. During this time, we embarked on exciting explorations, starting with the old flower district of Copenhagen. The district had undergone a remarkable transformation into a bustling urban area, yet it retained its environmentally friendly design, fostering biodiversity alongside human activity (**Fig. 16**).

Figure 16. Copenhagen urbanized flower district.

Our adventures also led us to the old Carlsberg brewery and gardens, renowned for their impressive architecture and rich history. Delving into urban planting concepts was a highlight, as we observed firsthand the effectiveness of various strategies (**Fig. 17**). The visit to Poul Petersen's nursery was equally captivating, offering a glimpse into the diverse array of grasses he cultivates (**Fig. 18**). On Saturday, I had the unique opportunity to immerse myself in the life of Anja, a team leader at Gunnar Christensens Planteskole, who was close to my age. From accompanying her to the mall and supermarket to cooking dinner together, I gained valuable insights into a life of someone my age in Denmark.

However, the pinnacle of my experience, that took place on my last day in Denmark, was undoubtedly stepping into the shoes of a team leader at Gunnar Christensens Planteskole for a day (**Fig. 19**). It was an unforgettable experience that left me feeling deeply honored. The organization's excellence in plant production and management was matched by their remarkable leadership and communication skills, creating a positive work environment admired by all employees. Collaborating with the team members not only brought me contentment but also instilled a sense of pride in their work and the company they represented. Inspired by this experience, I aspire to one day provide a similar enriching environment as a nursery owner.

Figure 17. Effectiveness of various urban planting concepts.

Figure 18. Poul Petersen's nursery demonstration garden.

Figure 19. Gunnar Christensens Planteskole. From left to right (Bent, Teagan, Anja).

CLOSING REMARKS

I am overwhelmed with gratitude for the incredible opportunity I had to embark on this journey. Every aspect of it filled me with wonder and optimism for the future. I witnessed breathtaking sights, encountered remarkable individuals, delved into horticultural practices, and forged memories that will forever hold a special place in my heart. This experience was truly invaluable, a once-in-a-lifetime gift provided to me by IPPS.

I express my gratitude to the IPPS-SRNA board for choosing me as a Vivian Munday winner in both 2021 and 2022. Additionally, I appreciate their decision to select me as the Early-Career Professional International Exchange Program awardee for the 2023 IPPS European Region Conference. Special thanks go to my European hosts Louise, Bent, and Anja, whose incredible hospitality made my visit immensely enjoyable and welcoming.

Combined Proceedings IPPS 74:496-500. 2024.

Prenatal Care: Healthy Stock for Healthy Cuttings

Ben Averitt

Willoway Nurseries, 4534 Center Road, Avon, Ohio, USA

baveritt@wwnic.com

Keywords: sensors, substrate water, EC, growing degree days

Summary

At Willoway, we cultivate approximately 3,500 SKUs, primarily consisting of trees and shrubs with an additional propagation production exceeding three million cuttings per year, predominantly sourced from current production plants. We adhere to stringent protocols to ensure healthy stock, which in turn results in superior and healthy cuttings. This document aims to describe some of the methodologies that contribute to our ongoing success.

INTRODUCTION

There are many factors that contribute to a continuous successful outcome for future production needs. Nutrient management is a critical component in our ongoing efforts to produce quality liners on a consistent basis. Controlled release fertilizers are utilized as both a top dress and as a component to the propagation mix. We often vary the application of fertilizers based upon the specific crop, and timing during the year. It is clear that fertilizer applications in May are significantly different from those of August or September even for the same plant. Also, when using control release fertilizers, the timespan or the longevity of the particular product is of major importance as is the

IPPS Vol. 74 – 2024

release rate for nutrients such as nitrogen. We use fertigation on some crops to augment the dry formulations. The application of fertilizers is a critical component and with many individual crops such applications must be tailored specifically for those crops. One size does not fit all.

When developing a fertilizer program, it is essential to consider multiple factors. Three primary criteria are cost, plant outcome, and environmental considerations such as runoff, stream and body of water effects, and regulations governing nutrients like nitrogen and phosphorus entering waterways. Environmental stewardship is a responsibility we should all embrace with vigor. Willoway we have multiple interconnected ponds each with its own pump and either Anderson Injectors or Bauerle Precision fertigation system. There are several individual stock nutrients capable of being injected into the outgoing system either alone or in combination. These include nitric acid, phosphoric acid, $CaNO_3$, KNO_3, $NHNO_3$, $MgSo_4$ and a minors package.

We established new protocols in 2024 which include daily monitoring of inputs and metrics, tightened nutrient levels with more defined seasonal levels and changes, and a strong grower accountability. The outcome has been better mother plants and superior cuttings.

Water management, as in most nursery operations, is a continuous issue. At our Avon, OH farm, we have no ground water while there is limited access to Lake Erie at our Huron, OH facility. We recapture approximately 95% of our water and recycle it with city water available for targeted clear water or specialized fertilizer applications.

Our soilless media has a high-water holding capacity for nursery stock, and the use of water has to be tailored to specific plants. Adjustments, of course, are weather dependent along with product mix in the production areas. This can lead to a tendency to overwater in some situations. But we are making improvements in the mix ratios and in water monitoring and application.

We realize that improved accountability necessitates revised responsibilities, a regular targeted training program with improved tools such as a grower's manual, implementation of best management practices (BMPs) and technological advances.

Data for moisture, pH and substrate conductivity can be gathered at the plant level for specific crops using available instruments like Bluelab Pulse meters and probes (Bluelab Website | Bluelab USA).

Data generation and logging at Willoway is an important on-going process. We use a HOBO data system that has data loggers at each pumphouse to record pump activity and nutrient data. Data is available on dashboards for system management **(Fig. 1)**.

Figure 1. Data collection stations at Willoway nursery.

Pumphouse data is tracked over each 24-hour period. It provides both flow and pressure data as well as sensor data for pH and nutrient monitoring. Set points are available that trigger alarms and emergency cuttoffs (**Fig. 2**).

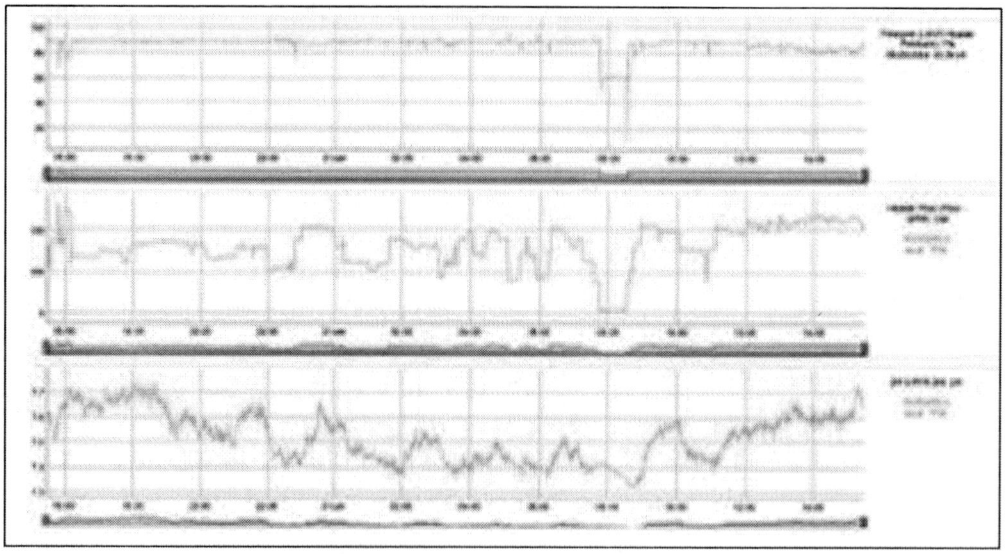

Figure 2. Example of pumphouse data output for 24 hours.

Substrate moisture can be measured at the container level using moisture sensors like Echo probes that connect to a wireless data logger to provide EC and container substrate moisture.

Results from our management efforts have proven themselves to be promising. It is still early but there are encouraging signs indicating reduced scap numbers and decreased pest and disease pressure. We have also observed improved rooting for difficult rooting crops like *Berberis* with a 60% rooting increase and 25% increase in *Hydrangea*.

Another technology we are exploring is using growing degree days gain insights into pest management and to schedule cutting harvest and improve rooting (Balteel, 2025 ; Barnes, 2005; Castillo and Castillo, 2001). Growing degree days use daily temperatures relative to a base temperature to predict a specific aspect of pest development or plant growth.

The relationship between growing degree days (GDD) to harvest cuttings and rooting percentages are illustrated for a *Berberis* (**Fig. 3**) and a *Hydrangea* crop (**Fig. 4**). It is clear that rooting percentage is significantly affected by the physiological state of the mother plant and in the case of *Berberis* Crimson Cutie rooting occurs best before the spring flush whereas in the case of the *Hydrangea* Incrediball optimum rooting has two peaks, one early on at close to 1000 GDD and the other at 2100 GDD.

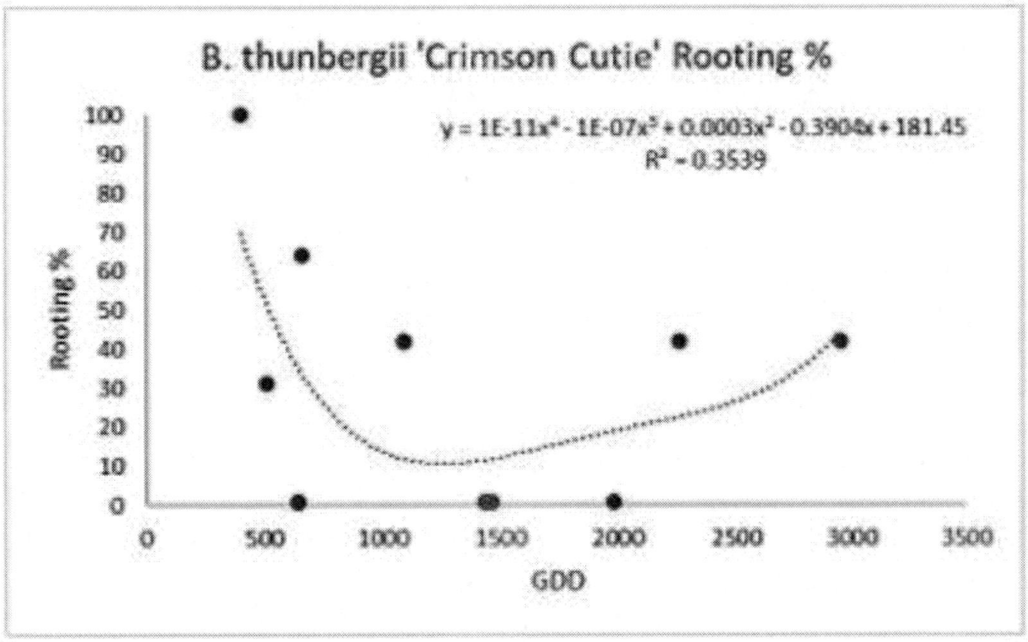

Figure 3. Rooting percentage in *Berberis thunbergii* Crimson Cutie in cuttings harvested after various growing degree days.

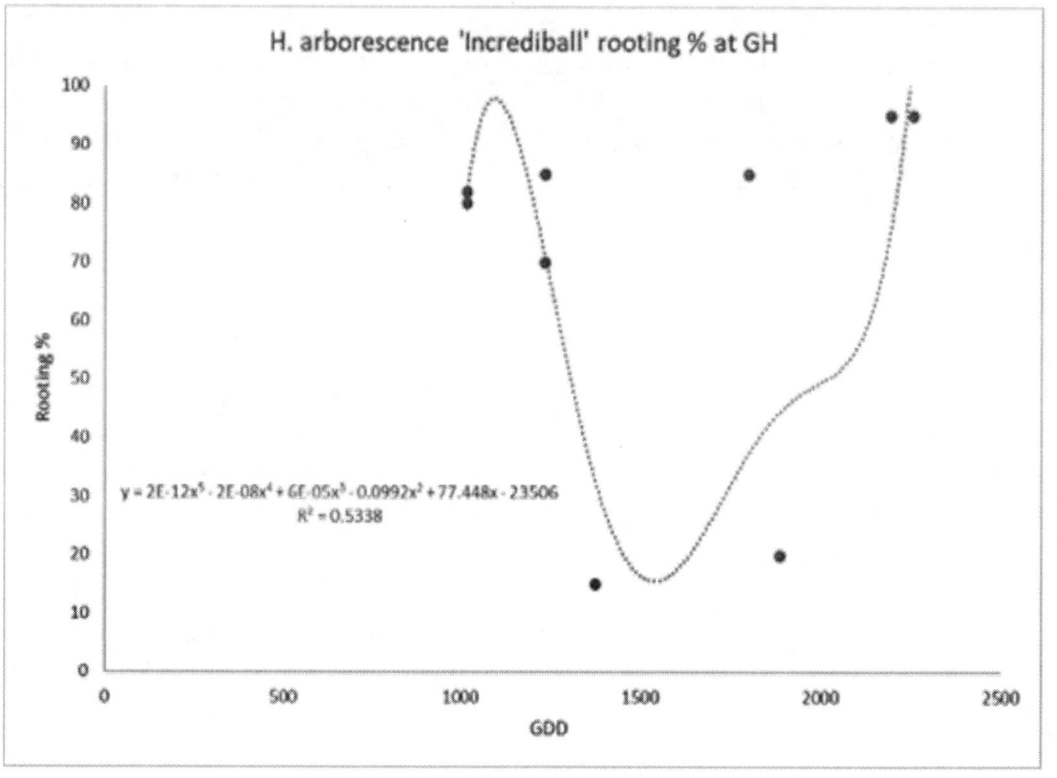

Figure 4. Rooting percentage in *Hydrangea arborescence* Incrediball in cuttings harvested after various growing degree days.

Conclusions

Improved plant health can improve numerous aspects of nursery production including stock plants by having a rigorous nutition and water management program. Growing degree days also offer another tool to improve cutting management. It needs to be specific for each crop and the models will get better as more data becomes available. Data may also be used from historical published propagation data if cutting harvest dates and rooting percentages are available. In conclusion, technology is an omnipresent component of our lives and work. Putting it into practice can yield significant improvements in any operation.

LITERATURE CITED

Baltell, B. 2017. Understanding growing degree-days. Michigan State University. https://www.canr.msu.edu/news/understanding_growing_degree_days

Barnes, H. W. 2005. Correlation of growing degree days and the timing of cuttings. Comb. Proc. Intern. Plant Prop. Soc. *55*:.386-392.

Castillo, Francisco, and Alfredo Castillo, 2001, How degree days effect the rooting percentage of *Cotoneaster acutifolia* cuttings. Comb. Proc. Intern. Plant Prop. Soc. *51*: 537.

Combined Proceedings IPPS 74:501-503. 2024.

An Overview of the Ant Plant (*Hydnophytum*)

Anna Baloh and Robert Geneve

Department of Horticulture, University of Kentucky, Lexington, Kentucky 40506 USA

anna.baloh@outlook.com

Keywords: ant plant, morphology, tuberous stem, *Hydnophytum*

Summary

Many plants have interactions with other species. Some plants have mutualistic relationships that benefit both species, while others are parasitic relationships. Ants are a common species that many plants harbor interactions with. Ants have been known to guard peony buds for their sugars, farm aphids on plants for nutrition, and even living in plants for habitat. In the following paper, ant relations with Rubiaceous ant plants will be explored. One of the major types of ant plant genera, *Hydnophytum*, is investigated in more detail. Ant plants feature a large tuberous stem that has many internal chambers that ants find hospitable. In return for habitat, ants provide protection to the plant from other herbivores. This relationship provides advantages to both species, making it a mutualistic interaction.

INTRODUCTION

Plants can form symbiotic or mutualistic relationships with other species for a variety of reasons. The madder family (Rubiaceae) has several species that form relationships with ants. The ants provide protection to the plant while the plant provides a habitat (domatia) for the ants. The ant plant features a large, tuberous stem with internal cavities, making it the perfect habitat for ants. However, these cavities are not produced by ant tunneling. The plant forms these cavities regardless of ant activity (Huxley, 1978).

Ant plants are epiphytic plants that are native to Southeast Asia (Kew Gardens, 2023). There are five genera of ant plants in Rubiaceae. Genera of ant plants include *Myrmecodia*, *Hydnophytum*, *Myrmephytum*, *Squamellaria*, and *Anthorrhiza* (Biopower Plants, 2024). The focus of this study was to describe early cavity formation in *Hydnophytum* seedlings.

Hydnophytum is one of the major genera of Rubiaceous ant plants. There are many known species of *Hydnophytum*, and they are well represented in gardens across the world (Biopower Plants, 2024). This genus features a large tuberous stem that houses the ant inhabitants (Fig. 1). Species can range in size and shape of the caudex (Biopower Plants, 2024).

Hydnophytum forms a mutualistic relationship with several ant species, but the most common inhabitant are the *Philidris* ants (Biopower Plants, 2024). The ants live in the large tuberous stem of the ant plant and provide protection in return for a habitat. The caudex forms by creating a series of internal chambers and cavities when the plant is still young. A small hole at the base of the tuberous stem can be found after a few months of development (**Fig. 1**).

Figure 1. A) *Hydnophytum* before and B) after a portion of the swollen domatia removed.

Depending on the size of the caudex, there can be only one plant that the ant colony occupy, or the colony can spread across several smaller, nearby ant plants (Biopower Plants, 2024). However, ant plants do not require an ant colony for growth. Many ant plants can grow, and still form chambers, without having any inhabitants. This is commonly seen in ant plants in greenhouses outside of the native range. Plants that occupy an ant colony have been found to be more vigorous (Biopower Plants, 2024.).

LITERATURE CITED

Biopower Plants. (2024). Frank's Ant-Plants. https://www.franksantplants.com/ (accessed Nov. 2024).

Huxley, C.R. (1978). The ant-plants *Myremcodia* and *Hydnophytum* (Rubiaceae), and the relationships between their morphology, ant occupants, physiology and ecology. *New Phytologist,* 80: 231-268.

Kew Gardens. (2023). Hydnophytum formicarum. Kew Royal Botanical Gardens.

Combined Proceedings IPPS 74:504-508. 2024.

Propagation of Northern Bayberry (*Morella pensylvanica* 'Bobzam') Through Tissue Culture

Mark Brand and Jackie Edgett

Department of Plant Science and Landscape Architecture, University of Connecticut Storrs, Connecticut 06269 USA

Mark.brand@uconn.edu

Keywords: native plants, micropropagation, woody plant

Summary

Northern bayberry is difficult to propagate from stem cuttings which limits its availability in the nursery industry. It was shown that cuttings taken in July and treated with IBA rooted at only about 20%. Micropropagation is an alternative to cutting propagation. An efficient micropropagation protocol was developed where microcuttings rooted at 98% and were successfully acclimated to greenhouse conditions.

IPPS Vol. 74 – 2024

INTRODUCTION

Northern bayberry (*Morella pensylvanica*) is a valuable native shrub found primarily in coastal areas of the northeastern United States from Virginia to the Canadian Maritimes. It is a suckering, colonizing, semi-evergreen shrub that is useful in landscape situations that present challenging conditions. Bayberries can tolerate salt spray, full sun, reflected light, drought, infertile sandy soils and cold winter temperatures. Its rhizomatous habit allows it to regenerate well following canopy damage resulting from snow clearing activities and people pressure. It is an ideal plant for parking lot islands, seaside plantings and bank stabilization.

Bayberry is a dioecious species, and female plants produce significant amounts of grayish white small fruits that offer seasonal interest, are a source of food for many songbirds and are covered with aromatic wax that can be used for candles. Female plants are more desirable than male plants due to their fruits and their more compact habits. Most bayberry plants are seed propagated because stem cutting propagation is difficult, or at least inconsistent. Plants offered for sale are essentially a 50%:50% mix of males and females, so obtaining a majority of female plants for landscape purposes is frequently not easy to accomplish. Reliable vegetative propagation of northern bayberry has remained elusive for the nursery industry but would provide great advantages over the current seed propagation that is used.

To confirm the general knowledge that northern bayberry is difficult to propagate from stem cuttings, we conducted a rooting study using the female cultivars 'Bobzam' and 'UConn Compact' and an unnamed male genotype. We collected cuttings at the end of June, end of July and end of August and stuck them individually in 2" pots in one part sphagnum peatmoss, one part horticultural-grade fine perlite, and one part horticultural-grade medium vermiculite under intermittent mist. Cuttings were wounded on one side and dipped in 2000, 4000 or 8000 ppm indole-3-butyric acid (IBA) in 50:50 water and ethanol, along with a 0 ppm IBA control. The male clone could be rooted at 100% using June cuttings treated with 2000 or 4000 ppm IBA and above 70% using 2000 ppm or no IBA. Both female cultivars rooted relatively poorly. 'Bobzam' rooted between 55% and 35% using July cuttings treated with between 0 and 4000 ppm IBA. The 'UConn Compact' cultivar reached a maximum rooting of only 20% using July cuttings and 2000 ppm and exhibited overall dismal rooting ability across all times and IBA concentrations. Our findings confirm that stem cutting propagation is likely an insufficient method of vegetatively propagating female northern bayberry.

Studies were conducted to determine if micropropagation of northern bayberry in vitro might be a way to efficiently vegetatively propagate a desirable female clone. New shoots (4-6 cm long) were harvested from 'Bobzam' plants forced in a greenhouse in late winter and surface sterilized using a 10% bleach solution for 15 min. Murashige and Skoog (MS) and Woody Plant (WP) media were tried in combination with benzyladenine (1 mg/l), or meta-topolin (5 mg/l), or thidiazuron (0.1 mg/l), or zeatin (4 mg/l). All explants on MS media turned black and failed, but many of those on WP with zeatin remained green and showed promise. A second shoot initiation was conducted using more advanced 10-14 cm shoots with better formed axillary buds (**Fig. 1**) and they were placed on WP medium containing 0, 2, 4, or 6 mg/l zeatin. Explants on 4 mg/l zeatin performed best. They were very slow to acclimate to in vitro culture taking 7 subculture cycles of 3 weeks each (total time of 147 days) to miniaturize and develop a steady 2X shoot multiplication rate (**Fig. 2**). With additional time in culture (beyond 190 days) the multiplication rate increased to 3X every 21 days and eventually reached 4X multiplication.

Figure 1. *Morella pensylvanica* 'Bobzam' Bobbee™. A) Well-developed, 10-14 cm shoots that were used to provide starting explants and trimmed shoot tip segment introduced into in vitro culture; B) Initiated shoot tip explants beginning to produce new shoot growth in vitro from apical meristems.

Figure 2. *Morella pensylvanica* 'Bobzam' Bobbee™. A) In vitro shoots developing from lateral buds on nodal explants following subculture; B) In vitro shoot developing from the apical meristem on tip explants following subculture; C) Shoot cultures that have miniaturized and acclimated to in vitro growth at 147 days following culture initiation.

On WP medium with 4 mg/l zeatin, 'Bobzam' cultures produced nicely elongated microshoots with well-expanded leaves. The shoots that were produced were ideal for use as microcuttings for rooting, or they could be cut into 3-5 node basal and apical pieces that could be recultured onto fresh multiplication medium. Apical explant pieces mostly produced shoot expansion only from the apical meristem, developing elongated straight shoots. Basal nodal pieces typically produced 2-5 shoots from axillary buds, but shoots were shorter than those from apical growth (**Fig. 2**).

Microcuttings were easily rooted (98%) in vitro on WP medium containing 1 mg/L indole-3-butyric acid (IBA). Root primordia were visible on the basal portions of microcuttings between day 10 and day 14 following sticking. At this point, pre-rooted microcuttings were transferred to one part sphagnum peatmoss, one part horticultural-grade fine perlite, and one part horticultural-grade medium vermiculite in clear deli tray humidity chambers that were placed under LED lighting providing an 18 hr photoperiod. Microcuttings could also be rooted easily (100%) by directly sticking unrooted shoots from tissue culture into deli tray humidity chambers (**Fig. 3**). Prior to sticking, the basal ends of directly stuck microcuttings were dipped in Hormodin 1 IBA powder (1000 ppm). During the initial 3 weeks of the rooting period, it was important to keep light levels low (below 30 uM/m2/sec) to avoid light stress that tended to turn microcutting leaves red or bronzy-black. Microcuttings rooted after 3-4 weeks and were then potted to 50 plug trays containing screened nursery mix (4:2:1 pine bark:peat moss:sand) and were covered with a clear plastic dome (Fig. 3). Acclimation to greenhouse conditions was accomplished by gradually increasing the number of vent holes in the clear domes to decrease humidity and removing shade cloth to increase light levels.

Figure 3. *Morella pensylvanica* 'Bobzam' Bobbee™. A) Well-rooted microcuttings that were directly stuck into peat moss-based rooting medium in humidity chambers; B) Acclimated bayberry plantlets growing in a 50-plug tray in the greenhouse.

Once acclimated to the greenhouse, micropropagated plants were potted into quart containers with nursery mix and then potted up to 1-gallon containers prior to being moved to outdoor, full sun growing conditions. Outdoors plants were provided with controlled-release fertilizer and grew rapidly into normal, well-branched and highly marketable plants (**Fig. 4**). Micropropagated plants were superior in quality (vigor and branching) to stem cutting propagated plants.

Figure 4. *Morella pensylvanica* 'Bobzam' Bobbee™ micropropagated plants after growing for three months in one-gallon containers outdoors.

Combined Proceedings IPPS 74:509-516. 2024.

Redesigning the Future of Plastic Plant Containers

Marie Chieppo

Echo Plant Plans LLC, Naushon RD, Sagamore Beach, Massachusetts 02562 USA

mariechieppo@me.com

Keywords: recycle, pots, alternative containers, policy, environment

Summary

Plastic containers are common in the greenhouse and nursery industry. To avoid contributing to plastic pollution and its ecological impact, it would be ideal if plastic containers were recycled. However, over 90% are sent to landfills. There are several reasons for this limited recycling and the current paper will discuss the current state of container use by the industry and potential alternatives to plastics for growing plants.

INTRODUCTION

Plastic plant containers have become ubiquitous in the gardening, horticulture, and landscaping industries, offering convenience, durability, and a cost-effective means for growing and selling plants. Plastic's strong molecular chains allow it to be pressed, stretched and molded into every conceivable shape (**Fig. 1**).

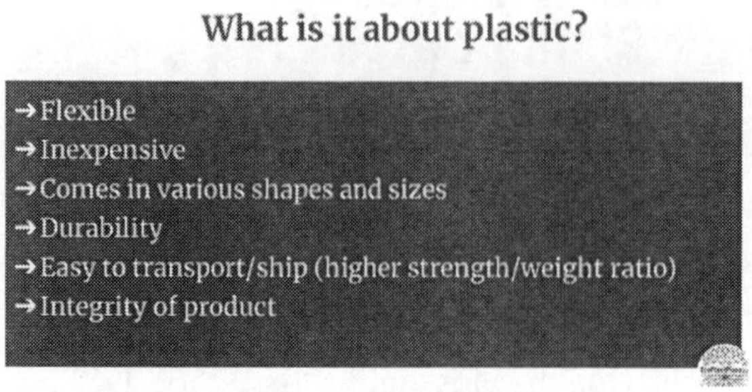

Figure 1. Attributes of plastic.

When growers' only options for containers consisted of materials that were prone to breaking and proved cumbersome, their markets were quite limited. The plastic container's popularity vastly expanded growers' markets resulting in one of the fastest growing industries in modern history. However, the widespread use of plastic plant containers has raised concerns about their environmental impact, particularly plastic pollution and resource depletion. The horticulture industry is being faced with significant pressures from the public and regulatory bodies to reduce the amount of plastic used in our products.

In 2019, I contracted with the Association of Professional Landscape Designers to research the end of life of plastic plant containers. For years, we have been disposing of them at recycling facilities, attempting to return them to nurseries or throwing them out with the trash. In 2020, I authored the white paper entitled "Plastic Pots and the Green Industry: Production, Use, Disposal and Environmental Impacts." My discovery that 95-98% of them go to landfills was sobering and quite frankly, surprising.

Plastics that are used in the horticulture industry largely include low density polyethylene (LDPE #4), polypropylene (PE #5) and polystyrene (PS #6). High density polyethylene (HDPE #2) is used for larger plants like trees and shrubs. Perennials, annuals and other small plants are typically grown and sold in #s 4, 5, and 6. Our containers use a tremendous amount of plastic: in 2009, the U.S. ornamental plant industry used 1.66 billion pounds of plastic. **(Fig. 2)**.

Figure 2. Plastic use in horticulture per year in 2009.

The overarching reason for why our containers and other plastic products go to landfills, is our recycling system is not adequately prepared to handle the vast quantities that are produced. Its production has increased 200-fold since 1950. In 1980 the world produced 98 million tons of plastic. By 2019, 450 million tons were produced. In 2019 only 8.5% of plastic was recycled (**Fig. 3**).

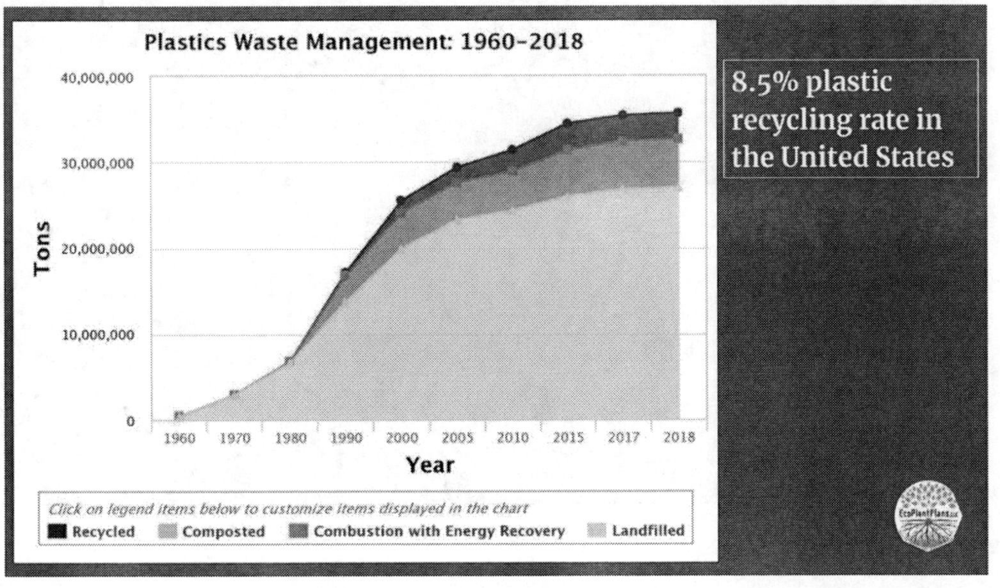

Figure 3. Plastic recycling rates between 1960 and 2018.

The numbers have not improved since. For years the United States depended on China to handle our waste. In 2017, China imposed the Sword Policy prohibiting the exportation of our waste to their borders. We were ill-prepared to handle the vast quantities of trash we had relied on them to handle.

One of the largest overarching reasons containers are not recycled, is the recycling system itself is not capable of handling the majority of common products. Aside from not having enough recycling facilities, the technology to process our waste is archaic. Collection is very inadequate. State by state residential recycling rates range between 9-37%.

Effective recycling requires precise sorting of different plastic types. Each plastic is composed of different chemicals and properties so it is difficult to recycle them uniformly.

Plastic often becomes contaminated with food residue, oils, and other material that can hinder the recycling process rendering the recycled product less viable.

The cost can sometimes exceed the value of the recycled material, especially if oil prices are competitive with recycled plastic.

Plastic can degrade each time it is recycled, limiting the number of times it can be reused before it is no longer suitable for production.

In addition to these barriers, plastic plant containers pose unique challenges. These include: optical readers cannot detect the color black, soil and chemicals can contaminate the recycling stream, they are composed of mixed plastics or weak plastics with little to no market value, it is cost prohibitive to collect and clean them, it is extremely difficult to identify the resin code located at the bottom of the container, there is an insufficient access to recycling facilities, and the design of the containers can vary widely, complicating the sorting process (**Fig. 4**).

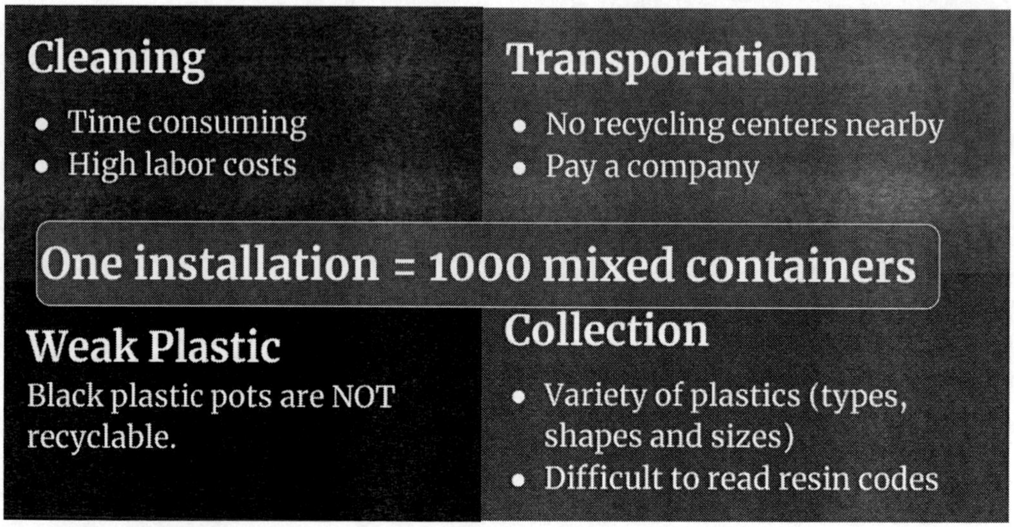

Figure 4. Complications in recycling plastic.

Plastics are notoriously difficult to break down in the environment. It can take anywhere from 500 to 1,000 years to decompose. Instead, they persist in the environment, where they pose a range of environmental risks. Chemicals leach into waterways and groundwater.

When plastics enter aquatic ecosystems, they break down into microplastics, which are consumed by marine life. This contamination affects everything from plankton to fish and seabirds and can ultimately impact human food sources (**Figs. 5 and 6**).

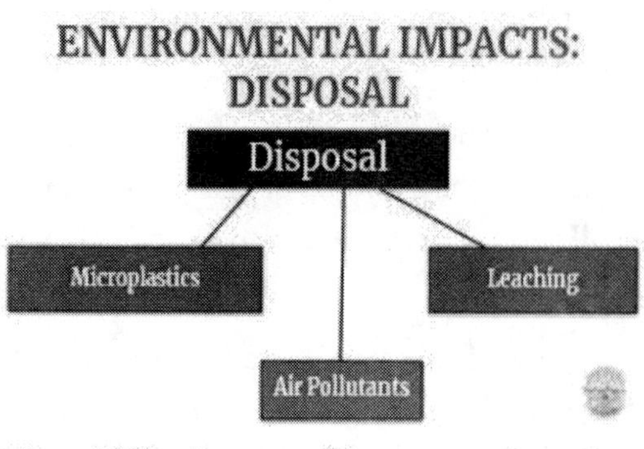

Figure 5. Environmental impacts on disposal.

Figure 6. Environmental impacts on manufacturing.

Plastics, especially those that break into smaller pieces, can be ingested by animals leading to blockages, injuries, and even death. The MacArthur Foundation estimates that humans ingest the equivalent of a credit card's worth of microplastic per week. Recent studies have found nano plastics in sperm and placentas.

For these reasons, there is growing concern over the negative impacts plastics are having on our planet. In response, a great deal of attention is being given to plastics that are used for short periods of time and discarded. Otherwise known as single use plastics, the majority of them are produced by the packaging industry. According to the Environmental Protection Agency, containers and packaging are defined as products that are assumed to be discarded the same year the products they contain are purchased. Plastic plant containers fall into this category.

A growing outcry against single-use plastics and the burgeoning quantity that is impacting our planet has the public demanding that manufacturers be responsible for the products they produce. In 2022, the IPSOS surveyed 20,000 people across 28 countries. 85% of the respondents want manufacturers and retailers to be held responsible for reducing and recycling plastic packaging. A USA public opinion poll indicated that 80% of voters are in favor of holding companies accountable for plastic waste and 82% favor products with less plastic packaging (**Fig. 7**). In July 2024, the U.S. government said it would phase out its purchases of single use plastics (**Fig. 10**). This was a significant step considering it is the largest buyer of consumer goods in the world. This is not a phase, but a new movement demanded by environmentally conscious consumers.

Figure 7. USA public opinion poll on plastics.

Horticulture Industry Responds

In the United States there are a growing number of facilities accepting horticultural containers. We can close the loop by capturing the material before it is disposed of.

East Jordan Plastics, a container manufacturer, works with growers to collect used containers they then recycle and use for new products. A line of their products is composed of 100% recycled content (**Fig. 8**). Five years ago, Pride's Corner in Lebanon CT created a program with the States of Maine and Vermont to collect containers when making deliveries to nurseries. At present, it is the only large collection hub in New England.

Figure 8. Recycling plastic containers.

We can "redesign" our way out of concerns over plastic by minimizing the amount we use in our containers. The horticulture industry does have alternative products that can be used (**Fig. 9**). The majority of them are either plantable or compostable thus negating concerns over disposal. A number of them have been on the market for quite some time, however the majority of growers are not familiar with them.

Figure 9. Types of alternative containers.

Understandably, the reliability of plastic and the unproven benefits of switching has slowed the course. Recently, a host of alternative containers have been tested including the new bioplastic containers made from cellulose. The excitement behind this material is it does "behave" similarly to plastic but will biodegrade over a period of 5-6 years without leaving toxic chemicals and can be produced on existing equipment. Large quantities require industrial composting. Another new product from Blackmore, uses a milk carton-like material to contain the plant that decomposes (**Fig. 10**). Like the other alternative materials (wood pulp fiber, rice hulls, coconut coir fiber, etc), what is left behind returns to the soil. New studies will be published soon comparing some of these alternative containers' performance.

Figure 10. Alternative milk carton-like containers.

Regulatory Activity

For years Europe has been utilizing extended producer responsibility laws, extending responsibility for products on producers rather than municipalities and taxpayers. California was the first state to invoke the extended producer responsibility regulations (SB54), followed by Oregon, Washington, Maine, Colorado and New Jersey. To improve recycling a handful of states have adopted Extended Producer Responsibility (EPR) laws. Governments and industry organizations are beginning to adopt Extended Producer Responsibility policies, which place the onus on manufacturers to manage the disposal of their products after they reach the end of their life. Under such programs, companies that produce plastic plant containers would be required to take responsibility for the collection and recycling of these items.

A synchronous attempt to reduce plastic are laws requiring a minimum percentage of recycled plastic in each product. In New Jersey for example, a law requires that products be composed of at least 10% post consumer resin (PCR). The PCR must come from commonly recycled products not industrially recycled. The ratio in this state and others will incrementally increase each year. By 2025, it will rise to 50%.

Truth in Labeling Laws (California SB), have been created to accurately inform consumers of how to handle their plastic purchases. Replacing the typical arrows surrounding the plastic type number, a triangle without arrows is now being stamped onto products.

The main drive behind these changes are consumers looking for sustainable options. Many products purchased by big box stores for example, are required to be third party certified as sustainably produced with minimal impact on the environment.

At present, the Horticultural Research Institute's Plastic Task Force is defining the challenges by tackling the issue from a number of different angles. Education of recent and upcoming regulations is paramount. Sharing knowledge of where to source PCR that works with a brand's color. This is real and being felt by producers. Container manufacturers and growers need reliable sources that are affordable and readily available. Create standards to ensure compliance and transparency. Involving consumers would behoove growers and companies. Getting a seat at the table to ensure plastic plant containers are on the radar as new recycling infrastructures are built would be in our best interest (**Fig. 11**).

There is a call to action for the green industry to work together and advocate for healthier practices and products. As Dr. Charles Hall from states: "Economic data shows that there is a payback for investment in sustainability. It is worth it. "

While plastic plant containers have become an integral part of the gardening industry, their contribution to plastic pollution is a growing environmental concern. By adopting sustainable alternatives and steadily reducing the amount of plastic used in the manufacturing process, the horticulture industry can contribute to a greener, cleaner future for both plants and people.

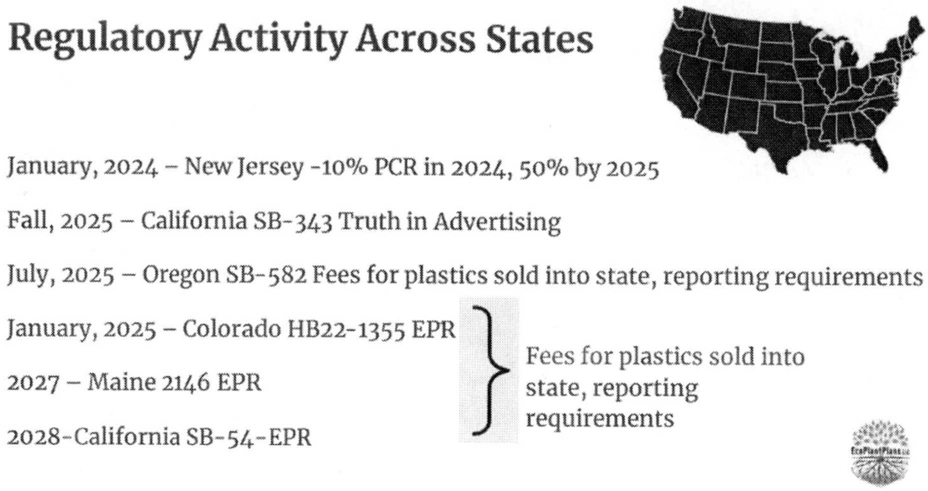

Figure 11. Pending or enacted regulatory action in several States from 2024 to 2028.

Combined Proceedings IPPS 74:517-519. 2024.

Propagation of *Tsuga canadensis*

Daniel M. Dlugos[1], Jennifer L. Koch[2], and David J. Burke[1]

[1] Holden Arboretum, Kirtland, Ohio 44094; [2] USDA Forest Service, Northern Research Station, Delaware, Ohio 43015 USA

ddlugos@holdenfg.org

Keywords: hemlock, native plants, cuttings, air layering, juvenility

Summary

Canadian hemlock is native to northern North America. It is under threat from the hemlock woolly adelgid, and breeding and selection programs are searching for potential resistance to this invasive pest. An efficient clonal propagation system would be required to study and eventually multiply resistant or tolerant selections. Preliminary results are reported for cutting and air layering propagation.

INTRODUCTION

There has been a renewed interest in propagation of *Tsuga canadensis* as part of breeding efforts aimed at developing resistance to hemlock woolly adelgid (HWA), an invasive insect that has affected hemlock stands throughout the eastern US. Efforts in conservation breeding have brought a focus on cloning cultivars or individuals that may exhibit evidence of potential resistance under field conditions. Two separate techniques were attempted for both mature and juvenile *T. canadensis*.

Cutting propagation success was tested with dormant, fall (November) cuttings using mature *T. canadensis* from natural areas of The Holden Arboretum (THA), and five-year-old potted trees grown from seed collected in natural areas of THA. Since much of the published literature is on success of cuttings with *T. heterophylla*, we incorporated one-year-old seedlings of this species obtained from the Franklin H. Pitkin Forest Nursery, Moscow, Idaho, into the experiment as a third tree type. Cuttings were a uniform 2.5 inches cut from the growing tip at 45° angles. The lower 1 inch of needles was stripped from each cutting prior to treatments. Applied treatments were a full-factorial combination of IBA (Advocate, Fine Americas Inc.) at 0%, 0.5%, and 1%, and fungicide (3336F, Cleary Chemicals LLC.) at 0 or 20 ounces per 100 gallons. Cuttings were treated with IBA during a 30 second basal dip, and fungicide with a 15-minute submersion. After allowing to dry, they were stuck in tray pots filled with Sunshine #1 potting mix (Sun Gro Horticulture). Ten cuttings were stuck in each tray, with a tray constituting a replicate. There were six pots for each of six treatments in the two blocks (bottom heated vs unheated) totaling 216 pots or 2,160 cuttings. Trays were located inside a greenhouse in a polyethylene chamber covering a bench top and receiving mist at the rates of 1 minute every 2 hours or 2 minutes every 1 hour, depending on outside conditions. The bottom-heated portion was set to 70° F.

Cuttings were harvested in March (4 months) and May (6 months). In March, there was no rooting in the non-bottom heated portion of the chamber. In the bottom-heated portion, success was highest in the juvenile *T. canadensis*, slightly higher than *T. heterophylla*, with success of up to 63% in the 0.5% IBA + Fungicide treatment. This was significantly greater than the highest success of the mature *T. canadensis* at 16%, achieved in the 1% IBA without fungicide treatment. In May, the unheated portion also began to show similar rooting success for all tree types except the mature *T. canadensis*, which showed none. Juvenile cuttings of both *T. canadensis* and *T. heterophylla* reached 80% success while mature *T. canadensis* peaked at only about 20%. At 6 months, rates were very similar across treatments for a given species.

Air layering propagation was initiated in August 2023, with three treatments: control treatments receiving no rooting compounds; a 1:5 dilution of liquid Dip 'N Grow; and powder Hormodin #3. There was replication of 6 for each treatment on each of two tree types: mature *T. canadensis* from natural areas of THA, and five-year-old potted trees grown from seed collected in natural areas of THA. All were harvested in November 2023. Mature *T. canadensis* had no rooting success. On juvenile trees, there was rooting on one air layer and callous formation on two air layers, all treated with Hormodin #3. All three were potted. The rooted air layer survived the one-year mark to August 2024.

Parent age is shown to present an obstacle to rooting success in both cuttings and air layering, necessitating more investigation into overcoming this issue. For cuttings, the necessity of bottom heat, especially for mature *T. canadensis*, is of significance to the industry. While not necessary, IBA appeared beneficial in many cases, but there was variability between tree type, time since sticking, and bottom heat in relation to the best concentration to use. Air layering of *T. canadensis* is possible, and to the best of our knowledge, this is the first documented instance of such.

Combined Proceedings IPPS 74:520-526. 2024.

Clematis and *Hydrangea*: An Avid Collector Shares a Passion for Plants

Deborah Hardwick

Hardwick Hall, Delaware, Ohio 43015 USA

Deborah.hardwick@icloud.com

Keywords: plant evaluation, pruning, consumer preference, native plants

Summary

Clematis and *Hydrangea* are popular garden plants. New selections are commonly being made available to gardeners. Also, both *Clematis* and *Hydrangea* require different cultural practices depending on the selectin being grown and this can be confusing to novice gardeners. Criteria for plant evaluation and growing tips are provided in this paper.

INTRODUCTION

In this paper, I will attempt to share my experiences with *Clematis* and *Hydrangea* evaluation and production.

Clematis

I have evaluated thousands of *Clematis* selections over the past 25 years. My mission has become to simplify their evaluation and bust myths surrounding *Clematis* plants. For *Clematis*, I will present information about objectively scoring plants for garden performance and breeding potential. I will also discuss plant pruning and the enthusiasm for native species.

Clematis performance is objectively evaluated on a scale of 1 to 10 using the following criteria.

1. Flower quality of color: Is the color consistent? No extreme fading with more than 6 hours of sun? Are flowers resistant to discoloration?
2. Flower form: Is the form consistent? No disfiguring due to early weather aberrations?
3. Flower durability: do flowers last better than other comparable flowers when blooming in a garden?
4. Flower quality: Are sepals self-cleaning by dropping off before they become brown and unattractive?
5. Plant weatherliness: Are plants winter hardy to at least zone 5 and do they handle periods of high heat well?
6. Plant foliage quality: Is foliage durable and not prone to any type of discoloration?
7. Plant: Is the plant sterile without fruit production or do they produce attractive persistent fruit? Do plants spread by volunteer seedlings?
8. Plant size: Are plants compact or medium sized? (compact being under 5 ft, medium 5 to 9 ft).
9. Flower performance: Does the plant grow and flower well with minimal fertilizer to no fertilizer?
10. Flower and plant remontancy: Do plants regrow fast and flower in less than 7 weeks after a mid-season cut?

There are also several negative aspects of *Clematis* growth that need to be evaluated. These include:

1. Excessive lateral spreading plant growth by rhizome or stolon extension (except native plant C. *socialis*). Large size plants over 15 ft height at maturity.
2. Shy to flower, not many flowers for the volume of the plant or only producing terminal flowers.
3. Untidy central boss that is shedding filaments when the sepals are still intact, or other issues with unattractive center of the flower.

By assigning one point to each positive and negative attributes, *Clematis* selections can receive an objective score between one and ten. There are a good number of high point winners in our evaluations, but way too many low scoring plants on the North American market **(Figs. 1, 2, and 3)**.

Figure 1. Two high scoring *Clematis* selections. A. 'Perel D/Azure' B. 'Mazurek' scored 10 points in the evaluation.

Figure 2. *Clematis ochroleuca* was a North American native plant with high potential (10-point score) as a garden plant and for plant breeding.

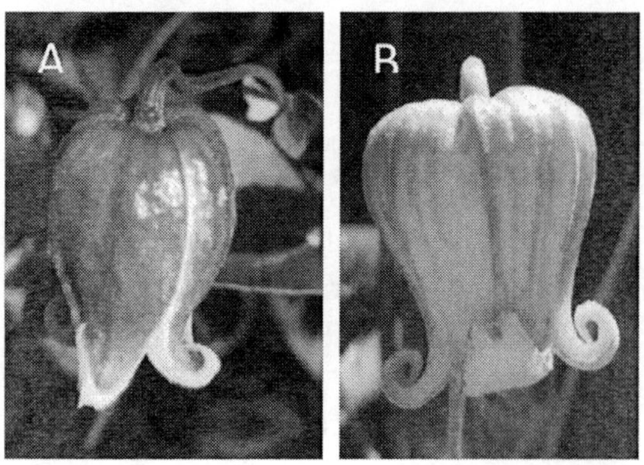

Figure 3. *Clematis* species that were low scoring plants for potential breeding due to their size and habit included A. *Clematis viorna* and B. *Clematis pitcheri*.

True *Clematis texensis* is one of the strongest native plants that you can grow (**Fig. 4**). However, *C. texensis* are very difficult to propagate. Available plants are variable from open pollinated seeds and hybrids have been sold as species.

Figure 4. *Clematis texensis*.

Clematis Pruning

Choosing a pruning method for *Clematis* can be simplified using the "Green – Yellow – Red" method. "Green" means to prune plants hard. Plants in this group bloom on new wood and have abundant large flowers appearing later in the season. This also includes non-vining types. They can be pruned multiple times (every three to four weeks) during the season. "Yellow" means to slow down and show caution. Wait until plants have finished flowering or only prune to remove excess plant growth. "Red" indicates stop or have limited pruning because these plants bloom on old wood. These include the earliest flowering species such as *Clematis montana* or plants in the *Clematis* Atragene group.

Clematis Production

There are some misconceptions about producing *Clematis* that can influence the choice to grow and produce plants. There is the assumption that trellising *Clematis* is needed to sell plants, but this can create headaches that are not profitable. Also, there is the perception that *Clematis* propagation may be trickier, and more plant material required for vegetative pieces. *Clematis* have a bad reputation as difficult with many losses in the garden that can limit their popularity.

Can our native *Clematis* fit into the mix and stand up to their non-native brethren? Do native North American plants fit into the trends in landscape and what customers want? My recent experience in consulting with companies selling hybrids or species of North American species is that *Clematis* is has a bit of cult fervor, but market size needs to be considered. Indeed, *Clematis ochroleuca* is a harder sell than *Clematis* 'Perel D/Azur' or 'Mazurek' (**Figs. 1 and 2**).

There has been numerous examples where North American species and their hybrids that have become available on the market. These include commercial breeding efforts and new introductions (**Figs. 5 and 6**).

Figure 5. A commercially successful *Clematis* hybrid is 'Stand by Me' from Has Hansen at Walters' Garden using A.) *Clematis integrifolia* from Eurasia and *Clematis fremontii* from North America.

Figure 6. Native North American *Clematis* species and hybrids. A. *Clematis viorna* hybrid; B. *Clematis crispa* hybrid; C. *Clematis gattingeri* from Tennessee; D. *Clematis addisonii* from the shale barrens in Virgina.

***Hydrangea* – The New Frontier**

The world of *Hydrangea* has changed forever since grandma's days. Some of the old fashioned yet reliable types found just about everywhere have been surpassed by a plethora of outstanding new varieties. With over 70 known species and some 600+ varieties dealing with *Hydrangea* can be quite the morass.

Some edification is definitely needed to sort things out especially when it comes to knowing how to prune different *Hydrangea* species.

***Hydrangea* Pruning Made Simple**

It can seem overwhelming in the spring to face your garden and take the simple steps to set your *hydrangeas* up for maximum flowering. In a nutshell, pruning of any plant always comes back to when the plant flowers, on what type of stem, (old or new) as well as size considerations. My pruning method solves the confusion by breaking the genus into pruning groups green, yellow, and red.

Sometimes called the stoplight guide.... everyone already knows that green means go, yellow means caution and red means stop. These groups area based on plant genetics and explains how to prune your *Hydrangea* which is everyone's biggest dilemma in the spring garden.

Green means go...these hydrangeas usually flower on new stems so we prune them hard in the spring, meaning remove the old stems in early winter or spring. If the stems are not cut hard, some may break a little growth on the bottom part of the stems. While you can cut to just above a growth break, removing the stems down low will create a better shaped plant as well as a better flowering impact. Types of *Hydrangea* that are green pruned are *Hydrangea arborescens*, *Hydrangea paniculata* and selections include Annabelle, Incrediball, Hass Halo, Limelight, Berry White, and Bobo.

Yellow means caution…these hydrangea bloom mostly on old stems. In the spring cut the stems to just above a nice fat bud break. Some reshaping can be done in the spring if you use a lower bud break on outer stems to create better plant support and silhouette. Types of hydrangea that are yellow pruned are *Hydrangea macrophylla* (big leaf), *Hydrangea serrata* (mountain hydrangea), *Hydrangea involucrata*, and *Hydrangea aspera* and selections include Bloomstruck, and Summer Crush.

Red means stop! These hydrangeas need old stems to generate spring regrowth and flower on the old stems. Old flowers that might linger can be deadheaded off, but other than that stop! No pruning needed. Pruning these hydrangeas will not kill the plant, but you will not have flowers from the new growth until the following year. Types of hydrangeas that are red pruned are *Hydrangea schizophragma, Hydrangea dichroa, Hydrangea quercifolia, Hydrangea anomala petiolaris* and selections include Munchkin, Ruby Slippers, Little Honey.

Easing the Confusion

The ideal role of the propagator and producer in simplifying things for gardeners include reducing misinformation, labeling and tagging plants correctly, using tags or QR codes to expand information especially about how to grow plants well in the home garden.

An example of the impact of misinformation is that *Clematis terniflora*, an Asian species had historically been referred to by the syn. paniculata, (not species *paniculata*). Plants were tagged in the USA by propagators as paniculata, therefore able to ship to the restricted states who had banned the plant *Clematis terniflora* 'Sweet Autumn' due to the invasive spread. *Clematis terniflora* is considered an invasive (and poisonous) species, particularly in the East, including North Carolina, Alabama, Delaware, Florida, Georgia, Illinois, Maryland, New Jersey, South Carolina, Tennessee, Colorado, Washington, Oregon. In the right climate, escaped plants spread invasively along riverbanks and roadsides. Plants with the *paniculata* tag can be found in nurseries to this day. *Clematis paniculata* the species would be in italics, is diecious, and is not hardy in North America. Clearly NOT what is sold tagged as paniculata. 'Sweet Autumn' is likely a hexaploid variant of *terniflora. (C. flammula robusta*) and propagation should be stopped. https://www.invasive.org/alien/pubs/midatlantic/clte.htm

Back to the passionate and positive side …Pruning *Hydrangea* Could propagators influence the confusion with pruning right from the get-go, so the customer knows how to handle *Hydrangea* and *Clematis* before they purchase the plant or immediately afterwards?

In Poland, one propagator sells *Clematis* in colored pots to correspond with flower color. This was deemed a success because flower color with *Clematis* is a number one interest and made it easier for the customer since *Clematis* cannot often be displayed in flower, and many tags do not have photos or QR codes. For *Hydrangea* and *Clematis* could plant tag color or pot color correspond to the pruning style? Other variables and info can be grouped with pruning color (**Table 1**).

Table 1. General care of *Hydrangea* by pruning color.

	Green Prune	Yellow Prune	Red Prune
Sunlight	Up to 6 hours	At least 4 hours	At least 4 hours
Winter Mulch	2 to 3 inches	Up to 6 inches	Normal garden level
Soil, Watering	Neutral to acidic, water from trunk to dripline	Neutral to acidic, water from trunk to dripline	Neutral to acidic, water from trunk to dripline
Plant size	Large to colossal 8 ft plus	Small to medium 2.5 to 5 ft	Medium to large 5 ft plus
Flower size	Large 5 inches plus	Medium 2.5 in to 5 in	Medium to large 2.5 in to 5 in

The Right Fit for Native Plants

Hydrangea's native North American plants are less interesting and may have a hard fit to find commercial viability. Flowers are smaller, plants rangier... can demand be created for them against the plethora of show stopping hydrangea varieties on the market?

Clematis have a similar struggle for native plants to find their way into commercial popularity. We have to face the fact that like Ultra Processed Foodstuffs, customers vote with their purchases and for big sales it is usually big flowers.

Outliers like me, and maybe you...are not the customer.

Combined Proceedings IPPS 74:527-529. 2024.

Propagation in Ash for Emerald Ash Borer Resistance

Rachel Kappler

Great Lakes Basin Forest Health Collaborative Coordinator, Holden Forests and Gardens, Kirtland, Ohio 43015 USA

rkappler@holdenfg.org

Keywords: pest management, breeding urban forests, *Fraxinus*, EAB

Summary

The Great Lakes Basin Forest Health Collaborative (GLB-FHC) is a partnership network supported by the United States Forest Service and Holden Forests & Gardens, an NGO near Cleveland, Ohio. I am coordinating the GLB-FHC network of partners (state, fed, tribal, NGO, private landowners, citizens, etc.), that are interested in coordinating and discussing various parts of participatory pest/disease resistance breeding activities that support tree health. Our current goals involve increased communication across organizations with activities involving ash (*Fraxinus pennsylvanica, F. americana),* American elm, (*Ulmus americana*) American beech (*Fagus grandiflora*), and eastern hemlock species, (*Tsuga canadensis*). Our collaborative works towards assisting each other in activities that improve and increase our ability to breed pest/disease resistant tree species such as: forest monitoring, tree reporting, tree selection, sample or seed collection, tree propagation and planting/orchard development. In this position I also serve as a technology transfer liaison, providing training for partners wanting to be involved in tree breeding activities for pest/disease resistance, sharing knowledge and answering questions.

IPPS Vol. 74 – 2024

INTRODUCTION

The EAB resistant ash tree breeding (*Fraxinus sp.*) program work flows breakes it down into 6 steps (**Fig. 1**). These steps are important in propagating EAB resistant ash because the order ensures that the best candidate trees are tested, tested trees let us know specific resistance levels as we progress into producing seed orchards. Planting trees into seed orchards allows for preservation of the individuals while producing seeds important for future plantings in the forest.

Figure 1. The 6 simplified steps needed in breeding EAB resistant ash.

One of the areas that slowed progress in EAB resistance work was a lack of available grafters. With the addition of some employees and grant funding we now have more locations that can assist with lingering ash grafting. Let me share some specifics on how the Forest Service grafts lingering ash clones or ramets. It can be dificult to retreive scion from some trees, but there are multiple tools that can help gather scion for grafting, for example, a slingshot, ropesaw, pole pruner extensions, and tree climbers. Trees that are lingering are not always in the best condition and it is important to check that you collect branches that have healthy buds on them. Research has found that hot-callus grafting gives the highest success rate (Carey et al. 2013). This means we will either top graft or side venir graft and add heat to the join area afterwards for around 6 weeks during the coldest time of the year. Before placing it near heat we coat the top of the graft in parafin wax to help it retain moisture. At the Forest Service Northern Research station trees are placed standing up in pots with a heating cable applied to 2 x 4 inch beams in order to keep the heat close to the join. Then surrounding the join there is insulation typically used for window sealer, it's thin size and short width allows for easy application to the beams. Openings are cut in the insulation to allow for placement of the tree and then the opening is taped over.

Temperatures are monitored to keep the heated area no warmer than 75°F (24°C). The room remains cold to keep buds from opening early. Future cloning for green ash can now include rooted cuttings, likely from the grafts, to increase ramets(clonal selections) of individuals (Aletta Doran, Holden Forests & Gardens, "personal communications" 10/15/24).

After grafted trees have grown into saplings individuals are tested for resistance via an EAB egg application bioassay, where those proving to show resistance successfully kill EAB larvae; however variations in resistance indicates only some are able to kill enough larvae to prevent their own mortality (Stanley et al. 2023). Other ramets of resistant individuals are planted out in a test orchard to examine over time how individuals cope with EAB in a more natural setting. Currently, green and white ash(Fraxinus pennsylvanica, Fraxinus americana respectively) from the southern Michigan and Ohio regions are set up in test orchards in hopes of converting them to seed orchards in the future. It's been between 8-15 years since they were planted and they have started to make seed. Plans are now underway to plant these open pollinated seeds out and assess how well they perform in a test orchard. In order to do this we have to collect and propagate ash from seed.

Collection of seed is done in September and sorting is required to remove empty seed or those infested with ash seed weevil larvae, a naturally occurring issue. Setting seed into cold storage for up to four weeks induces the larvae to remove themselves from the seed and leave a clear exit hole. If not propagating that year ash seed can be dried to 30% relative humidity and stored well for up to 5 years and longer if dried and frozen. Stratification for ash takes 3-4 months of cold, moist and dark conditions (Burns et al. 1990). We place our seeds in bags of moist sphagnum moss in cold storage for this timing and then place the bag in the greenhouse to start germination. Emerging seeds can then be potted, but other more direct sowing is an option.

Future work includes research plantings of open pollinated seed from EAB resistant ash plantings, which will likely begin in a year or two. GLB-FHC also has a goal of assisting areas across the Great Lakes Region with adding new lingering clones to new plantings. This will help preserve lingering trees as original trees may die in the forest before EAB resistance testing can be done. Preserving these trees locally allows for a greater remaining gene pool for resistance breeding programs.

LITERATURE CITED

Burns, R.M. and Honkala, B.H., [Technical coordinators]. (1990). Silvics of North America: Volume 2. Hardwoods. United States Department of Agriculture (USDA), Forest Service, Agriculture Handbook 654.

Carey, D.W., Mason, M.E., Bloese, P. and Koch, J.L. (2013). Hot callusing for propagation of American beech by grafting. HortScience. 48:620-624.

Stanley, R.K, Carey, D.W., Mason, M.E., Doran, A., Wolf, J., Otoo, K.O., Poland, T.M., Koch, J.L., Jones, A.D. and Romero-Severson, J. (2023). Emerald ash borer (*Agrilus planipennis*) infestation bioassays and metabolic profiles of green ash (*Fraxinus pennsylvanica*) provide evidence for an induced host defensive response to larval infestation. Front. For. Glob. Change 6:1166421.
doi: 10.3389/ffgc.2023.1166421

Combined Proceedings IPPS 74:530-539. 2024.

Propagation of *Packera aurea*, *Eutrochium fistulosum* and *Sagittaria latifolia* at Kind Earth Growers

Josheph LaMent

Kind Earth Growers, 2503 Mountain View Dr., Ottsville, Pennsylvania 18942 USA

incLaMent@gmail.com

Keywords: native plants, seed germination, dormancy, ragwort, Joe-pye weed, duck potato

Summary

Kind Earth Growers specializes in growing native plants from seeds. Production is mainly in cell trays. This paper describes the propagation and production strategies for three different crops including golden ragwort (*Packera*), Joe-pye weed (*Eutrochium*) and duck potato (*Sagittaria*).

INTRODUCTION

Kind Earth Growers (**Fig. 1**) is a wholesale nursery operation in Bucks County Pennsylvania; growing native perennials, largely from seed, in plug trays. As I outline how we grow these species, it is important to factor in our water quality and location as these conditions can change how plants respond in the greenhouse environment. Kind Earth pulls its water from a single well, pH 7.0-7.6, hardness >260mg/L. The water is injected either with 93% sulfuric acid at a rate of .6oz per 100 gallons or Sprint 330 Iron chelate at a rate of .025oz per 100 gallons, applied at grower discretion. Bottom heat is the main heat source for early stages of growth and runs from January through

April. Seeds are germinated at 72 F and plants are grown on from 55-60 F. Our methods are centered around manipulation by hand. Seeds are sown over 288 trays with graded vermiculite carrier and a set of sieves. Seedlings are singled down from this tray into larger sizes by our transplanters (see photo to left). Plants start in a standard peat-based propagation mix and are transplanted into a peat based plug mix or a peat free compost and pine fines mix depending on species and time to availability.

Figure 1. The Kind Earth growers' team in 2024.

We incorporate slow-release fertilizer at the low rate given by the manufacturer, and supplement that weekly as-needed with various liquid salts, typically 150 ppm N maximum. Because of our water quality we use an acidic complete fertilizer two weeks in a row, followed by a neutral or basic fertilizer for the third week. We inoculate religiously with *Trichoderma* products, and I attribute that to our lack of significant root disease. Our IPM program leans heavily on beneficial insect releases.

Finally, I would like to mention that we focus on growing a plant that is not only healthy, and able to survive direct planting into the landscape, but capable of propagating and supporting its community for years to come.

***Packera aurea* (Golden ragwort)**

Packera is a low growing semi-evergreen perennial in the Asteraceae and this picture is the original patch where seed was collected in our first year of operation 2016 (**Fig. 2**).

Figure 2. *Packera aurea* seed production stock.

We grow two distinct variants, our open pollinated variety from Bucks County and a *Packers aurea* which is solely vegetatively propagated, which we procure via tissue culture. The open pollinated variety has a smaller habit in almost all parameters, purple coloration on the leaf underside is less intense, it grows much better in full sun and produces viable seed. Some argue the vegetative variety is a better garden plant being more handsome, and because the seed it produces does not germinate. It is believed that the low seed fecundity of the vegetative form is due to the common characteristic of most Asteraceae to be self- sterile. Both plants are vigorous growers and excellent groundcover.

The first set of ripe seed is on or around May 14th. It's ready when the light lime green seeds begin to turn beige at their core, the seed will ripen fully to a ruddy brown and the pappus will fluff out all at once near the end (**Fig. 3**). Keeping your eye on the weather can be important as the seed is dispersed quickly in rain or wind. The first round is the best, I suspect the beautiful and diminutive, white-crossed seed bug may be the culprit, eating the late seed more thoroughly. As many as three rounds of ripening can occur, usually by May 21st.

Figure 3. Initial seed ripening stage.

We have found that managing a garden plot with many of these plants can dramatically increase the volume, quality, and reliability of the seed they produce. Rub the seed on a sieve to remove the pappus and winnow it before storage. Stratify seed for 30 days, up to two years before losing viability.

Seeds take 4-8 days to germinate. This is a cool season crop, so lower light levels and slightly lower nighttime temperatures don't cause the finish time to fluctuate dramatically. Using peat free media pushes finish time to 4-5 weeks. Our last week to transplant into a finished size is week 40. The most prominent insect pest is aphids, check early and often.

Should you miss your seed or struggle to germinate it, the many tips can be pulled off plants as young as 12 weeks old (**Fig 4a**). The youngest reliable division from seed grown plants is pictured above. Side shoots will sprout from the base of the cutting so get that towards the center of the plug.

We also use tissue culture plants to multiply plants (**Fig. 4b**). It's best to get good contact between the very base of the plantlet without burying the small basal leaves. We plant them in pl128 trays. The biggest issue with vegetative propagation is *Phytophthora* crown rot and botrytis. Treat them until sale.

Once the tissue culture plants are inserted into the tray, they are held in our humidity tent for hardening off. Cut the mist down as quickly as possible. Allow the plantlets to flag between mist events from day 2 onward. We often use a silicone surfactant on day 1 at labelled cutting rate.

Table 1 shows the times to finish a seed, division and tissue culture crop.

Figure 4. Vegetative propagation in *Packera*. A. Side shoot divisions. B. Plants arriving from the tissue culture vessel. C. Individual plantlet. D. High humidity tent for tissue culture plantlets.

Table 1. The time to finish a seed, division and tissue culture crop in *Packera aurea*.

Starting size	Finish size	Weeks to finish
Sown seed	288	3-4
288 from seed	128	2-3
288 from seed	72	3
128 from seed	Dp50	3-4
Divisions	Dp50	3-4
Stage 3 tissue culture	128	3-4
128 from tissue culture	Dp50	3-4

***Eutrochium fistulosum* (Hollow stem Joe-pye weed)**

Joe-pye weed (aside from having a great first name) is a beautiful meadow and roadside weed, often to 9 feet tall with whorled leaves (**Fig. 5**).

Figure 5. *Eutrochium fistulosum* in flower.

The seed ripens around September 15th and can hang on the plant into October. The pappus does not separate easily from the seed so don't bother. The seed will germinate without treatment or ideally with a stratification of 1 to 15 days to help it imbibe moisture. Longer stratification times often reduce viability. Caterpillars and pollinators love it as well as aphids, thrips, and mites. Late summer weather, or overwatering, will bring leaf spot.

Joe-pye germinates in stages over 5-9 days. Seedlings grow vertically with such enthusiasm that they need to be pinched before the plug is ready for transplant. The plugs at left aren't quite rooted through but have begun stretching. To the right is the proper trim height to promote a sturdy habit in the finished plant. Get the crown buried slightly at transplant to ensure a strong connection with the media.

Figure 6. Seedling plug production in *Eutrochium*. A. Young seedlings beginning to stretch. B. Pinched plants.

An exclusive selection of double flowering *Eutrochium fistulosum* is 'JoJo' (**Fig. 7**).

Figure 7. 'JoJo' *Eutrochium fistulosum*.

'JoJo' plants are shorter than the species, topping out at 5' while retaining its vigorous habit. The leaves are dark green, and the stems have red/purple tint. Color saturation in the flowers is excellent especially in full sun. The flowers will persist into October as they will not get pollinated or produce seed. Strikingly, the flowers and stems turn black at dormancy providing good winter interest. While it may not provide nectar or pollen it attracts caterpillars, aphids, thrips, mites, and their associated predatory insects.

Eutrochium fistulosum's hollow stem and whorled arrangement cause some irregularities when making cuttings and keeping stock plants. To combat these issues, I grow the mother pot tall and harvest cuttings down the stem rather than up. Three-year-old 'JoJo' stock plants are forced in the propagation house. The goal is to leave 7-9 solid nodes and pinch the tips off. These tips are not good for rooting, their hollow stem cause them to warp and bend under the substate rather than develop healthy callus. The plants average 5 productive stems per plant (**Fig. 8**).

After pinching, we remove the large leaves from the top to allow light in, as the top two nodes will push cuttings. The expanding shoots will be ready between day 14 and 20 from pinch. Each node, being whorled, will produce 3-7 cuttings. This first flush amounts to 100 cuttings per stock pot, conservatively.

Take cuttings before stems get too large. If you let them get long, they will need their leaves trimmed, which is time consuming.

Take the cuttings before this happens! The leaves expand incredibly fast, and furthermore they will continue to expand after you cut them, so go a little earlier than you think is ideal.

We stick them in 128's and under the mist tent. Mist requirements are medium-low. Cuttings will take 2-4 weeks to root.

Figure 8. 'JoJo' three-year-old stock plants. A. Growth after nine weeks. B. Pinched plants with large top leaves removed. C. Shoots developed after pinching. D. Shoots ready to take cuttings.

You can now prepare the mother stock plants for another round of cuttings (**Fig. 9**). Method, time, and quantities will be similar to the first flush. Transplant into larger size quickly, when there is something like 6 nodes pinch the plant and root the tips. We aim to finish crops before mid September to make sure the plants have enough time to create winter buds. Usually our last dp50's get planted week 34.

Figure 9. Stock plants cut back to induce a second round of cuttings.

Table 2. Production schedules for *Eutrochium fistulosum* from seeds or cuttings.

Seed Propagation		
Starting size	Planted into:	Weeks to finish
Sown seed	288	4-5
288 from seed	128	2-3
288 from seed	72	3
128 from seed	Dp50	3-4
'JoJo' Cutting Propagation		
Mother plant	1st flush cuttings	11-12
Mother plant	2nd flush cuttings	14-15
Unrooted cuttings	128	2-4
Unrooted cuttings	tip cutting	7-8
128 from cuttings	Dp50	3-4

Sagittaria latifolia (Duck potato)

Wetland plants are at the heart of what we grow at Kind Earth and play a crucial ecological role such as nutrient and sediment capture, soil stabilization, groundwater recharge, floodwater mitigation, food, and cover for wildlife.

Wetland habitats are not currently preserved, protected, or managed with the care that they need to be to maintain the health of our land and water.

Sagittaria is an excellent plant for use in wetland restoration and stormwater management (**Fig. 10**). It is still able to tolerate dry spells, possibly due to its (edible) underground corms. White flowers with butter yellow centers bloom as early as July but can appear as late as September. The seed ripens September 15th through October. Stratify the seed for 60 days up to two years and we find a short freeze helps break seed dormancy. *Sagittaria* should be given plenty of fertilizer, a characteristic of many wetland plants such as *Typha*. Aphids are the main pest in a greenhouse setting.

Figure 10. Duck potato (*Sagittaria latifolia*) growing in dry earth cracks can tolerate wet or dry soils.

The edible corms are also excellent propagation material (**Fig. 11**). These corms vary in size considerably, often growing larger or smaller depending on the size of the pot, quantity of fertilizer, and how late in the season they began to form. A single *Sagittaria* plant will produce 3-12 corms on long stolons, 10-16 weeks from germination. Once they have formed, main plants have a habit of going dormant a little while after, which is problematic for sale. Corms may resprout the same season from the terminal claw-shaped bud, but it takes some time. Corms that have formed fully can be plucked off the stolon, put damp in a bag and stored in the fridge, 60 days to 9 months. Keep an eye out for fungal growth in the bags. To sprout; float them in warm water until the terminal grows. Once roots have begun to expand, break the plant off, plant it, and put the corm back in the water to resprout. This next set of plants will be smaller and more delicate, plant them with the corm this time. We sprout them March through July. Sprouting before March will result in dormancy as early as May.

Figure 11. Sprouting *Sagittaria* corms.

Sagittaria germinates best in fully saturated soil, we grow many wetland species in ebb and flood benches simply plugged up to hold water. One could put the propagation tray into a solid flat to hold enough water in the media. Trays can be drained but kept wet once germination is complete. Seedlings are small, which contributes to a lengthy stay in the prop tray. Manage the inevitable algae by disturbing the soil surface with a fine mist nozzle operated by hand.

Because of the role these plants play in the environment they respond well to fertilizer. Seedlings are fertilized weekly at 100 ppm N. Once transplanted into larger sizes you can apply 200 ppm N twice weekly.

Higher incidences of fertilizer will finish your plugs faster but the plants will be "softer" which is not ideal when they will be planted in a colder or drier landscape. When you're growing wetland plants, don't dry them out.

Table 3. Production schedule for *Sagittaria latifolia* from seeds.

Starting size	Finish size	Weeks to finish
Seed sown	288	4-5 (5-8d to germ)
288	72	2-4
Seed sown	Mature corms	10-16
Fridged corm	Terminal plant (1st)	5-7
Fridged corm	Sprouted corm (2nd)	8-10

Combined Proceedings IPPS 74:540-542. 2024.

Optimizing Rooting for Efficient Nursery Production

Olivia Liebing and W. Garrett Owen

Department of Horticulture and Crop Science, The Ohio State University, Columbus Ohio 43201 USA

Liebing.4@osu.edu

Keywords: daily light integral (DLI), root-zone heating, foliar rooting hormone, ornamental woody nursey propagation

Summary

Propagation can be a significant operational and production expense. Controlled environment management can improve rooting efficiency and the final quality of the liner. These include light management, root-zone temperature and foliar auxin levels.

INTRODUCTION

Major barriers to the ornamental woody nursery plant industry include increasing operational and production expenses, workforce availability, and year-round accessibility to high-quality rooted propagative materials (liners). Ornamental nursery woody plants are often propagated by rooting cuttings, which requires extensive hand labor to dip cuttings in rooting hormones, stick cuttings into media, and the capital

IPPS Vol. 74 – 2024

and operational investment of facilities to promote adventitious root development (**Fig.1**).

Figure 1. *Taxus* cutting with healthy, well-formed, and evenly distributed adventitious roots.

To achieve this goal, it is proposed to improve rooting by investigating how environmental management components of controlled environment production systems and rooting hormone applications can be integrated to improve rooting and overall quality of ornamental nursery liners. The objectives are to identify optimal propagation (1) DLI, (2) root-zone temperatures, and (3) rates of foliar rooting hormone that improve adventitious rooting of ornamental nursery cuttings. Upon completion of this project, efficient, sustainable, and profitable propagation strategies to root ornamental woody nursery cuttings will be identified while also creating resources for propagators to effectively apply recommendations to key taxa.

Despite these steps, rooting failure and cutting loss can still impede economic success (**Fig. 2**). When plants are propagated without optimal conditions or rooting hormones, uniform rooting is difficult to achieve.

Figure 2. Non-uniform rooting in *Buxus* cuttings. The rooting success and quality of this *Buxus* crop varied greatly even when all cuttings were harvested and placed into propagation on the same day. If optimal conditions were achieved in propagation, then rooting would be more synchronized amongst the cuttings.

Greenhouses offer technologies to empower propagators to precisely control the environment, yet optimal propagation conditions and foliar rooting hormone applications to hasten adventitious rooting is largely unknown for ornamental woody nursery cuttings.

Thus, there is a critical need to identify cost-effective strategies such as manipulating the daily light integral (DLI), providing root-zone heating to improve rooting of ornamental nursery cuttings, and applying foliar rooting hormones to promote efficient, sustainable, and profitable propagation regimens; therefore, this critical need is the overall goal of the project (**Fig. 3**).

Figure 3. Foliar spray applications of rooting hormone using booms reduces hand labor and creates opportunities for easy repeat applications over time.

Combined Proceedings IPPS 74:543-545. 2024.

Substituting Hemp Hurd Fiber for Peat in Plant Production

Jessica Lubell-Brand

Department of Plant Science and Landscape Architecture, University of Connecticut
Storrs, Connecticut 06269 USA

jessica.lubell@uconn.edu

Keywords: *Cannabis*, transplants, petunia, substrate, media

Summary

Fiber products have become an interesting alternative to traditional substrate components. Hemp hurd fibers are a biproduct of other hemp uses and could be an alternative organic source for container production.

INTRODUCTION

Growers are interested in sustainable alternatives to peat moss for nursery plant container production. Substrates including coir, biochar and wood fiber have been the subject of recent research trials. Hemp (*Cannabis sativa*) farming has seen a resurgence since its legalization in 2018. Hemp stems consist of long bast fibers and short hurd fibers. The bast is desirable for the textile industry, however the hurd has fewer identified uses and is a byproduct of hemp production. Novel uses for hurd include hempcrete, wood paneling and animal bedding. I am conducting research to evaluate hurd as a substitute for peat moss in container production of horticultural crops.

Hemp hurd fiber is available from a limited number of suppliers in a few different blends based on particle size. The first product tested had a 10 mm diameter particle size. Unfortunately, this product was rich in tannins that had to be washed from the substrate prior to potting, which was labor intensive. In addition, the product was contaminated with hemp seeds that germinated 1 to 3 days post transplanting. Seedings could be easily removed by hand; however, this was undesirable due to legal restrictions on growing hemp and added labor.

The second hurd product tested had 2 mm diameter particle size and was clear of tannin and seed contaminants. In experiments with petunia, hurd was substituted at three different rates (low, medium and complete replacement) for the peat portion of a 1:1 peat:vermiculite container medium **(Fig. 1)**. The porosity of the four media tested ranged from 77% to 84%.

Figure 1. The four substrate treatments using different levels of hemp hurd.

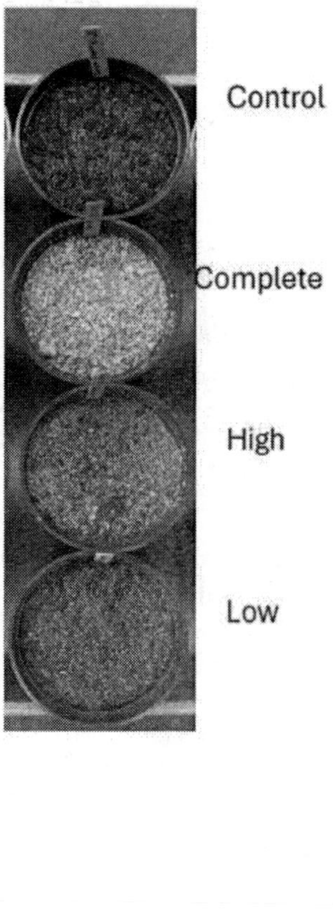

In the first petunia experiment, plants received a low rate of controlled release fertilizer and constant liquid fertigation with a 20-10-20 product that was acidifying. After one week the pH of all media, except the complete replacement treatment, had dropped well below the recommended pH range for petunia. Upon changing to a 13-2-13, strong basifying fertilizer, the pH rebounded to acceptable range for the medium and complete replacement treatments and close to recommended range for the low and control treatments. Petunia plants grown in control and low treatments performed similarly, were visibly indistinguishable, and larger than plants grown in medium and complete replacement treatments **(Fig. 2)**.

Figure 2. Petunia plant growth during experiment one.

In the second petunia experiment, a 15-5-15 basifying fertilizer was used and pH over time was more constant for all treatments; however, pH was below recommended range for the low and control treatments as in experiment 1. Performance of the medium treatment for some measured parameters had improved to match the low and control treatments in experiment 2.

These initial findings for petunia indicate that hurd may be a viable alternative for peat, but additional research is needed to evaluate fertilizer formulation and rate of delivery. Future work will evaluate hurd for other greenhouse crops like geranium and tomato and nursery crops including woody shrubs and herbaceous perennials.

Combined Proceedings IPPS 74:546-552. 2024.

Sowing Seeds of Change

Gale Martin

Natives in Harmony Nursery, 4652 Township Road, Marengo, Ohio USA

gale@nativesinharmony.com

Keywords: native plants, nursery production, Ohio, prairie

Summary

The native plant production and education program at Natives in Harmony nursery are described. This includes plant selection, propagation and nursery production. In addition, their educational outreach program is described.

INTRODUCTION

Most of my work has been centered around the Sandusky Plains, a small residual plains area in Ohio that is threatened by development and "progress" on all sides. It encompasses Wyandot, Green Camp and Upper Sandusky municipalities in Ohio. This area is well within reach of the I-80 corridor and its associated businesses.

Many endemic native plants abound around Trella Romine Prairie with some being quite rare. *Asclepias tuberosa* of various colored forms, *Echinacea purpurea, Baptisia alba* , *Silphium species*, and numerous grasses can be found in this environment . It is important to know that plants are found in communities and rarely is a plant an island, the community aspects are varied but one key ingredient is a shard microflora heavily dependent upon mycorrhizal associations.

The Claridon Prairie, in Marion County, Ohio is one of the few surviving remnants of the once extensive grass land prairies that were part of pioneer Marion County (**Fig. 1**). An isolated strip between a railroad track and a highway has over 75 species of significant prairie grasses and forbs.

Figure 1. Claridon Prairie.

Remnant prairies are often fragile and are susceptible to numerous threats in addition to road, railroad and general construction activities.

An example of this is the royal catchfly (*Silene regia*) in the 95 West Remnant prairie that was destroyed by the illicit dumping of driveway debris and asphalt which eliminated the very small population found on the railroad remnant of the prairie (**Fig. 2**)

Figure 2. Royal catchfly before being eliminated by dumping debris on the site.

Remnant prairies are often refuges for rare plants. Reasons for this vary, but part of the occurrence of rare plants is an interdependence upon specific soils, water relationships and as mentioned earlier the shared use of mycorrhizal affiliations. This of course does not take away from the competitive effects of woody plant intrusion which is significant in the alteration of water relationships, removal of critical nutrients and the alteration of fungal populations. Prairies are unique eco-systems and require diligent efforts of observation, understanding and preservation.

Across the road from an industrial plant on 95 West Remnant prairie (**Fig. 3**) a brief survey turned up a myriad of rare species including:

Purple rattlesnake root - *Nabalus racemosus* (*Prenanthes racemosa*)
Savannah blazing star- *Liatris scariosa var. niewlandii*
White false indigo - *Baptisia alba*
Culvers root- *Veronicastrum virginicum*
Michigan lily- *Lilium michganense*
Virginia mountain mint - *Pycnanthemum virginicum*
Wild hyacinth- *Camassia scilloides*
Stiff goldenrod- *Solidago rigida*
Prairie phlox- *Phlox pilosa*
Canada anenome- *Anenome canadense*
Golden alexander- *Zizia aruea*
Black eyed Susan – *Rudbeckia sp.*

Purple rattlesnake root

Savannah Blazing Star

White False Indigo

Culvers root

Michigan lily

Virginia mountain mint

Figure 3. Native species in the 95 West Remnant prairie.

Prairies Are Not Just Homes for Plants

Prairies are reservoirs for not only plant species but also for the myriad of insects, other arthropods, small and larger mammals, reptiles and in some cases amphibians. It is essential to recognize prairies for the enormous potential they represent, and it is not sufficient to just restrict our activities to observation and recognition. The much troubled Monarch butterfly (*Danaus plexippus*) is but one of many examples of the value of a prairie ecosystem. *Asclepias* species are its primary food source, and prairies often serve as the stockpile for those plants **(Fig. 4)**.

Figure 4. Monarch butterfly on *Asclepias* (above) and American Lady (lower) caterpillars on sweet everlasting (P*seudognaphalium obtusifolium*) production material.

Propagation Facilities

Natives in Harmony Nursery was established to address some of the shortages of endemic native plants that have been forsaken by the mainstream horticultural world **(Fig. 5)**. We do extensive research and scouting for native populations with a keen eye towards the collection of seed for propagation. We take care to ensure that sufficient quantities of seed are left in place to allow the endemic population to prosper.

Figure 5. Production facilities for propagation of native species.

As our endeavors increased, we saw a need for closer observation and a means for harboring certain species so that they would not meet the fate of the *Silene regia*.

We decided to implement a grow your own program starting with a raised bed (**Fig. 6**). This allowed us to see how our plants do when planted out and we have the bonus of more available seeds to collect, always a plus.

Figure 6. Raised bed planting for plant evaluation and seed production.

Germination may be variable with some seeds germinating immediately while others germinate irregularly over a period of weeks, months or even years.

This reproductive strategy is advantageous for a wild plant because offspring are dispersed over time, a better strategy for dealing with weather fluctuations.

As is the case with many northern U.S. and Canadian species stratification is a must to achieve significant seed germination (**Fig. 8**). Again, conditions vary from species to species, some requiring more elaborate treatment than others.

Seeds arrive from all across Ohio

This endangered striped gentian is just one of the state-listed species we grow from seed acquired via plant enthusiasts who want to help us preserve Ohio genetics.

Figure 7. Striped gentian (*Gentiana villosa*).

Figure 8. Cooler facilities for seed storage and chilling stratification.

We have grown significantly since our early days. This includes product diversity and the ability to deliver plants (**Fig. 9**).

Figure 9. Plant delivery.

Educational Outreach

We hold a number of educational events; we have to spread to word. Teaching children about the native world and what can be done to help it is an ongoing process (**Fig. 10**). And an important one too!

Figure 10. Educational outreach.

Future Directions

Hugelkultur or Hügelkultur is a Germanic word describing an old European system that tried to mimic some of the natural processes one can find in the forests.

This concept also applies to prairies and prairie management. The quickest way to really "see" a plant is to go look for yourself. Nothing surpasses on the spot observation and analysis. If you want to know how a plant grows, go find where and how it grows. It could well be an eye opener and will assist greatly with learning how to produce it.

Combined Proceedings IPPS 74:553-558. 2024.

Optimizing Hemp Growth: Harnessing the Synergy of Aquaponics through a Split Root System

Casey McKay, Yuta Inoue, Victor Zayas, Neil Mattson, Mark Bridgen

School of Integrative Plant Science, Cornell University, Ithaca, New York 14583 USA

cam575@cornell.edu

Keywords: *Cannabis*, hydroponics, greenhouse production

Summary

This study investigated the potential of integrating aquaponics and hydroponics using a split root system to optimize hemp (*Cannabis sativa L.*) growth. The results demonstrated that the single bucket hydroponic system provided the most optimal growing environment measured by fresh weight and kite measurements which outperformed the other treatments. The aerobically digested fish water treatment had significantly lower plant growth. A multiloop system was developed that included the combination of coupled and decoupled aquaponics. Additional work will determine combinations enhance nutrient availability for the plants.

IPPS Vol. 74 – 2024

INTRODUCTION

Integrating hemp cultivation with aquaponics presents a sustainable and efficient approach to enhancing plant growth while supporting fish cultivation (Yep, 2019). The split root system, which divides plant roots into two zones to access distinct nutrient environments, has been shown to improve nutrient uptake and overall plant health (Shen, 2004). Since hemp's legalization in the 2018 Farm Bill, there has been little research on indoor cultivation of hemp using aquaponics. This study investigates implementing a split root system with aquaponics and hydroponics for hemp cultivation to optimize nutrient absorption and overall plant health. It also explores research on the possibility of a dual decoupled and coupled aquaponics growing technique, also known as a multiloop system. The split root system involves dividing the roots into two halves and submerging one half into one treatment bucket and the other half into another treatment bucket (**Fig. 1**).

Figure 1. Split root system design.

In the split root system, each hemp plant is positioned in a net pot, with its roots divided into two separate sections. One half of the roots is submerged in a bucket containing one treatment, while the other half is submerged in a second bucket containing a different treatment.

There are two control treatments using single buckets: one with Hydroponic water and another with Fish water–water enriched with fish waste. Throughout the experiment, plant growth was measured through kite measurements–canopy development, pH, EC, and chlorophyll content. Post-harvest measurements such as fresh weight, dry weight, root weight, and root area were taken. A single cultivar, TJ's CBD, will be used to analyze the effects of different root zone treatments. Since hemp recently became legal in the United States, there is almost no research on how hemp is cultivated. This study aims to investigate how well hemp grows in aquaponic treatment.

METHODS AND MATERIALS

This experiment was conducted in the Kenneth Post Lab greenhouse at Cornell University. Twenty-four one-gallon buckets and six two-gallon buckets equipped with 30 aero stones were used to grow hemp using a split root system integrating aquaponics and hydroponics. Eighteen Rockwool cubes, each placed in plastic net pots, housed the rooted hemp cuttings, with the roots divided between the two respective treatments that were randomly chosen for that plant. The cuttings were treated with RootX, a rooting hormone, then were placed in an aeroponics chamber to promote rooting for two weeks.

Then the cuttings were placed in a Rockwool cube, put under a humidity dome, and were in a water bath to encourage longer roots. After two weeks, the roots were long enough to be put into the split root system and their respective treatments **(Fig. 2)**.

Figure 2. A comparison of the split root fish water and aerobically digested fish water (bottom right plant) and its size comparison to the single bucket hydroponic treatment (bottom left plant).

The three treatments are composed of fish water, hydroponic water and aerobically digested fish water (**Fig.** 3). Aerobically digested fish water refers to nutrient-rich water that is typically wasted from a fish tank that has been biologically processed by microorganisms in the presence of oxygen, breaking down fish waste into stable, beneficial forms for plant growth. Fish water refers to water that is directly drained from the fish tank with solids still in it. The six combinations of treatments include single bucket fish waste, split root fish water and aerobically digested fish waste, single bucket hydroponics water, split root hydroponics water and aerobically digested fish waste, split root hydroponics water and aerobically digested fish waste, and split root hydroponics and hydroponics. High Pressure Sodium (HPS) grow lights provided consistent lighting, and environmental conditions were carefully controlled. Weekly measurements of plant height, width, and chlorophyll index were recorded, while post-harvest measurements included fresh weight, dry weight, root weight, and root area after three weeks. Data from these measurements were analyzed to compare growth and health outcomes between the different cultivation methods.

Figure 3. Hemp plants were subjected to six treatments. H (Hydroponics only) , HA (Hydroponics- Aerobically digested fish waste), HH (Hydroponics-Hydroponics, FA (Fish waste and Aerobically digested fish waste), HF (Hydroponics-Fish), and F (Fish only).

Multi-Loop Aquaponics System:

The multi-loop system built for this study includes both decoupled and coupled aquaponics setups to compare their effects on hemp growth (**Fig. 4**). In the decoupled setup, nutrients supplied by the fish do not meet the nutrient threshold for hemp, and separating the water systems prevents chemical exposure to the fish while allowing precise control over the plants' nutrient intake. Conversely, in the coupled setup, the water recirculates between the fish tank and the plant grow beds, creating a single-loop system. This design allows for the direct use of fish waste as nutrients for plants. Due to the project's time constraints, the decoupled system was implemented with a drip-to-drain setup, while the coupled system maintained a recirculating loop. This dual approach aims to optimize environments for both the fish and the hemp, and to compare the outcomes between the two systems.

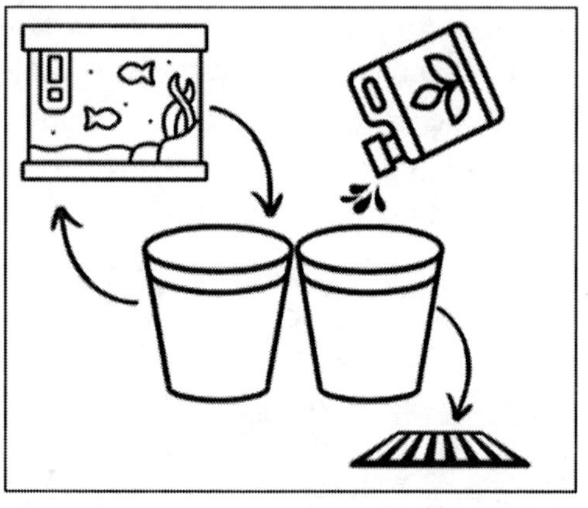

Figure 4. The multi-loop aquaponics system design. The left bucket demonstrates a closed loop between the bucket and the fish tank, while the right bucket demonstrates an open system with additive fertilizer that is drip to drain.

RESULTS AND DISCUSSION

The hydroponics single bucket (H) treatment resulted in the highest mean fresh weight after three weeks of growth significantly outperforming other treatments (**Fig. 5**). Plants in the split root fish waste and aerobically digested fish waste (FA) treatment had the lowest mean fresh weight. Treatments F (Fish only), HA (Hydroponics-Aquaponics), HF (Hydroponics-Fish), and HH (Hydroponics-Hydroponics) showed intermediate results. Statistical analysis indicated that treatments labeled with the same letter are not significantly different from each other at a p-value threshold of 0.05. This data suggests that the hydroponic system, whether coupled with aquaponics or used independently, tends to support better plant growth in terms of fresh weight compared to fish-based treatments.

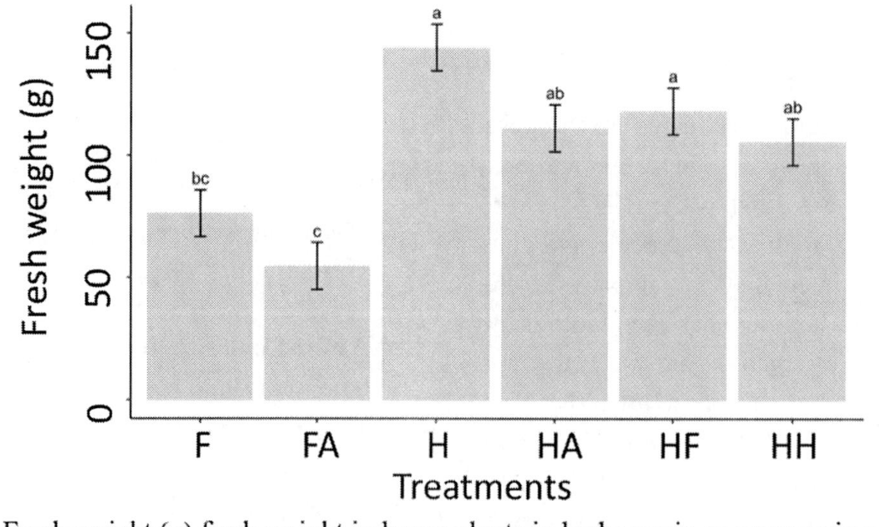

Figure 5. Fresh weight (g) fresh weight in hemp plants in hydroponic or aquaponic solutions. Treatments sharing the same letter above the bars are not significantly different from each other ($p > 0.05$), as determined by a post-hoc test.

There is no significant difference in kite measurements among the various treatments, after three weeks of growth in their respective treatments as indicated by the identical letter 'a' above all bars (**Fig. 6**). This suggests that the different treatments do not have a statistically significant impact on the kite measurements of the hemp plants.

The kite measurements across all treatments range from approximately 900 to 1300 units, with no clear pattern or substantial variation among the treatments. This uniformity indicates that, within the parameters of this experiment, the type of nutrient system—whether it involves fish waste, hydroponics, or a combination thereof—does not significantly affect the kite measurement of the plants.

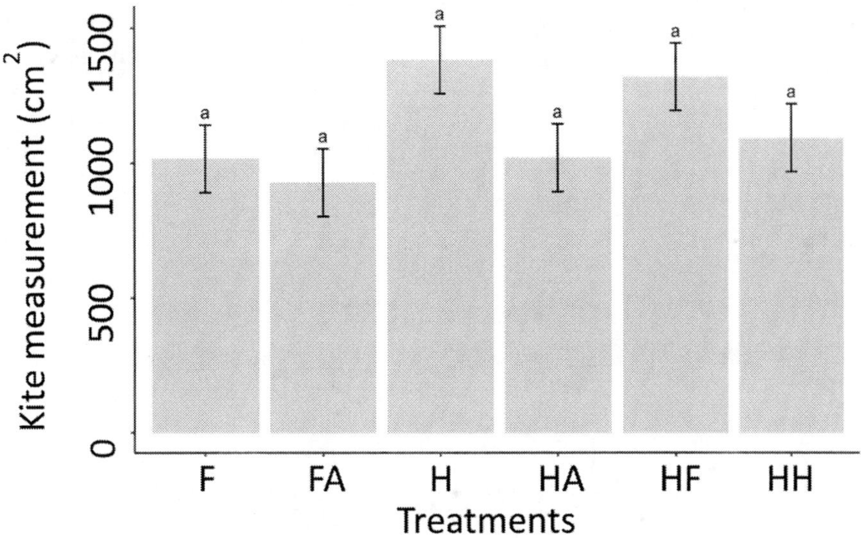

Figure 6. Kite measurements in hemp plants in hydroponic or aquaponic solutions. Treatments sharing the same letter above the bars are not significantly different from each other ($p > 0.05$), as determined by a post-hoc test.

Conclusion:

This study investigated the potential of integrating aquaponics and hydroponics using a split root system to optimize hemp (*Cannabis sativa L.*) growth. The results demonstrated that the single bucket hydroponic system provided the most optimal growing environment measured by fresh weight and kite measurements which outperformed the other treatments. The aerobically digested fish water treatment demonstrated significantly lower plant growth. This result highlights the challenges in the use of organic aquaponics as a nutrient source for hemp instead of non-natural alternatives like fertilizer hydroponics. The investigation of a multiloop system, the combination of coupled and decoupled aquaponics, shows challenges for precise nutrient management and a labor-intensive design, but it allows for the diversification of income and semi-organic production. Further refinement of the design system is needed in order to enhance nutrient availability for the plants.

These findings underscore the potential of hydroponics as a reliable method for hemp cultivation while revealing the complexities of integrating aquaponics into commercial hemp production. Future research should focus on optimizing fish safe nutrient formulations for aquaponic systems, investigating longer cultivation periods, aquaponics' effect on cannabinoids, and exploring the scalability of multiloop systems. This study contributes valuable insights into sustainable hemp cultivation practices and serves as a foundation for further exploration into innovative agricultural systems.

LITERATURE CITED

Shen, J., Li, H., Neumann, G., and Zhang, F. (2004, November 18). Nutrient uptake, cluster root formation and exudation of protons and citrate in Lupinus albus as affected by localized supply of phosphorus in a split-root system. Plant Science 168:837-845. https://doi.org/10.1016/j.plantsci.2004.10.017

Yep, B. and Zheng, Y. (2019). Aquaponic trends and challenges – A Review. Journal of Cleaner Production 228:1586-1599. https://doi.org/10.1016/j.jclepro.2019.04.290

Combined Proceedings IPPS 74:559-567. 2024.

Tree Root Management Trials at NVK Nurseries

Ryan Munroe

NVK Nurseries, 1155 Highway 5, Dundas, Ontario, Canada L9H 5E2

ryan@NVKnurseries.com

Keywords: transplanting, nursery production, containers

Summary

Root growth, especially during container growth is critical to transplanting success. Various container designs were evaluated for their ability to modify root systems in crops like oaks, honeylocust, and conifers. Studies compared traditional solid wall plastic containers with alternative paper and fiber containers.

INTRODUCTION

NVK Nurseries is a wholesale grower of trees, evergreens, shrubs and perennials operating on over 1,200 acres of field and container farms about an hour southwest of Toronto, Ontario, Canada. Our nursery is committed to continually improving our growing practices, and in the last few years we have pursued several growing methods to improve the tree root systems in our propagation and container field-liner departments. The primary motivations behind these efforts are to: 1) improve transplant success of 'hard to transplant' tree species; 2) eliminate major root deformities, such as stem-girdling roots and matted, pot-bound root systems, and; 3) save labour costs associated with trying to correct (e.g. slice, shave, tease apart) these deformities.

While bare root field-liners can easily be assessed for root defects before planting, container-grown trees can possess a range of root issues that are not easily observed. The root quality of container liners from suppliers can be variable, so we are beginning to develop our own container tree liner production system to ensure a higher and more uniform quality. Our testing over the last 2 years has been highly focused on tree species known to be difficult to bare root transplant; for example, most conifers and tap-rooting deciduous trees, which are commonly sold/grown in containers to be lined out. Our efforts have involved the testing of 2 different air-pruning systems in propagation (tulip bulb crates and AirTrays) and 4 different air-pruning/woven fabric containers (Pioneer Pots, RediRoot pots, RootPouch fabric pots, and Ercole pots).

Traditionally, the only way to control the growth of tree roots in production was through pruning. Since almost all trees in the first three quarters of the 20th century were field-grown, this meant that roots could either be hand pruned before planting, during field-growing with implements, or after harvest again. As more tree production has steadily shifted to container growing, various techniques have been employed to manage root growth within the constricted space of a pot. This has included chemical control (lining pots with copper hydroxide), entrapment (using woven natural and synthetic fibres to trap and stop root growth) and air-pruning (allowing air to desiccate root tips). Since most of our trialing has revolved around newer air-pruning tray and pot designs, I'll provide a bit more detail here. Air-pruning is simple in concept: allow enough air to regularly interact with the periphery of the root system in order to stop roots from growing, while initiating fine root development within the existing root ball. However, in practice, it is challenging to implement in a nursery production setting, since it necessarily requires greater exposure of the root system to desiccating air. Striking the balance between allowing enough air to consistently 'prune' roots, while avoiding too rapid and sustained dry-out periods, is critical for success at a large commercial scale. Since root quality starts as early as seed propagation, that's where I'll begin.

In the spring of 2023, we began testing a couple different air-pruning systems for our tree seedlings. First, we used readily available tulip bulb crates, of which we have hundreds, compliments of our perennials department. The slotted sides and bottom of the crates is ideal for allowing for consistent air-root pruning, and we kept the crates elevated off of the floor to help with this. We sowed mainly *Quercus* spp. in the bulb crates in the fall and overwintered them in a minimally heated greenhouse. We left them in these crates and allowed them to grow over much of the growing season (**Fig. 1**). We found this method to work very well for producing well-structured root systems with a moderate amount of fine absorptive roots, especially due to the reliable and consistent air-root pruning of the seedlings' taproots. Importantly, dry out speed was moderated by the large volume of potting media held in the bulb crates. Another positive aspect of this method is that it is inexpensive and simple to execute; however, the seedlings are still considered bare root upon removal, which does limit the timing and ease of potting up/lining out.

Figure 1. The tulip bulb crate method, showing the entire crate of *Quercus* spp. seedlings on the left, a young seedling in the middle and the root system of a 1-year-old seedling on the right. Notice the brown, desiccated bottom of the tap root of the middle photo, which helped initiate early growth of lateral fine roots along the tap root.

We also wanted to test an air-pruning plug system, which led us to AirTrays®. These are a relatively new line of propagation trays. In 2023, we trialed *Quercus rubra, Gleditsia triacanthos* and *Pinus strobus* in the deep, 18-count deep trays pre-filled with 60x120mm Ellepots (FP paper), as seen in Figure 2. In this particular design of AirTray®, the only contact between the tray and the plug is a small plastic spike at the very bottom which suspends the plug (**Figs. 2 and 3**). This allows for air-pruning because it elevates each plug several centimeters off the tray and provides a sizable air gap around the sides. Although there was an adjustment required to manage irrigation properly in this open-tray design, the use of an even-watering boom system allowed us to provide the consistent moisture needed to avoid patchy dry-out. The results were very positive overall, with growth and root quality in all three trial species deemed excellent by our head propagator. Importantly, the AirTrays® strike a good balance between allowing substantial air flow, but still retaining relatively high humidity within each cell of the tray.

Figure 2. *Quercus rubra* seedlings grown in 18 count deep AirTrays® in 60x120mm Ellepots. Notice the desiccation of the tap root in the middle photo, as well as the brown root tips around the plug in the far-right photo, which contains root growth largely within the plug.

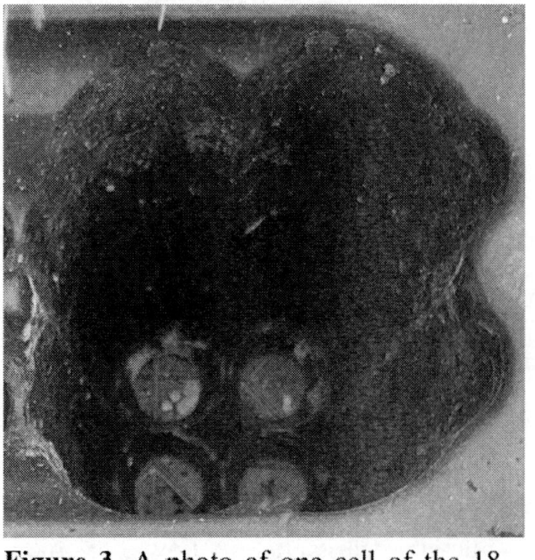

Figure 3. A photo of one cell of the 18-count deep AirTray®. Notice the 4 large holes at the bottom of the cell. If you look closely at the center of the bottom, you can see a spike which elevates the plug completely off of the bottom of the tray. The 4 ridges along the sides of the cell also ensure that there is minimal contact between the plug and the side walls.

The only modification that we wished to make for our 2024 propagation trials was to replace the pre-loaded Ellepots in the AirTrays® with alternative cell liners that we could loose-fill with our own growing media. To do this, we used FertilPots, which are 100% biodegradable wood fibre pots. After requesting samples of different pot sizes, we found that the 517.C 7x12cm pot fit very well into the 18 count deep AirTray® (**Fig. 4**).

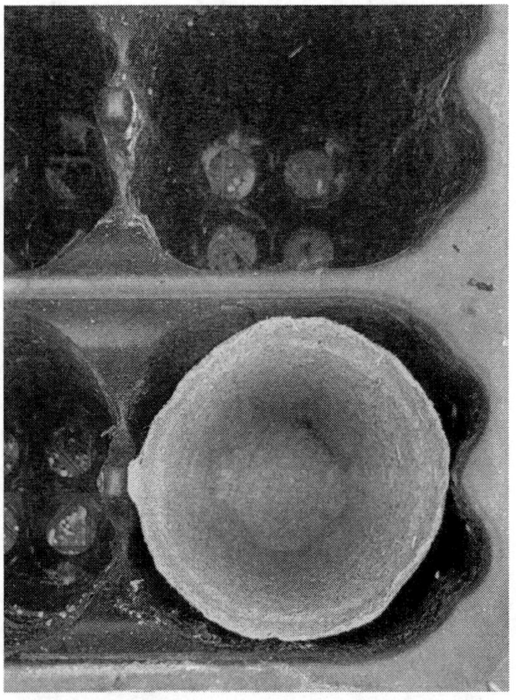

Figure 4. The 517.C 7x12cm wood fibre pot fit nicely as a replacement insert for the Ellepot in the 18-count deep AirTray® to allow us to loose fill our own potting mix.

The FertilPots also functioned well (Figure 5). The benefit of both the Ellepot and the Fertilpot is that their woven, porous structure allows roots to grow through them, as long as they are not too dry. We found that both worked well for developing good root structure, but that the FertilPots had the added benefit of capturing irrigation water more effectively, and more clearly showing differences in moisture (as is visible in **Fig. 5**).

During the 2023 growing season, we were also trialing different air-pruning systems and woven fabric pots in our container field-liner department. Our aims were to determine which pots would produce the highest quality roots, while also considering factors such as: dry-out, growing media temperatures, relative costs, ease of production, and reusability of different pots. The 3 alternative pots we used in our 2023 trials were the Pioneer Pot with holder, the RediRoot pot, and the RootPouch pot (Figure 6). We also included a standard solid-walled pot as a control. We used drip emitters to ensure equal irrigation volumes.

Figure 5. *Picea* spp. seedlings potted from smaller plugs and grown for several months in FertilPots in an 18 count deep AirTray®. Roots easily grow through the porous wood fibre pots and then air-prune once humidity drops.

Dry-out was an important factor to consider in this trial, and we found that the Pioneer Pot with holder performed similarly to the standard solid-walled pot in terms of having slower dry-out than the RediRoot and RootPouch pots. Although it is not depicted in **Figure 6**, the Pioneer Pot system nests the air-pruning pot inside of a holder, which was designed to maintain higher humidity levels and slow down dry-out. Growing media temperatures were another consideration, and we found that temperatures were consistently lower in the Pioneer Pot with holder and RootPouch pots in comparison to the RediRoot and standard pots. While the Pioneer Pot was significantly more expensive than the other pots, its higher manufacturing quality gave it a longevity advantage. While the RootPouch pot only allows for a one-time use, the Pioneer pots can be reused up to 10 growing cycles, on average (personal communications with several growers across North America). RediRoot, in contrast, we were told by a number of growers, lasted only between 3-5 years.

The RootPouch pot performed well on laterally rooting, fibrous tree species, but sat too wet and didn't effectively manage root growth on the bottom of the pot in the taprooting tree species we grew. Both the Pioneer and RediRoot pots performed much better in this regard due to their elevation of root systems off the ground. Ultimately, based on the criteria of root quality, costs of production and reusability, we decided to advance with the Pioneer and RediRoot pots for our more extensive 2024 trials.

Figure 6. **A**) Pioneer pot with holder (holder not depicted above); **B**) RediRoot pot; **C**) RootPouch pot; **D**) Standard solid-walled pot. We used these pots in our root quality trials to grow several different tree species in 2023 in our container field-liner department. Pots range in size from 2 to 4 gallons.

In 2024, we scaled up our research and development efforts, in part thanks to support from the Barborinas Family Fund Tree Fund Grant, which we were awarded in Spring 2024. We continued testing the Pioneer and RediRoot pots with over two dozen different tree species and cultivars, and we also included the 5L Ercole pot into our trialing, since FertilPots' largest 580.C (18cm x 16cm) offering fit very nicely (**Fig. 7**).

The perforated sides of the Ecole pot, in addition to the small air gaps separating the FertilPot from the Ercole pot, provided an ideal air root-pruning design along the sides. Unfortunately, the bottom of the Fertilpot still rested on a solid pot bottom. The results showed much faster dry-out in the RediRoot pot when compared to the Pioneer pot with holder, with the FertilPot inside the Ercole pot performing somewhere in between.

Figure 7. The 580.C Fertilpot (17x16cm) (Pictured by itself on the right) fits perfectly into the black 5L Ercole pots. *Fagus sylvatica* plugs were potted 3 months prior to this photo.

Fagus sylvatica was one of the species we used in these pot trials, and we photo-monitored the progress of their growth in the different pots across time. In **Figure 8**, we show *F. sylvatica* seedlings that we started growing in the spring of 2023. Only the solid-walled pot produced root systems with significant root deformities after one season in 2023.

Figure 8. Root systems of 2023 grown *Fagus sylvatica.* Trees were grown for one season in the following pots from left to right: Pioneer with holder, RediRoot, RootPouch, and Standard. Notice the lack of visible roots on the exterior of the 3 air-root pruning and fabric pots and the circling and descending roots on the exterior of the standard pot.

In **Figure 9**, we show the same trees in a Pioneer and standard pot at the end of the 2024 growing season. By 2024, the visual difference between the two root systems is even more stark. In the standard container grown trees, you can see an abundance of circling and matted roots. In the Pioneer pot-grown trees, while you can't see many roots on the periphery of the root ball, it is packed with fine roots, which is the result of two seasons of air-pruning.

Figure 9. *Fagus sylvatica* trees grown for two seasons in a Pioneer Pot with holder (left) standard smooth-walled plastic pot (right). Notice the matting and circling roots on the tree grown in the standard smooth-walled plastic pot compared to the lack of visible roots on the sides of the Pioneer Pot-grown tree.

In general, when we compared the growth and health of the many trees grown in the different pots during our 2024 trials, we found few obvious differences. Some exceptions occurred, such as when trees in RediRoot pots, which were prone to faster dry-out, were under irrigated and resulted in decreased tree growth. Although we didn't take quantitative measurements of stem or root growth, we were able to draw these conclusions from regular observation and extensive photo monitoring.

Conclusions

These series of trials have been very useful for us to test out not only the impacts of different air-pruning designs on root quality, but also to help answer bigger picture questions around economic viability, production efficiency and overall plant performance. We will be using our observations from 2023 and 2024 to guide our growing practices moving into the spring of 2025. In propagation, we will scale up the use of the AirTrays with FertilPots because of the flexibility of this system, both in terms of producing plugs with superior root systems, and because of the ease of use with our existing irrigation boom to efficiently manage dry-out. Our results from the previous years will be useful in knowing which species to plant in the FertilPots, because we found that the highly degradable Fertilpots, if used for longer than 6 months, disintegrate and require a fibrous root system to hold the plug together. Therefore, certain slow-growing, heavy taprooting trees such as *Carya glabra*, which we grew in 2024 trials, will not work well in this system.

In our container field-liner department, we will move forward in 2025 with the #2, #5, and #7 Pioneer Pots because of their superior longevity, the high quality root systems they produce and less severe dry-out experienced when utilizing the holders. With the Pioneer Pots, we are trying to create our own system that replaces the need to purchase the pot holder, making production less expensive, while still retaining the benefits of the microclimate created by some form of protective holder.

We acknowledge that the implementation of a new practice or new technology always requires adjustments; some large and others small. Overall, the benefits of growing high quality root systems that can establish and root out faster in their growing next stage continues to motivate us to work through the challenges associated with these production changes.

At NVK, we are always striving to innovate better growing practices, and we look forward to see what 2025 holds.

Combined Proceedings IPPS 74:568-572. 2024.

Starting and Sustaining a Beneficial Insect Program

Freedom Shelley

Spring Meadow Nursery, 12601 120th Ave, Grand Haven, Michigan 49417 USA

freedom@springmeadownursery.com

Keywords: mites, aphids, thrips, greenhouse,

Summary

Having a beneficial insect program provides effective pest control but also helps prevent future pest outbreaks while reducing chemical treatments and chemical resistance. When starting a beneficial insect program essential components to consider include consulting with experts, training staff, and implementing a detailed.

IPPS Vol. 74 – 2024

INTRODUCTION

In 2018, Spring Meadow Nursery encountered significant challenges of pest resistance in two-spotted spider mite (TSSM (*Tetranychus urticae*), a key pest in the operation. Despite applying chemicals one to two times per week, TSSM outbreaks persisted, and the effectiveness of traditional spray methods diminished. In response, a trial program was initiated by incorporating beneficial insects into the integrated pest management strategy. Beneficial insect companies suggested targeting the most affected crops with the highest TSSM populations, the roses. A small-scale trial was conducted on approximately two acres of rose greenhouse production. The successful reduction in TSSM populations with minimal chemical sprays in this area led to the expansion of the beneficial insect program across the entire 50+ acres of greenhouse production. This integration of beneficial insects has resulted in improved pest control, particularly for TSSM, a reduction in chemical pesticide use, and valuable insights on how to establish and optimize a beneficial insect program in a large-scale greenhouse operation.

Starting a Biological Insect Program

Spring Meadow Nursery began its beneficial insect program in 2018, and learned many lessons along the way. Several key considerations are crucial when it comes to setting up a program:

1. Identification of target pests: Identifying the primary pest(s) to target and select the most appropriate beneficial insect(s) for control of the pest. It was identified that *Tetranychus urticae* was the primary pest at Spring Meadow Nursery due it its chemical resistance (Maffly, University of Utah, 2023) and its detrimental effects on plant quality.

2. Selection of beneficial insects: Over the past six years, Spring Meadow Nursery has trialed various beneficial insect species for TSSM, including *Neoseiulus californicus, Phytoseiulus persimilis, Amblyseius andersoni,*and *Neoseiulus cucumeris.* Among these species, it was found that *P. persimilis*, commonly known as the spider mite destroyer **(Fig. 1)**, proved to be the most effective in controlling TSSM outbreaks under optimal environmental conditions (temperatures between 70-80°F and relatively humidity above 60%). In addition, it was found that supplementing with *N. californicus* and *A. andersoni* during temperature extremes in the beginning and end of the growing season enhanced the program's success, as these species are able to tolerate a broader range of temperature and humidity conditions compared to *P. persimilis.*

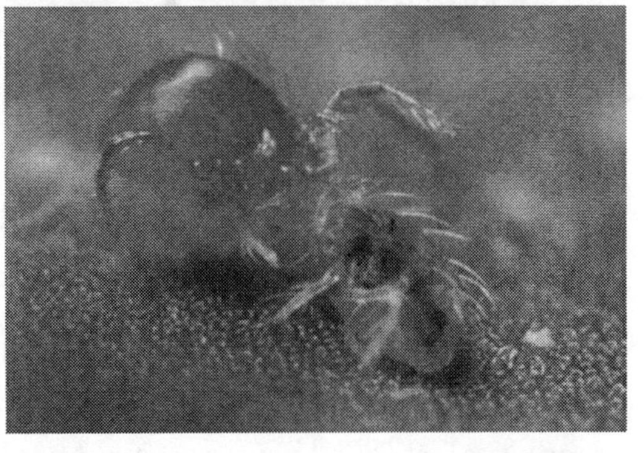

Figure 1. *Phytoseiulus persimilis,* commonly referred to as the "spider mite destroyer," preying on *Tetranychus urticae* (two-spotted spider mite). Image courtesy of Biobee.

3. Targeting secondary pests: In addition to TSSM, Spring Meadow Nursery addressed secondary pests such as aphids, thrips, and other soft-bodied insects. To do so, a portable, self-watering banker plant system (**Fig. 2**) was designed, which is easily moved to areas experiencing pest outbreaks. This system serves both as a protective habitat and as an alternate food source for beneficial insects. The following beneficial insects are commonly released on these banker cart systems: *Aphidius colemani,* a parasitic wasp that feeds on aphids (**Fig. 3**), *Dicyphus hesperus,* which feeds on soft-bodied insects, and *Orius insidiosus,* which preys on thrips.

Figure 2. A portable, self watering banker cart system, custom built at Spring Meadow Nursery, containing *Triticum aestivum* (winter red wheat grass), *Lobularia maritima* (sweet alyssum), *Verbascum spp.* (mullein), *Portulaca grandiflora* (portulaca), and *Capsicum annuum* (purple flash pepper). Image courtesy of a Spring Meadow Nursery grower.

4. Quantifying beneficial insect requirements: To determine the optimal quantity of beneficial insects needed for Spring Meadow Nursery, growers collaborated with insect suppliers to establish insect application rates based on square footage.

The growers, with the help of insect suppliers, developed a calculator that allows growers to input the growing space, select the desired insect species, and determine the required quantity of insects. This calculator serves as a tool that ensures precise, data-driven decisions and effective pest management.

Figure 3. Parasitized aphids, a result of *Aphidius colemani* parasitism. Image courtesy of a Spring Meadow Nursery grower.

Sustaining a Biological Insect Program

While initiating a beneficial insect program is critical, its long-term success depends on dedication and consistent maintenance. At Spring Meadow Nursery, growers have identified several essential strategies for sustaining a successful biological control program over the course of the growing season:

1. Regular applications: Weekly or biweekly applications of beneficial insects are necessary to maintain pest control effectiveness. Without a consistent food supply, beneficial insect populations decline rapidly (within 5-7 days). Regular applications not only replenish the beneficial insect populations but also facilitate their reproduction and growth, ensuring a continual cycle of pest control.

2. Use of Technology: To enhance the efficiency of insect applications, growers at Spring Meadow Nursery integrated modern technologies, such as the BugFlow system from Biobee (**Fig. 4**). This system enables even distribution of beneficial insects across large greenhouse bays, reaching up to 12 feet across from the application point. This has reduced the limitations associated with hand applications, which only covered a limited area and often resulted in uneven distribution of beneficial insects.

Figure 4. (left) Bug flow system developed by BioBee, (right) Freedom Shelley, grower at Spring Meadow Nursery, holding an assembled bug flow unit. Image courtesy of Biobee and Spring Meadow Nursery grower.

3. Chemical Compatibility: One challenge in maintaining a biological control program is the potential of adverse effects of chemical pesticides on beneficial insects. Certain chemicals can kill up to 75% of beneficial insect populations, compromising pest control efforts. To mitigate this, growers at Spring Meadow Nursery prioritize the use of "softer" chemical sprays that have minimal impact on beneficial insects (typically 25% mortality). They also utilize resources such as the spray compatibility tools provided by Koppert and Biobest, which allows them to assess the impact of specific chemicals on beneficial insect populations. This helps ensure that pesticide applications do not undermine the effectiveness of the beneficial control program.

Conclusion

In conclusion, starting small, carefully selecting the appropriate beneficial insect species, consulting with experts, training staff, and implementing a detailed plan are essential components for establishing a successful and sustainable beneficial insect program. Having a beneficial insect program not only provides effective pest control but also helps prevent future pest outbreaks while reducing chemical treatments and chemical resistance. Lastly, a key advantage of using beneficial insects is that pests cannot develop resistance to a predator that consumes them, ensuring no resistance, long-term effectiveness and sustainability of an integrated pest management program.

Resources

Biobee: www.biobee.com

Biobest: www.biobestgroup.com/side-effects-app

Koppert – Side Effects: Sideeffects.koppert.com/side-effects/

Maffly, B. (2023). Research unravels how spider mites quickly evolve resistance to toxins. The University of Utah. https://attheu.utah.edu/research/research-unravels-how-spider-mites-quickly-evolve-resistance-to-toxins/

Combined Proceedings IPPS 74:573-579. 2024.

Assessing Variation in Photosynthetic Performance of Soybean Using MultispeQ Phenotyping

Harmeet Singh-Bakala[1,2], Francia Ravelombola[1], Cheryl Adeva[1], Jessica Argenta[1], Grover Shannon[1] and Feng Lin[1]

[1]University of Missouri – Fisher Delta Research, Extension, and Education Center, Portageville, MO 63873, USA; [2]Division of Plant Science and Technology, University of Missouri, Columbia, Missouri 65211, USA

hsmyz@missouri.edu

Keywords: photosynthesis, plant breeding, non-destructive

Summary

Soybean, a globally important crop, requires improved yield potential to meet the rising demand without expanding production areas. Photosynthesis is fundamental to plant growth and a key target for yield enhancement. This study aimed to evaluate the variation in photosynthetic efficiency among 54 elite soybean breeding lines using the non-destructive MultispeQ tool. Field experiment was conducted in 2024 at Lee Farm, Portageville, Missouri. Photosynthetic traits, including SPAD, Light intensity (PAR), PSII efficiency (FvP/FmP), maximum quantum yield (Phi2), non-photochemical quenching (NPQt) and linear electron flow (LEF), were measured at the R2-R3 stage. Significant phenotypic variation was observed across the breeding lines for key traits, such as SPAD (mean 43.3), Phi2 (mean 0.36), and LEF (mean 211 μmol m²/s), highlighting genetic diversity in photosynthetic performance. Correlation analyses revealed positive associations between FvP/FmP and Phi2 ($r = 0.47$), as well as LEF and PAR ($r = 0.80$), indicating a strong relationship between PSII efficiency and light-use efficiency. Conversely, NPQt negatively correlated with Phi2 ($r = -0.45$), illustrating a trade-off between energy dissipation and photosynthetic performance. These findings provide insights into the genetic potential of photosynthetic traits for breeding programs. Future analysis of seed yield will further elucidate the role of these traits in improving soybean yield, supporting targeted breeding efforts to enhance crop performance under varying environmental conditions.

INTRODUCTION

Soybean is one of the world's most important food crops and serves as a crucial source of vegetable protein and oil. To meet the growing global demand for soybean, the rate of yield improvement must double to avoid further expansion of production areas (Zhu et al 2010; Rains 2011). Photosynthesis is the fundamental physiological process driving plant growth and development. Future improvements in crop yield will largely depend on enhancing net photosynthesis and energy transduction efficiency (Ainsworth et al 2008; Parry et al 2011). Leveraging natural variation in photosynthetic capacity offers an opportunity to breed genotypes with enhanced carbon assimilation (Parry et al. 2011; Faralli and Lawson 2020). Multiple studies have shown that improving photosynthetic (PS) efficiency is a promising strategy for increasing soybean yield potential, especially in the context of climate change (Zhu et al 2010; Long et al 2006;). Other than genetic characteristics of the plant (genotype), photosynthesis also depends upon the influence and constraints of environmental parameters including light, temperature, CO_2, moisture etc.

Therefore, it is important to evaluate the variation in photosynthetic performance in natural and breeding populations to utilize this variability to improve seed yield in soybean (Araus et al 2016; Sakoda et al 2016). Traditionally, different methods have been in practice to capture photosynthetic data such as via gas exchange measurements, which measure CO_2 assimilation rates and stomatal characteristics of the leaves. Another major technique involves chlorophyll fluorescence (CF) which provides insights into photosynthetic performance by assessing light energy capture, distribution, and photosystem II (PSII) efficiency in both controlled and field conditions. Screening diverse and elite soybean germplasm for photosynthetic traits is critical for capturing valuable variation and elucidating the genetic basis of PS efficiency (Montez et al 2022; Ort et al 2020; Dhanapal et al 2016).

Non-destructive phenotyping methods are required for effective screening breeding materials in field conditions (Araus et al 2014; Meacham-Hensold et al 2020). Understanding the relationships between photosynthetic traits and seed yield is crucial for leveraging these phenotypic correlations in a breeding program. The first objective of this study was to assess and characterize the variation in multiple PS-traits among soybean breeding lines using a non-destructive phenotyping tool named 'MultispeQ'. In addition, the interrelationships among various photosynthetic traits will be identified and analyzed.

MATERIALS AND METHODS

Fifty-four soybean breeding lines (Maturity group IV late) from advanced yield trials were evaluated in 2024 at Lee Farm, Portageville, Missouri. The experiment was laid out in a randomized complete block design (RCBD) in 2 replications. MultispeQ data collection was done at full flowering (R2-R3) stage on a clear day (between 10.00 AM and 2.00 PM) using upper canopy leaf (**Fig.1**) for different photosynthetic traits including SPAD, Light intensity (PAR), PSII efficiency (FvP/FmP), PSII maximum quantum yield (Phi2), Non-photochemical quenching (NPQt), Linear electron flow (LEF), leaf thickness, leaf angle and leaf temperature. Descriptive statistics (Mean, SD, Min and Max) were calculated for all traits to capture variation within breeding lines. Analysis of variance (ANOVA) were performed to test the significance of differences among lines for each photosynthetic trait. Then, Pearson Correlation coefficients were computed to investigate the interrelationships among measured PS traits, providing insights how these photosynthetic traits interact in our advanced breeding material. The statistical analyses were performed in R-studio software using agricolae, ggplot and dplyr packages (R core team 2022).

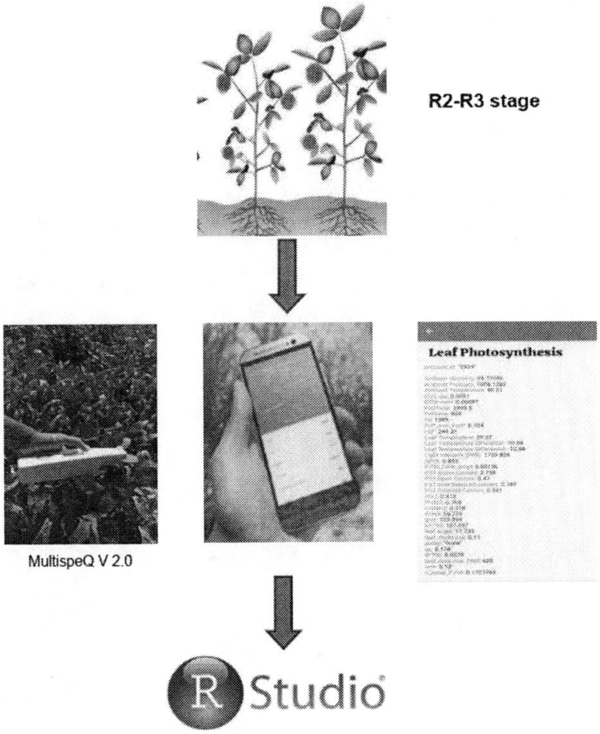

Figure 1. Experimental workflow of photosynthetic data collection and analysis.

RESULTS AND DISCUSSION

To investigate the variation in photosynthetic performance using high throughput non-destructive phenotyping tool MultispeQ 2.0, we used 54 advanced breeding lines (MG IV). Good phenotypic variation was observed in multiple photosynthesis related traits including FvP/FmP (0.70), SPAD (Mean 43.3), Phi2 (mean 0.36) indicating differences in relative chlorophyll content, photochemical efficiency and overall photosynthetic performance of the advance breeding lines. Notably, lines with higher Phi2 and LEF (Mean 211 μ moles/m^2/sec) suggested greater energy conversion efficiency and higher productivity potential. Variation in NPQt (mean 1.10) highlights the differences in heat dissipation mechanisms, with some lines showing enhanced photoprotection than others.

Table. 1. Summary statistics for major PS-traits in elite soybean breeding lines.

TRAIT	MEAN	SD	MIN	MAX
SPAD	43.3	4.37	31.4	53.4
PAR (μ moles/m^2/sec)	1338	366	177	2111
FvP/FmP	0.70	0.04	0.548	0.776
Phi 2	0.36	0.062	0.229	0.546
NPQt	1.10	0.427	0.41	3.02
Leaf thickness (mm)	0.53	0.4	0.04	2.02
Leaf temp (°C)	31.2	1.08	28.2	35.5
LEF (μ moles/m^2/sec)	211	45.7	43.6	324

The variability in key photosynthetic traits in advanced breeding lines is also shown in the density plots **(Fig 2)**. The SPAD density plot shows quite wide distribution, indicating a substantial variation ($P < 0.001$) in relative chlorophyll content across soybean breeding lines. The FvP/FmP plot ($P < 0.05$) displays relatively narrower distribution, suggesting differences in the maximum quantum efficiency of PSII photochemistry. Leaf thickness also exhibits significant variability ($P < 0.05$), with most lines clustering around thinner leaves but some lines also have much thicker leaves. The Phi2 plot ($P < 0.05$) highlights variation in the quantum yield of PSII, pointing to differences in light energy conversion efficiency. These multiple density plot distributions suggested the diverse photosynthetic efficiencies within the advance breeding lines, which can be valuable for selecting traits linked to improved seed yield and stress tolerance.

We also observed several interesting correlations among multiple photosynthetic traits measured with MultispeQ in elite soybean breeding lines at R2-R3 stage. The strongest positive correlations are observed with LEF and light intensity-PAR i.e. $r = 0.80$, indicating the increased light intensity directly boosts the electron flow, enhancing photosynthetic performance of the plants. Similarly, FvP/FmP is positivity correlated with Phi2 ($r = 0.47$), which suggests that lines with higher quantum yield of PSII exhibit greater PSII efficiency downstream. Also, SPAD positively correlated with FvP/FmP ($r = 0.42$), implying that higher chlorophyll content may support greater PSII efficiency.

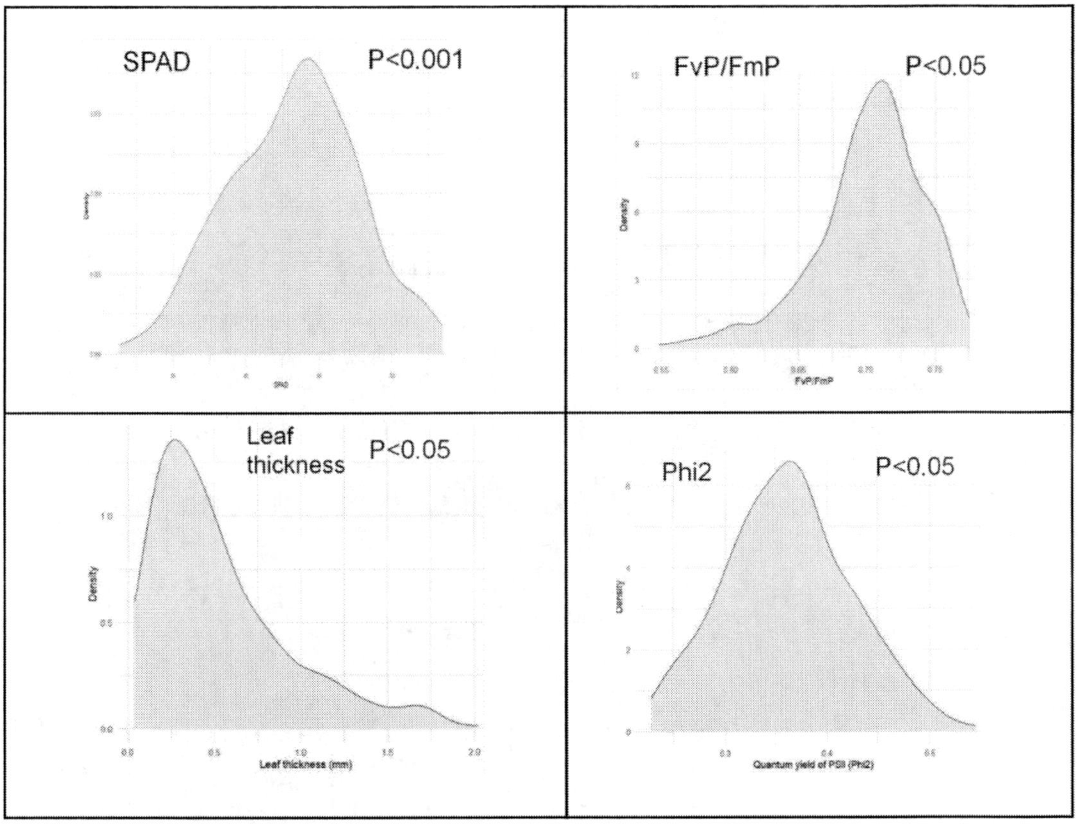

Figure 2. Density plots showing variability for various MultispeQ based traits.

Conversely, NPQt shows a notable negative correlation with Phi2 (r = -0.45), indicating that as energy dissipation through non-photochemical quenching increases, the quantum yield of PSII decreases, reflecting a trade-off between energy use for photosynthesis and dissipation to prevent photo-damage. Leaf thickness has minimal relationships with most of the other traits.

Interestingly, light intensity (PAR) also shows a strong negative correlation with NPQt (r = -0.73), showing that under high light conditions, non-photochemical energy dissipation is reduced. This allows more energy to be directed towards photosynthesis. So, these interrelationships provide insights into how different soybean lines respond to light conditions and regulate photosynthetic efficiency. This information can further be used to correlate with the seed yield performance to select lines with optimal photosynthetic traits for breeding purposes.

Conclusion

In a nutshell, substantial variations were observed for multiple photosynthetic traits in elite soybean materials (Fig.2; Table.1), showing their genetic potential for selection for improvement in relevant traits. Using the MultispeQ v2.0, we successfully found strong and significant correlations **(Fig. 3)** among multiple PS traits like FvP/FmP, NPQt, PAR and SPAD in our breeding materials. In the near future, final seed yield will also be analyzed along with all PS-traits to study multiple correlations, aiming to pinpoint valuable photosynthesis related traits for targeted breeding efforts, ultimately improving soybean yield performance.

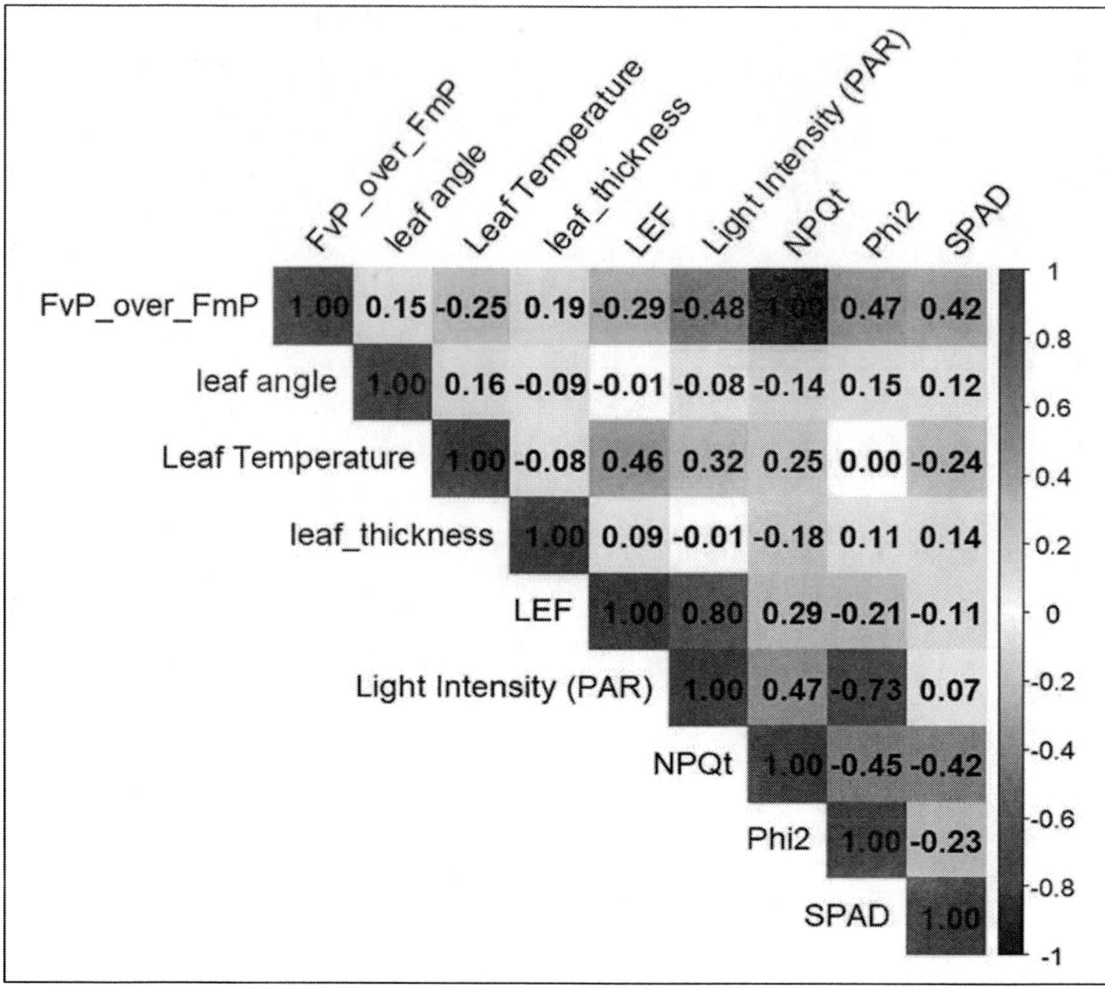

Figure 3. Correlation matrix for PS-traits in elite lines.

LITERATURE CITED

Ainsworth, E.A., Rogers, A. and Leakey, A.D.B. (2008). Targets for crop biotechnology in a future high CO_2 and high O_3 world. Plant Physiol. *147*:13–19.

Araus, J.L. and Cairns, J.E. (2014). Field high-throughput phenotyping: the new crop breeding frontier. Trends Plant Sci. *19*:52–61.

Dhanapal, A.P., Ray, J.D., Singh, S.K., Hoyos-Villegas, V., Smith, J.R., Purcell, L.C. and Fritschi, F.B. (2016). Genome-wide association mapping of soybean chlorophyll traits based on canopy spectral reflectance and leaf extracts. BMC Plant Biol. *16*.

Faralli, M. and Lawson, T. (2020). Natural genetic variation in photosynthesis: an untapped resource to increase crop yield potential? Plant J. *101*:518–528.

Long, S.P., Zhu, X.G., Naidu, S.L. and Ort, D.R. (2006.) Can improvement in photosynthesis increase crop yields? Plant Cell Environ. *29*:315–330.

Lopez, M.A., Xavier, A. and Rainey, K.M. (2019). Phenotypic variation and genetic architecture for photosynthesis and water use efficiency in soybean (*Glycine max* L. Merr). Front. Plant Sci. *10*.

Meacham-Hensold, K., Montes, C.M., Wu, J., Guan, K.Y., Fu, P., Ainsworth, E.A., Pederson, T., Moore, C.E., Brown, K.L., Raines, C., et al. (2020). Plot-level rapid screening for photosynthetic parameters using proximal hyperspectral imaging. J. Exp. Bot. *71*:2312–2328.

Montes, M.C., Fox, C., Sanz-Sáez, A., et al. (2022). High-throughput characterization, correlation, and mapping of leaf photosynthetic and functional traits in the soybean (*Glycine max*) nested association mapping population. Genetics *221* iyac065. https://doi.org/10.1093/genetics/iyac065

Ort, D.R., Merchant, S.S., Alric, J., Barkan, A., Blankenship, R.E., Bock, R., Croce, R., Hanson, M.R., Hibberd, J.M., Long, S.P., et al. (2015). Redesigning photosynthesis to sustainably meet global food and bioenergy demand. Proc. Natl. Acad. Sci. U.S.A. *112*:8529–8536.

Parry, M.A.J., Reynolds, M., Salvucci, M.E., Raines, C., Andralojc, P.J., Zhu, X-G., Price, G.D., Condon, A.G. and Furbank, R.T. (2011). Raising yield potential of wheat. II. Increasing photosynthetic capacity and efficiency. J. Exp. Bot. *62*:453–467.

R Core Team. (2022). R: A language and environment for statistical computing. R Foundation for Statistical Computing, Vienna, Austria. URL https://www.R-project.org/

Raines, C.A. (2011). Increasing photosynthetic carbon assimilation in C3 plants to improve crop yield: current and future strategies. Plant Physiol. *155*:36–42.

Sakoda, K., Tanaka, Y., Long, S.P. and Shiraiwa, T. (2016). Genetic and physiological diversity in the leaf photosynthetic capacity of soybean. Crop Sci. *56*:2731–2741.

Zhu, X.G., Long, S.P. And Ort, D.R. (2010). Improving photosynthetic efficiency for greater yield. Annu. Rev. Plant Biol. *61*:235–261.

Combined Proceedings IPPS 74:580-584. 2024.

Taxus canadensis: Can We Do Without Synthetic Auxins and a Mist Tent?

Hilary Sowa and Brian Maynard

Department of Plant Sciences and Entomology, The University of Rhode Island, Kingston, Rhode Island 02881 USA

Hilary.Sowa@uri.edu

Keywords: Canada yew, auxins, IBA, restoration, sustainability, water

The authors wish to acknowledge a student research grant to H.S. from the International Plant Propagator's Society Eastern Region Foundation (2024).

Summary

In this preliminary study, we investigated whether the propagation of *Taxus canadensis* (Canada yew) from cuttings could be made more sustainable by eliminating the use of synthetic auxins and a mist tent. What we found was: (1) softwood cuttings with or without the application of auxins did not root without the use of a mist tent; and (2) hardwood cuttings taken in early January and propagated in a mist tent rooted equally well with or without auxins. Plans to extend this research include examining whether hardwood cuttings taken in early January can be successfully rooted without the use of auxins in a humidity tent.

INTRODUCTION

Taxus canadensis (Canada yew) is a native, evergreen, understory shrub found in bogs and swampy forests in USDA hardiness zones 2-6 westward from Newfoundland to southeastern Manitoba, southward to northern Iowa and New York, and further into Tennessee and Virginia. This species prefers acidic and moist soil conditions and reaches heights of 6 feet with a spread of up to 10 feet. Insignificant yellow blooms appear in the spring, followed by red berries that ripen in the late summer and early fall. *T. canadensis* has a sweeping, plagiotropic, and open habit and is often used as a ground cover in heavily shaded sites (Cullina, 2002). Because seed propagation of this species has been found to be challenging and often requires multiple warm and cold stratification periods (Dirr & Heuser, 2018), seedlings, not surprisingly, rarely appear in nature with propagation typically achieved through layering (Allison, 1990).

The U. S. Forestry Service (n.d.) considers *T. canadensis* to be an important species for conservation and restoration and finding ways to more sustainably propagate this species would support the goals of such ecologically focused projects. Synthetic auxins, including indole-3-butyric acid (IBA), have long been relied upon in propagation, and while generally considered safe in small doses, toxicity if swallowed (U. S. Environmental Protection Agency, 2015), neurological and immunological effects in animals (Yilmaz & Celik, 2009), and environmental hazards (U. S. Environmental Protection Agency, 2015) are among some of the identified risks. High auxin levels also have negative implications for subsequent development in cuttings, including reduced bud break and leaf drop (Sun & Bassuk, 1993). Mist tents, another mainstay of the propagation industry, come with setup and operating costs including equipment, space, and water use, but perhaps more importantly, a continued reliance on a resource that may be scarce in many areas of the country. In totality, eliminating the use of auxins and mist tents would make the propagation process more environmentally and economically sustainable.

In this study, we began our investigation into whether *T. canadensis* cuttings could be propagated without the use of synthetic auxins and/or a mist tent in order to minimize environmental impacts and costs associated with this process. We compared the effects of different hormone treatments against using no hormones at all, time of year cuttings were taken, and the efficacy of using a humidity tent versus a mist tent on rooting success.

MATERIALS AND METHODS

Softwood cuttings were collected in late June. Hardwood cuttings were collected at two time periods: early January and mid-February (**Fig. 1**). While hardwood cuttings have previously been shown to root successfully with high auxin concentrations (Dirr and Heuser, 2018; Hartmann et al., 1996), we were also interested in testing the efficacy of using a humidity tent for softwood cuttings.

Treatments

C: Control (no IBA)
H2: Hormodin H2 (0.3% IBA; 3000ppm)
H3: Hormodin H3 (0.8% IBA; 8000ppm)
H45: Hormex H45 (4.5% IBA; 45,000ppm)
DNG (Dip n Grow): (1% IBA; 10,000ppm)
KIBA: diluted to 0.15% IBA; 1500ppm)

Figure 1. Collecting hardwood cuttings.

Softwood Cuttings

The bottom 1" of each cutting was stripped of foliage and lightly wounded to stimulate root development. Cuttings were then randomly assigned to one of four treatment groups (C, H2, H3, and H45, with 5 replicates per treatment group), stuck in a sharp sand and peat mix (Dirr & Heuser, 2018; Hartmann et al., 1996), and placed in a poly humidity tent.

Hardwood Cuttings

The bottom 1" of each cutting was stripped of foliage and lightly wounded to stimulate root development. Cuttings were then randomly assigned to one of four treatment groups (C, DNG, H45, and KIBA, with 5 replicates per treatment group), stuck in a 1:3 peat:perlite mix, and placed in a mist tent. The mist interval was set to mist for 20 seconds every 20 minutes, and bottom heat (70°F) was used.

RESULTS

Softwood cuttings in all treatments in a humidity tent did not root after approximately 2 months. Hardwood cuttings taken in early January and placed in a mist tent rooted equally well, with or without the application of auxins. Cuttings were scored approximately 2 months after sticking, and rooting success was measured both in terms of mean number of roots per rooted cutting and mean percentage rooted.

The mean number of roots per rooted cutting in the control group (4.0) was comparable to those in the auxin treatment groups (DNG 3.7, H45 5.0, and KIBA 3.2) for cuttings taken in January (**Fig. 2A**). In contrast, when harvesting of cuttings was delayed to mid-February, the number of roots per rooted cutting in the control group (1.0) was far fewer than those found in the DNG and H45 treatment groups (5.0 and 4.8, respectively) (**Fig. 2B**).

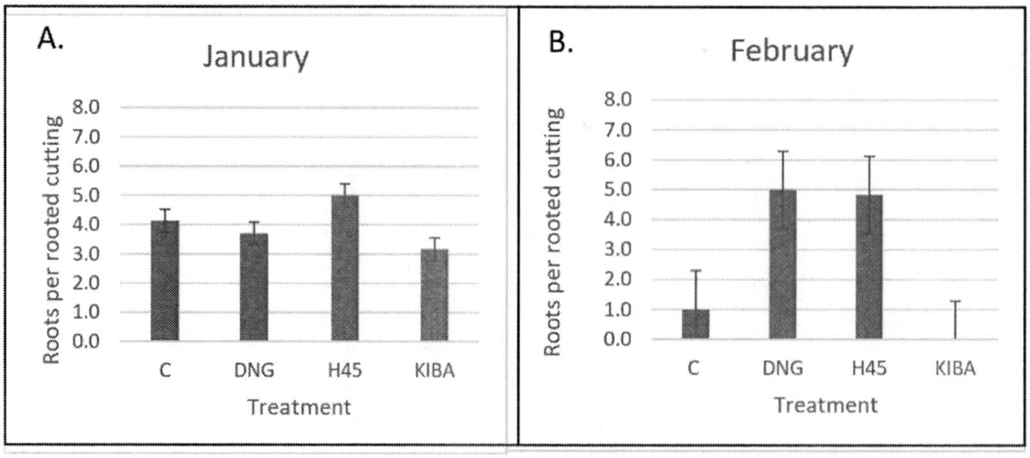

Figure 2. Roots per rooted cutting in untreated or IBA-treated softwood cuttings taken in January or February.

The mean percentage of cuttings that rooted successfully in the control group (53.3%) was also comparable to those found in the auxin treatment groups (DNG 46.7%, H45 20.0%, and KIBA 60.0%) for cuttings taken in January **(Fig. 3A)**.

In contrast, when harvesting of cuttings was delayed to mid-February, the mean percentage of cuttings that rooted successfully in the control group (20.0%) was lower than those found in the DNG and H45 treatment groups (30.0% and 46.7%, respectively) **(Fig. 3B)**. In February, cuttings treated with KIBA failed to root.

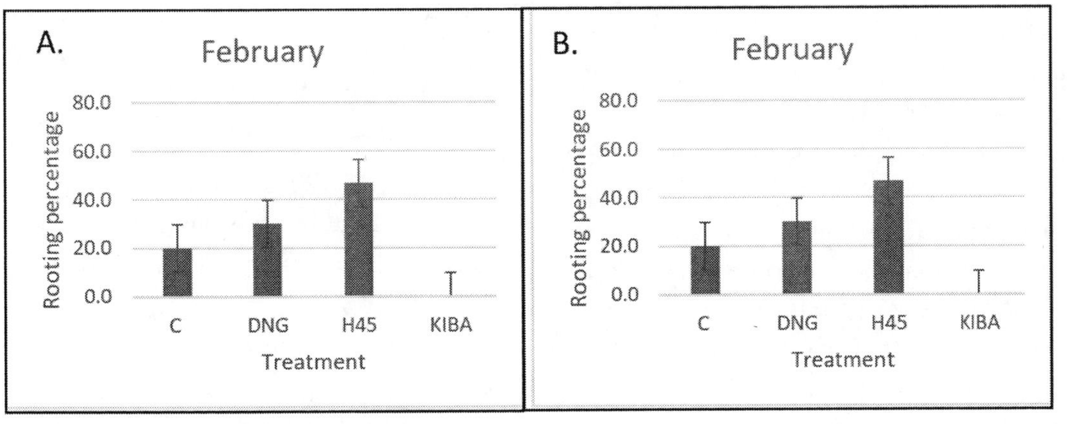

Figure 3. Rooting percentages in untreated or IBA-treated softwood cuttings taken in January or February.

A representative example of a cutting that rooted successfully without the use of auxins is shown below **(Fig. 4)**.

Figure 4. Rooted hardwood cutting without the use of auxins.

DISCUSSION

In the current study, *T. canadensis* softwood cuttings did not root in a humidity tent regardless of treatment, providing further support that *Taxus* species are best propagated from hardwood cuttings (Dirr & Heuser, 2018; Hartmann et al., 1996).

Hardwood cuttings taken in early January in a mist tent, however, rooted successfully without the use of auxins. These results support the idea that not every species requires auxins to root and that auxin use should be considered on a species-by-species basis rather than simply as a default (Barakat and Draie, 2024; Maynard, 2012). In the current study, targeting the correct time of year to harvest cuttings (i.e., early January) had a significant impact on rooting success.

Given the small sample size in this study, these results should be considered preliminary; it would be valuable to repeat this study using a larger sample size to determine whether results could be replicated. The current results would, however, suggest that cuttings should be taken in early January as this time period appears to be optimal for rooting without the use of auxins.

A follow-up study could examine whether hardwood cuttings taken in early January could be propagated without auxins and without a mist tent. Reducing auxin and mist tent use has both environmental and economic benefits including eliminating contact with a potentially toxic substance, reducing reliance on a sometimes scarce natural resource, and decreasing overall propagation costs. The overall goal of continued research in this area would be to find ways to make the propagation process more sustainable.

LITERATURE CITED

Allison, T. D. (1990). The influence of deer browsing on the reproductive biology of Canada yew (*Taxus canadensis* marsh.): I. Direct effect on pollen, ovule, and seed production. Oecologia *83*:523–529.

Barakat, H. and Draie, R. (2024). Effect of indole butyric acid (IBA) concentration on rooting of some ornamental plants. International Research Journal of Innovations in Engineering and Technology *8*:173-180. https://doi.org/10.47001/IRJIET/2024.811019

Cullina, W. (2002). Native trees, shrubs, and vines: A guide to using, growing, and propagating North American woody plants. Houghton Mifflin Company, Boston, MA, USA.

Dirr, M. A., and Heuser, C. W. Jr. (2018). The reference manual of woody plant propagation: From seed to tissue culture (2nd ed.). Timber Press, Portland, OR, USA.

Hartmann, H. T., Kester, D. E., Davies, F. T. Jr., & Geneve, R. L. (1996). Plant propagation: Principles and practices (6th ed.). Prentice-Hall, Boston, MA, USA.

Maynard, B. (2012). Bringing the art and science of cutting propagation back to the bench. Proc. Intern. Plant Prop. Soc. *62*:249-253). https://ipps.org/uploads/docs/62_057.pdf

Sun, W.-Q., and Bassuk, N.L. (1993). Auxin-induced ethylene synthesis during rooting and inhibition of budbreak of 'Royalty' rose cuttings. J. Amer. Soc. Hort. Sci. *118*:638- 643.

U. S. Environmental Protection Agency (2015). IndoleButyric Acid Technical. https://www3.epa.gov/pesticides/chem_search/ppls/062097-00027-20150715.pdf

U. S. Forest Service (n. d.). *Taxus canadensis*. Retrieved on September 30, 2024 from https://www.fs.usda.gov/database/feis/plants/shrub/taxcan/all.html

Yilmaz, Z. and Celik, I. (2009). Neurotoxic and immunotoxic effects of Indole-3-butyric acid on rats at subacute and subchronic exposure. Neurotoxicology *30*:382-385. doi: 10.1016/j.neuro.2009.03.002

Combined Proceedings IPPS 74:585-589. 2024.

Sand Bed Propagation

Kevin Zuidervliet

Octoraro Native Plant Nursery, 6126 Street Road, Kirkwood, Pennsylvania 17536 USA

kevin@octoraro.com

Keywords: acorns, sowing, mist, seed, oak

Summary

Octoraro Native Plant Nursery has been successfully propagating a variety of species of plants in their sand bed system for years. It is a low-cost system that allows dense planting of seeds. This method is especially effective with acorns, small seeds, and plants that have a low germination rate. Below you will find a detailed description on how to build the sand beds, the technique for sowing seeds, and an in-depth guide on how to harvest oak trees bare root while they are in leaf.

INTRODUCTION

Octoraro Native Plant Nursery is a wholesaler of native trees and shrubs in Lancaster County Pennsylvania. We produce +350,000 native plants per year that are primarily used for riparian buffer plantings and reforestation. We have a propagation facility that sits on a very small footprint and because of that we need systems that allow us to propagate a lot of plants very efficiently.

The sand bed system that we use was developed in conjunction with Bill Barnes who consulted with us as we constructed our new propagation facility. It is made up of several layers which you can see being constructed in **Figure 1**. The layers are as follows:

1. Greenhouse floor - concrete or tamped stone.
2. Weed fabric.
3. Biotherm heat tubing - Biotherm heat tubing or a similar product creates an ideal temperature for seeds to germinate.
4. Round pebbles - To protect the heat tubing from not being punctured.
5. Aluminum flashing - Acts as a heat dispersal layers as well as a protects the heat tubing when you are digging plants.
6. Weed fabric - This keeps plant roots from going deeper than this layer.
7. Pure sand - High quality sand that does not have any additives or chemicals is necessary to create plants that have a robust root system.
8. Mist nozzles - These hang over the sand beds to create a humid environment.

Figure 1. Sand bed construction consists of layers of materials.

Sowing and Harvesting

Sowing seeds is very simple for plants that have small seeds or are wind dispersed. Simply sprinkling them evenly over the surface of the sand is all that is needed. Seeding too thickly is preferred as you can always cull plants to a proper spacing. Acorns are planted using your hand as a plow to create a straight row at approximately one inch depth. The acorns are dropped into this furrow at .5"-1" spacing and another furrow is made. This second furrow pushes sand in both directions which is what covers the first furrow. As the acorns germinate, they push themselves slightly above the sand (**Fig. 2A**). Once the plants are fully germinated, they do not appear in rows but create a canopy (**Fig. 2B**).

Figure 2. Oak propagation in the sand bed. A. Germinating acorns. B. Oak seedlings.

We have developed a specialized process for harvesting plants in leaf. This is especially effective for oaks. We harvest plants in leaf so that we can "flip" our sand beds as many times a year as possible. The more times you can sow and harvest from your sand beds the more efficient space is. This is essential for any propagation facility no matter its size.

Twenty – four hours prior to harvesting we apply BioPlex Transplant Concentrate and Plant Enhancer. We do this in a bucket with a sump pump and a hose at a rate of 5 fl. oz. per 10 gallons. When we harvest plants, we do so very gently so as not disturb that fine root system that the plants have developed in the sand. The plants go straight into a bucket which contains a hydration gel and mycorrhizal slurry (**Fig. 3**). From here the plants are potted. As soon as one tray or one plant is potted it is placed on a wagon and mist is applied. This keeps the plant from transpiring which is essential as the plant begins to establish itself in the pot. As wagons are filled, they are brought into a greenhouse and placed on the floor under mist nozzles which are run at a rate to keep 100% humidity in the space (**Fig. 4**). This rate depends on the weather and size of the greenhouse. At this point, BioPlex is reapplied. Immediately after potting and while the plants are transitioning it is important to keep the leaves wet for 1 week as the plant establishes. After that week the mist can be slowly scaled back until it is off. This weaning off period typically takes 7-10 days.

Figure 3. Harvesting oak trees into buckets with a mycorrhizal and gel slurry.

Figure 4. A greenhouse that is being filled with potted oak trees under mist.

Conclusion

Using the sand bed method can be an incredibly space efficient tool that allows propagators to produce 1000's of plants on a small square footage. The technique used to produce plants on that space is specialized but as the system is used, propagators will learn best methods which have been outlined in this paper. When sowing seeds, it is important to do so densely as the more plants in the sand that germinate the more finished product you will get at the end. Using tools such as a mycorrhizal slurry, mist systems, and Bioplex you are able to harvest plants in leaf which ensures multiple crop rotations. This system is excellent for growers who are operating on a small square footage and are trying to produce the most plants possible.

Combined Proceedings IPPS 74:590-593. 2024.

Native Seed Cleaning: The Oregon Seed Blower

Claire Zuidervliet

Mt. Cuba Center, 3120 Barley Mill Rd, Hockessin, Delaware 19707 USA

czuidervliet@mtcubacenter.org

Keywords: native plants, seed cleaning, debearder

Summary

Mt. Cuba Center has explored the use of the Oregon Seed Blower in native seed cleaning which has proven to be effective for cleaning native seeds with pappus or calyx structures. The blower successfully removes lighter debris and unviable seed from the most samples, resulting in clean seed for storage and sowing. While not suitable for all species and situations, the Oregon Seed Blower can be a useful tool for operations that store and/or propagate small to medium amounts of seeds. This paper describes the function of the Oregon Seed Blower and shows the process for cleaning *Solidago rugosa var. apsera.*

IPPS Vol. 74 – 2024

INTRODUCTION

Mt. Cuba Center is public garden in Hockessin, DE whose mission is to inspire an appreciation for the beauty and value of native plants and a commitment to protect the habitats that sustain them. Mt. Cuba's greenhouse facility focuses on conservation-related propagation, plant production, and research. At Mt. Cuba Center, we typically handle smaller samples of seed compared to a large commercial nursery and tend to store seed for longer periods of time.

The Oregon Seed Blower, produced by Hoffman Manufacturing Inc., is a simple yet effective machine that removes small, lightweight debris from seed samples and separates unviable seeds from viable seeds. It is described as an "economical laboratory" machine, meant for operations that need to separate seeds by mass or have a laboratory-like setting. There are two models available: the original freestanding model and the compact tabletop model. The blower is operated with a 5-minute timer. It utilizes airflow generated by a blower motor that pushes air through a column (**Fig. 1A**). The air lifts lightweight debris or empty seeds into the removable cups on each side of the column (**Fig. 1B**), while heavier, viable seeds stay in the bottom cup (**Fig. 1C**). Airflow is controlled by an adjustable cap at the top of the air column (**Fig. 1D**). The Oregon Seed Blower is essentially using the winnowing method for seed cleaning in a controlled environment.

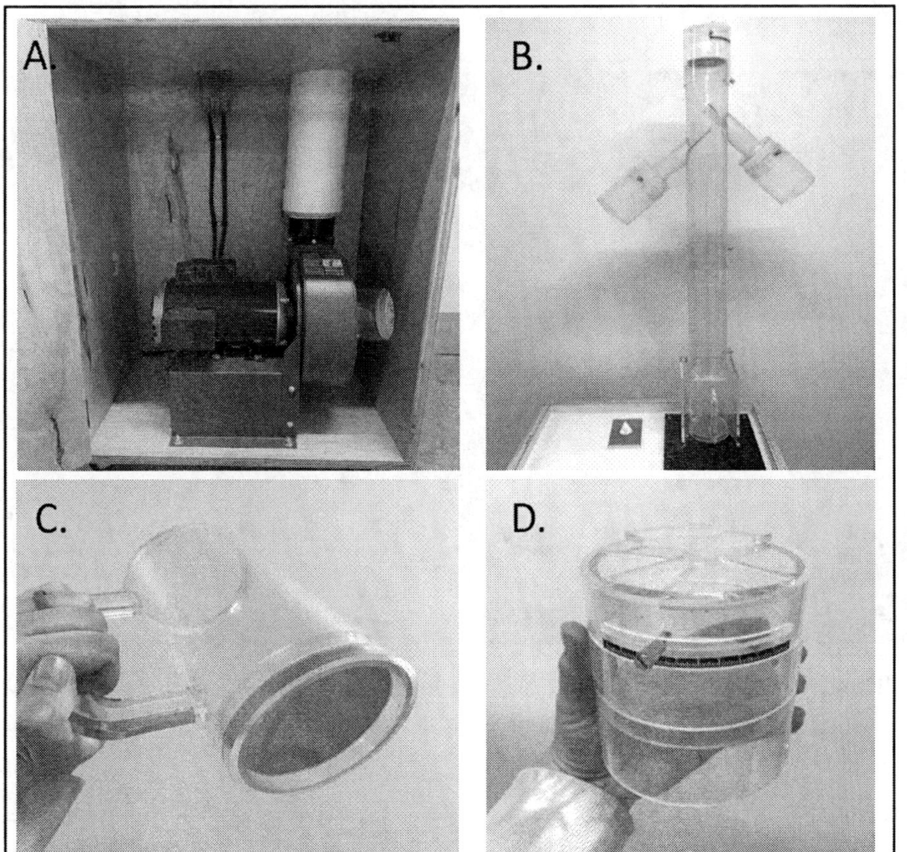

Figure 1. Oregon Seed Blower. A) Blower motor. B) Air column. C) Bottom cup with mesh. D) Air flow adjustment cap.

Through much experimentation, I have learned there are two ideal seed types to clean with the Oregon Seed Blower. Seeds in the *Asteraceae* family that have a pappus structure are a great candidate for this machine due to the light weight of the fluffy debris. *Asteraceae* species that I have cleaning using the blower include: *Soldiago, Symphyotrichum, Vernonia, Eutrochium, Liatris,* and *Cirsium*. The other seed type that works well with this machine is species that contain seeds inside of a calyx or other papery covering. This includes: *Pycnanthemum, Monarda, Verbena, Rumex,* and others. On the Hoffman Manufacturing website, they mention that hemp growers and grass seed farmers may also use this machine.

Before utilizing the Oregon Seed Blower, it is necessary to perform bulk cleaning of the seed samples. Pappus and calyx structures must be removed from the seeds. The two tools I find most effective for this process include the Debearder, also made by Hoffman Manufacturing Inc., and various screens/sieves (**Fig. 2**).

The Debearder consists of a wooden trough and a paddle, both lined with texturized rubber. Placing the seed sample into the trough and gently passing the paddle over it removes pappus as well as papery coverings. Screens and sieves can be used to release seeds from a calyx and are especially useful when still attached to the flower heads (**Fig. 2B**). It is important to note that the Oregon Seed Blower will not always remove all debris from a seed sample and may require multiple runs to fully clean the seeds. Even still, there may be remaining debris that is the same weight as the seeds and therefore unaffected by the airflow.

Figure 2. A) Hoffman debearder and pan sieves. B) Removing calyx parts from *Pycnanthemum*.

Cleaning *Solidago rugosa var. aspera*

One seed cleaning protocol I have developed involving the Oregon Seed Blower is for *Solidago sp*. In this example, I cleaned a sample of *Solidago rugosa var. aspera*. After harvest, the unclean seed was allowed to dry for several months. First, the seed was manually plucked and shaken from the stems (**Fig. 3A**). Next, the sample was placed in the Debearder to remove pappus and crush large debris into smaller particles (**Fig. 3B**). The seeds and debris were placed in the bottom cup of the Oregon Seed Blower. The blower ran for 2 minutes at the lowest airflow setting. The airflow from the blower successfully floated the pappus and empty seeds into the removable side cups on the column while the viable seeds remained in the bottom cup.

After running through the blower, the seeds were screened with a #20 sieve to remove any heavy debris that was smaller than the seeds. The remaining seeds were mostly clean, with only a small amount of debris remaining that was the same size as the seeds and not light enough to be removed with the blower (**Fig. 3C**).

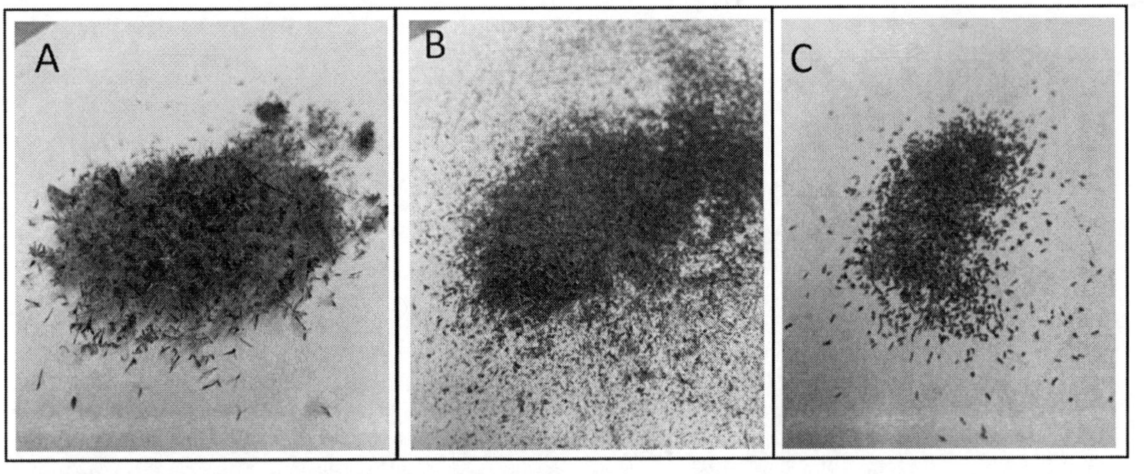

Figure 3. Seed cleaning in Solidago. A) Seeds manually plucked for seed heads. B) Debearded seeds. C) Seed sample after using the seed blower.

Ten of the cleaned seeds were cut in half to observe viability and all the seeds contained full, white embryos with no evidence of pests, assuming 100% viability in this sample. By Mt. Cuba's standards, this sample is considered clean enough for sowing and long-term storage.

Conclusion

Native seed cleaning is an art as much as it is a science. Developing effective cleaning techniques for each species requires trial and error, combining all available tools at different capacities until the seed is sufficiently clean. A cleaning method that works for one operation may not work for another due to the size of samples being cleaned or the end goal for the seed. At Mt. Cuba Center, The Oregon Seed Blower has proven to be effective at removing debris and empty seeds from viable seeds in the Asteraceae family with pappus, or seeds within a calyx or other papery covering.

Using tools such as a mycorrhizal slurry, mist systems, and Bioplex you are able to harvest plants in leaf which ensures multiple crop rotations. This system is excellent for growers who are operating on a small square footage and are trying to produce the most plants possible.

Combined Proceedings IPPS 74:594-607. 2024.

New Plant Forum 2024 – Eastern Region IPPS

Shelby French, Moderator and Program Chair

The Farm at Doe Run, Coatesville, Pennsylvania 19320 USA

sfrenchdoerun@gmail.com

Keywords: breeding, plant introduction, genetics

Summary

New plants for 2024 are highlighted and described. This year six IPPS-ER breeders presented herbaceous and woody perennial plants.

PRESENTER

Olivia Bamaszewski, Bailey Nuseries, St. Paul, Minnesota 55119 USA
Olivia Bamaszewski@baileynursery.com
Contact alec.Charais@baileynursery.com

***Vitex agnus-castus* 'Bailtexfour'** PP36,066
First Editions® Violet Mist™

Violet Mist® Chasetree is a highly compact cultivar, maturing to only 3.5' tall and wide. Drought and salt-tolerant, Violet Mist® has a heavy load of blue-purple flowers in early summer. It will naturally rebloom after 60 days when deadheaded, adding to the floral display.

The colorful flower spikes are a pollinator favorite, providing nectar all season long. In production, Violet Mist® is a vigorous grower that fills liners quickly, making the plant salable in record time. Bred at Bailey Innovations™.Winter-hardy to USDA Zones 6-9.

Figure 1. Violet Mist chastetree plant habit and flowers.

***Hydrangea arborescens* 'BAlful'** PP35,613
First Editions® FlowerFull™

FlowerFull® Smooth Hydrangea boasts superior genetics that produce significantly bolder blooms and sturdier upright stems. With two-to-three times more blooms per season than other *Hydrangea arborescens*, FlowerFull® is a major improvement to the industry standard. That proliferation of blooms is held high on strong stems that don't flop in the wind and rain. With fewer touch points from day one, this low-maintenance marvel has improved resistance to bacterial leaf spot and makes for an outstanding focal point in smaller landscapes and commercial settings alike. Maturing at 3-4' tall and 4-5' wide, FlowerFull® fits well into a modern landscape. Winter-hardy to USDA Zones 3-8.

Figure 2. Flowerfull smooth hydrangea plant habit and landscape use.

Gardenia jasminoides **'Baildeniaone'** PP36,284
First Editions® Big Beauty Gardenia™

Big Beauty® stands out with double blooms and strong fragrance. Massive numbers of pure white flowers cover the plant in summer, allowing the fragrance to spread through the landscape. Its naturally round habit eliminates the need for pruning, keeping it low maintenance. At six-to-seven feet tall and wide, this larger gardenia is ideal for hedges or foundation plantings. Introduced by Bailey Innovations™.Winter-hardy to USDA Zones 7-9.

Figure 3. Big Beauty gardenia plant habit and semi-double flowers.

PRESENTER

Justin Cartmel
Anthony Tesselaar Plants
327 Monbuld Road, Silvan, Victoria, Australia 3795.
contactus@tesselaar.com

Rosa x damascene
Flower Carpet® shrub roses

Flower Carpet Shrub Roses have large blooms with an old world look and are intensely fragrant. The foliage is mid green, lush and is highly disease resistant.

Will thrive in Zones 4-10 and best planted in well drained soils with some degree of organic matter. The plants will tolerate dry conditions once established. Typical growth depending on variety is around 3 feet tall.

Figure 4. Flower Carpet® shrub rose 'Berry' has rich dark pink-red fragrant blossoms.

Figure 5. Flower Carpet® shrub rose 'Lipstick' has reddish-pink fragrant long-lasting flowers that continue throughout the summer.

Figure 6. Flower Carpet® shrub rose ‘Champagne’ has creamy orange-pink fragrant blooms that continue to unfurl petals.

PRESENTER

Brent Horvath
Intrinsic Perennial Gardens, Inc.
170702 Seaman Road, Hebron, Illinois 60034 USA
BrentH@intrinsicperennialgardens.com

Scriptus pendulus ‘Stars and Stripes’ PPAF

Scirpus 'Stars and Stripes' is a grass-like plant native to North America, known for its easy and fast growth. It thrives in well-drained soil, requires low fertility, and has minimal pest issues. It is best planted in early spring and prefers medium moisture. The plant is intolerant of shade and needs at least 80 frost-free days. It blooms in mid-spring and has a long lifespan with slow-spreading rhizomatous roots. Plants grow approximately 24 inches tall and 18 inches wide and are hardy in USDA Zones 3-9.

Growing conditions

- Easy, fast grower
- Best in well-drained mix, allowed to dry between watering
- Low fertility requirement, max pH 7.0, low salinity tolerance
- No significant pest or disease pressure

Planting and care

- Splits in a cool greenhouse in February/March
- Plant plugs in pots or just below the surface of the growing media

- Can be trimmed to control height
- Best planted when temperatures are at least 65 degrees
- Avoid transplanting when not actively growing

Fertilization and media

- Performs well with control-release fertilizer for perennials
- Adapted to all soil textures, grows in conventional peat or bark-based media
- Low drought tolerance, needs medium moisture

Garden transplanting

- Trade gallons recommended for success
- Well-developed root mass critical for survival
- Planting bare root not recommended

Additional information

- Bred by Intrinsic Introductions
- Intolerant of shade, needs 80 frost-free days
- Blooms mid-spring, fruiting June-August
- Long lifespan, slow-spreading rhizomatous roots

Figure 7. *Scriptus pendulus* 'Stars and Stripes'.

PRESENTER

Jeff Kurts
Spring Meadow Nursery
12601 120th Ave., Grand Haven, Michigan 49417 USA
jeff@springmeadownursery.com

***Bignonia capreolata* 'SMNBFW'**
Proven Winners® Dressed to Thrill™

Climbing plants grow 12-18 feet tall and 5-8 feet wide. Orange-red trumpet shaped flowers are attractive to hummingbirds. It is the heaviest flowering bignonia in our trials. Winter-hardy to USDA Zones 6-9.

Figure 8. Dressed to Kill bignonia as a container plant and in full bloom.

***Styrax japonica* 'RNI-RIXRED'**
Proven Winners® Stairway to Heaven™

Plants grow 12-18 feet tall and 5-8 feet wide. They have glossy foliage and red new growth. Plants have white pendent flowers will re-bloom. Winter-hardy to USDA Zones 5-9.

Figure 9. Stairway to Heaven snowbell plant habit and flower close-up.

Diervilla x **'SMNDSN'**

Proven Winners® Kodiak Jet Black™

Plants grow 3-4 feet tall and wide. They produce jet black foliage that lasts all summer. Bright yellow flowers are produced in summer. Winter-hardy to USDA Zones 3-8.

Figure 10. Stairway to Heaven snowbell plant habit and flower close-up.

PRESENTER

Angela Treadwell-Palmer
Plants Nouveau
Box 40125, Mobile, Alabama 36640 USA
angela@plantsnouveau.com

Paeonia 'Smithopus10'
Candy Apple™

Candy Apple is the first commercially available double red ITOH peony on the market from breeder Don Smith. Plants grow 3feet tall and wide. Truly red flowers and the plant is perfectly round with the flowers are on stems held tight to the leaves for an overall rounded look. All the Garden Candy® ITOHS make perfect foundation and middle-of-the-border plants. Each flower is 4-6" wide and full of crimson red petals that last for weeks. Maroon fall color gives an added season of interest. Like all the Garden Candy® peonies, Candy Apple is deer and mildew-resistant. Tissue culture propagated. Winter-hardy to USDA Zones 4-9.

Figure 11. Candy Apple ITOH peony plant habit and flower close-up.

***Thuja standishii x plicata* 'Junior Giant'**

Junior Giant is a fast-growing evergreen with dark-green, fragrant, feathery foliage. Junior Giant will mature at 15-20 feet tall and 8-10 feet wide, making it a much better fit for sites where space is limited. However, it is as long-lived as its predecessor, Green Giant. It is appropriate for hedges, screening or as specimen plants. Winter-hardy to USDA Zones 5-8.

Figure 12. 'Junior Giant' arborvitae plant habit and landscape use.

***Picea* 'Kolmschagi'**

Spruce It Up^{TM}

Spruce It Up^{TM} is a year-round evergreen that does not need a lot of trimming. Plants will mature at 7 feet tall and 3 feet wide. Use it in foundation plantings, a mixed border, or anywhere you'd like a cone-shaped, upright accent. It grows more quickly than other selections, but no taller, so you won't have to wait as long for this dwarf evergreen to mature. The branches are just a little more open, allowing airflow between needles, so it gets no mites. It can take extreme heat without burning, does not revert, and has not been known to get needle cast. Winter-hardy to USDA Zones 3-7.

Figure 13. Spruce It Up plants in production and its use in the landscape.

PRESENTER

H.William Barnes
Barnes Horticultural Services LLC
New Castle, Pennsylvania 16105 USA
bill@barnhortservices.com

Fraxinus sogdiana

Fraxinus sogdiana is a medium size tree from the "Stans" region of Asia Minor and western China (**Fig. 14**). Glossy green compound leaves are about 3 inches long and 1.5 wide and taper to a point in the middle of the leaf. Being an ash chances are each plant from seed is either a male or female. Fall color is yellow with tints of red.

Culture is basically the same as other *Fraxinus*. Tolerant of most soils and shows no preference for pH differences or for variable moisture conditions. Full sun is preferrable but will tolerate some shade. Cold hardy easily in zone 6, probably can extend well into zone 5. Hardy in Denver, Colorado. Since Emerald Ash Borer rarely attacks non-blooming trees, it is not known if it is resistant to EAB.

Propagation can be from seed, cuttings and presumably grafting on suitable root stock. Efforts underway to ascertain the potential for grafting. Cuttings root well under mist.

Figure 14. A) *Fraxinus sogdiana* in an urban setting in Bologna, Italy fall 2024. Some yellow fall color just beginning to form. B) and C) Close-up of leaves. D) Fully rooted cutting from material stuck in mid-June.

Elaeagnus multiflora (Cherry silverberry)

Elaeagnus multiflora is amongst 10 species of *Elaeagnus* found in China and the *Elaeagnacea* in China has attracted considerable interest as potential food sources (Cheng et al. 2022) It is a medium to large deciduous shrub with heavy flowering and subsequent fruit set (**Fig. 15**). It is not demanding with soils preferences ranging from moist to considerably dry. Soil pH is not a particular limiting factor and can range from 6 to 8. It can tolerate some shade but does perform better in full sun. In general, it is documented to zone 6 hardiness and observations here in Western Pennsylvania suggest that hardiness could well extend into Zone 5, but studies affirm that Zone 4 is impractical.

Figure 15. *Elaeagnus multiflora*. A) Heavy fruit crop. B) Flowering shrub and C) flower close-up.

At least 4 cultivars are recognized, *Elaeagnus multiflora ovata*, *E. multiflora* "Sweet Scarlett", a Ukrainian selection (McCree, 2016) and "Red Gem" (Planting Justice, 2024). The Nursery for Uncommon Plants lists an Austrian form as SSP (Carya, 2024). Heppy.org (2024) lists a cultivar with soft spines as "Tillamook".

A closely related species, *Elaeagnus longipes,* is sometimes considered to be a variant of *Elaeagnus multiflora* (Plants of the World, 2024).

Work at BHS has shown that seed is useful for propagation but unless reliably fresh can be slow to germinate. Suggestions by McCree (2016) indicate that a warm moist period of at least 4 weeks followed by a cold moist period of 12 weeks is effective. McCree goes on to say that both hardwood cuttings taken in fall and allowed to remain in situ for close to a year will root. He indicates that softwood cuttings will also root but I have not found that to be true, at least in a traditional mist system as they seriously resent mist, although the close cousin *Elaeagnus umbellata* roots readily from cuttings in summer. *Elaeagnus multiflora* will graft readily to *Elaeagnus umbellata* as a standard side veneer graft in late summer.

Cheng et al, (2022) discuss the nutritional value of the fruits produced by this plant and related species.

Culturally *Elaeagnus multiflora* is an easy plant to "farm", and it has its own mechanisms for nitrogen fixation. Interspecific hybrids of the various Chinese species have been reported and has attracted much interest as a potential food crop. One member of the group, *Elaeagnus umbellata* is closely related to *Elaeagnus multiflora* and is under investigation by the USDA (Carya, 2024). Dr. Dick Zimmerman (2024) requested of me some 20 years ago seed of a large, fruited *E. umbellata* from the Berkshire Mountains in Massachusetts for research into it as a food crop. He suggested that it was found to have the highest Vitamin C content of any known fruit. While many members of the tribe including *Elaeagnus multiflora* have high concentrations of Lycopene, vitamins A and E and essential fatty acids (Heppy, 2024).

Elaeagnus multiflora has a great potential as a food crop either as fresh fruit, dried or as jams or jellies. Even greater potential is possible if crossed with *Elaeagnus umbellata*. With new hybrids already under development as a food crop. The heavy fruit load characteristics of *Elaeagnus umbellata* when bred to *E. grandiflora* could greatly enhance the fruit bearing qualities of a hybrid.

To date there have been no reports of *Elaeagnus multifora* as being invasive. It is presumed that a hybrid of *E. multiflora x E. umbellata* will not be invasive opening the door to a useful food crop.

Literature Cited

Cheng, C., Fan, S., Wang, C., Ye, L., Wang, Z. and Huang, H. (2022). Phylogenetic analysis of *Elaeagnus* L. in China: A basis for genetic improvement of a berry crop. Frontiers of Plant Science *13* doi.org/10.3389/fpls.2022.899079

Carya Nursery for Uncommon Plants. (accessed 2024). Cornusmas.eu

Heppy. (accessed 2024). Goumi (*Elaeagnus*). Heppy.org

McCree, Todd. 2016. Greatescapefarms.com

Planting Justice, 2024,Goumi (*Eleagnus multiflora*) Plantingjustice.org

Plants of the World Online. (accessed 2024). *Elaeagnus longipes.* powo.science.kew.org

Zimmerman, R. (2004). Personal communication.